Smart Material Systems

FRONTIERS IN APPLIED MATHEMATICS

The SIAM series on Frontiers in Applied Mathematics publishes monographs dealing with creative work in a substantive field involving applied mathematics or scientific computation. All works focus on emerging or rapidly developing research areas that report on new techniques to solve mainstream problems in science or engineering.

The goal of the series is to promote, through short, inexpensive, expertly written monographs, cutting edge research poised to have a substantial impact on the solutions of problems that advance science and technology. The volumes encompass a broad spectrum of topics important to the applied mathematical areas of education, government, and industry.

BOOKS PUBLISHED IN FRONTIERS IN APPLIED MATHEMATICS

Smith, Ralph C., *Smart Material Systems: Model Development*

Iannelli, M.; Martcheva, M.; and Milner, F. A., *Gender-Structured Population Modeling: Mathematical Methods, Numerics, and Simulations*

Pironneau, O. and Achdou, Y., *Computational Methods in Option Pricing*

Day, William H. E. and McMorris, F. R., *Axiomatic Consensus Theory in Group Choice and Biomathematics*

Banks, H. T. and Castillo-Chavez, Carlos, editors, *Bioterrorism: Mathematical Modeling Applications in Homeland Security*

Smith, Ralph C. and Demetriou, Michael, editors, *Research Directions in Distributed Parameter Systems*

Höllig, Klaus, *Finite Element Methods with B-Splines*

Stanley, Lisa G. and Stewart, Dawn L., *Design Sensitivity Analysis: Computational Issues of Sensitivity Equation Methods*

Vogel, Curtis R., *Computational Methods for Inverse Problems*

Lewis, F. L.; Campos, J.; and Selmic, R., *Neuro-Fuzzy Control of Industrial Systems with Actuator Nonlinearities*

Bao, Gang; Cowsar, Lawrence; and Masters, Wen, editors, *Mathematical Modeling in Optical Science*

Banks, H. T.; Buksas, M. W.; and Lin, T., *Electromagnetic Material Interrogation Using Conductive Interfaces and Acoustic Wavefronts*

Oostveen, Job, *Strongly Stabilizable Distributed Parameter Systems*

Griewank, Andreas, *Evaluating Derivatives: Principles and Techniques of Algorithmic Differentiation*

Kelley, C. T., *Iterative Methods for Optimization*

Greenbaum, Anne, *Iterative Methods for Solving Linear Systems*

Kelley, C. T., *Iterative Methods for Linear and Nonlinear Equations*

Bank, Randolph E., *PLTMG: A Software Package for Solving Elliptic Partial Differential Equations. Users' Guide 7.0*

Moré, Jorge J. and Wright, Stephen J., *Optimization Software Guide*

Rüde, Ulrich, *Mathematical and Computational Techniques for Multilevel Adaptive Methods*

Cook, L. Pamela, *Transonic Aerodynamics: Problems in Asymptotic Theory*

Banks, H. T. , *Control and Estimation in Distributed Parameter Systems*

Van Loan, Charles, *Computational Frameworks for the Fast Fourier Transform*

Van Huffel, Sabine and Vandewalle, Joos, *The Total Least Squares Problem: Computational Aspects and Analysis*

Castillo, José E., *Mathematical Aspects of Numerical Grid Generation*

Bank, R. E., *PLTMG: A Software Package for Solving Elliptic Partial Differential Equations. Users' Guide 6.0*

McCormick, Stephen F., *Multilevel Adaptive Methods for Partial Differential Equations*

Grossman, Robert, *Symbolic Computation: Applications to Scientific Computing*

Coleman, Thomas F. and Van Loan, Charles, *Handbook for Matrix Computations*

McCormick, Stephen F., *Multigrid Methods*

Buckmaster, John D., *The Mathematics of Combustion*

Ewing, Richard E., *The Mathematics of Reservoir Simulation*

Smart Material Systems

Model Development

Ralph C. Smith
North Carolina State University
Raleigh, North Carolina

siam
Society for Industrial and Applied Mathematics
Philadelphia

10 9 8 7 6 5 4 3 2 1

Library of Congress Control Number: 2005920039

ISBN 0-89871-583-0

Contents

Foreword

With this volume, the Frontiers series continues its mission of publishing "cutting edge" treatments on applied mathematical aspects of emerging scientific and technological issues. This is the sixth book to appear in the general area of smart materials and structures since 1992 when the field grew to the point where a book treatment of the subject was in order. Some of the early monographs were sparse in mathematical detail, consisting of either collections of the authors' papers or collections of chapters written by several different authors embodying several points of view. One exception was the 1996 volume co-authored by the current author in which a first attempt at a rigorous and detailed mathematical treatment was given. The field has now had an additional decade of maturation and innovation and the timing is right for a new "frontiers" account.

In the present volume, Professor Smith discusses mathematical and modeling foundations for all of the basic types of material systems that make up the field of smart materials and structures. Notable features of the present volume include careful attention to nonlinear physically-based models exhibiting significant hysteresis (and the resulting damping) as well as treatment of numerical approximation techniques appropriate for both estimation and control design.

The field of smart materials and structures is relatively new and is known by several names: adaptronics, adaptive structures, intelligent material systems and structures, smart materials and structures and combinations of these words. Several conferences a year have been devoted to these topics since the first one was held approximately fifteen years ago. Professor Smith has been a significant and prolific contributor to the largest of these conferences, eventually becoming one of the Chairs. He also serves as an associate editor for one of the major journals serving this community. In this sense, his career has developed along with this discipline, placing him in an excellent position to write this text.

In this volume, Smith discusses current and future applications of advanced smart material systems. The text provides a comprehensive development of both linear and nonlinear models required to characterize these materials in a manner that facilitates design and control development. While the focus is primarily on piezoelectric, magnetic and shape memory material systems, the text also includes applications exploiting ionic polymers, magnetorheological compounds and fiber optic sensors. Three classes of nonlinear models are discussed, all of which provide unified characterization frameworks for the broad class of combined smart material systems. The book includes an extensive development of structural models based on the nonlinear constitutive models as well as a chapter on numerical techniques for approximating solutions to the structural systems. In summary, this monograph will surely be of significant use to researchers and students in the engineering and material sciences as well as those in applied mathematics who are interested in modeling smart materials and structures.

Daniel J. Inman
Virginia Tech
Blacksburg, VA
December 1, 2004

Preface

Writing a book on smart materials bears a certain semblance to purchasing a high-end laptop — due to rapid advances in the field, it is tempting to postpone the decision every six months in order to include the latest technological development. At some point, however, one must simply focus on the present technology and realize that upgrades will be necessary in the future.

The field of smart materials has advanced rapidly in the last 15 years due to an increasing awareness of material capabilities, the development of new materials and transducer designs, and increasingly stringent design and control specifications in aerospace, aeronautic, industrial, automotive, biomedical, and nano-systems.

Equally important for the advancement of the field is the development of models, numerical approximation techniques, and control designs which accommodate the hysteresis and constitutive nonlinearities inherent to the materials. The majority of initial investigations focused on linear analysis based on the approximately linear material behavior observed in low to moderate drive regimes and obtained with certain amplifier and feedback designs. When applicable, linear models and control designs are certainly advantageous and should be considered first. However, for an increasing number of applications, the attributes which provide smart materials with unique actuator and sensor capabilities are inexorably due to physical mechanisms that produce hysteresis — hence these mechanisms must be incorporated in models and control designs to achieve the unique performance capabilities offered by the compounds. For example, the damping provided by shape memory alloy tendons in civil and aerospace structures is proportional to the area of the hysteresis loop. Hence optimal vibration attenuation requires optimal hysteresis which is diametrically opposite to the strategy of operating in approximately linear regimes.

In this monograph, we provide a unified development of linear and nonlinear models for smart material systems as a prelude to model-based control design. We focus significant attention on the physical mechanisms that provide the materials with unique transducer capabilities but yield hysteretic and nonlinear behavior, and, to the extent possible, we use the underlying physics to guide the development of unified constitutive frameworks for quantifying the dynamics of a broad class of ferroic compounds. These constitutive relations are subsequently employed to construct structural models for a range of transducer constructs and geometries. Finally, we address the development of numerical approximation techniques which adhere to the physical principles used to construct models and are appropriate for transducer and control design.

In Chapter 1, we summarize a range of present and projected smart material applications to illustrate both the scope of the field and issues that must be addressed in models. Chapters 2–5 address the development of linear and nonlinear models for ferroelectric, ferromagnetic and ferroelastic compounds. In each of these chapters, we discuss three classes of nonlinear models — Preisach models, domain wall models, and homogenized energy models — since they comprise unified frameworks for the combined class of ferroic compounds, as illustrated in Chapter 6, and are amenable to control design. Whereas all three frameworks encompass physical principles, we focus primarily on the third since the combination of lattice-level energy principles and stochastic homogenization to incorporate material nonhomogeneities provide it with extreme flexibility in a range of smart material applications and operating regimes. It is illustrated in Chapters 2 and 4 that the homogenized energy framework provides an energy basis for extended Preisach formulations. It is thus anticipated that the symbiotic investigation of energy-based and extended Preisach models will strongly contribute to the growth of unified characterization frameworks with the former providing strengths inherent to the energy basis and the latter providing a mature mathematical framework for model analysis and identification.

The unified constitutive models developed in Chapters 2–6 provide the nucleus for the linear and nonlinear rod, beam, plate and shell models detailed in Chapter 7. In this manner, relevant physics is incorporated in the system models to augment accuracy and improve efficiency as required for real-time implementation. Chapter 8 summarizes numerical approximation techniques that are appropriate for both simulations and control design.

The topic of model development for smart material systems is highly interdisciplinary and this book was written with the goal of making it accessible to scientists from a broad range of disciplines with backgrounds ranging from students entering the field to experts within the constituent disciplines. To aid the understanding of both mathematical and physical terminology, we have included numerous definitions throughout the text and have provided an extensive glossary of terms in an appendix. We have also placed mathematical proofs in an appendix so they are available for those who are interested but do not deter readers focusing primarily on physical aspects of the theory. Finally, open research questions are indicated at various points to encourage investigations deemed necessary to advance the state of knowledge in this rapidly evolving field.

Whereas the text does not include exercises, preliminary versions have been used in graduate courses on smart material systems and much of the detail provided throughout the book was motivated by feedback from students in those classes.

Various resources will be maintained at the website `http://www.siam.org/books/fr32` to augment the text and provide a mechanism to update the material. To aid the reader in the implementation of certain models, we have included MATLAB® m-files for the homogenized energy framework used to characterize hysteresis in ferroelectric, ferromagnetic and shape memory alloy compounds as well as well as rod, beam and plate codes. Whereas these are research rather than commercial-level codes, they illustrate facets of the models and will provide a template for simulating the behavior of various smart material systems. We will also maintain a list of errata and certain updates to the material at this website.

The reader will note that while a number of smart materials are discussed in Chapter 1, the model development in subsequent chapters focuses almost exclusively on ferroelectric, relaxor ferroelectric, ferromagnetic and ferroelastic compounds. Although this includes a large cross-section of presently employed materials, it neglects a number of established and emerging compounds such as ionic and amorphous polymers, MR and ER fluids and solids, and fiber optic sensors. This omission is dictated solely by the unified nature of ferroic compounds, in addition to length limitations, and should not be construed as an indication of material merit. Numerous references are provided in Chapter 1 to guide readers investigating applications which exploit alternative materials.

A significant body of this research resulted from collaborations with students, postdocs and colleagues, and credit to their contributions is liberally given through the citations. Special thanks are also extended to collaborators whose comments on preliminary versions of the manuscript have significantly improved the exposition and reduced the number of typos by several orders of magnitude. Specifically, sincere thanks are given to Brian Ball, Lynn Boyles, Tom Braun, Marcelo Dapino, Andrew Hatch, Emily Lada and Stefan Seelecke for their attention to detail and honest feedback when reading parts of the manuscript. Andrew Hatch and Jordan Massad are credited with writing and editing much of the software that is included at the text website to render it in a form that is friendlier to users.

The support provided by several federal funding agencies has been instrumental both for the research summarized here and the writing of this text. These agencies include the Air Force Office of Scientific Research (Dynamics and Control Program), the DARPA Mosaic Program, NASA Langley Research Center, and the National Science Foundation (Division of Civil and Mechanical Systems). Finally, I would like to thank Lisa Briggeman, Simon Dickey, Elizabeth Greenspan and April Schilpp from SIAM for their assistance and encouragement throughout the process of writing this book.

Ralph C. Smith
North Carolina State University
Raleigh, NC
December 1, 2004

Notation

Symbol	Meaning	Page
	Material Properties and Models	
α	Coupling coefficient in effective field models	86
α	Index for austenite or martensite	264
α_1, α_2, α_3	Direction cosines between magnetization and crystallographic axes	163
β	Standard deviation in Boltzmann probably	102
β_A, β_M	Chemical free energies for austenite, martensite	256
$\dot{\gamma}$	Shear strain rate for ER or MR fluids	34
ε	Strain	427
$\overline{\varepsilon}$	Local average strain	266
ε_R	Remanence strain	426
ε_0	Spontaneous strain	256
ϵ	Dielectric permittivity	87
ϵ_0	Permittivity of a vacuum (8.854×10^{-12} F/m)	55
η	ER or MR fluid viscosity	34
η	Local inverse susceptibility $\frac{dE}{dP}$	100
η_α	Phase-dependent entropy constants	256
$\theta_c, \theta_i, \theta_r$	Incidence and reflection angles for an optical fiber	38
λ	Thermal conductivity of medium	126
$\lambda_0, \lambda_s, \lambda$	Magnetostriction	423
λ_B	Bragg wavelength for an optical fiber	40
$\lambda_{s\parallel}$, $\lambda_{s\perp}$	Spontaneous magnetostriction measured parallel and perpendicular to applied field	174
λ_{100}, λ_{111}	Saturation magnetostrictions in the $\langle 100 \rangle$ and $\langle 111 \rangle$ directions	174
Λ	Grating period for a Bragg optical sensor	40

Symbol	Meaning	Page
	Material Properties and Models (Continued)	
μ	Measure in Preisach models	82
μ	Boltzmann probability function	102
μ	Magnetic permeability	166
μ_0	Permeability of free space ($4\pi \times 10^{-7}$ H/m)	167
ν, ν_1, ν_2	Densities in Preisach and energy models	112
ρ	Material or charge density	135
$\bar{\rho}^e$	Average electrical resistivity	126
σ	Stress	427
σ_A, σ_M	Critical austenite and martensite stresses	266
$\sigma_{A_s}, \sigma_{M_s}$	Austenite and martensite starting stresses	246
σ_c	Coercive stress	412
σ_e	Effective stress	260
σ_I	Interaction stress	260
σ_R	Relative stress	269
τ	Shear stress	34
ϕ	Electric potential	135
φ	Conjugate field	282
Φ	Average exchange energy	75
Φ_0	Energy required to reorient single dipole or moment in an ordered lattice	75
χ	Dielectric or magnetic susceptibility	423
$\chi_A, \chi_{M^+}, \chi_{M^-}$	Characteristic functions	267
ψ	Helmholtz energy	62
ω	Frequency at which dipoles or moments attempt to switch	219
Ω	Transducer surface area	126
A	Austenite phase	22
$\mathcal{A}$	Cross-sectional area	126
A_f, A_s	Austenite finish and start temperatures	22
$\mathbf{B}, B$	Magnetic induction or flux	422
B_s	Saturation magnetic induction	165
c_A, c_M	Phase-dependent specific heat capacities	254
$\bar{c}$	Average specific heat for a material	268
c	Kelvin-Voigt damping coefficient	68
d_{31}, d_{33}	Piezoelectric strain constants	55

Symbol	Meaning	Page
	Material Properties and Models (Continued)	
$\mathbf{D}, D$	Electric D field	414
D_e	Effective D field	91
e	Order parameter	423
$\mathbf{E}, E$	Electric field	47
$\widehat{E}$	Electric field data	119
E_c	Coercive electric field	412
$\overline{E}_c$	Local average coercive field	113
E_e	Effective electric field	86
E_h	Bias electric field	76
E_I	Interaction electric field	111
$\mathcal{E}_\pi$	Energy for 180° domain walls	90
E_0	Depolarizing electric field	46
g_{31}, g_{33}	Piezoelectric stress constants	28
G	Gibbs energy	62
h_α	Specific enthalpy	271
h_c	Heat transfer coefficient	126
$\mathbf{H}, H$	Magnetic field	166
H_c	Coercive magnetic field	412
$\overline{H}_c$	Local average coercive field	199
H_e	Effective magnetic field	180
H_h	Bias magnetic field	186
H_I	Interaction magnetic field	161
$\mathcal{J}$	Exchange integral	179
k	Boltzmann's constant (1.38×10^{-23} J/K)	75
k_p	Irreversible loss constant in domain wall models	91
k_s	Preisach kernel	81
k_{31}	Electromechanical or magnetomechanical coupling factor	67
K, K_1, K_2	Anisotropy constants	180
$L(t)$	Interface line in the Preisach plane between $S^+(t)$ and $S^-(t)$	194
$\mathbf{m}, m$	Magnetic moment	185
$\mathbf{M}, M$	Magnetization	422
$\mathbb{M}$	Set of finite, signed Borel measures	82

Symbol	Meaning	Page
	Material Properties and Models (Continued)	
$\mathcal{M}$	Actuator mass	126
$\overline{M}$	Local average magnetization	203
M_{an}	Anhysteretic magnetization	411
M_f, M_s	Martensite finish and start temperatures	22
M_{irr}	Irreversible magnetization	199
M_{rev}	Reversible magnetization	199
M_R	Remanence magnetization	422
M_s	Saturation magnetization	426
M_0	Spontaneous magnetization	426
$\langle M_+ \rangle$, $\langle M_- \rangle$	Average magnetization due to positively and negatively oriented moments	218
M^+, M^-	Martensite phases	244
MPB	Morphotropic phase boundary	48
n_1, n_2	Refractive indices for optical fibers	37
n_e	Modal index for a Bragg optical sensor	40
N_+, N_-	Number of positively and negatively oriented dipoles or moments	74
p_{+-}, p_{-+}	Likelihood of dipole, moment switch from + to − or conversely	102
p_0	Dipole strength	74
$\mathbf{P}, P$	Polarization	425
$\overline{P}$	Local average polarization	79
$\widehat{P}$	Polarization data	119
P_{an}	Switching polarization	85
P_{irr}	Irreversible polarization	91
P_{rev}	Reversible polarization	92
P_R	Remanence polarization	425
P_s	Saturation polarization	426
$\langle P_+ \rangle$, $\langle P_- \rangle$	Average polarization due to positively and negatively oriented dipoles	102
P_0	Spontaneous polarization	426
Q	Heat	58
Q	Electric charge	135
R	Correlation between cells of Mg cations in PMN	146
s	Compliance	56

Symbol	Meaning	Page
	Material Properties and Models (Continued)	
s_i	Dipole, spin or moment orientations	74
s_1, s_2	Thresholds in Preisach models	191
S	Entropy	58
S	Preisach plane	81
S_A, S_M	Specific entropies for austenite and martensite	254
S_-	Index set of negative dipole states	106
$S^+(t)$, $S^-(t)$	Sets in the Preisach plane of positively and negatively oriented moments	194
$S_\triangle$	Compact Preisach plane	193
S_{irr}	Entropy production due to irreversible processes	59
t	Time	82
t_s	Switching time	82
T	Temperature	58
T_c	Curie point (transition temperature)	413
$\overline{T}_c$	Local Curie point	148
T_E	Temperature of surrounding environment	126
T_f	Freezing temperature	142
T_g	Glass transition temperature	30
T_R	Temperature of a reference state	254
T_0	Curie temperature	414
$\mathcal{T}$	Relaxation time	104
U	Internal energy	58
U_A, U_E, U_λ, U_M	Anisotropy, exchange, magnetomechanical and magnetostatic energies	163
V	Voltage	68
V	Control volume	102
W	Work	58
x_A, x_M	Fractions of austenite, martensite variants	273
x_+, x_-	Fractions of positively and negatively oriented dipoles or moments	102
X	Density of neighboring B' sites containing Mg cations in PMN	146
Y_A, Y_M	Young's moduli of austenite, martensite	256
$Y^P_{ijk\ell}$	Young's modulus at constant polarization	57

Symbol	Meaning	Page
	Numerical Approximation	
DP_n	Degree of precision	376
E_n	Quadrature error	376
$\mathbb{K}$	Stiffness matrix	386
L_1, L_2, L_3	Area coordinates	383
$\mathbb{M}$	Mass matrix	386
$\mathbb{Q}$	Damping matrix	386
	Operators and Spaces	
$\mathcal{C}$	Observation operator	119
$C^n[a,b]$	Space of functions that are n-times continuously differentiable on the interval $[a,b]$	412
$C(0,T;X)$	Space of X-valued continuous functions on $[0,T]$	412
$H^p(a,b)$	Sobolev space of functions whose first p distributional derivatives exist in $L^2(a,b)$	419
$H_0^p(a,b)$	Subset of $H^p(a,b)$ whose elements satisfy essential boundary conditions	419
$\mathcal{K}$	Parameter-to-observation map	119
$\mathcal{K}^\dagger$	Moore–Penrose generalized inverse of $\mathcal{K}$	119
$L^2(a,b)$	Space of square integrable functions	421
$L^2(0,T;X)$	Space of X-valued square integrable functions on $(0,T)$	421
$\mathcal{Q}$	Parameter space	119
$\mathcal{R}(\mathcal{K})$	Range of the operator $\mathcal{K}$	119
$T(q)$	Least squares functional	119
$T_\alpha(q)$	Regularized least squares functional	120
$\mathcal{Y}$	Observation space	119
	Structural Models	
γ	Air damping coefficient	316
ε_x	Strain at thickness coordinate z (also $\varepsilon_\beta, \varepsilon_{\alpha\beta}$, $\varepsilon_{\beta\alpha}$ where $\alpha, \beta = x, y, \theta$)	329
$\theta_\alpha, \theta_\beta, \theta_x, \theta_y$	Neutral surface rotation	344
κ	Curvature	318

Symbol	Meaning	Page
	Structural Models (Continued)	
ρ	Density	307
σ_x	Normal stress (also σ_β where $\beta = y, z, \theta$)	329
σ_{xy}	Shear stress (also $\sigma_{\alpha\beta}, \sigma_{\alpha z}$ where $\alpha, \beta = x, y, \theta$)	329
σ_1, σ_2	Sesquilinear forms	366
χ_{pe}	Characteristic function	324
Ω	Structural region	325
$\mathcal{A}$	Action integral	312
$\mathcal{A}$	System matrix	369
c	Kelvin–Voigt damping parameter	307
c_ℓ	Boundary damping coefficient	308
e_x	Reference surface strain (also $e_\beta, e_{\alpha\beta}, e_{\beta\alpha}$ where $\alpha, \beta = x, y, \theta$)	329
$\mathcal{H}$	Total energy	311
H_D	Dome height	360
I	Moment of inertia	319
k_ℓ	Boundary stiffness coefficient	308
K	Kinetic energy	311
$\mathcal{L}$	Lagrangian	311
m_ℓ	Boundary mass coefficient	308
M	Bending moment (also $M_\alpha, M_\beta, M_{\alpha,\beta}, M_{\beta,\alpha}$ where $\alpha, \beta = x, y, \theta$)	327
N	In-plane force (also $N_\alpha, N_\beta, N_{\alpha,\beta}, N_{\beta,\alpha}$ where $\alpha, \beta = x, y, \theta$)	328
Q	Shear force (also Q_α, Q_β where $\alpha, \beta = x, y, \theta$)	327
R_α, R_β	Radius of curvature	342
u, u^N	Longitudinal displacements	308
U	Potential energy	311
v, v^N	Circumferential displacements	327
V, V^N	Spaces of test functions	310
$\mathcal{V}$	Product space of test functions $\mathcal{V} = V \times V$	368
w, w^N	Transverse displacements	316
X	State space	310
$\mathcal{X}$	Product state space $\mathcal{X} = V \times X$	368
Y	Young's modulus	307
z_n	Neutral line or surface	306

Elements and Compounds

Element/Compound	Name	Designation
Al	Aluminum	
Au	Gold	
Ba	Barium	
$BaTiO_3$	Barium titanate	BT
$CaTiO_3$	Calcium titanate	CT
Cd	Cadmium	
Co	Cobalt	
Cu	Copper	
CuZn	Brass	
Dy	Dysprosium	
Fe	Iron	
Fe_3O_4	Iron oxide (magnetite)	Loadstone
Ga	Galium	
Ho	Holmium	
K	Potassium	
KH_2PO_4	Potassium dihydrogen phosphate	KDP
Mg	Magnesium	
Mn	Manganese	
Ni	Nickel	
Ni_xTi_{1-x}	Nickel titanium	Nitinol
Nb	Niobium	
O	Oxygen	
P	Phosphorus	
Pb	Lead	
$Pb(La,Zr,Ti)O_3$	Lead lanthanum zirconate titanate	PLZT

Element/Compound	Name	Designation
$Pb(Mg_{1/3},Nb_{2/3})O_3$	Lead magnesium niobate	PMN
$Pb(Mg_{1/3},Ta_{2/3})O_3$		PMT
$Pb(Sc_{1/2},Ta_{1/2})O_3$		PST
$PbTiO_3$	Lead titanate	PT
$PbZrO_3$	Lead zirconate	
$Pb(Zr,Ti)O_3$	Lead zirconate titanate	PZT
	polyvinylidene fluoride	PVDF
Sc	Scandium	
Si	Silicon	
SiO_2	Silicon oxide	Glass
Sr	Strontium	
$SrTiO_3$	Strontium titanate	ST
Ta	Tantalum	
Tb	Terbium	
$Tb_xDy_{1-x}Fe$		Terfenol-D
Ti	Titanium	
Zn	Zinc	

Abbreviations for Units

Symbol	Units	Quantity
μm	micrometer (10^{-6} m)	Length
A	ampere	Electric current
Å	angstrom (10^{-10} m)	Length
C	coulomb	Electric charge
^{o}C	degree Celsius	Temperature
dB	decibel	Sound pressure level
F	farad (1 C/V)	Capacitance
GPa	gigapascal (10^9 Pa)	Pressure
H	henry	Inductance
Hz	hertz	Frequency
J	joule	Energy
kg	kilogram (10^3 g)	Mass
ksi	kilo-pounds per square inch	Pressure
K	kelvin	Temperature
m	meter	Length
mil	10^{-3} inch	Length
MJ	megajoule (10^6 J)	Energy
MPa	megapascal (10^6 Pa)	Pressure
ms	millisecond (10^{-3} s)	Time
nm	nanometer (10^{-9} m)	Length
N	newton	Force
Pa	pascal	Pressure
psi	pounds per square inch	Pressure
s	second	Time
T	tesla	Magnetic induction
V	volt	Electric potential
W	watt	Power
Wb	weber	Magnetic flux

Chapter 1

Smart Material Applications

1.1 Smart Material Systems

Increased demands for high performance control design in combination with recent advances in material science have produced a class of systems termed smart, intelligent or adaptive systems. While subtle differences may be associated with the individual terms, smart systems are generally defined as ensembles whose dynamics can be monitored or modified by distributed sensors and actuators, in accordance with an integrated control law, to accommodate time-varying exogenous inputs or changing environmental conditions. Specific choices for the actuators, sensors and control laws are dictated by the design requirements for the system.

For aeronautic and aerospace systems, control transducers must be lightweight and should typically have minimal effect on the passive system dynamics. Furthermore, actuators must provide the required strain or force inputs using the available power supplies which, in certain aerospace structures, may require the scavenging of power from other components in the system. Restrictions on size and weight also dictate that transducers in some regimes must be capable of multiple roles. For example, transducers which monitor and control vibrations in an aircraft fuselage may also be required to act as inputs and sensors for health monitoring or nondestructive evaluation of the structure. The limitations on the mass and size of transducers are often relaxed in industrial applications but output requirements may be more stringent. For example, magnetostrictive transducers employed in the cutting head of a milling machine can weigh several pounds but are required to achieve cutting tolerances on the order of 1 μm while operating in kilohertz regimes.

Actuators and sensors comprised of smart or active materials can meet many of these criteria. Like the definition for smart systems, the definitions of smart or active materials can vary between fields. We define active actuator materials as those which convert electrical, magnetic or thermal energy to mechanical energy whereas sensor effects are provided by the opposite conversion of energy. In a similar vein, we define smart material transducers as fully integrated devices which employ smart materials in concert with the hardware required to generate the electric, magnetic, thermal, or stress fields required for actuation or measured during sensing. The difference between the constituent smart materials and smart material transducers is illustrated by a magnetostrictive transducer design presently employed for high accuracy, high speed industrial milling. As detailed in Section 1.3, a Terfenol-D rod provides the input/diagnostic capabilities whereas the full transducer is additionally comprised of a prestress mechanism, wound-wire solenoid and surrounding magnet which are used to generate or measure magnetic fields and shape the flux for optimal actuator/sensor design.

A review of the literature or consideration of conference programs in 1990 would reveal that smart materials research at that time was focused primarily on piezoelectric materials with emerging emphasis on shape memory alloys (SMA), magnetostrictive compounds, electrorheological (ER) and magnetorheological (MR) fluids, PVDF films, polymer gels and fiber optic sensors. Whereas the nonlinear and potentially hysteretic constitutive behavior of most of these compounds was recognized and being quantified, the majority of models and model-based control designs for systems utilizing smart material actuators and sensors were linear.

In the subsequent time period, research on all of these materials has burgeoned and advances in materials science have produced compounds such as ionic polymers, electrostrictive polymers, and ferromagnetic shape memory alloys (FSMA) which have the potential for providing unique transducer capabilities for high performance aerospace, aeronautic, automotive, industrial and biomedical applications. Figure 1.1 summarizes a number of present and projected applications and the material behavior which must be accommodated when designing smart material transducers for these systems.

Piezoelectric and ferroelectric materials and transducers are widely considered for smart structure design due to the fact that they are lightweight and compact, relatively inexpensive, and exhibit moderately linear field-strain relations at low drive levels. They also exhibit broadband drive capabilities and very high set point accuracy which, to date, has made them the material of choice for stages in nanopositioners. Finally, the converse and direct piezoelectric effects provide these materials with both actuator and sensor capabilities. Due to the noncentrosymmetric nature of ferroelectric materials, they exhibit hysteresis and constitutive nonlinearities at all drive levels. For low input regimes, these deleterious effects can typically be mitigated through feedback mechanisms. In high drive regimes, however, it is necessary to employ charge or current control, or models and control designs which incorporate hysteresis to achieve the tolerances required for nanopositioning or high accuracy tracking.

In certain applications, electrostrictive transducers constructed from relaxor ferroelectric materials are advantageous over piezoelectric materials due to the fact that they exhibit minimal hysteresis when employed in the diffuse transition region near the material's bulk Curie point. This makes them advantageous in applications ranging from sonar transduction to deformable mirror design. Unlike piezoelectric materials, electrostrictive compounds are not poled and hence exhibit few aging effects. However, their highly temperature-dependent and nonlinear saturation behavior must be accommodated when designing control systems which incorporate these compounds.

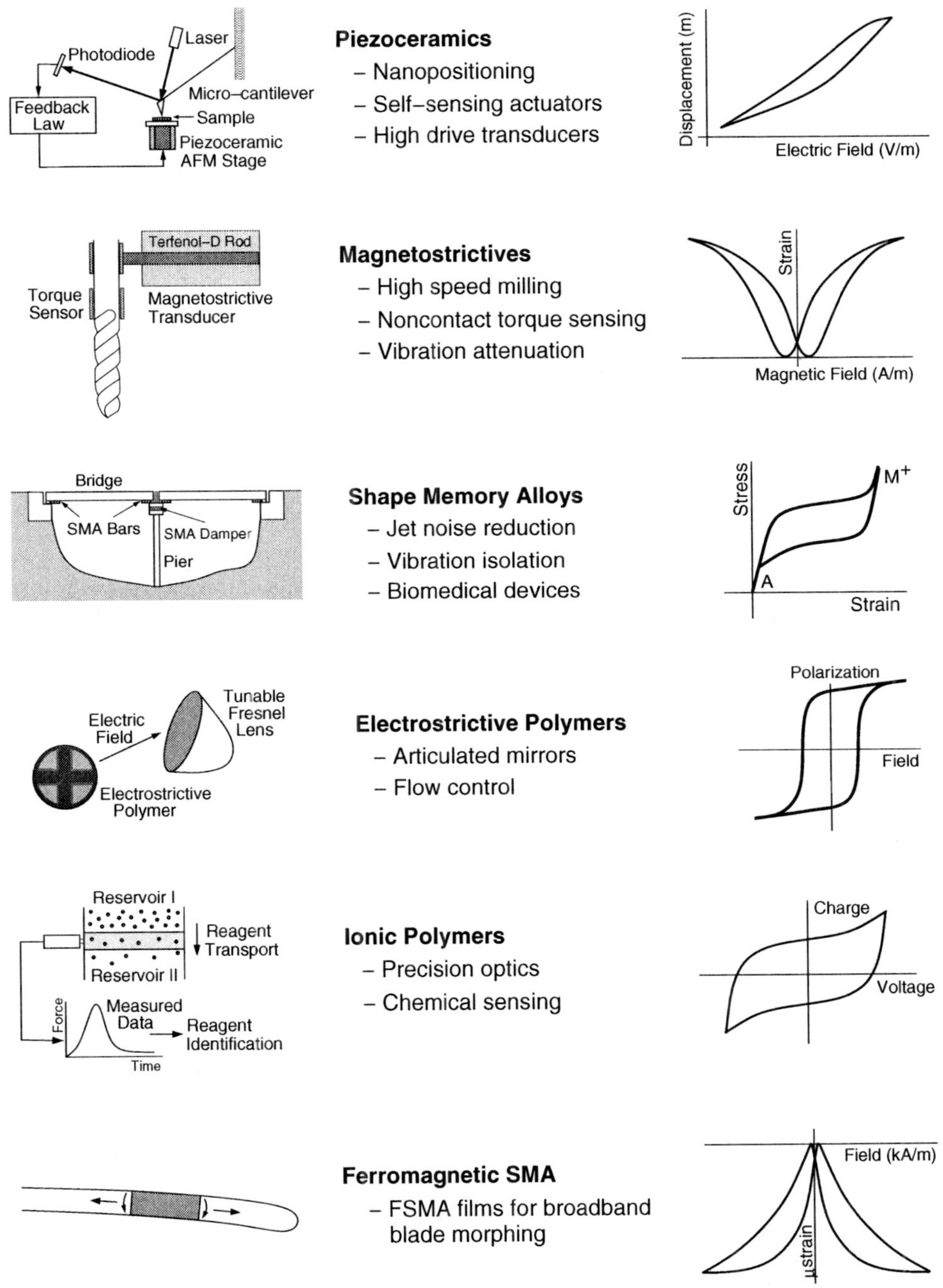

Figure 1.1. *Materials under consideration for aerospace, aeronautic, industrial and biomedical applications and the inherently nonlinear and hysteretic constitutive behavior exhibited by these compounds.*

The magnetic analogue of electrostrictive compounds are magnetostrictive materials which convert magnetic energy into mechanical energy and conversely. Due to the circuits required to generate the driving magnetic fields, transducers which utilize magnetostrictive cores are larger and more massive than piezoelectric or electrostrictive patches. The ruggedness, giant forces, and moderate strains generated by the transducers, however, make them advantageous in certain industrial systems — such as transducers for high speed milling — and as material properties and transducer designs are refined, the scope of their application should rapidly increase. From a modeling and control perspective, the nonlinearities and hysteresis inherent to the materials at moderate to high drive levels must be accommodated before the materials can be utilized to their full potential.

Shape memory alloys (SMA) are being increasingly considered for civil, aeronautic, aerospace and industrial applications which require significant passive damping or utilize the high work output densities exhibited by the materials. Because the energy dissipated by the materials is proportional to the area of the hysteresis loop, pseudoelastic operating regimes which *maximize hysteresis* are required when employing SMA as tendons to attenuate earthquake or wind-induced vibrations in buildings or as fibers to eliminate vibrations in articulated antennas or membrane mirrors. The utilization of temperature-induced phase transitions to provide actuator capabilities is under intense investigation in the context of microelectromechanical systems (MEMS), thin film SMAs, and microactuator applications since surface area to volume ratios in these geometries promote rapid cooling and hence higher frequency drive capabilities.

The goal of obtaining higher drive frequencies while maintaining high work densities has motivated the development of ferromagnetic shape memory alloy (FSMA) materials which rely on magnetic field-induced phase transitions to provide actuator inputs. Initial investigations focused on FSMA films have demonstrated drive frequencies on the order of 5 kHz with potential for reaching 10–15 kHz as compared with SMA films which presently have maximum operating frequencies on the order of 100 Hz.

Electrostrictive polymers including PVDF, polyimides and polymeric elastomers are lightweight, highly flexible and malleable, and provide both actuator and sensor capabilities. For these reasons, they are being investigated for use as remote lens cleaners for aerospace missions, acoustic pressure sensors, flow control actuators, synthetic jets, artificial muscles for robotic units, backing material for membrane mirrors and tunable lenses, and actuator implants to stimulate tissue and bone growth. However, their success is predicated on the quantification of constitutive nonlinearities and hysteresis in a manner which promotes real-time control design.

Ionic polymers differ from PVDF, polyimides and polymeric elastomers in that electromechanical coupling in ionic polymers is produced through the transport of charged and uncharged ions within the polymer matrix. This provides them with both mechanical actuator and sensor capabilities and the potential for chemical and biological sensing.

The design of smart structures which utilize these materials requires both the characterization of their constitutive properties and the development of cou-

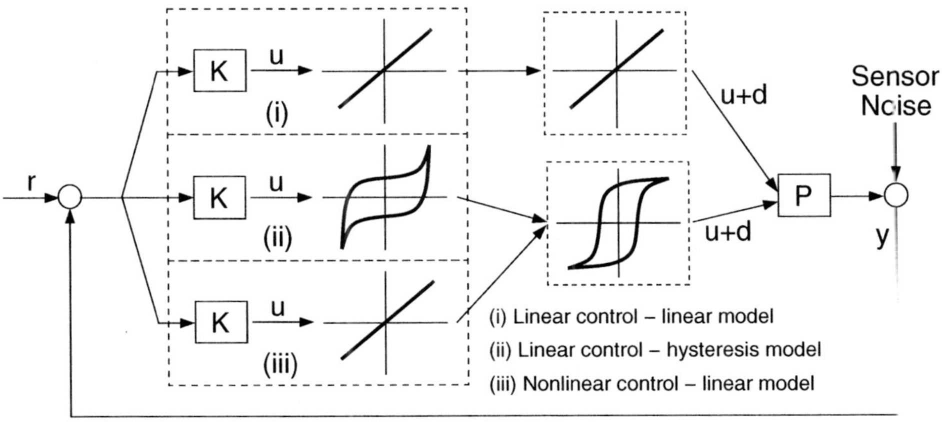

Figure 1.2. *(i) Linear robust control design employing a linear model. (ii) Linear robust control design utilizing a model-based inverse filter for hysteretic transducers, and (iii) Nonlinear control design employing a linear model for hysteretic transducers.*

pled models which quantify their interaction with underlying systems. Control laws must be compatible with the properties of the sensors and actuators as well as the mechanisms through which they interact with the system. For example, a number of the previously mentioned actuators yield unbounded (discontinuous) input operators in the mathematical formulation of the control problem. The extension of control theories to this regime has been been completed in certain applications but is lacking in general. Moreover, all of the active materials exhibit nonlinear dynamics and hysteresis at high drive levels. This must be incorporated in both the models and control methods before the materials can be utilized to their full capability in smart structure design. As depicted in Figure 1.2, control design for nonlinear and hysteretic transducers yields a hierarchy of algorithms ranging from linear control algorithms utilizing approximate inverse models to fully nonlinear control laws which directly accommodate the hysteresis and constitutive nonlinearities inherent to the transducer materials. While aspects of the nonlinear designs have been addressed, the state of the theory lags far behind that for the linear case.

Because the dynamics of a smart structure are dependent upon the attributes of the constituent active materials, it is necessary to consider the development of linear, nonlinear and hysteretic constitutive relations, their incorporation in coupled system models, numerical approximation of these models, and subsequent control design in a concerted manner. This facilitates the development of models amenable to control design, and the formulation of control laws which are compatible with physical attributes of active sensor and actuator materials. By incorporating known physics into the models, the degree to which control laws must attenuate unmodeled dynamics is reduced thus improving their performance in high performance applications utilizing smart material transducers.

1.2 Piezoelectric and Electrostrictive Applications

1.2.1 History

To illustrate the evolution of piezoelectric materials and ferroelectric toward smart material applications, we summarize briefly their history. A detailed treatment of this history can be found in the classic texts [73, 234, 243, 325].

Piezoelectricity, which literally means "pressure electricity" from the Greek word "piezo" for pressure, was discovered by Pierre and Jacque Curie in 1880 while Pierre was investigating the relationship between pyroelectricity and certain crystal symmetries. In studies initially focused on tourmaline and later extended to quartz, cane sugar, and Rochelle salt, the Curie brothers were able to demonstrate the generation of electric charge in response to applied pressure or stresses. This is the direct piezoelectric effect which, in present materials, can produce voltages ranging from a fraction of a volt to several thousand volts. The converse effect, which constitutes the generation of strains or displacements in the material in response to applied fields, was subsequently justified using thermodynamic principles. Both effects are due to the noncentrosymmetric nature of certain ceramics, polymers and biological systems and it is this property which also produces the switching-induced hysteresis and constitutive nonlinearities inherent to ferroelectric and piezoelectric materials.

An initial *linear* characterization of the electromechanical properties of piezoelectric materials was published in 1910 in Voigt's *Lehrbuch der Kristallphysik* [500] which established the notation still employed in linear piezoelectric models. In 1916, Paul Langevin developed a transducer comprised of a quartz crystal sandwiched between two metal plates which functioned as an ultrasonic submarine detector [326]. Sound waves in water were produced when the quartz was used to oscillate the plates and a second quartz device was employed to measure rebounding waves. The distance to the reflecting source was computed from the time between the emitted and received acoustic signals. This was a forerunner of modern sonar and exemplifies the dual use of actuator/sensor capabilities exploited in present smart material designs.

From 1916 through the 1950's, research focused both on material development and the use of new materials to create novel technologies, with a number of the contributions made by Walter Cady. During this time period, the KDP (potassium dihydrogen phosphate) family was discovered in 1940 and the first piezoceramic compound $BaTiO_3$ (barium titanate) was produced. $BaTiO_3$ was the first of the perovskite family of piezoceramic materials to receive widespread usage and it served as a precursor to lead zirconate titanate (PZT) and lead magnesium niobate (PMN) — discovered in the 1950's — which presently are among the most widely employed ferroelectric compounds due to their high dielectric and piezoelectric strengths, moderate costs, and broad range of operating temperatures. The technologies based on these materials during this time period included the development of crystal phonograph pickups, microphones, wave filters for multichannel telephone design, and radio communication advances based on quartz crystals — over 50 million quartz crystals were employed by the United States during WWII. As detailed by Mason [325, 326], a substantial portion of both the material development and invention of technologies based on these materials was performed during this period at Bell Laboratories.

During the 1960's and 1970's, research focused both on the development of novel technologies utilizing $BaTiO_3$, PZT and PMN, and the investigation of new compounds to eliminate limitations due to the rigid nature of piezoceramics. This latter goal was achieved in 1969 by Kawai when he discovered that polyvinylidene fluoride (PVDF) had piezoelectric coefficients approximately ten times larger than previously discovered polymers [261]. PVDF has subsequently been widely considered for both sensor and actuator applications and to date has been the most widely used piezoelectric polymer.

From 1980 through the present time, research on both material development and the development of high performance aerospace, aeronautic, industrial and biomedical applications, based on ferroelectric and piezoelectric compounds, burgeoned. Materials research has focused on the development of single crystal materials which exhibit strains approaching 1% while extensive polymer research led to the production of polyimides [337], elastomeric and amorphous polymers [209, 468] and biological polymers [169, 170]. Applications utilizing the direct piezoelectric effect include gas igniters, accelerometers employing PZT disks which play a central role in automotive airbag systems, and mode-specific sensors based on geometrically-configured PVDF films. Commercial actuator applications include dot matrix printer heads, auto-tracking devices for VCR's which avoid magnetic noise, shutter mechanisms for cameras, and the PZT-based TEMS (Toyota Electronic Modulated Suspension) which was produced in 1989 to augment shock absorber capabilities (see [489] and included references). Piezo-actuators have also played a pivotal role in nanotechnology starting with their use as positioning elements in scanning tunneling microscopes (STM) in 1982 and atomic force microscopes (AFM) in 1985 and continuing to the present in essentially all nanopositioner applications.

In addition to the development of novel materials and technologies which has occurred in the last 20 years, three research areas have become increasingly important — (i) the development of nonlinear and hysteretic constitutive relations and construction of fully integrated system models based on linear and nonlinear constitutive relations, (ii) full and reduced-order approximation techniques for discretizing these system models, and (iii) model-based control designs which incorporate known physics to achieve the stringent design criteria dictated by present and projected applications. Furthermore, it has been increasingly recognized that these components should ideally be investigated in concert to achieve the novel design capabilities provided by these materials.

1.2.2 Prototypical Applications

The breadth of systems employing piezoelectric transducers precludes a survey and we instead focus on several prototypical applications which illustrate issues to be addressed when developing constitutive and system models, full and reduced-order approximation techniques and model-based control designs.

Structural Shape Modification

Many of the initial applications utilizing both the sensing and actuating capabilities of piezoelectric materials focused on structural shape modification or vibration control. This provided both a testbed for theoretical development and a source of emerging technologies.

To illustrate a theoretical prototype, consider the thin beam with surface-mounted piezoceramic patches depicted in Figure 1.3. Through the input of diametrically out-of-phase voltages to the patches, bending moments are created in the beam thus producing transverse displacements. This provides actuator capabilities for the configuration. Sensor capabilities are provided by the direct piezoelectric effect in which stresses in the patch produce charges and corresponding voltages. A truly integrated smart structure utilizes either self-sensing actuators or sensor-actuator pairs to provide the observations and inputs required for vibration attenuation [11, 103, 135].

Theoretical issues which have been investigated in this context are the development of linear and nonlinear constitutive relations which quantify the patch inputs to the structure as well as the passive effects contributed by the patch material to the structural dynamics. This configuration provides a natural prototype for investigating numerical approximation techniques since beam models are sufficiently complex to encompass a number of issues associated with more complex structures but avoid difficulties associated with shear locking and 2-D analysis which make the approximation of plate and shell models difficult. From the perspective of control design, this has been an important prototype for investigating the ramifications of the unbounded (discontinuous) control input operators which result from the piecewise input regions provided by the patches.

A technological prototype illustrating the use of piezoceramic patches to provide shape or structural changes to enhance system performance is illustrated by the flap assembly from [267] which is depicted in Figure 1.4. In this design, the

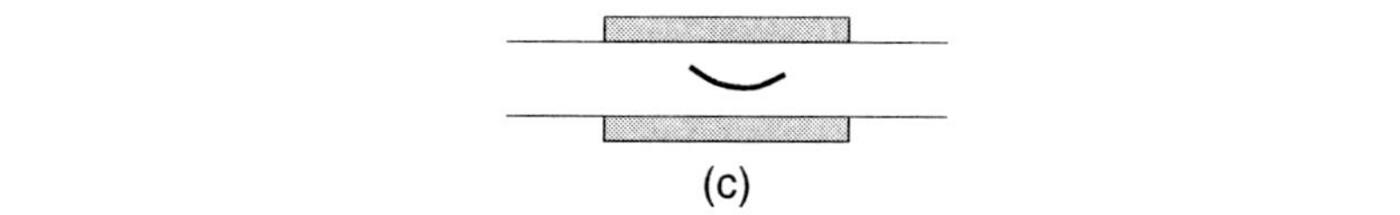

Figure 1.3. *(a) Thin beam with surface-mounted piezoceramic patches. (b) Bending moments generated by out-of-phase voltages to the patches. (c) Identification of structural damage using the dual actuator/sensor capabilities of PZT transducers.*

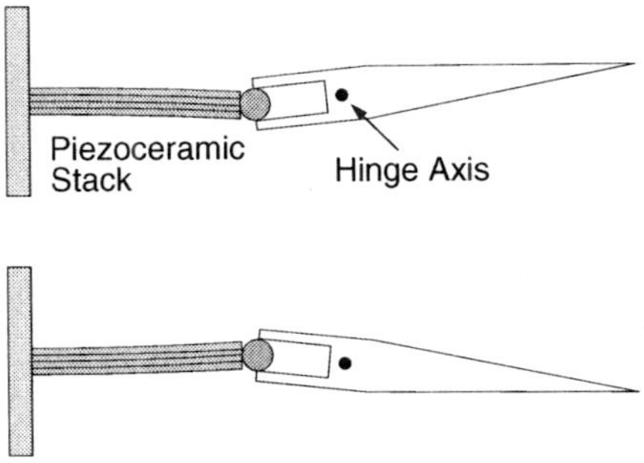

Figure 1.4. *Piezoceramic stack employed for flap rotation (after [267]).*

strains produced by stacked piezoelectric actuators are amplified by a rod and cusp assembly to provide the stroke required to rotate flaps. This research is directed at modifying the aerodynamic properties of aircraft wings or helicopter rotor blades.

Structural Health Monitoring

The dual sensor and actuator capabilities of piezoelectric and electrostrictive materials also provide the possibility for health monitoring in smart structures. As detailed in [33], this can be accomplished by driving the structure using the actuator facility of the material and sensing its response using the direct piezoelectric or electrostrictive effects. The status of the structure can then be evaluated using various criteria. The simplest strategy is to compare the response to baseline data for the original structure. If significant deviations are detected, more detailed analysis can be performed. If sufficiently accurate models are employed, this can include the identification of defects of the type depicted in Figure 1.3(c) through least squares fits to the data. Details regarding these procedures are provided in [33, 235] and references therein. To date, essentially all structural health monitoring techniques exploiting piezoceramic or electrostrictive transducers have been based on low drive dynamics for which linear models provide sufficient accuracy.

Structural Acoustic Systems

A second class of smart structures in which piezoceramic actuators and sensors have played a fundamental role are those involved in structural acoustic systems. Structure-borne noise arises in settings ranging from aircraft and automobiles to fields emanating from high voltage transformers. In all cases, the unwanted noise is generated by structural vibrations produced by an adjacent source (e.g., aircraft engines, impinging flowfields, vibrating machinery, electromagnetic cores). Smart materials provide the capability for reducing structure-borne noise by modifying the structural dynamics to regimes which couple less effectively with acoustic fields. This capability can be enhanced through the use of models which predict sound power levels as a function of structural displacements and velocities. In this case, piezoceramic patches can be utilized to sense the structural strains and produce bending moments in accordance with a structural acoustic model-based control law [99].

The sound pressure levels originally considered in aircraft, automotive and industrial systems were typically linear, and linear PZT models were employed for characterization and control design. As noted in [99], feedforward methods are currently employed for many structural acoustic applications. Because these techniques rely on superposition principles, they are inherently linear and hence require either linear models or linear filtering techniques such as inverse compensation.

Current and future structural acoustic applications are beginning to focus on regimes which are highly nonlinear and will require the development of nonlinear models and control techniques. For example, the sound pressure levels produced in a space launch vehicle payload fairing during liftoff or impinging on an aircraft bay at supersonic speeds can exceed 150 dB [149, 404, 513]. Smart materials employed in these applications will quite likely be operating in highly nonlinear ranges thus necessitating the use of nonlinear constitutive and hysteresis models of the type developed in Chapter 2.

A second area of active research is the development of full and reduced-order approximation techniques suitable for structural acoustic configurations such as the prototype depicted in Figure 1.5. This necessitates the approximation of shell models coupled to 3-D acoustic fields with general boundary conditions which are significantly more complex than modal approximations derived under the assumption of cylindrical geometries and simply-supported boundary conditions as often considered for initial theoretical or experimental prototypes.

Finally, optimal control designs for fully coupled structural acoustic systems with piezoceramic actuators has spawned extensive research on the propagation of discontinuous inputs for coupled hyperbolic/parabolic systems. This analysis has focused primarily on linear systems with linear input operators, and significant extensions to the theory are required to accommodate the constitutive nonlinearities and hysteresis present in high drive regimes.

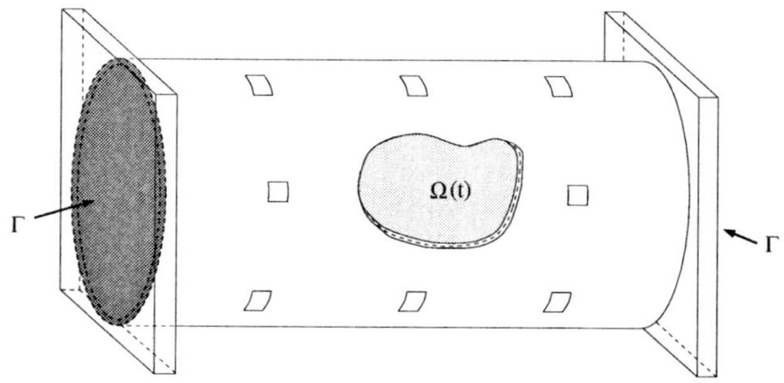

Figure 1.5. *Thin shell with surface-mounted piezoceramic patches which encloses an acoustic cavity.*

RAINBOW and THUNDER Transducers

A number of transducer designs have been developed, or are under investigation, to augment strain, force, or drive level capabilities of the constituent piezoelectric or electrostrictive materials through curvature enhancement, prestress augmentation, or geometrical coupling with surrounding materials or strain enhancement mechanisms. This includes RAINBOW (Reduced And InterNally Biased Oxide Wafer) [200, 201, 233, 294], THUNDER (THin layered UNimorph Driver and sEnsoR) [348, 350, 470, 509], and Lipca [349].

To illustrate, consider the THUNDER design depicted in Figure 1.6(a). As detailed in [77, 231, 350], present THUNDER designs are typically comprised of a piezoceramic wafer, a metallic backing layer, hot melt adhesive layers, and optional metallic top layers. During the manufacturing process, the assemblage is heated under pressure to temperatures in the proximity of the Curie point for PZT and then cooled to room temperature. During the cooling phase, the adhesive solidifies and internal stresses are developed in the constituent materials due to differing thermal properties. This produces the characteristic curved shape and prestresses which align dipoles to enhance strains generated by an applied field. Additionally, the backing layer provides robustness which allows the generation of large strains without damaging the transducer. The combination of robustness and curvature/prestress enhancement provides THUNDER with sufficiently large displacement capabilities to give great potential for applications including high speed valve design, synthetic jets for flow control, shape modification in space structures such as configurable mirrors, configurable shape modification of an airfoil to control flow characteristics, and linear motor design for microrobotics [372].

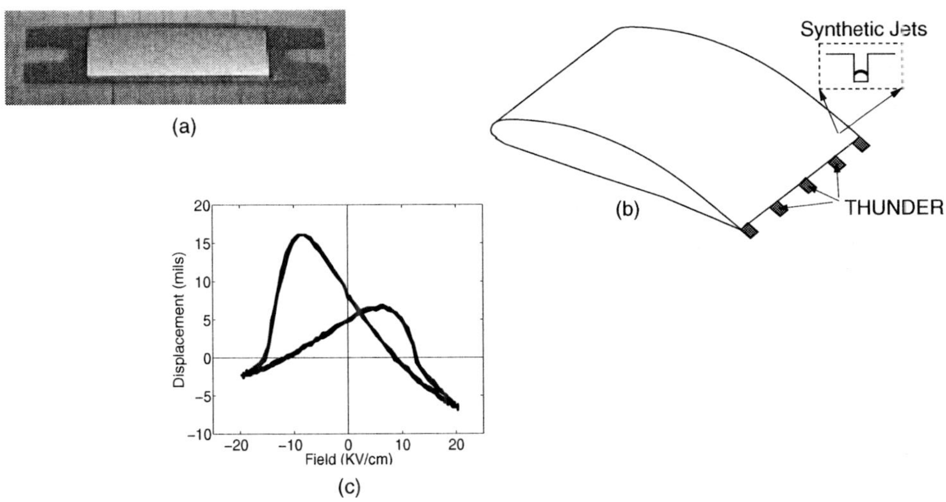

Figure 1.6. *(a) THUNDER transducer considered for (b) flow control and synthetic jet design. (c) Stress-dependent electromechanical behavior exhibited by THUNDER transducers.*

However, high drive levels in combination with stress-enhanced electromechanical coupling yield field-displacement relations which are nonlinear, hysteretic, and asymmetric as shown in Figure 1.6(c). This necessitates the development of nonlinear constitutive relations which incorporate both hysteresis and full electromechanical coupling, the construction and approximation of system models based on these constitutive relations, and the investigation of commensurate control designs to achieve the high drive capabilities provided by THUNDER.

Flextensional Design, Hybrid Transducers, and Ultrasonic Motors

A second class of composite transducers are flextensional designs which utilize PZT or PMN-PT-BT drivers [110, 221]. Prototypical designs are depicted in Figure 1.7, and the reader is referred to Section 1.3 and [118] for discussion about analogous transducers which employ magnetostrictive drivers. This class of transducers was originally developed for sonar projection but due to its wide range of hydrodynamic response, it has recently been used for applications including oil exploration and underwater imaging. For high drive levels and electrostrictive drivers, nonlinear models must be employed to characterize the drive dynamics. A second area of modeling and optimal design involves the determination of cap geometries which produce desired frequency responses for both sending and receiving hydrostatic signals. Because the transducers have both tunable actuator and sensor attributes, they provide a unique capability for certain adaptive smart structure applications.

A related class of transducers are the inchworm actuators which utilize piezoelectric, electrostrictive or magnetostrictive materials as drive elements and clamps [118]. A hybrid design utilizing piezoceramic and magnetostrictive components is illustrated in Figure 1.16 of Section 1.3. Such designs provide the capability for essentially unlimited displacements with speeds currently on the order of 1 mm/s. As the designs and drive electronics improve, the utilization of such devices in smart structure applications should continue to increase.

Finally, we note that a related, and very active area of research, is the design of high efficiency ultrasonic motors in which electrical inputs are converted to

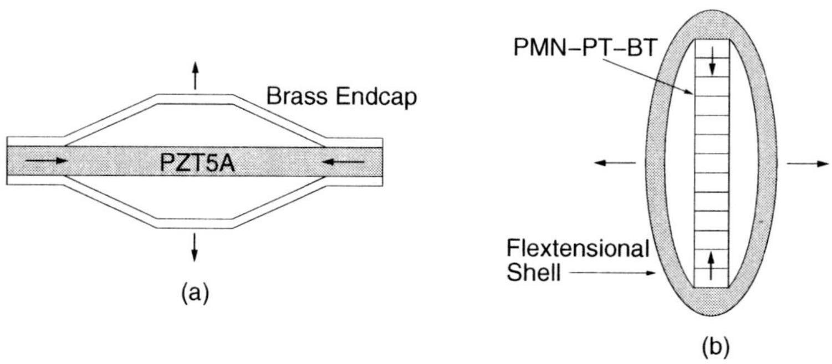

Figure 1.7. *Flextensional piezoceramic and electrostrictive transducers. (a) Piezoceramic cymbal actuator (after [110]) and (b) PMN-PT-BT transducer (after [221]).*

mechanical outputs using piezoelectric actuators. This provides the potential for eliminating the variability and aging associated with conventional hydraulic and electromagnetic motors. Rather than provide a partial summary of this research, we direct the reader to the work of Uchino [489] and references therein.

Nanopositioning Stages

Since the inception of the scanning tunneling microscope (STM) in 1982 and atomic force microscope (AFM) in 1985 [206], piezoceramic actuators have played a fundamental role in the design of stages for micro- and nano-positioning. This role has continually expanded as instruments have evolved to achieve increasingly stringent speed and accuracy specifications, and PZT-based stages are considered for applications ranging from nanoconstruction to the development of nuclear magnetic resonance microscopes (NMRM) with the goal of detecting single electron spins [398, 516].

To illustrate issues associated with both device design and subsequent model and control development, consider the prototypical AFM design depicted in Figure 1.8 and detailed in [206, 445, 446]. To ascertain the 3-D surface structure of a sample, the sample is moved laterally along an x-y grid using a PZT transducer and displacements in an adjacent microcantilever are monitored using a photodiode. Corresponding forces are determined via Hooke's law and a feedback law is used to compute z-displacements — also provided by a PZT actuator — which maintain constant forces. A complete scan in this manner provides a surface image of the compound as depicted in Figure 1.8(b).

The degree of accuracy provided by the PZT actuators when laterally and vertically positioning the sample is crucial to the resolution of the final image. Two representative transducer designs are illustrated in Figure 1.9. The first is comprised of a stage in which d_{33} inputs from PZT rods are used to provide lateral motion and

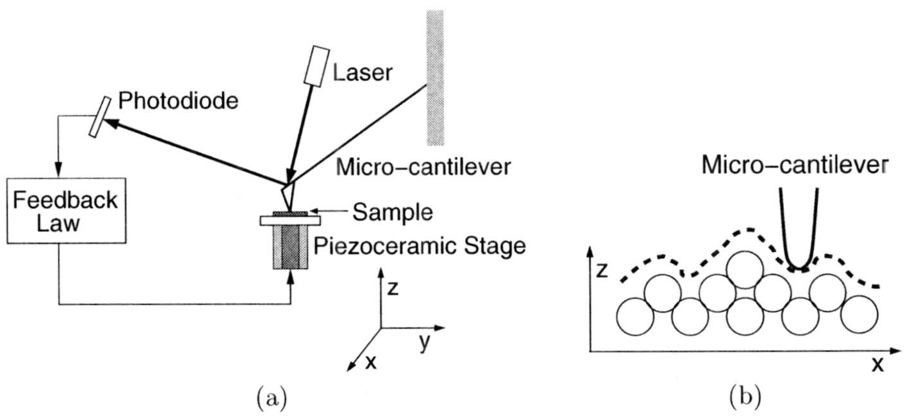

Figure 1.8. *(a) Configuration of a prototypical AFM. (b) Surface image determined by one lateral sweep.*

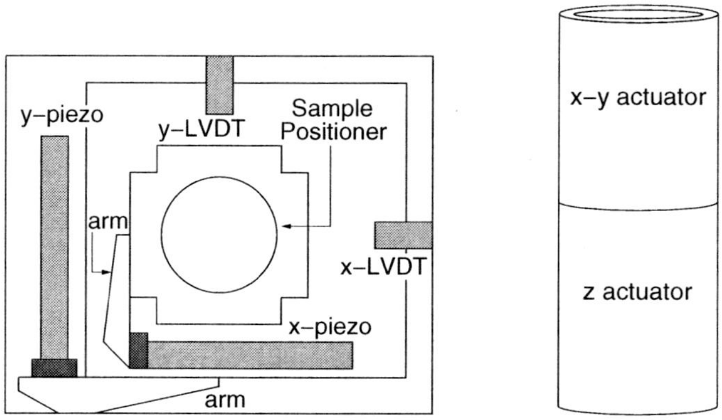

(a) (b)

Figure 1.9. *Actuator configurations employed for sample positioning in an AFM. (a) Stacked actuators employed as x- and y-stages and (b) cylindrical PZT transducer.*

a separate stack provides vertical or transverse displacements. In the second design, d_{31} and d_{33} inputs from a PZT shell respectively generate transverse and lateral motion. The stacked actuator stages offer the advantage of simple construction and reduced hysteresis whereas the shell configurations are better suited for isolation from exogenous vibrations and small stage construction. While both configurations provide highly repeatable and accurate set point placement, the relations between input fields and generated displacements exhibit hysteresis and constitutive nonlinearities as illustrated in Figure 1.10 with data collected from an AFM employing a stacked actuator.

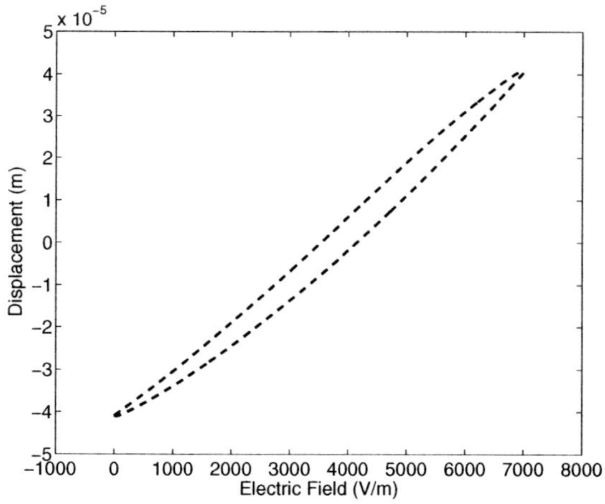

Figure 1.10. *Quasistatic relation between the input field E and displacements generated by a PZT stacked actuator in an AFM stage (from [436]).*

For certain drive regimes, the hysteresis and constitutive nonlinearities can be mitigated through either the drive electronics or feedback loops incorporated in the software. As detailed in [315, 316], the use of charge or current controlled amplifiers can essentially eliminate hysteresis. However, this mode of operation can be prohibitively expensive when compared with the more commonly employed voltage-controlled amplifiers, and current control is ineffective if maintaining DC offsets as is the case when the x-stage of an AFM is held in a fixed position while a sweep is performed with the y-stage. For low scan rates, PID or robust control designs can be employed to accommodate hysteresis [108, 401]. However, at the high scan rates required for real-time product diagnostics (e.g., semiconductor chips) or monitoring of biological processes (e.g., protein unfolding), increasing noise-to-data ratios and diminishing high-pass characteristics of control filters preclude a sole reliance on feedback laws to eliminate hysteresis. This motivates the development of control designs which incorporate and approximately compensate for hysteresis through model inverses employed either in feedback or feedforward loops. The models and inverse models must fully accommodate transient dynamics and be applicable to either the rod or shell geometries employed in present stage designs. This also necessitates the development of commensurate numerical approximation techniques and reduced-order numerical methods suitable for real-time implementation.

1.3 Magnetostrictive Transducers

1.3.1 History

The investigation of magnetostrictive materials began in 1842 when James P. Joule measured a change in the length of an iron sample when it was subjected to a magnetic field. Termed the Joule effect, the change in dimension due to a change in magnetization is phenomenologically similar to the converse piezoelectric effect and is the mechanism employed in most magnetostrictive actuators. The reciprocal behavior in which stresses produce a change in magnetization was discovered soon after this. Known as the Villari, magnetostrictive, or magnetomechanical effect, this mechanism is phenomenologically analogous to the direct piezoelectric effect and is commonly exploited in magnetostrictive sensors. Two additional magnetic phenomena which provide transducer capabilities are the Wiedemann effect, which is manifested as a twisting in the sample due to a helical magnetic field, and the inverse Wiedemann, or Matteucci effect, which has been exploited to produce magnetic torque sensors. Some of the initial magnetostrictive devices and technologies developed in the latter half of the 19th century are detailed in the monograph by Hunt [228].

During the first half of the 20th century, applications utilizing magnetostriction included telephone receivers, torque sensors, fog horns, and scanning sonar [118]. These transducers typically employed nickel, cobalt, and alloys of these compounds, which exhibited saturation magnetostrictions or strains on the order of 50×10^{-6} or 50 μL/L.

In 1963, "giant" magnetostrictive alloys comprised of the rare earth elements terbium and dysprosium were demonstrated to produce strains on the order of

10,000 μL/L but only at cryogenic temperatures. The combination of these rare earth alloys with the transition metal iron was independently and nearly simultaneously demonstrated by Clark and Belson of the Naval Ordnance Laboratory [92] — now the NSWC (Naval Surface Warfare Center) — and Koon, Schindler and Carter of the NRL (Naval Research Laboratory) [266] to produce alloys having giant magnetostrictive capabilities at room temperatures. This led to the development of Terfenol-D (terbium: Ter, iron: Fe, Naval Ordnance Laboratory: NOL, dysprosium: D) which exhibits room temperature magnetostriction up to 1600 μL/L with values up to 3600 μL/L achievable at resonance. While produced primarily as monolythic rods, Terfenol-D ($Tb_xDy_{1-x}Fe$) can also be incorporated in polymer matrix composites [138, 403] and thin films [47, 143, 306]. As detailed in [116–118, 391], applications which utilize Terfenol-D alloys include active vibration and noise control, micropositioning in high force regimes, medical and industrial ultrasonics, noncontact torque sensors, and tuned vibration absorbers.

Within the last 10 years, intense research has focused both on the development of novel materials having increased performance capabilities and technologies which exploit these materials. To address the goal of reducing hysteresis and material anisotropies, the inclusion of holmium has produced the alloy $Tb_xDy_yHo_zFe_{1.95}$ [518] whereas the replacement of terbium by gallium has yielded the transducer alloy Galfenol [93, 95]. A second area of current research is the development of ferromagnetic shape memory alloys (FSMA) which employ magnetic fields to produce phase transitions with the goal of achieving the strain capabilities of SMA with the dynamic response attributes of magnetostrictives. Present FSMA candidates include NiMnGa alloys, and details regarding this avenue of materials science and transducer design can be found in [90, 156, 244, 257, 367, 373, 455, 519].

1.3.2 Applications

We focus on a few selected applications which illustrate the issues associated with model development and control design rather than providing a survey of recent technologies which employ magnetostrictive actuators and sensors. A comprehensive summary of industrial, automotive and commercial applications employing Terfenol-D transducers can be found in [116–118, 391].

Terfenol-D Transducer Designs

Whereas the design of Terfenol-D transducers is still evolving as the capabilities of the materials become better understood, a common design for numerous applications is the piston transducer depicted in Figure 1.11. Strains in the Terfenol-D rod are produced when fields generated by current through the solenoid cause the rotation of magnetic moments in the manner illustrated in Figure 1.12. The sensor capabilities of the transducer are produced by the opposite effect in which stresses to the rod realign moments which produces subsequent changes in the magnetization. The prestress mechanism serves two roles: it increases the percentage of moments which are oriented perpendicular to the rod axis and it maintains the rod in a state of compression. The bias fields required to attain bidirectional strains

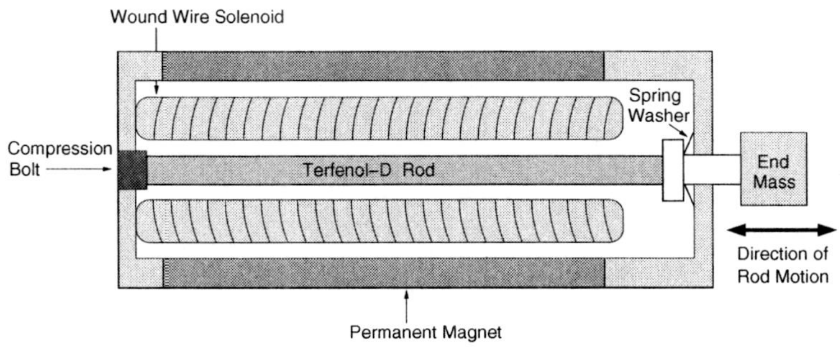

Figure 1.11. *Cross section of a prototypical Terfenol-D transducer.*

and improved linearity are provided either by a surrounding permanent magnet or through a prescribed DC current to the solenoid. Despite the fact that the permanent magnet adds bulk, it is typically employed since it reduces ohmic heating to the coil and provides additional flexibility for shaping the flux path to optimize performance. Modifications to this design include the use of laminated rods to reduce eddy current losses and the inclusion of tubing so that water flow around the solenoid can be used to regulate operating temperatures.

Typical H-M and H-ε relations, where H, M and ε respectively denote the magnetic field, magnetization and strain, are plotted in Figure 1.13 to illustrate the hysteresis and constitutive nonlinearities which must be incorporated in models and accommodated by model-based control algorithms. Whereas approximately linear strain and force outputs can be maintained at low drive levels through feedback loops, the inherent hysteresis and saturation nonlinearities produce delays and a degradation in accuracy at high stresses if neglected in the control design. Additionally, eddy current losses must be accommodated either through transducer design (e.g., laminated rods) or inclusion in the models. For certain applications, models must also incorporate full magnetomechanical coupling to characterize stress-induced changes in the magnetization analogous to those depicted in Figure 1.6 for PZT-based THUNDER actuators.

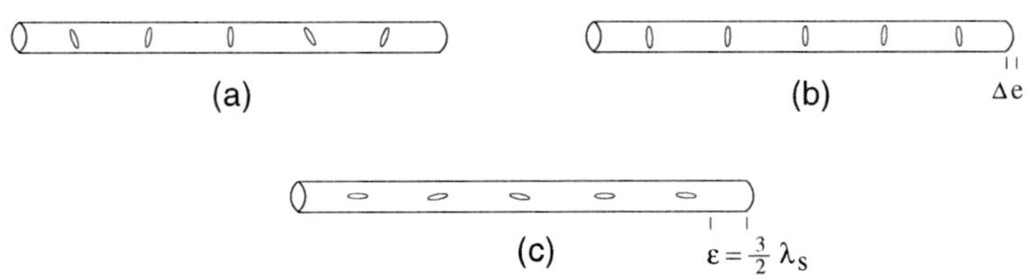

Figure 1.12. *Magnetic domains in a Terfenol-D rod: (a) orientation of unstressed rod in absence of applied magnetic field, (b) orientation of pre-stressed rod with no applied field, and (c) orientation of pre-stressed rod when field is applied in direction of longitudinal rod axis.*

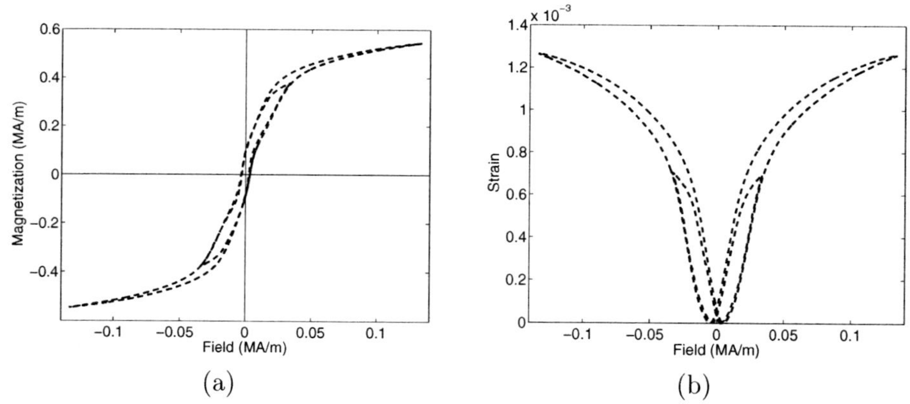

Figure 1.13. *Hysteretic data measured in a Terfenol-D transducer as reported in [119]: (a) field-magnetization relation, and (b) field-strain relation.*

Structural, Acoustic and Industrial Applications

Examples illustrating the manner through which magnetostrictive transducers can be employed in structural, acoustic and industrial systems are provided in Figure 1.14. The configuration (a) provides a testbed for investigating the theoretical

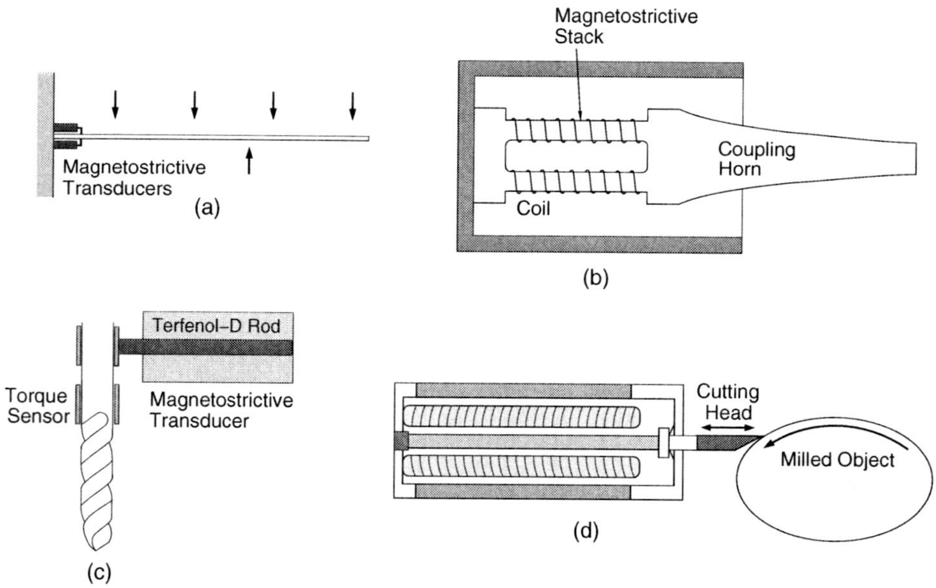

Figure 1.14. *Applications utilizing magnetostrictive transducers. (a) Vibration sensing and attenuation, (b) ultrasonic horn (after [118]), (c) torque sensing and vibration attenuation in a milling machine, and (d) high speed, high accuracy milling.*

development of actuator/sensor models and model-based control designs as well as an initial experimental testbed for model and control validation. Figure 1.14(b) illustrates the coupling of a Terfenol-D transducer with an ultrasonic horn for applications ranging from the cleaning of intricate or inaccessible machinery to the catalysis of chemical reactions. The industrial application (c) illustrates a setting in which the large forces generated by the transducer can be used to attenuate vibrations while the inverse Wiedemann, or Matteucci, effect is employed in a noncontact torque sensor. A second milling application is depicted in (d) where a transducer is employed for cutting out-of-round automotive parts at speeds of 3500 rpm and cutting tolerances of 1-2 μm [432,451]. A typical cutting trajectory is plotted in Figure 1.15(a) to illustrate that following an initial ramp and hold to bring the cutting head adjacent to the ingot, the actuator must periodically drive the cutting head while maintaining the specified tolerances. Results from [358, 359] are plotted in Figures 1.15(c) and (d) to illustrate that robust control design alone cannot achieve the specified tolerance due to saturation nonlinearities and hysteresis-induced delays

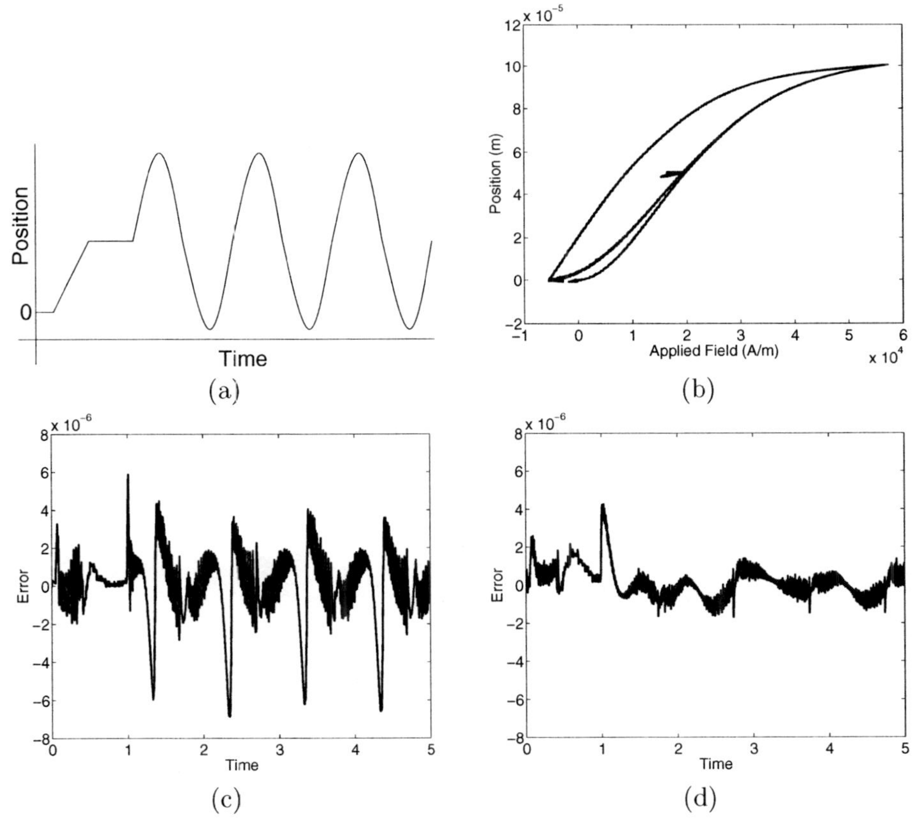

Figure 1.15. *(a) Trajectory to be tracked by the cutting head, and (b) hysteretic relation between H and the displacement y. (c) Error in the cutting head position obtained with an $\mathcal{H}_2$ design, and (d) error obtained with an $\mathcal{H}_2$ design incorporating inverse compensation.*

whereas robust control designs employing model-based inverse filters can maintain a tracking accuracy of 1-2 μm once cutting commences — even though the transducer is operating in the hysteretic and nonlinear regime depicted in Figure1.15(b).

Additional applications utilizing Terfenol-D transducers include sonar, geological tomography, bone conduction hearing aids, linear or rotational motors in addition to torque, displacement and force sensor designs [116–118, 391].

Hybrid Transducers

A second class of magnetostrictive transducers includes flextensional devices analogous to those described in Section 1.2.2, inchworm devices and rotational motors. The flextensional designs are similar to those depicted in Figure 1.7 but utilize a magnetostrictive core rather than PZT or PMN stacks. The magnitude of generated forces in combination with the dual actuating and sensing capabilities of the materials makes the transducers potentially advantageous in hydrostatic applications including sonar transduction and underwater imaging.

Moreover, there is an increased focus on the design of hybrid transducers which utilize complementary properties of constituent materials as illustrated in Figure 1.16 by designs employing magnetostrictive drive rods and piezoceramic clamps. In addition to the forces provided by the magnetostrictive materials, the 90° phase shift between Terfenol-D and PZT or PMN is being utilized to construct hybrid transducers having improved energy efficiencies and frequency bandwidths [70, 71, 136, 246].

A comparison of magnetostrictive transducers with the piezoelectric and electrostrictive devices discussed in Section 1.2.2 indicates that while they exhibit superior performance in certain applications, they are bulkier than their ferroelectric counterparts due to the required solenoid and housing. Hence they are currently employed in regimes where large forces or strains are required but weight is not a limiting factor. The range of applications will almost surely increase, however, as materials such as magneto-composites are perfected.

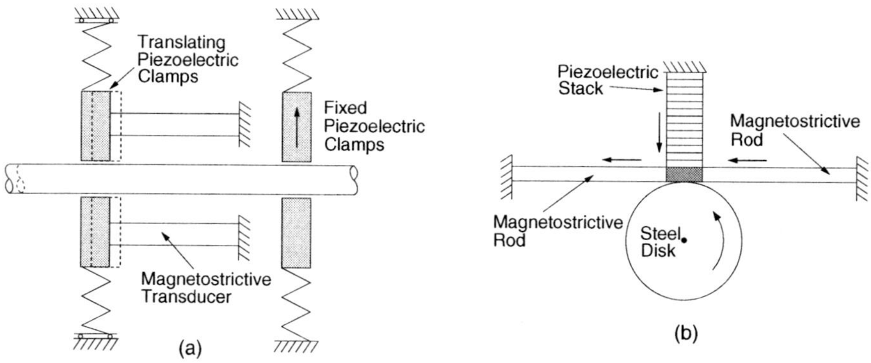

Figure 1.16. *Hybrid magnetostrictive-piezoceramic transducers. (a) Inchworm actuator (after [340]) and (b) Rotational motor (after [495]).*

1.4 Shape Memory Alloys

1.4.1 Material Behavior

When compared with piezoelectric, electrostrictive and magnetostrictive compounds, shape memory alloys (SMA) are relative newcomers to the arena of smart materials. In 1932, Swedish physicist Arne Ölander observed that a deformed AuCd alloy could be returned to its original shape when heated [368], while in 1938 the temperature-dependent nucleation and disappearance of martensite phases in brass (CuZn) was reported. Chang and Read in 1951 used x-ray analysis to establish some of the mechanisms in the AuCd transformations and showed that systems exploiting the shape memory effect could perform work [82]. However, the first sustained research on shape memory alloys is typically attributed to William Buehler and his colleagues at the Naval Ordnance Laboratory starting in 1961 [62–64]. This research focused on nickel-titanium alloys including the equiatomic compound NiTi. Like Terfenol-D which was developed in the same facility, the generic name Nitinol (Nickel Titanium Naval Ordnance Laboratory) reflects the acronym of its origin. Subsequently developed copper-aluminum-nickel (CuAlNi), copper-zinc-aluminum (CuZnAl) and iron-manganese-silicon (FeMnSi) alloys also exhibit shape memory effects but Nitinol remains the most widely used shape memory alloy.

Shape memory alloys derive their unique transducer capabilities from the fact that they can recover from up to 10% strains through temperature and stress-induced transformations between high temperature austenite and low temperature martensite phases. The austenite phase exhibits a cubic structure whereas shear deformations in the crystallographic structure — characterized by changes from 90° to approximately 96° in the lattice — produce 24 martensite variants in 3-D having either twinned or detwinned forms. For uniaxial configurations, there are only two martensite variants which, in the detwinned form, produce positive and negative shear strains as depicted in Figure 1.17. As detailed in [185], the martensite and austenite phases have very different mechanical, thermal, electrical, acoustic[1] and optical properties due to the differing crystallographic properties of the two phases — this necessitates the consideration of phase-dependent material coefficients when constructing models for SMA undergoing phase transitions.

The thermodynamic stability of the austenite and martensite phases provides SMA with the capability for "remembering" various shapes constructed in the austenite phase. To illustrate the origins of this shape memory effect, we

[1] Due to its regular atomic structure, austenite rings in response to an impact whereas martensite responds with a thud.

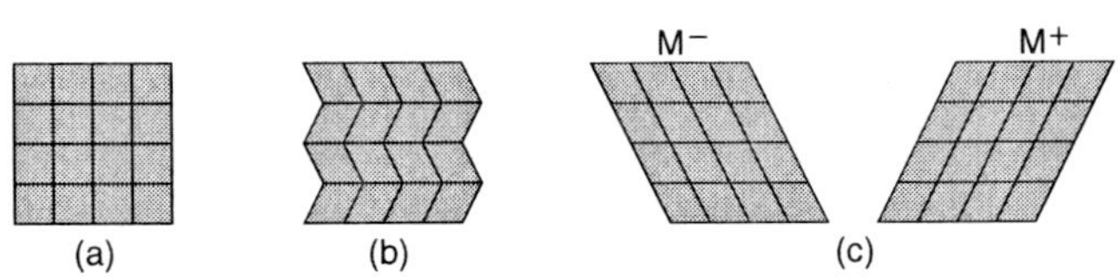

Figure 1.17. *Phases in uniaxial SMA: (a) austenite, (b) twinned (self-accommodated) martensite, and (c) detwinned martensite.*

consider first the effect of temperature-induced phase transformations. In the absence of an applied stress, the material transforms from austenite to twinned (self-accommodated) martensite as the material is cooled and returns to the austenite phase when reheated. The temperature at which the martensite and austenite transformations commence are termed M_s and A_s, and M_f, A_f denote the temperatures when the transformations are completed. Due to the twinning, which is energetically favorable in the absence of a load, macroscopic strains are negligible.

Phase transformations can also be induced through applied loads. If the temperature of the material is above A_f, the material exhibits a nearly linear stress-strain relation as the stress is increased until σ_{M_s} when a transition to the detwinned M^+ phase commences as depicted in Figure 1.18(b).[2] Similarly, the material again exhibits nearly linear behavior in the martensite phase until the stress is lowered to σ_{A_s} and the reverse process begins. In this high temperature regime, full shape recovery is observed upon unloading and the phenomenon is termed *pseudoelastic* or *superelastic*. For temperatures below M_f, the material retains a residual strain when unloaded as depicted in Figure 1.18(c). When heated above A_f, the deformation is recovered and the material returns to its original shape. This low temperature be-

[2]It is conventional in many smart material and mechanical engineering applications, where transducers operate in the mode of displacement control, to plot strains as the abscissae. In the materials science literature, it is more common to plot stress on the abscissa in accordance with mathematics and physics conventions.

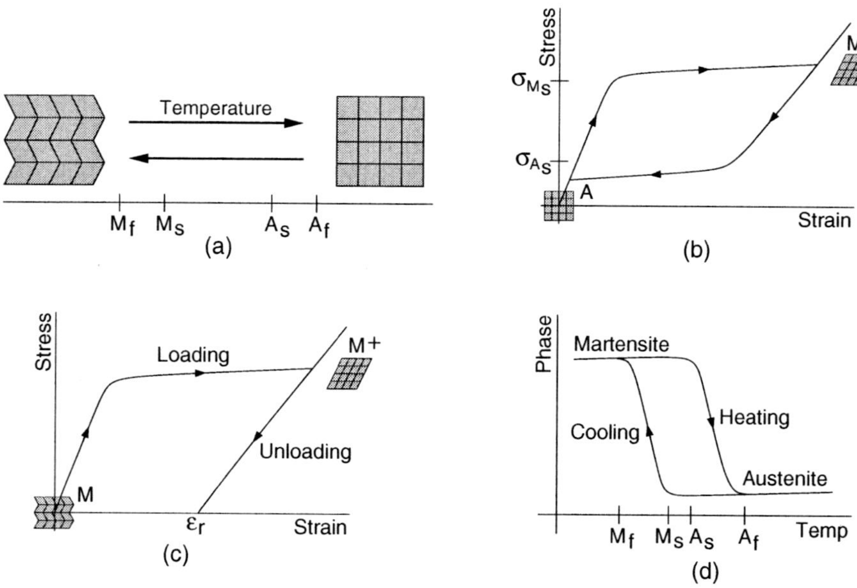

Figure 1.18. *(a) Temperature-induced phase transformations between austenite and twinned martensite in the absence of an applied load. (b) Stress-induced phase transformation and pseudoelastic behavior for $T > A_f$. (c) Quasiplastic behavior and residual strain ε_r generated when $T < M_f$. (d) Recovery of the residual strain which produces the shape memory effect when SMA is heated above $T > A_f$.*

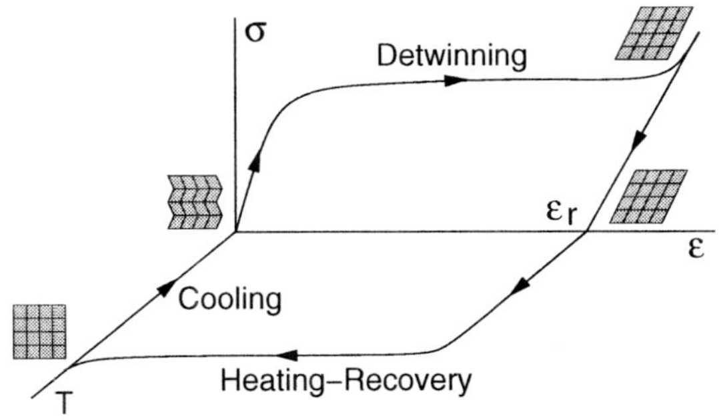

Figure 1.19. *Shape memory effect in a uniaxial SMA in which a residual strain ε_r is recovered through heating.*

havior is sometimes termed *quasiplastic* and the recovery of stress-induced residual strains through heating constitutes the *shape memory effect (SME)*. The strains associated with martensite transformations under an applied load are illustrated in Figure 1.18(d), and generation and recovery of residual strains ε_r through the detwinning and heating process are depicted in Figure 1.19. Details regarding these phenomena are provided in Chapter 5 as a prelude to model development.

1.4.2 Applications

In a number of early applications, the shape memory properties of SMA were exploited to create innovative and efficient fasteners, connectors, and clamps. For example, commercially available pipe couplers have been manufactured by employing a Nitinol sleeve whose inner diameter is smaller than that of the adjoining pipe. The sleeve is stretched in the low temperature martensite phase and the coupled joint is then heated to the austenite regime to recover the residual strain and produce a highly effective seal. Specifically, Nitinol couplers have been used to join hydraulic lines in F-14 fighter planes since the late 1960's. The alloys are also biocompatible for most individuals which has led to their use in biomedical applications including orthodontic wires which produce tooth movement as they revert from the high stress martensite phase to the low stress austenite phase, anchors with Nitinol hooks to attach tendons to bone, and Nitinol guides for the insertion of catheters into blood vessels. More recently, SMA have been considered as robotic actuators, fire detection, sprinkler and gas shutoff valves which sense and respond to elevated temperatures, and highly flexible cell phone antennas which exploit the pseudoelastic properties of SMA.

To illustrate modeling and control issues associated with SMA in present investigations, we discuss in detail a civil engineering application based on the pseudoelastic SMA behavior, an aerospace design that relies on temperature-induced phase transformations, and thin film and microactuator designs which exploit increased surface area to volume ratios to achieve improved dynamic capabilities. The reader is referred to [113, 171, 177, 245, 336, 413, 458] for additional examples illustrating a wide range of other applications utilizing shape memory alloys.

Vibration Attenuation in Civil Structures

Shape memory alloys possess a number of attributes which make them ideal candidates for vibration and displacement attenuation in civil structures, including (i) large elastic strains, (ii) significant energy dissipation capabilities due to the hysteretic martensite transformations, (iii) strain hardening in the martensitic phase, and (iv) excellent fatigue and corrosion resistance. Among the first to exploit these properties for civil applications were Graesser and Cozzarelli who investigated their use for seismic isolation [194]. A number of subsequent investigations have focused on vibration attenuation in bridges and buildings using Nitinol wires, rods and composites.

These applications exploit the pseudoelastic (superelastic) behavior depicted in Figure 1.18(b), due to stress-induced phase transformations in materials constructed so that the austenite phase occurs within the operating temperature range. Because the energy dissipated by the SMA device is proportional to the area of the hysteresis loop, the transducers are designed to *maximize* the hysteresis, thus necessitating the development of models and design packages which accommodate this nonlinear behavior and the resulting energy losses. A significant advantage provided by SMA tendons or gaskets is the fact that they dissipate energy without stiffening the structure which in turn can increase fatigue due to ensuing high frequency dynamics.

The use of SMA transducers in a multispan bridge design is illustrated in Figure 1.20. As detailed in [133], SMA restraining bars can be retrofitted to reduce hinge and abutment displacements — e.g., to prevent unseating during an earthquake — whereas Wilde et al. [510] illustrate that the addition of SMA bars to laminated rubber bearings provides substantially improved damping capabilities as well as displacement control due to the hardening of the bars in the high-stress, martensite phase.

Parallel investigations are focused on the development of SMA tendons to attenuate earthquake or wind-induced vibrations in buildings. Figure 1.21 depicts a structural testbed employed in a joint US-Japan program [7] which employs transducers comprised of Nitinol wires wrapped around cylindrical posts to reduce earthquake-induced dynamics by a factor of eight.

The use of SMA transducers to enhance passive damping capabilities also has significant potential for large-scale space structures — such as mirrors, antennas,

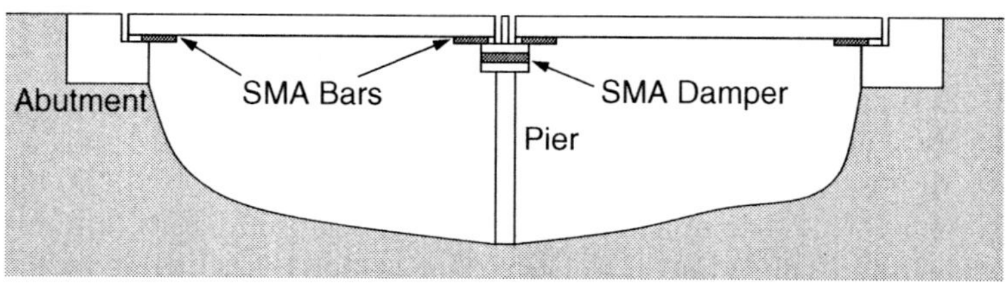

Figure 1.20. *Employment of SMA bars to reduce lateral displacements and SMA dampers for vibration attenuation on a multispan bridge.*

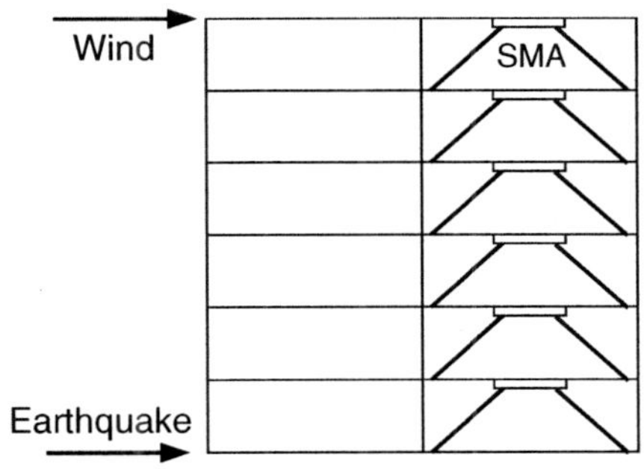

Figure 1.21. *Six story test frame used to investigate the use of SMA tendons to attenuate earthquake or wind-induced vibrations in buildings (after [7]).*

and solar arrays — due to the increased emphasis on lightweight polymers and composites which provide minimal internal damping. This is illustrated in Figure 1.22 by the depiction of an electrostrictive membrane mirror employing SMA tendons both for deployment and vibration attenuation.

It is anticipated that the exploitation of pseudoelastic SMA behavior for vibration attenuation in civil aerospace and aeronautic systems will burgeon as the technology matures. An important component of this technology is the combined development of hierarchical multiscale characterization frameworks ranging from micromechanic models necessary for material design to low-order macroscopic models that are sufficiently efficient for system design and real-time control implementation.

SMA Designs for Improved Flight Characteristics

The high work density-to-weight ratio of shape memory alloys makes them ideal candidates for a variety of aerospace applications. In the 1995 Smart Wing Program, SMA torque tubes were employed to modify aerodynamic properties of an airfoil to increase lift and rolling moments [273, 274]. This illustrated the potential of bulk SMA to improve flight characteristics within the constraint mandated by low switching frequencies.

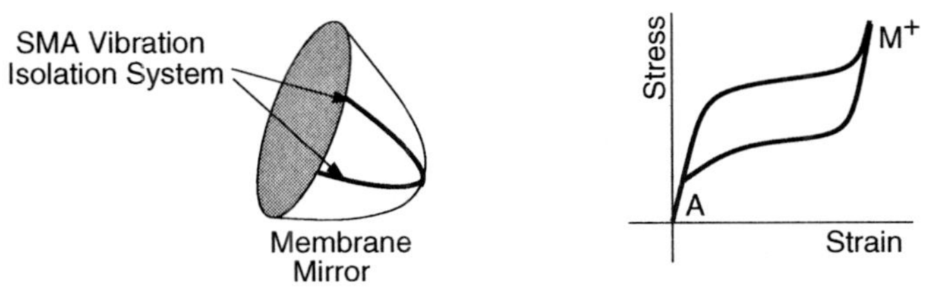

Figure 1.22. *SMA tendons employed to attenuate vibrations in a membrane mirror by optimizing the pseudoelastic damping (area of hysteresis loop).*

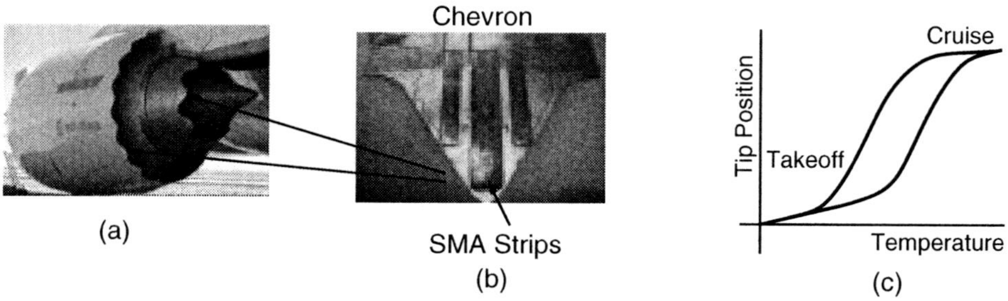

Figure 1.23. *(a) Jet engine and (b) SMA-driven chevrons employed to reduce jet noise and decrease drag. (c) Temperature-dependent behavior of 2-D or 3-D SMA flaps as a function of flight characteristics.*

Present investigations at Boeing are focused on the use of SMA flaps to reduce jet noise through improved mixing while maintaining efficient flight profiles [74, 311]. To illustrate the strategy and related research issues, consider the chevron depicted in Figure 1.23(b) which is located at the outlet of a jet engine, as shown in Figure 1.23(a). The strategy is to employ SMA flaps to position the chevron in the flow to improve mixing and decrease jet noise during takeoff and then to retract the chevron during cruise to reduce drag. This action utilizes the shape memory effect in the manner depicted in Figure 1.23(c). While proof-of-concept experiments performed at Boeing have demonstrated a 4 dB reduction in jet noise — a 3 dB reduction is measured if one of the two engines is turned off – the optimal design and control of SMA-driven chevrons requires the development of comprehensive models and model-based control techniques. This provides a significant challenge and is a present research direction since the SMA flaps are necessarily 2-D or 3-D whereas the majority of existing SMA models which are sufficiently low-order for optimal design and real-time model-based control algorithms are uniaxial or 1-D.

SMA Microactuators

SMA films, membranes, thin wires and microcantilevers are increasingly investigated for use as microactuators in applications ranging from infrared imaging to biomedical instruments for minimally-invasive surgery. Their advantage derives from a number of factors including large achievable strains, low drive voltages, high power-to-weight ratios and improved switching frequencies. As noted in [323, 324] SMA films heat in milliseconds in response to low voltage Joule heating ($\sim$ 5V), generate force and strokes on the order of 100 mN and 1%, and exhibit output work densities on the order of 10 MJ/m^3. For comparison, it is noted in Table 1 of [271] that thermopneumatic microactuators exhibit a work density of approximately 1.2 MJ/m^3 whereas that of PZT is approximately 0.12 MJ/m^3 — and for the fitness advocate, we note that the work density of muscle is approximately 0.02 MJ/m^3. Finally, the small thermal inertia and increased surface-to-volume ratios of SMA films and thin wires facilitate fast cooling and higher switching frequencies than their macroscopic counterparts with present designs having the potential to achieve 100 Hz [310, 426].

A number of investigations have focused on the design of SMA microgrippers which have the following advantages over conventional miniaturizations: they do not emit any lubricants or particles, they can be integrated into existing microassemblies, they exhibit clean room suitability and are biocompatible, and they permit fine-scale resolution of gripping forces. A biomedical grasper design discussed in [140] is depicted in Figure 1.24. This microinstrument is fabricated from a 0.3 mm diameter SMA tubing and is under investigation for clot removal in the brain or retrieval of coils used to seal aneurysms that have subsequently broken free into the bloodstream.

A survey of issues and applications associated with SMA thin films can be found in [236]. As detailed there and [323], thin film actuator designs include micropumps [521], microvalves [265], micromirrors, microswitches, micromechanical energy storage devices, and vibration dampeners in microelectronics packaging [223].

Biomedical applications exploiting the pseudoelastic and shape memory effects of SMA are summarized in the survey articles by Duerig [140] and Duerig, Pelton and Stöckel [141]. A number of these devices are being designed to operate in arteries, veins and even capillaries which necessitates miniaturization. For example, thin film TiNi stents are being developed for small blood vessels in the brain (on the order of 1 mm) whereas Nitinol guides used to redirect catheters or needles presently have diameters of 1–2 mm. On a slightly larger scale, SMA servo-actuators play a fundamental role in "active endoscope" designs being developed to improve diagnostics while reducing discomfort and risk to the patient. The resulting microtechnology has potential application to catheter design and remote inspection devices for aeronautic, aerospace, automotive and industrial processes.

Fundamental requirements for all miniaturized devices employing SMA are accurate characterization techniques and control designs capable of achieving specified tolerances. This necessitates the development of nonlinear models (often 2-D and 3-D), which quantify phase transition-induced hysteresis, and model-based control techniques that can be implemented in real-time. Whereas components of requisite models and control designs are presently in place, the extensions and implementation issues associated with 2-D and 3-D SMA designs pose significant challenges which have not yet been adequately addressed.

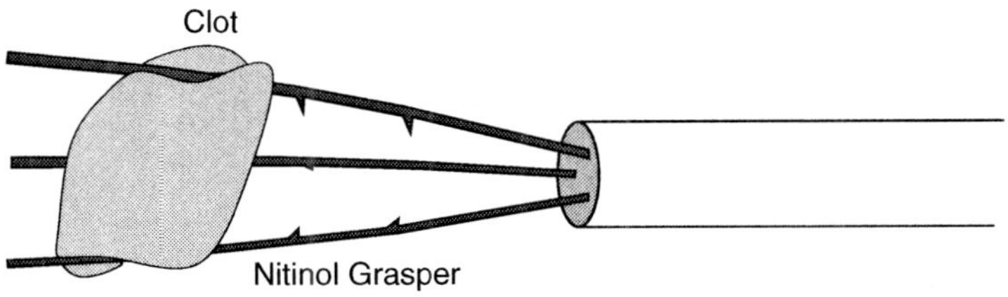

Figure 1.24. *Nitinol grasper designed to remove blood clots or retrieve embolized occlusion coils from the bloodstream (after [140]).*

1.5 Piezoelectric, Electrostrictive and Ionic Polymers

Piezoelectric, electrostrictive and ionic polymers are similar to piezoceramic compounds due to the fact that they exhibit electromechanical coupling which provides them with sensor and actuator capabilities. However, their polymer nature provides them with unique dielectric and mechanical properties which prove advantageous in a number of smart material applications. To illustrate these capabilities, we focus individually on semicrystalline, amorphous, and ionic polymer compounds.

1.5.1 Semicrystalline Polymers

The first piezoelectric polymer to be widely utilized and the most popular to date is polyvinylidene fluoride (PVDF) whose piezoelectric properties were discovered by Kawai in 1969 [261]. Other semicrystalline polymers include copolymers of PVDF with trifluoroethylene (TrFE) and tetrafluoroethylene (TFE), odd-numbered polamides or nylons, liquid crystal polymers, and biopolymers. Due to its prevalence in applications, we focus on PVDF and refer the reader to [209, 503] for details regarding the morphology and usage of other semicrystalline polymers.

The electromechanical properties of PVDF are provided by a polar crystalline phase resulting from the spatially symmetric location of fluorine and hydrogen atoms along a polymer chain. This morphology provides PVDF with a high dielectric constant and excellent polarization stability at room temperature. The fundamental, linear, piezoelectric properties of PVDF are compared in Table 1.1 with those of PZT. While the piezoelectric constant $d_{31} = 22 \times 10^{-12}$ m/V is larger than that of most other polymers, its magnitude is significantly less than that of PZT thus making the latter preferable for high drive actuator designs. However, the piezoelectric stress constant g_{31} is nearly a factor of 20 higher than that of PZT, thus providing the polymer with excellent sensor capabilities. PVDF also offers the advantage of being highly flexible and easily formed to irregular surfaces as contrasted with PZT

Property	Symbol	Materials: PVDF	Materials: PZT	Units
Strain Constant	d_{31}	22×10^{-12}	-175×10^{-12}	(m/V)
	d_{33}	-30×10^{-12}	400×10^{-12}	(m/V)
Stress Constant	g_{31}	216×10^{-3}	-11×10^{-3}	(Vm/N)
	g_{33}	-330×10^{-3}	25×10^{-3}	(Vm/N)
Coupling Factor	k_{31}	0.14	0.34	
Relative Dielectric Const	ε_r	12	1700	
Max Operating Temp		80	150	(^{o}C)
Density	ρ	1780	7600	(kg/m^3)
Young's Modulus	Y	2×10^9	7.1×10^{10}	(N/m^2)

Table 1.1. *Piezoelectric material properties for standard PVDF and PZT compounds.*

which is inflexible due to its ceramic nature. The sensor capabilities of PVDF are augmented by the low stiffness and density which provides the compound with a high voltage sensitivity to stress. Moreover, the fact that the polymer exhibits a high dielectric breakdown means that PVDF actuators can withstand much higher drive fields than their PZT counterparts. Whereas these high field capabilities prove advantageous in certain transducer designs, it also means that significant hysteresis in the E-P relation, as illustrated in Figure 1.25, must be incorporated in models and control designs. Finally, while the polarization behavior of PVDF is stable at room temperature, its performance degrades significantly above approximately 80 ^{o}C which limits its high temperature utility.

Rather than attempt to catalogue applications utilizing PVDF actuators and sensors, we summarize here only a few representative examples and refer the reader to [177, 209, 503] for more comprehensive discussion on this topic. The flexibility, light weight, low acoustic and mechanical impedance, and facility for being cut in complex patterns or bonded to irregular surfaces made PVDF films immediate candidates for structural and structural acoustic sensing [96, 98]. There has been significant discussion regarding the possibility of designing distributed modal sensors through the shaping and placement of PVDF films on vibrating structures [66, 197, 288, 341, 488]. While the concept is intriguing, Clark and Burke [97] illustrate that there are stringent practical limitations including the necessity of requiring that boundary conditions on 2-D structures satisfy orthogonality constraints between the sensor aperture and both structural modes and their curvature. An alternative for providing distributed sensing capabilities is the differential poling of PVDF to provide gradients or regions having differing electromechanical sensitivities. While again a promising idea, technological issues remain to be resolved and the concept is still under investigation.

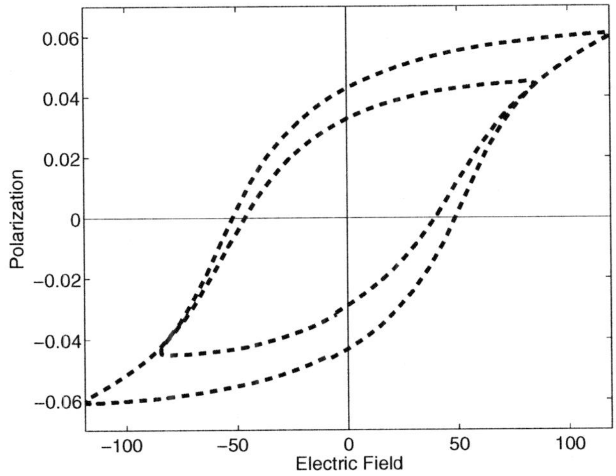

Figure 1.25. *PVDF data from [443] collected at 20 mHz with input fields having maximum values of 85 MV/m and 120 MV/m.*

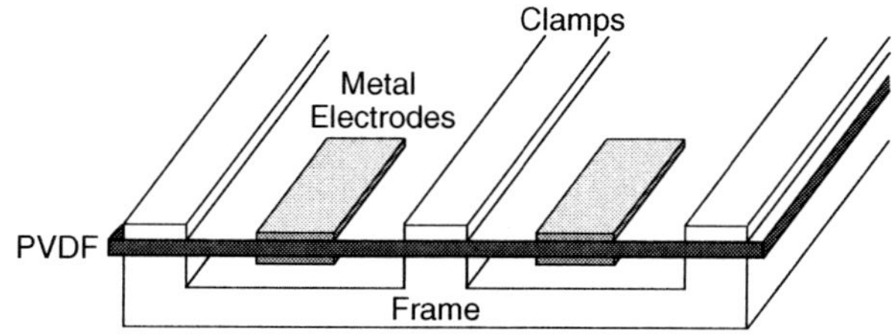

Figure 1.26. *Contactless keyboard employing PVDF after [503].*

The direct piezoelectric effect has been utilized in electromechanical devices such as the contactless keyboard depicted in Figure 1.26. As detailed on page 322 of [503], the application of pressure to metal electrodes stretches the film thus producing a charge which initiates the desired action. The advantages provided by such designs are low cost and durability due to the reduction in moving parts.

Finally, the biocompatibility, conformability and impedance of PVDF and other electroactive polymers make them excellent candidates for a wide range of biomedical applications. This includes actuator implants to stimulate bone and tissue growth, sensors to monitor vascular stints and grafts, and artificial muscle actuators [320].

1.5.2 Amorphous Polymers

Amorphous polymers are similar to semicrystalline polymers in the general sense that they exhibit permanent molecular dipoles which reorient in response to applied fields to produce actuator effects with converse actions yielding sensor capabilities. As detailed in [209, 371], however, the electromechanical mechanisms for the two classes differ quite substantially. Specifically, PVDF and the previously mentioned polymers operate in a state of thermal equilibrium whereas amorphous polymers exhibit piezoelectric properties in a quasi-static state due to the freezing-in or locking of dipoles below a glass transition temperature T_g. This provides amorphous polymers with higher temperature operating capabilities — 200 ^{o}C for (β-CN) APB/ODPA polyimide as compared with 80-100 ^{o}C for PVDF — but at the expense of reduced dielectric and remanence properties. For example, d_{31} for polyimides is roughly one quarter that of PVDF which has limited their use in high drive actuator design.

Presently, polyimides are employed primarily as passive or protective components in smart material transducers. To illustrate, consider the MEMs thin film actuator design depicted in Figure 1.27. As detailed in [147], the actuator is comprised of a polyimide/gold/polyimide composite flap adjacent to an electroded substrate comprised of silicon, glass or sapphire. During the manufacturing process, differing thermal properties of the constituent materials cause the flap to curl due to the same thermal stress mechanisms which produce the curvature in THUNDER transducers as depicted in Figure 1.6. Actuation is achieved by applying voltage to the electroded flap and substrate causing the film to simultaneously flatten and uncurl due to attractive electrostrictive forces. When the voltage is terminated, the

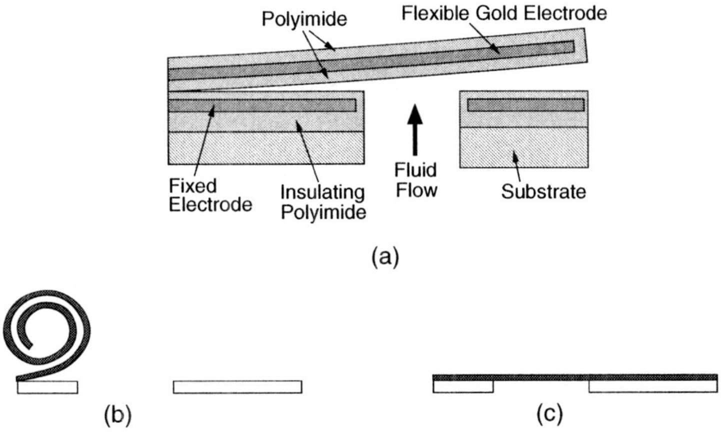

Figure 1.27. *(a) Electrostatic MEMs actuator comprised of gold and polyimide layers adjacent to an electroded surface employed as a microfluidic valve. (b) Open aperture in the absence of voltage and (c) closed aperture resulting from an applied voltage.*

flap recoils to is original curved state. By varying the geometry and thickness of the gold and polyimide layers, one can modify the effective aperture, while achieving mechanical displacements on the order of 100 μm at actuation speeds less than 100 μs and frequencies greater than 5 kHz. Moreover, the actuators require little power with individual devices using less than 50 μW. This thin film MEMs design is presently being considered for applications which include electrical relays and switches, integrated infrared choppers for night vision cameras, optical shutters, and microfluidic valves.

To date, the use of amorphous polymers for high performance transducer design lags significantly behind PVDF. However, as the material properties and characterization of amorphous polymers matures and high temperature operating requirements become increasingly stringent, their use in biomedical, aerospace, and robotic applications will likely grow.

1.5.3 Ionic Polymers

Building on research from the 1940's [260, 275], ionic polymers were investigated in the 1960's for use in water desalinization and chemoelectric power generation. Developments in materials science during the 1980's led to improvements in mechanical strength and ionic conductivity which were capitalized on in the 1990's by Orguro, Kawami and Takenaka [366], Sadeghipour, Salomon and Neogi [399], Segalman et al. [414], and Shahinpoor [418–420] who established the use of ionic polymers for actuating and sensing in a broad range of structural, robotic and biomimetic applications as summarized in [421, 522].

Ionic polymers differ from the previously described PVDF, polyimide and polymeric elastomers in the manner through which electromechanical coupling is generated and manifested. Specifically, electromechanical coupling in ionic polymers is produced through the transport of charged and uncharged ions within the polymer matrix. To illustrate, consider the configuration depicted in Figure 1.28, which consists of a teflon polymer backbone sulfanated with negatively charged side groups in a hydrated membrane with conductive surfaces. Application of an electric field across the membrane produces ionic migration toward the negative surface which combines with the redistribution of solvent to produce bending in the structure – this provides the membrane with actuator capabilities. Conversely, mechanical deformations produce a redistribution of charges and solvent which produces an electric field – this gives the polymer sensor capabilities.

The unique electromechanical properties of ionic polymers provide them with the following properties:

- Strain outputs greater than 1% (PZT: 0.1-0.3%)
- Operating voltages on the order of 1-10 V (PZT: 1-2000 V)
- Chemoelectric coupling properties
- Controllable charge and solvent transport properties.

The fact that electromechanical coupling is directly related to charge transport provides the materials with the potential for chemical sensing [302]. This latter capability is illustrated in Figure 1.29 where an ionic polymer coated with a chemical-specific permeable membrane is used to separate two reservoirs. Introduction of the chemical in Reservoir I will cause a concentration gradient which in turn produces diffusion from Reservoir I to Reservoir II. The accompanying charge flow generates a stress which can be measured to monitor the presence of the chemical.

As detailed in [361,421,469], original ionic polymers required hydration which severely restricted their use in most aerospace and aeronautic applications. However, recent investigations have illustrated that significant strain and stress outputs can be achieved by constructs in which active ionic polymers are encapsulated in polyimides to provide space durability and the possibility for operation in nonhydrated environments. This provides the potential for substantially expanding the utility of ionic polymers for a broad range of aerospace and aeronautic applications.

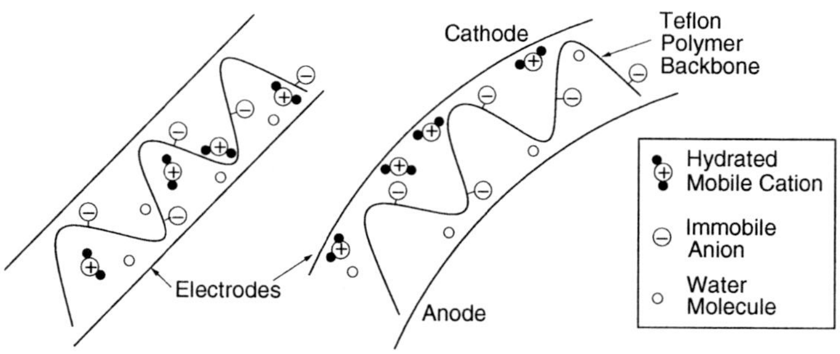

Figure 1.28. *Electromechanical transduction in ionic polymers in response to an applied field.*

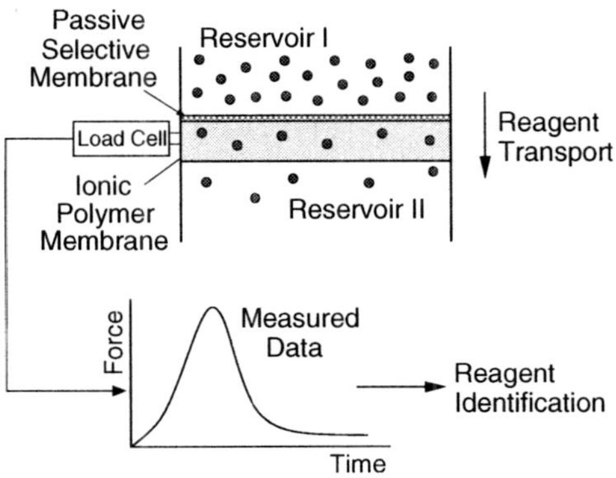

Figure 1.29. *Chemical detection using chemical-specific permeable membranes.*

The mechanisms which provide ionic polymers with unique electromechanical and chemomechanical transduction capabilities also endow the materials with an inherently nonlinear and hysteretic behavior as illustrated by the voltage-current relation plotted in Figure 1.30. The hysteresis is due to capacitive storage within the material whereas the saturation nonlinearities in the current near the peak potential of 0.5 V are indicative of solvent breakdown due to electrochemical instability. This nonlinear and hysteretic behavior, as well as other nonlinearities exhibited by the materials, must be quantified to provide accurate models for material characterization and design as well as model-based control designs. The development of models for ionic polymers dates back to the work of Grodzinsky and Melcher [195] and Yannas and Grodzinsky [523], and presently constitutes a thriving research area necessary both for advancements in material design and the development of applications utilizing these novel compounds — e.g., see [181, 360, 520].

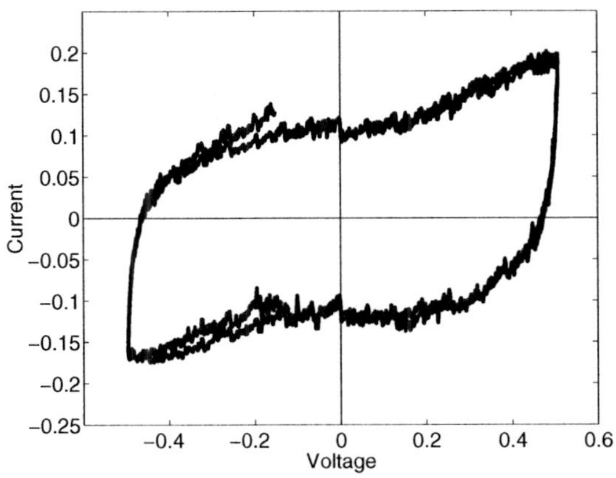

Figure 1.30. *Nonlinear and hysteretic voltage-current behavior exhibited by an ionic polymer.*

1.6 Electrorheological and Magnetorheological Compounds

1.6.1 ER and MR Fluids

Electrorheological (ER) and magnetorheological (MR) fluids are characterized by reversible changes in the rheology of the fluids when subjected to electric or magnetic fields. This capability for changing the yield stress has motivated their consideration in applications ranging from variable resistance mechanisms in exercise machines to active shock absorbers and clutches.

Initial investigations regarding the development of MR fluids for high speed transduction, including clutch design, were performed at the US National Bureau of Standards as reported by Rabinow [387] in 1948. At approximately the same time, Winslow reported on the use of ER fluids — with some mention of MR fluids — for clutch, valve, brake and loudspeaker designs [514, 515]. Hence the two technologies are equally mature with nearly similar research levels at the present time. Whereas there are significant differences between the electric and magnetic circuitry required for actuation, as well as the microscopic material behavior, the mesoscopic mechanisms producing changes in the fluid rheology are analogous and we summarize them simultaneously.

As illustrated in Figure 1.31, MR or ER fluids are comprised of particles, on the order of 1-10 μm in size, suspended in inert fluids such as mineral or silicon oils. In the absence of applied fields, the particles are uniformly distributed and, to a first approximation, the fluids are characterized as Newtonian. Under this assumption, the shear stress τ is modeled by the linear relation

$$\tau = \eta\dot{\gamma} \tag{1.1}$$

where η and $\dot{\gamma}$ respectively denote the viscosity and shear strain rate. The application of an applied electric field E or magnetic field H causes the particles to align in chains — called tendrils or fibrils — in the direction of the field. This provides one component of the mechanism which produces the field-dependent change in shear stress; other physical mechanisms include coulombic interactions in ER fluids and interparticle repulsion in MR fluids. From a phenomenological perspective, these

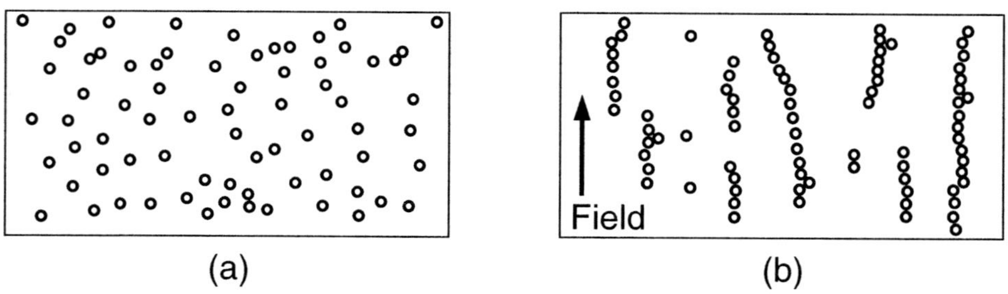

Figure 1.31. *MR or ER fluids comprised of particles suspended in an inert fluid medium: (a) no applied field and (b) applied magnetic or electric field.*

field-dependent mechanisms produce several facets of non-Newtonian behavior as illustrated for ER fluids in Figure 1.32. In the absence of motion, the fluids exhibit a field-dependent static yield stress $\tau_s(E)$ which decreases to $\tau_y(E)$ once the fluid is in motion. For moderate shear strain rates, the fluids exhibit approximately affine behavior

$$\tau = \eta\dot{\gamma} + \tau_y(E) \tag{1.2}$$

where $\tau_y(E)$ incorporates the measured functional dependence of τ_y on the field. Similar phenomenological behavior is observed for MR fluids.

Due to its simplicity, the Bingham model (1.2) has been employed for characterization and design in a number of ER and MR applications. However, it ignores the dependence of η on the field and strain rate, the mechanisms producing the field-dependence in τ_s and τ_y, the dynamics of the fluids, temperature-dependencies, and particle interactions, thus motivating the investigation of more sophisticated ER [48, 56, 69, 258, 375] and MR [308, 346, 428, 429, 456] fluid models. The incorporation of additional physical mechanisms, the construction of reduced-order models feasible for real-time implementation, and the development of model-based control designs which accommodate the inherently nonlinear fluid dynamics are open research topics necessary for the continued exploitation of MR and ER fluids.

It is a tribute to Rabinow and Winslow that a number of the MR and ER applications under present investigation were either motivated by or described in their pioneering work [387, 515]. These include vibration isolation devices such as shock absorbers, seat dampers, or engine mounts, magnetically or electrically triggered clutches, and automotive valves. As detailed in [78, 144, 177, 207, 254, 317, 458], additional applications include variable resistance exercise bikes and vibration attenuation devices for civil structures which provide semi-active damping of wind or earthquake-induced vibrations using minimal power — compare with the SMA devices discussed in Section 1.3 and depicted in Figure 1.21. Finally, the reader is directed to [68, 525] for discussion regarding the development of hydraulic actuators which exploit MR valves employed in combination with piezoelectric or Terfenol-D pumps. These hybrid devices provide durability, controllable damping, high output forces, and the facility for miniaturization; hence they exhibit great potential for a number of aeronautic, aerospace and industrial applications.

To illustrate the use of MR fluids for clutch construction, consider the two prototypical designs depicted in Figure 1.33. The concentric cylinder facilitates

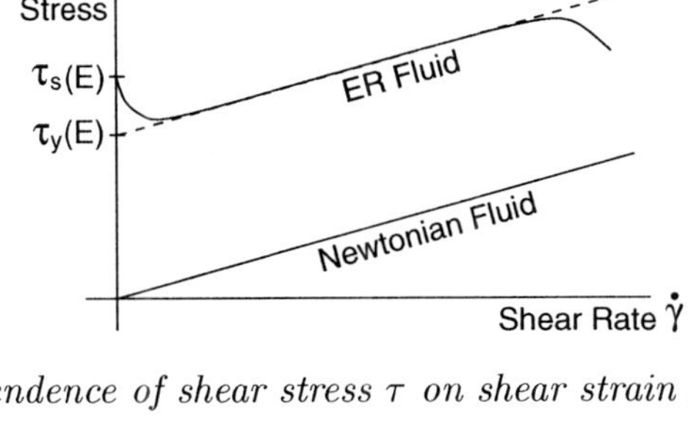

Figure 1.32. *Dependence of shear stress τ on shear strain rate $\dot{\gamma}$ for ER fluids.*

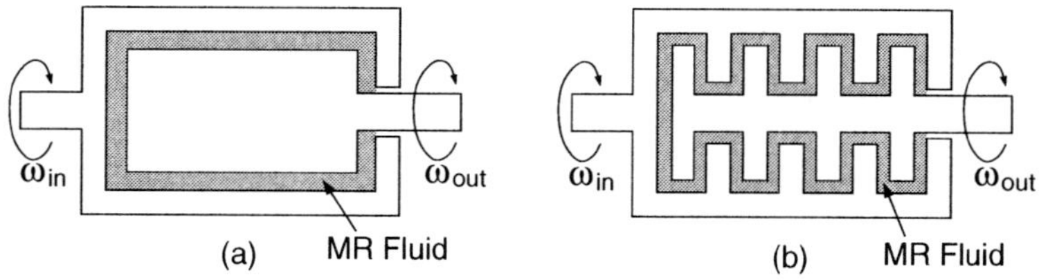

Figure 1.33. *Clutch designs employing MR fluids: (a) single plate and (b) multi-plate design (after [458]).*

manufacturing whereas the multi-plate construction provides higher torque capabilities for a given volume. In the absence of an applied magnetic field, the input shaft spins freely whereas power is transferred from the engine to drive train when applied fields increase the yield stress of the MR fluid. This application also motivates technological issues which delineate between the use of ER or MR fluids. As noted by Carlson and Weiss [80], present ER fluids do not have the capability for supporting the shear stresses ($\sim$ 14 kPa) required for current automotive clutches whereas the values of τ_y provided by MR fluids are within specified thresholds.

The complexity of ER and MR fluid behavior provides both highly unique design capabilities and significant challenges from the perspective of model development, numerical approximation, and control design. It is anticipated that the most rapid and significant advances will come from interdisciplinary research teams who investigate design, characterization, approximation and control issues in concert.

1.6.2 ER and MR Elastomers

Viscoelastic or solid analogues of ER and MR fluids are ER and MR gels and elastomers. The investigation and development of these compounds is relatively new compared with the fluids and they have not yet been incorporated in smart structure designs to the extent of the fluids. ER and MR elastomers offer the advantage of avoiding design issues required to prevent fluid leakage and in some applications may provide additional damping and elastic capabilities. Like ER and MR fluids, the elastomers derive the capability for changing yield behavior through chains of polarizable particles embedded in a carrier medium — in this case, typically silicon or natural rubber. The application of electric or magnetic fields enhances dipole moments associated with the particles thus producing a field-dependent shear modulus. Potential applications utilizing this mechanism include adaptive tuned vibration absorbers, tunable automobile suspensions, and tunable engine mounts. The reader is referred to [162, 176] for additional information on ER elastomers and [79, 124, 183, 255] for details regarding MR elastomers.

1.7 Sensor Technologies — Fiber Optics

Two necessary components of any feedback control system are sensors and actuators. When considered in isolation, an ideal sensor is one having the capability for monitoring necessary system properties while affecting the structural properties in a manner which enhances performance — e.g., improved load-carrying capacity. Hence sensors should be lightweight, flexible and capable of achieving the frequency bandwidth dictated by the application. Furthermore, they should be sufficiently robust so that the lifetime of the control system is not dictated by that of the sensor. For some control designs, it may also be advantageous for sensors and actuators to be collocated which motivates consideration of transducer materials capable of both functions.

Of the previously discussed transducer materials, piezoelectric, electrostrictive and magnetostrictive materials have been widely investigated for use as both sensors and actuators with ionic polymers and shape memory alloys more recently joining the list. All offer unique features which make them candidates for the design of self-sensing actuators [11, 135] or sensoriactuators [103] tailored to certain applications.

Present smart material control systems also employ a number of other sensor constructs including accelerometers, laser vibrometers for noncontact velocity measurement, proximity sensors for displacement measurement, strain gauges and fiber optics [113, 177]. While all of these choices offer advantages in various applications, we summarize here the properties of fiber optic sensors due to their unique capability for providing high accuracy, quasi-distributed strain measurements.

1.7.1 Fiber Optic Principles

Optical fibers operate by the principle of *total internal reflection* which occurs when light encounters an interface with a material having a lower refractive index. To illustrate, consider two materials having refractive indices n_1 and n_2 where $n_1 > n_2$. Snell's law, or the law of refraction, states that if a ray has angle of incidence θ_1 in the first material, then the angle of refraction θ_2 in the second material is quantified by the relation

$$\frac{\sin\theta_2}{\sin\theta_1} = \frac{n_1}{n_2}.$$

For the considered case $n_2 < n_1$, it thus follows that $\theta_2 > \theta_1$ as depicted in Figure 1.34. The critical angle of incidence θ_c is defined as that which yields $\theta_2 = \frac{\pi}{2}$ so that $n_1 \sin\theta_c = n_2$ or

$$\theta_c = \arcsin(n_2/n_1).$$

Finally, total internal reflection occurs for $\theta_1 > \theta_c$. In this latter regime, the rays also satisfy the law of reflection which states that

$$\theta_i = \theta_r$$

where θ_i and θ_r denote the incidence angle and angle of reflection.

Optical fibers are constructed from a high-index core surrounded by a cladding material having a slightly lower refractive index as shown in Figure 1.34(b). Both the core and cladding are typically comprised of glass (SiO_2) mixed with dopants to control the refractive indices. By choosing dopants so that $\theta_i > \theta_c$ for the geometry and strains under consideration, optical fibers thus act as wave or light guides in the manner depicted in Figure 1.34(c).

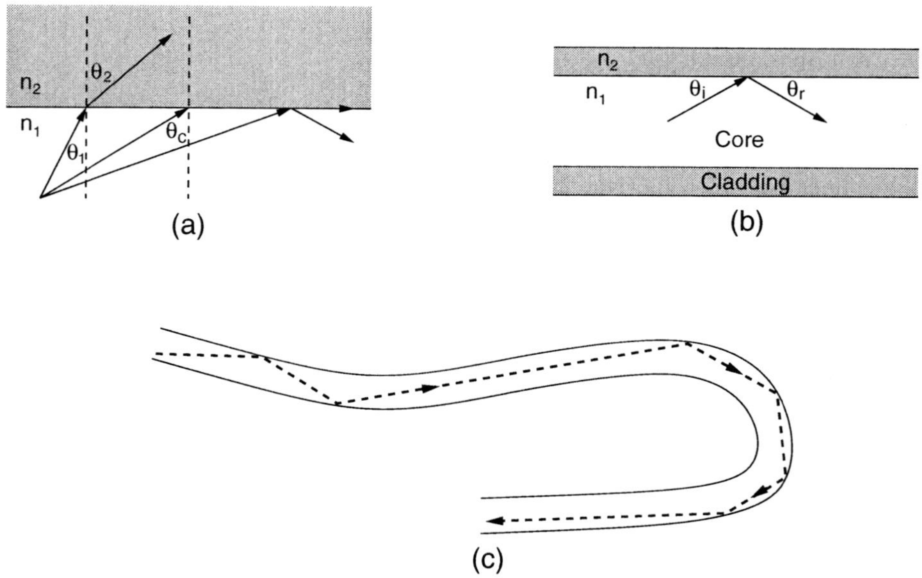

Figure 1.34. *(a) Refraction and reflection of light at the interface between materials with refractive indices n_1 and n_2, $n_1 > n_2$. (b) Single-mode optical fiber comprised of a high-index silicon core surrounded by a silicon cladding with a slightly lower refractive index. (c) Travel of light rays within an optical fiber.*

Single mode fibers have core diameters on the order of 5 μm whereas the diameters of multimode fibers range from 100 μm to 200 μm. From the perspective of structural design, optical fibers can withstand strains up to 8% which is on the order of those generated by shape memory alloys. Their Young's modulus (6.9×10^{10} N/m^2) is close to that of aluminum (7.3×10^{10} N/m^2) and PZT (7.1×10^{10} N/m^2) which facilitates impedance matching with these materials. Finally, it has been demonstrated that with the proper selection of adhesives and coatings (e.g., polyimides), optical fibers can operate in excess of 1 million cycles and, in general, have a longer operating life than conventional strain gauges [458].

1.7.2 Fiber Optic Sensor Designs

There exist a number of sensor designs employing optical fibers, each proving advantageous in certain applications. We summarize three designs: Mach-Zehnder interferometers, Fabry-Perot sensors, and Bragg grating sensors, and refer the reader to [112, 113, 262, 483] for details regarding additional sensor designs utilizing optical fibers. Further discussion detailing the use of fiber optic sensors for smart systems can be found in [100, 490].

Mach-Zehnder Interferometers

This class of fiber optic sensors exploits interference produced in light signals due to strains in the underlying structure. As depicted in Figure 1.35, a single

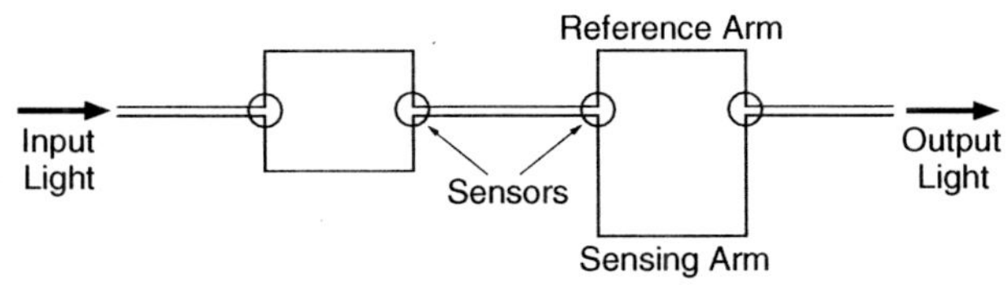

Figure 1.35. *Mach-Zehnder interferometer array which measures the interference in a beam of light split into a reference signal which is isolated from strains and a sensing signal subjected to strains.*

input beam is split to provide equivalent signals in two single mode fibers. The first fiber, which acts as a reference arm, is isolated from strains whereas the second (termed a sensing arm) undergoes deflections produced in the underlying structure. The differing path length in the strained fiber produces a phase shift which, when compared with the reference signal, can be used to deduce the strain amplitude.

Mach-Zehnder interferometers offer the advantage of high sensitivity and can be used to measure strains on the order of 10^{-8}. The requirement of a reference arm, however, increases the complexity and size of the sensors and limits the degree to which they can be localized or used to measure strains at a point.

Fabry-Perot Strain Sensors

Fabry-Perot sensors exploit the interference between reflected waves in the device to provide localized or pointwise strain measurements. The fiber optic components of the sensor consist of a single-mode fiber separated from a multimode fiber by an air gap on the order of 4 mm as depicted in Figure 1.36. When an incoming wave in the single-mode fiber encounters the glass-air interface at the beginning of the gap, part is reflected to provide a reference signal r_1 while a sensing component r_2 continues through the gap where it is reflected at the air-glass interface with the multimode fiber. Deformations in the underlying structure change the width of the gap thus altering the interference between the signals r_1 and r_2. This in turn produces a phase change which is measured by the device. In this manner, Fabry-Perot sensors provide the capability for obtaining highly accurate, localized strain measurements.

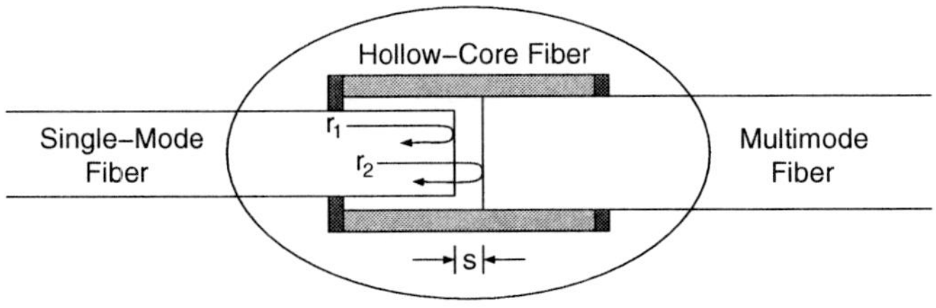

Figure 1.36. *Fabry-Perot strain sensor exploiting interference between reflected signals r_1 and r_2.*

An example illustrating the sensitivity of Fabry-Perot sensors is provided by Murphy et al. [354] in the context of measuring strains produced in an F-15 airframe during fatigue tests. These results demonstrate that Fabry-Perot sensors are capable of detecting strains down to 0.01 μstrain (0.01 μm/m) while exhibiting negligible hysteresis. Hence this technology offers strong potential for sensor design in smart material systems.

Bragg Grating Sensors

One of the most promising sensor designs exploiting optical fibers is based on the use of Bragg gratings to systematically perturb the refraction index of the fiber core in the manner depicted in Figure 1.37. The grating, which is formed by exposing the fiber to an intense optical interference pattern, acts as a step-band filter in the sense that it reflects certain wavelengths while allowing others to pass. As detailed in [215], the strongest interactions or reflections occur at the Bragg wavelength

$$\lambda_B = 2n_e\Lambda$$

where Λ is the period for the grating and n_e denotes the modal index.

The sensor capabilities for the device derive from the fact that any environmental changes that alter the Bragg spacing or modal index will also change λ_B, a quantity that can be measured with high accuracy. For example, both axial and transverse strains will compress or dilate the grating to produce shifts $\delta\lambda_B$ in the Bragg wavelength. As reported in [347], strain responses measured in this manner are highly accurate, linear, and exhibit no hysteresis at temperatures as high as 370 ^{o}C.

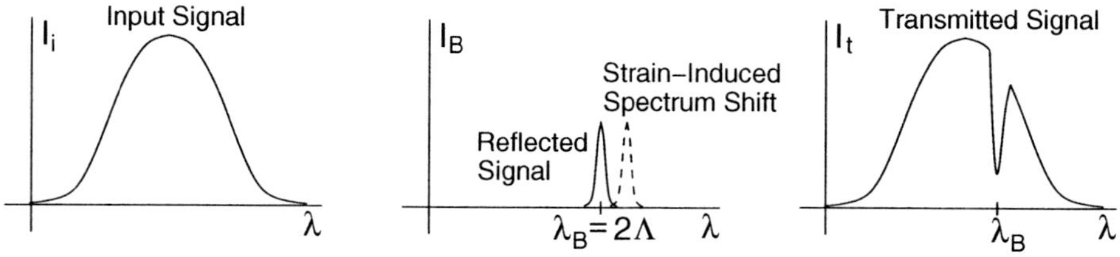

Figure 1.37. *A fiber Bragg grating (FBG) generating a reflected wave having the Bragg wavelength $\lambda_B = 2n_e\Lambda$. Changes in the modal index n_e or grating period Λ due to strains, temperature deviations, or changes in polarization produce corresponding shifts in λ_B which provides the device with highly sensitive sensor capabilities.*

The elasto-optic properties of the grated fibers also provide the capability for detecting pressure changes and measuring acoustic fields. Like strain measurements, large pressure changes can be monitored with simple readout devices. As noted in [215], however, the stiffness of the glass fibers necessitates the use of geometric or complex readout systems when detecting sound fields due to the low amplitude of generated transverse strains.

Fiber Bragg grating (FBG) sensors can also be used to measure temperature as well as electric and magnetic fields. The temperature sensitivity derives from the thermo-optic properties of the device and care must be taken to insulate against this effect if measuring other quantities in variable temperature environments. The ability to measure magnetic or electric fields is induced by coating the fibers with magnetostrictive or piezoelectric jackets. As with the acoustic sensors, the induced strains are small so fine-scale readout techniques are required to resolve the measurands.

The advantage of FBG sensors are augmented by the fact that the shifts in wavelength used to measure strains, temperature changes, and electric, magnetic and acoustic fields are independent of intensity and do not change as the light passes through adjacent fibers or connectors. This enhances the robustness of the devices. Finally, FBG technology is being heavily investigated for numerous other applications ranging from optical storage to telecommunications [215] which will continue to produce technological advances that benefit sensing for smart material systems.

Chapter 2

Model Development for Ferroelectric Compounds

Ferroelectric materials are defined as those which exhibit, at temperatures below the Curie point, a *domain structure* and *spontaneous polarization* which can be reoriented by applied electric fields. As detailed in Section 2.1, the domain morphology results from the alignment of dipoles to minimize electrostatic and elastic energy, and materials with this structure will exhibit varying degrees of hysteresis at *all drive levels*. This definition is analogous to the designations *ferromagnetic* and *ferroelastic* for compounds exhibiting magnetic and elastic domains, and the combined class of ferroelectric, ferromagnetic and ferroelastic materials will be collectively referred to as *ferroic* materials.

To simplify the discussion, we will focus primarily on the ferroelectric materials $BaTiO_3$ (barium titanate), $Pb(Zr,Ti)O_3$ (lead zirconate titanate or PZT), the film polyvinylidene fluoride (PVDF), and the electrostrictive material $Pb(Mg,Nb)O_3$ (PMN) at temperatures below the freezing point — however, the theory and models encompass a broad range of ferroelectric compounds. $BaTiO_3$ was the first piezoceramic material to be developed commercially and, due to its simplicity, it complements $PbTiO_3$ in providing an ideal prototype for describing the constituent dipole and domain processes. For present applications, however, it has been largely superseded by PZT due to the enhanced performance capabilities of the latter, and we focus the model development on PZT. At low temperatures, PMN exhibits ferroelectric behavior whereas it exhibits relaxor ferroelectric behavior at temperatures between the freezing point and Curie point — model development for PMN in this latter state is addressed in Chapter 3.

To achieve bidirectional strains in compounds such as PZT, it is necessary to pole the material and operate around this poled state. For low to moderate input fields, this yields approximately linear responses which motivated the original linear analysis of Voigt and the linear constitutive relations developed in Section 2.2. These are often referred to as the *piezoelectric relations* and materials operating in approximately linear regimes are frequently designated as *piezoelectric materials*. It should be emphasized that piezoelectric in this context has a *linear* connotation and refers to the coupled direct and converse piezoelectric effects exhibited by the materials. This can be somewhat of a misnomer since the "piezoelectric" material

PZT is inherently nonlinear and hysteretic due to its noncentrosymmetric nature. Moreover, there exist piezoelectric materials such as bone and quartz which are not ferroelectric and compounds such as PMN that are ferroelectric but not piezoelectric due to quadratic field-strain dependencies.

Chapter Organization

- Section 2.2 – Linear constitutive relations: Applicable for low to moderate drive regimes.
- Section 2.3 – Higher-order energy relations: Characterize hysteretic behavior of certain single crystal compounds and provide a basis for subsequent models.
- Section 2.4 – Preisach models: Phenomenological theory with a well established mathematical framework. Requires extensions to accommodate frequency, temperature and stress-dependencies as well as reversibility and noncongruency.
- Section 2.5 – Domain wall theory: Highly efficient for symmetric major hysteresis loops. Model does not enforce minor loop closure without *a priori* knowledge of turning points which precludes transient control design for certain applications.
- Section 2.6 – Homogenized energy theory: Characterizes hysteresis in a broad range of biased and symmetric operating regimes. Incorporates reversibility, noncongruency, and provides a framework which can accommodate rate, temperature and stress-dependencies. Efficient inverse formulation facilitates linear control design.
- Section 2.7 – Ginzburg–Landau energy relations: Quantify domain and domain wall properties but difficult to implement for transducer or model-based control design.

2.1 Physical Properties of Ferroelectric Materials

Barium Titanate and Lead Titanate

Barium titanate ($BaTiO_3$) and lead titanate ($PbTiO_3$) are isostructural with the mineral perovskite ($CaTiO_3$) and exhibit what is termed a perovskite structure comprised of a cubic form for temperatures above the Curie point T_c and tetragonal, orthorhombic, and rhombohedral forms for $T < T_c$ as illustrated for $BaTiO_3$ in Figure 2.1. While the specific temperatures and hierarchy of forms differ for $BaTiO_3$ and $PbTiO_3$, the fundamental mechanisms are analogous so we consider the two simultaneously when illustrating properties of perovskite compounds as a prelude to the more complex behavior of PZT.

Above T_c, the materials exhibit paraelectric behavior whereas they are ferroelectric for $T < T_c$. As illustrated in Figure 2.2, the paraelectric cubic structure is centrosymmetric whereas it is energetically favorable for the O^{2-} ions to be biased

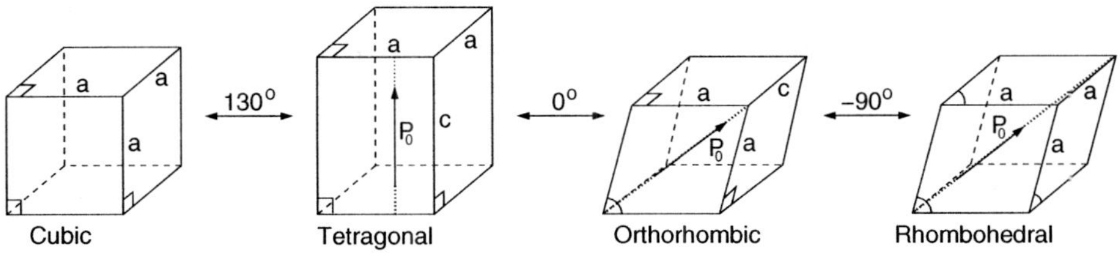

Figure 2.1. *(a) Cubic, tetragonal, orthorhombic and rhombohedral forms of perovskite compounds and approximate transition temperatures for $BaTiO_3$ (°C).*

slightly below face centers and Ti^{4+} ions are biased upward from the unit cell center in the tetragonal phase. The relative change in position of the Ti^{4+} and O^{2-} ions produces a spontaneous polarization P_0 as well as the noncentrosymmetric structure which is one source of hysteresis in the materials.

The energy profiles for the paraelectric and ferroelectric phases are plotted in Figures 2.2(c) and (d). For the high temperature phase, the internal energy has a unique minimum whereas it exhibits a double well profile for the low temperature tetragonal structure. The minima correspond to the equilibrium positions of the Ti^{4+} ions above or below the cell center. If sufficient electrostatic or elastic energy is provided to move the ion across the unstable equilibrium, it will enter the opposite potential well producing a dipole switch — at the macroscopic scale, this produces

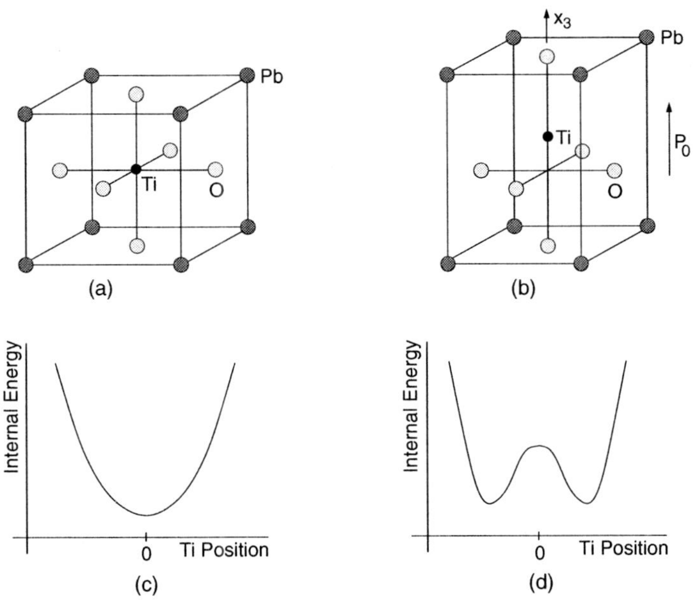

Figure 2.2. *(a) Perovskite structure of $PbTiO_3$ in the cubic form above T_c. (b) Tetragonal structure of $PbTiO_3$ for $T < T_c$ and resulting spontaneous polarization P_0. (c) Internal energy as a function of Ti position along the x_3-axis in the paraelectric phase $T > T_c$ and (d) in the ferroelectric phase $T < T_c$.*

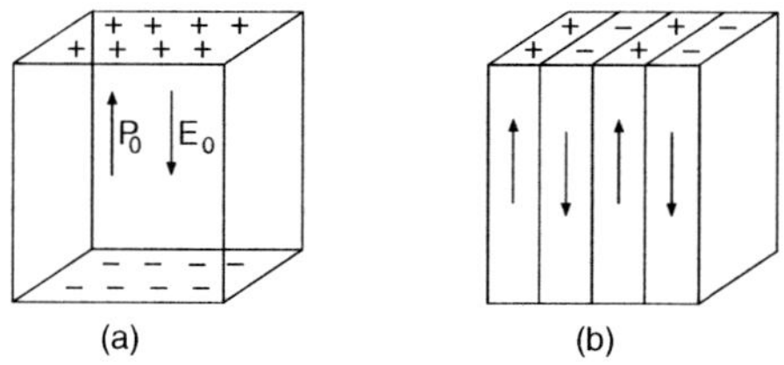

Figure 2.3. *(a) Surface charge due to a spontaneous polarization. (b) Twinned 180° domains which form to minimize electrostatic energy.*

a discontinuous jump in the polarization as experimentally illustrated for single crystal $BaTiO_3$ on pages 72-76 of [351]. The quantification of the internal energy, in a manner which accommodates the double to single well transition at T_c and the distortions in the energy profile due to applied fields, constitutes a fundamental component of the modeling frameworks discussed in later sections.

A direct consequence of the spontaneous polarization which forms as the material cools through the Curie point is the generation of surface charge and a depolarizing field E_0 as depicted in Figure 2.3. To minimize the resulting electrostatic energy, dipoles align in a 180° twinned domain structure.[3] The quantification of 180° domain wall dynamics forms the basis for the polarization models summarized in Section 2.5.

A second mechanism which occurs as the material is cooled from the paraelectric to ferroelectric phase is the generation of thermal stresses. Stresses in the direction of E_0 will have no effect on the 180° domain structures but instead lead to the formation of 90° domains to minimize ferroelastic energy. This produces a twinned structure of the form depicted in Figure 2.4.

Direct and Converse Piezoelectric Effects

The *converse piezoelectric effect* constitutes the linear and reversible strains generated in ferroelectric materials in response to an applied field. As illustrated in Figure 2.5(b), this is due to changes in the ionic structure which subsequently deform the material. The *direct piezoelectric effect* designates the opposite phenomenon in which low stress inputs produce changes in the dipole configuration or polarization.

[3]Domains refer to regions in which dipoles are aligned whereas domain walls denote the transition regions between domains.

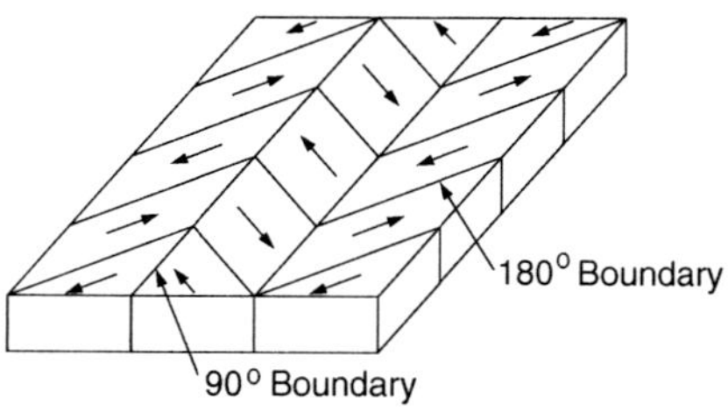

Figure 2.4. *Twinned 90° and 180° domain structure for $BaTiO_3$.*

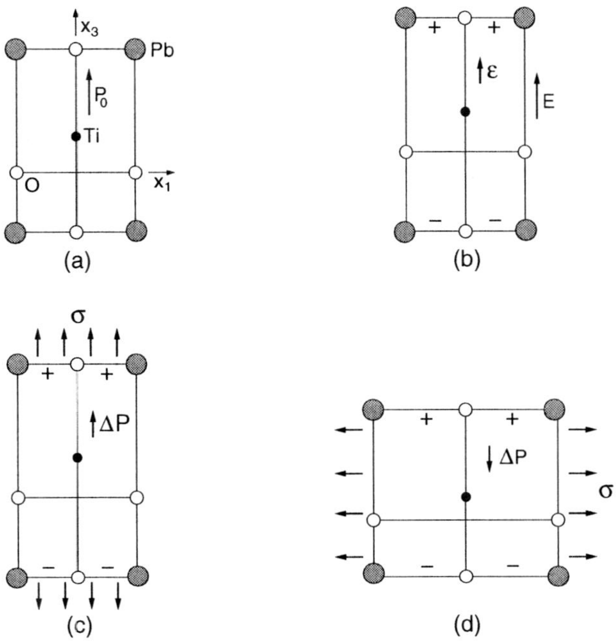

Figure 2.5. *(a) Tetragonal form of $PbTiO_3$ and the spontaneous polarization P_0. (b) Converse piezoelectric effect $\varepsilon = d_{33}E$. (c) Direct piezoelectric effects $\Delta P = d_{33}\sigma$ and (d) $\Delta P = d_{31}\sigma$ for $\sigma < \sigma_c$.*

The polarization changes produced by two tensile stress configurations are depicted in Figures 2.5(c) and (d). It is emphasized that both piezoelectric effects refer to *linear* and *reversible* electromechanical mechanisms which occur well in advance of ferroelectric or ferroelastic switching.

As detailed in Section 2.2, where linear models quantifying reversible effects are developed, the coefficients d_{nij} quantify the electromechanical coupling between input stress or electric fields and changes generated in the crystallographic structure. In general d is a tensor but for the specific geometry depicted in Figure 2.5, it can be reduced to the form d_{nm} where n denotes the direction of the polarization and m quantifies the direction of the applied stress. The application of a positive (tensile) stress σ_3 parallel to the dipole produces an enhancement of the spontaneous polarization whereas the application of a perpendicular tensile stress decreases P_0. To indicate the magnitude of these effects, we note that $d_{33} \approx 400 \times 10^{-12}$ m/V and $d_{31} \approx -170 \times 10^{-12}$ m/V for standard PZT compositions.

Ferroelectric and Ferroelastic Switching

The application of electric fields larger in magnitude than the coercive field E_c[4] causes the central Ti ion to have sufficient energy to overcome the barrier in the double well potential and move to the minimum which coincides with the

[4]The coercive field is defined as the electric field required to reduce the polarization to zero.

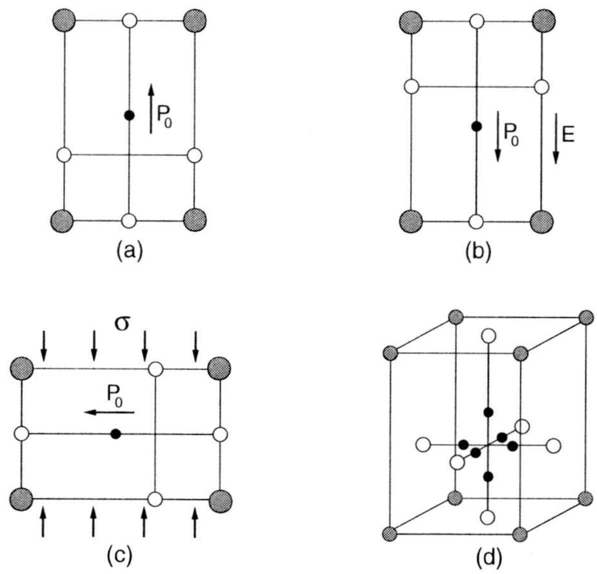

Figure 2.6. *(a) Spontaneous polarization P_0 for $PbTiO_3$. (b) Ferroelectric 180° polarization switch due to an applied electric field $E > E_c$ and (c) ferroelastic 90° degree switch due to a compressive force σ larger in magnitude than the coercive stress σ_c. (d) Six minimum energy sites for the tetragonal structure.*

applied field, thus producing a 180° ferroelectric polarization switch as depicted in Figure 2.6(b). Similarly, applied stresses larger than the coercive stress σ_c will cause the ion to reorient in the well closest to the direction of the stress. This produces 90° ferroelastic switching as illustrated in Figure 2.6(c). The six minimum energy sites for tetragonal structures are indicated in Figure 2.6(d) to delineate possible ferroelectric and ferroelastic switching mechanisms for applied fields and stresses.

Lead Zirconate Titanate (PZT)

It was noted previously that $Pb(Zr,Ti)O_3$ (lead zirconate titanate or PZT) alloys are the most common ferroelectric compound employed for smart material applications. These compounds are comprised of $PbTi_{1-x}O_3$ and $PbZr_xO_3$ with x chosen to optimize electromechanical coupling. The limiting case $PbTiO_3$ has a tetragonal structure similar to $BaTiO_3$ for $T < T_c$ whereas $PbZrO_3$ is orthorhombic below the Curie point — with 71° and 91° domains. A typical phase diagram, illustrating the locus of Curie points which delineate the transition from the paraelectric to ferroelectric phases and the morphotropic phase boundary (MPB) between the tetragonal and rhombohedral structures, is depicted in Figure 2.7(a). As detailed in [114, 362] and illustrated in Figure 2.7(b), strong electromechanical coupling and large dielectric constants are achieved near the MPB thus motivating why PZT is advantageous over $PbTiO_3$ for high performance transducer design. Depending on the molar fraction x, the Curie point T_c ranges from 240^oC to 480^oC, thus extending the operating temperatures as compared with $BaTiO_3$.

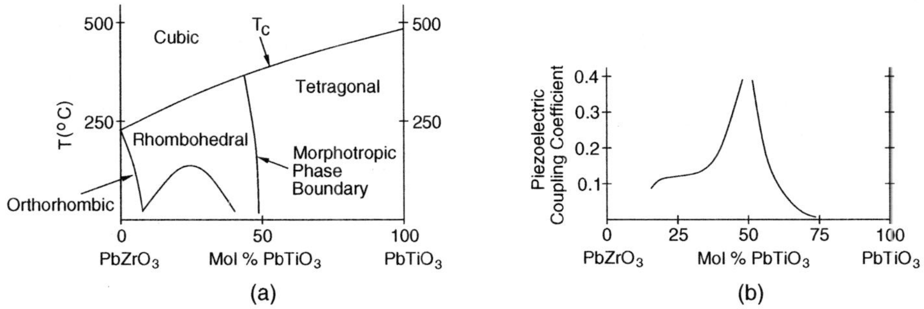

Figure 2.7. *(a) Phase diagram for Pb(Zr, Ti)O_3 illustrating the dependence of the Curie point T_c on the molar fraction x of Zr and the morphotropic phase boundary (MPB) separating the rhombohedral and tetragonal structures. (b) Dependence of the piezoelectric coupling coefficient on the the molar fraction x (after [243]).*

Poled Polycrystalline Materials

The discussion thus far pertains to single crystals whereas the ferroelectric compounds used in smart material applications are typically polycrystalline and in ceramic form. Whereas a single crystal is polar, an unpoled polycrystal exhibits zero net polarization due to the random orientation of grains and domains. The random orientation of domains also restricts the saturation polarization which can be obtained with the material. For example, it is detailed in [351] and illustrated in Figure 2.8 that the saturation polarization of polycrystalline $BaTiO_3$ is approximately half that for a single crystal due to limitations in 90° switching.

To generate the net polarization required to achieve bidirectional strains and general sensing capabilities, it is necessary to pole the materials before use or after manufacturing processes that take place at temperatures near or above the Curie point — e.g., THUNDER must be repoled after the heat treatment which yields its characteristic shape. To attain the energy required to reorient dipoles, poling is typically performed by applying high DC fields at elevated temperatures slightly

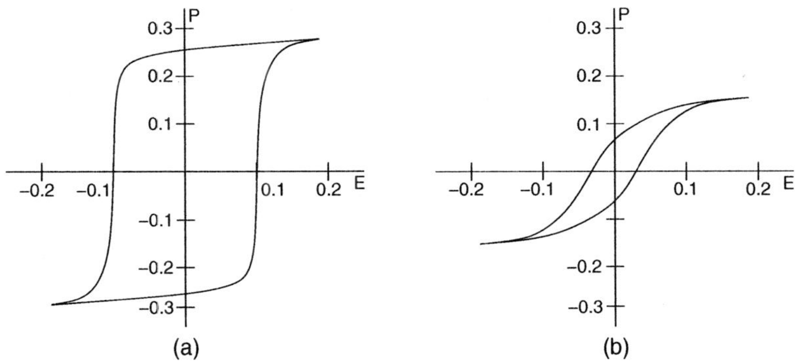

Figure 2.8. *Prototypical BaTiO_3 behavior: (a) single crystal and (b) polycrystal (after [351]).*

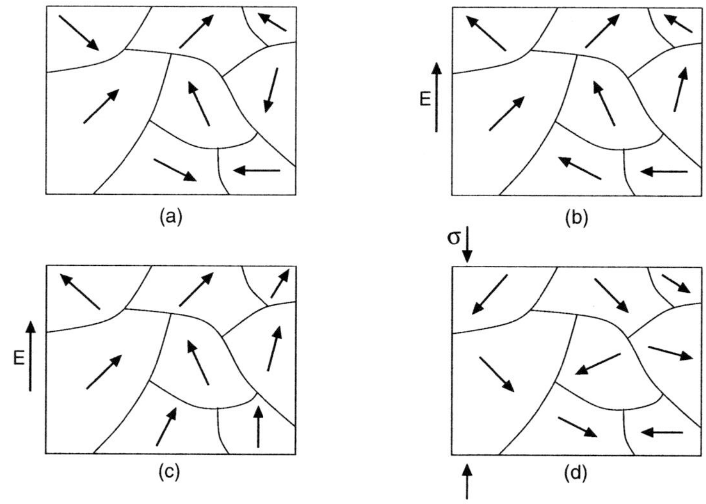

Figure 2.9. *(a) Unpoled material, (b) changes in 180° domains, and (c) changes in 90° and 180° domains due to poling. (d) Ferroelastic 90° switching due to an applied stress.*

below the Curie point. The dipole reorientation achieved during the poling process is illustrated in Figure 2.9. The typical convention is to align the resulting polar vector with the x_3-axis and we will employ this convention throughout subsequent model development.

We point out that the random orientation of grains must be accommodated when quantifying the constitutive behavior of general ferroelectric compounds. This directly motivates the stochastic homogenization techniques employed in the energy models developed in Section 2.6 and implicitly motivates the use of general densities and effective macroscopic parameters in the Preisach and domain wall models of Sections 2.4 and 2.5.

Hysteresis and Constitutive Nonlinearities

An important property of all ferroelectric materials is the presence of hysteresis and constitutive nonlinearities in the relation between input fields E and stresses σ and output polarization P and strains ε as illustrated in Figures 2.10 and 2.11. While influenced by a number of mechanisms, hysteresis is directly associated with the noncentrosymmetric structure of ferroelectric compounds and is observed to some degree at *essentially all drive levels.*

To illustrate switching mechanisms which produce hysteresis and saturation nonlinearities, we illustrate in Figure 2.10 prototypical E-P and E-ε relations for a ferroelectric compound. As detailed in [230, 256, 309], these are indicative of curves measured for lead lanthanum zirconate titanate (PLZT) and typify ferroelectric and ferroelastic properties common to a broad range of *soft* piezoelectric compounds.

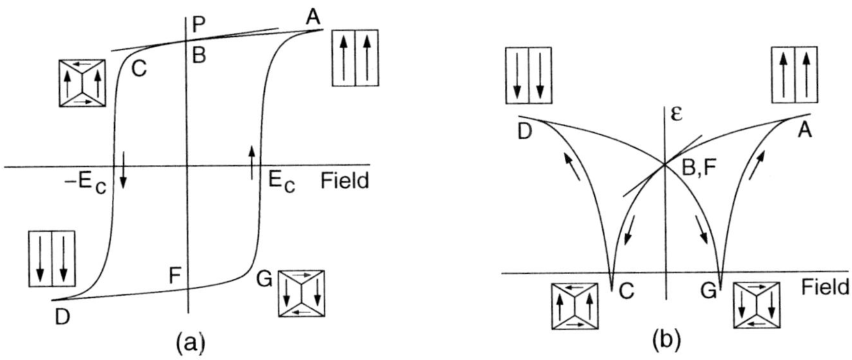

Figure 2.10. *(a) Hysteretic field-polarization relation and linear approximation where the remanent polarization P_R occurs at Point B. (b) Hysteretic field-strain relation and linear approximation where the remanent strain ε_R occurs at B, E.*

Point A: For sufficiently large positive fields, all dipoles are aligned with the field and the material acts as a single domain. Increasing the field beyond this point causes ion movement which stretches the unit cell in a reversible and approximately linear manner.

Point B (Positive Remanence:) At point B, the applied field is zero and the material exhibits a positive remanence polarization P_R and remanence strain ε_R. Small field inputs produce changes in P and ε which are approximately reversible and linear — it is within this regime that the linear direct and converse piezoelectric equations, developed in Section 2.2, are applicable.

Point C: As the field is reduced through the negative coercive field $-E_c$, the polarization begins to switch. The rapid transition in this "burst region" of the E-P curve is attributed to 180° dipole switching. This 180° domain switching does not affect ε and the negative strains observed at coercivity are attributed to 90° switching.

Point D: At Point D, all dipoles have oriented with the applied field and the material again acts as a single domain. The resulting polarization is opposite in sign to that at Point A whereas the strains are equal to those at A since distortions in the unit cell due to increasing negative fields are the same as those produced by positive fields when the material acts as a single domain.

Point F (Negative Remanence:) Increasing the field E to zero causes dipoles to reorient to the negative remanence polarization $-P_R$. The behavior of both P and ε for small field inputs is analogous to that at Point B.

Point G: Switching of 180° domains produces the burst region in the E-P relation whereas the excursion to negative strains in the E-ε curve is again due to 90° switching. The increase in E through the coercive field E_c to the saturation values, indicated by Point A, returns the material to single domain behavior.

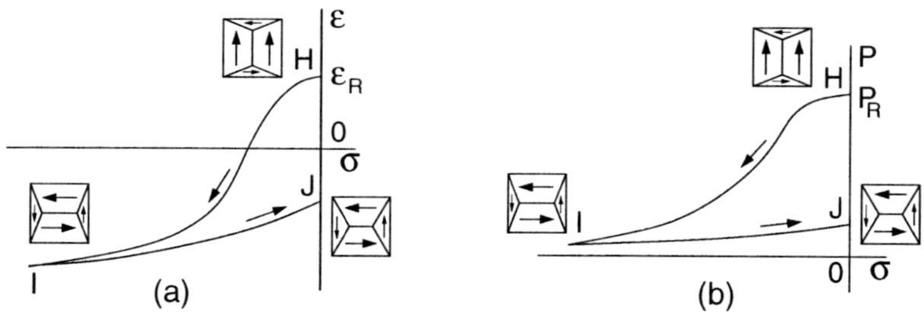

Figure 2.11. *Hysteresis in the (a) stress-strain relation and (b) stress-polarization relation due to 90° switching induced by a compressive stress.*

To illustrate ferroelastic switching effects due to stresses applied in the same direction as the field, consider the application of a negative (compressive) stress to a material as depicted in Figure 2.11. Experimental validation of these mechanisms for PLZT is detailed in [230, 309].

Point H (Positive Remanence): This corresponds with the remanence Point B in Figure 2.10 which occurs in the absence of applied fields or strains.

Point I: The application of a compressive stress causes 90° switching which in turn produces negative strains and reduces the polarization toward zero. At Point I, switching is nearly, but not totally, complete and the Young's modulus $Y = \frac{d\sigma}{d\varepsilon}$ is approximately linear.

Point J: The reduction of the applied stress to zero produces reversible and approximately linear changes in P and ε since the primary mechanism is elastic changes in the unit cell.

The combined effects of stresses and large applied fields for soft PZT compounds are plotted in Figure 2.12 to illustrate that the application of stresses to a poled soft ferroelectric material produces a *shape memory effect (SME)* analogous to that exhibited by shape memory alloys (SMA) as detailed in Section 1.4. The application of stress produces a relative change in strain ε_r which remains until the material is repoled through the application of an electric field as illustrated experimentally by Lynch for PLZT [309]. We plot this behavior in terms of $-\sigma$ and $-\varepsilon$ so that it can be directly compared with analogous SMA behavior plotted in Figure 1.19. We emphasize that the microscopic mechanisms producing the two shape memory effects differ substantially, with the SME in SMA due to phase transitions. However, similarities in the mesoscopic and macroscopic behavior permit the construction of unified modeling frameworks as detailed in Chapter 6.

These ferroelectric and ferroelastic switching mechanisms contribute strongly to the hysteretic and nonlinear E-P and E-ε behavior of ferroelectric materials but are not the sole mechanisms responsible for this complex phenomenon. As discussed by Lynch [309], the behavior is augmented by intermediate rhombohedral dipole reorientations for materials having compositions near tetragonal-rhombohedral morphotropic phase boundaries to enhance electromechanical coupling — see Figure 2.7.

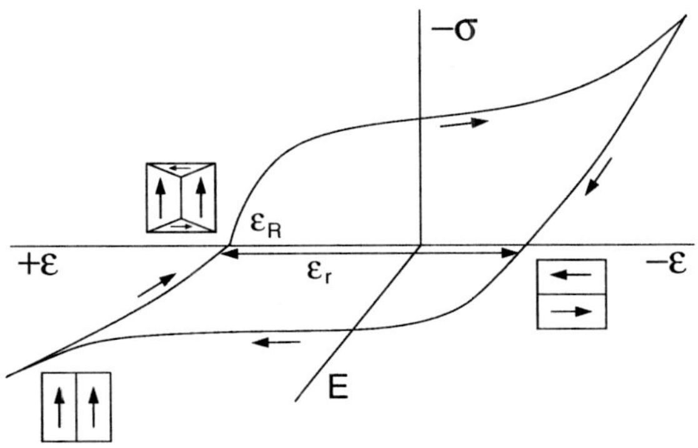

Figure 2.12. *Shape memory effect exhibited by soft PZT compounds due to 90° switching produced by a compressive stress. Repolarization is achieved for certain PLZT compounds with electric fields having coercive values on the order of half the initial coercive value. Axis orientation permits direct comparison with analogous SMA behavior in Figure 1.19.*

In polycrystalline materials, the hysteresis curve often has a sigmoid form analogous to that depicted in Figure 2.8 for $BaTiO_3$. In such cases, the remanence polarization, saturation polarization, coercive field, and slope through the burst region are decreased due to the random orientation of domains and grains[5] and the presence of intergranular stresses. Finally, it is observed that even single crystal compounds do not exhibit instantaneous transitions at E_c due in part to the restriction of domain wall movement by inclusions or pinning sites inherent to the materials.

For material characterization and model-based material design, it is necessary to characterize all of these mechanisms. However, the random nature of the phenomena make it difficult to incorporate all of them into energy formulations. In the domain wall theory of Section 2.5, the energy required to translate domain walls is derived for a uniform cell and *effective macroscopic parameters* are employed to incorporate random nonhomogeneities due to polycrystallinity, nonuniform crystals, and variable intergranular stresses. For the homogenized energy framework developed in Section 2.6, energy analysis for uniform lattices is used to construct *kernels* or *hysterons* quantifying the nonlinear hysteretic behavior for *homogeneous*, single crystal compounds. The effects of material nonhomogeneities, which produce pinning sites, and random domain and grain orientations inherent to polycrystalline compounds are subsequently incorporated through *stochastic homogenization techniques* based on the assumption that parameters such as the local coercive field E_c and interaction field E_I are manifestations of underlying distributions rather than fixed parameter values. This combined energy and stochastic homogenization technique also provides a basis for the Preisach models described in Section 2.4 which were originally derived solely from mathematical principles.

[5]Grains are defined as regions of uniform anisotropy delineated by transition zones termed grain boundaries.

Design Considerations to Reduce Hysteresis

Whereas ferroelectric materials are inherently hysteretic, the deleterious effects of this phenomenon can be reduced through several design considerations: (i) low to moderate operating regimes, (ii) use of feedback laws to linearize the response, and (iii) current or charge controlled amplifiers rather than voltage controlled amplifiers.

The first option is the most obvious. As illustrated in Figure 2.10, the E-P and E-ε relations are reversible and approximately linear in biased regimes with small inputs — e.g., near remanence — in which case linear models can provide sufficient accuracy. This has led to the widespread use of linear models for ferroelectric transducers in applications ranging from structural acoustic control to vibration isolation in elastic structures. However, the linear relations are not exact as illustrated by the low field AFM data plotted in Figure 1.10. For applications such as nanopositioning, which require high set point accuracy, the inherent hysteresis must be accommodated even when operating at low input fields or voltages.

The use of feedback laws is also widely employed and has led to the success of linear piezoelectric models in a number of the applications illustrated in Chapter 1. In high accuracy or high frequency drive regimes, however, increasing noise-to-signal ratios and diminishing control characteristics limit the degree to which feedback designs can solely be employed to linearize the response of piezoelectric transducers. To illustrate, consider the performance of the stacked and cylindrical PZT actuators depicted in Figure 1.9 for nanopositioning in present AFM designs. At low drive frequencies, high gain feedback laws are currently employed to mitigate hysteresis in the E-ε relation [16,107,108,115,400,401,405], thus leading to the phenomenal success of the instruments. However, at the higher frequencies required for large-scale product diagnostics — e.g., quality control of semiconductor chips — or real-time monitoring of biological processes — e.g., observation of protein folding dynamics — the efficacy of feedback laws is diminished by inherent thermal and measurement noise. Robust control design can be used to extend the frequency ranges of operation [401,408], but the loop shaping and gains required to attenuate hysteresis have the negative effect of accentuating high frequency noise.

As detailed in [305, 313–316], the use of charge or current controlled amplifiers can essentially eliminate hysteresis. However, this mode of operation can be prohibitively expensive as compared with the more commonly employed voltage controlled amplifiers, and current control is ineffective if maintaining DC offsets as required for numerous applications - e.g., the x-stage in an AFM must hold a specified position while a sweep is performed in the y-stage.

Finally, it is noted that procedures such as the addition of dopants or thermal annealing can be employed to reduce hysteresis. However, this can diminish performance and requires processing capabilities which may not be generally available.

In summary, all of these options exhibit limitations for general operating regimes and applications. This motivates the development of nonlinear hysteresis models which provide the accuracy required for fundamental material characterization and the efficiency necessary for system design and real-time implementation of model-based control algorithms.

2.2 Linear Piezoelectric Models

For low input field levels, the hysteretic E-P and E-ε relations can be approximated by the slope at remanence, or about other bias points, to yield linear constitutive relations quantifying the converse piezoelectric effect as illustrated in Figure 2.8. Similarly, the polarization and fields generated by the direct piezoelectric effect exhibit a nearly linear dependence on the stress for low inputs levels, thus motivating the use of linear constitutive models for sensor applications. The development of linear models relating the dielectric and elastic behavior of piezoelectric materials was initiated by Voigt [500], and the resulting coupled models are employed in numerous smart material applications which require low to moderate input fields or voltages.

2.2.1 Linear Constitutive Relations

When defining both linear and nonlinear constitutive relations, we can choose either the polarization P or D field (also termed the electric displacement) as the electric variable. It is argued by Ikeda [234] and Cady [73] that the choice of P offers certain theoretical advantages whereas the historical development of Voigt is formulated in terms of the D field. Furthermore, both are easily measured quantities; the normal component of the D field equals the charge per unit area on an electrode [230] while P can be measured with a Sawyer tower comprised of a reference capacitor placed in series with the sample. Because the final constitutive relations are analogous, the choice between D and P is not fundamental for what follows and can be based on the measurement capabilities available for a particular application. Since data used for model validation in later sections includes polarization measurements, we employ P and summarize analogous relations for D at various points in the discussion.

In the absence of an applied field, the polarization produced by a stress σ is

$$P = d\sigma$$

where d is the piezoelectric charge coefficient. For general materials, P is vector-valued and related to each component of the stress σ_{ij}. It thus follows that

$$\begin{aligned} P_n &= \sum_{i=1}^{3}\sum_{j=1}^{3} d_{nij}\sigma_{ij} \\ &= d_{nij}\sigma_{ij} \end{aligned} \tag{2.1}$$

where the second equality results from the use of the Einstein summation convention in which terms with repeated indices are automatically summed. Note that the 27 coefficients d_{nij} comprise a third rank tensor.

The second contribution to P is due to the field. For *linear* operating regimes, this component can be quantified by the polarizability relation

$$P_n = \chi^{\sigma}_{nm} E_m. \tag{2.2}$$

Here $\chi^{\sigma}_{nm} = \overline{\chi}^{\sigma}_{nm}\epsilon_0$ where $\overline{\chi}^{\sigma}_{nm}$ denotes the dielectric susceptibility measured at constant stress and ϵ_0 is the permittivity of a vacuum. We note that the relation (2.2) should be used with caution for ferroelectric materials since the susceptibility is not only nonlinear but is a multi-valued map at high field levels — see Figures 1.1, and 1.26, as well as Figure 2.10 for a depiction of the linear operating range.

The combination of (2.1) and (2.2) then yields

$$P_n = d_{nij}\sigma_{ij} + \chi^{\sigma}_{nm} E_m \tag{2.3}$$

as a linear model for the *direct piezoelectric effect.*

The *converse piezoelectric relation* combines Hooke's law or, in this case, its reciprocal, with the linear E-ε relation depicted in Figure 2.10(b) to express the strain as

$$\varepsilon_{ij} = s^E_{ijk\ell}\sigma_{k\ell} + d_{nij} E_n \tag{2.4}$$

where s^E is the fourth rank compliance tensor measured at constant field. The equivalence of the piezoelectric d coefficients follows from the thermodynamic treatment in Section 2.2.2 and has the limitations delineated in Remark 2.2.3.

The tensor descriptions in (2.3) and (2.4) provide a complete description of the linear behavior of general piezoelectric compounds. However, the number of coefficients can be reduced significantly by invoking elastic symmetries along with electric symmetries due to poling. When combined with a change in index, this provides a matrix system suitable for typical smart material applications.

We first note that for linear elastic materials, the stress and strain tensors are symmetric so that $\sigma_{ij} = \sigma_{ji}$ and $\varepsilon_{ij} = \varepsilon_{ji}$. The symmetry in stress and strain implies that d_{nij} is symmetric in i and j thus reducing the number of independent coefficients to 18.

To formulate the system as a matrix equation, the indices jk are replaced by a single index m according to the following convention:

11	22	33	23, 32	31, 13	12, 21	Tensor
1	2	3	4	5	6	Matrix

The coefficients d_{nij} are then written as $d_{nij} = d_{nm}$ when $m = 1, 2, 3$ and $d_{nij} = \frac{1}{2} d_{nm}$ when $m = 4, 5, 6$. The factor of 1/2 accommodates the symmetry in i and j. A similar convention is applied to the compliance coefficients $s^E_{ijk\ell}$ and is used to formulate the stress and strain tensors as matrices, thus yielding

$$\begin{bmatrix} \sigma_{11} & \sigma_{12} & \sigma_{31} \\ \sigma_{12} & \sigma_{22} & \sigma_{23} \\ \sigma_{31} & \sigma_{23} & \sigma_{33} \end{bmatrix} \longrightarrow \begin{bmatrix} \sigma_1 & \sigma_6 & \sigma_5 \\ \sigma_6 & \sigma_2 & \sigma_4 \\ \sigma_5 & \sigma_4 & \sigma_3 \end{bmatrix}$$

$$\begin{bmatrix} \varepsilon_{11} & \varepsilon_{12} & \varepsilon_{31} \\ \varepsilon_{12} & \varepsilon_{22} & \varepsilon_{23} \\ \varepsilon_{31} & \varepsilon_{23} & \varepsilon_{33} \end{bmatrix} \longrightarrow \begin{bmatrix} \varepsilon_1 & \frac{1}{2}\varepsilon_6 & \frac{1}{2}\varepsilon_5 \\ \frac{1}{2}\varepsilon_6 & \varepsilon_2 & \frac{1}{2}\varepsilon_4 \\ \frac{1}{2}\varepsilon_5 & \frac{1}{2}\varepsilon_4 & \varepsilon_3 \end{bmatrix}$$

(e.g., see [365, Chapter VII] or [114, 396]).

The number of piezoelectric coefficients d_{im} and stress-strain components of interest are further reduced when we consider a material poled in the x_3 direction.

As noted in [362], conical symmetry dictates that in this case, all of the piezoelectric coefficients are zero except $d_{31} = d_{32}, d_{33}$, and $d_{15} = d_{24}$.

The utilization of the structural and electric symmetries along with matrix re-indexing yields the system

$$\begin{bmatrix} \varepsilon_1 \\ \varepsilon_2 \\ \varepsilon_3 \\ \varepsilon_4 \\ \varepsilon_5 \\ \varepsilon_6 \\ \hline P_1 \\ P_2 \\ P_3 \end{bmatrix} = \left[\begin{array}{cccccc|ccc} s_{11}^E & s_{12}^E & s_{13}^E & 0 & 0 & 0 & 0 & 0 & d_{31} \\ s_{12}^E & s_{11}^E & s_{13}^E & 0 & 0 & 0 & 0 & 0 & d_{31} \\ s_{13}^E & s_{13}^E & s_{33}^E & 0 & 0 & 0 & 0 & 0 & d_{33} \\ 0 & 0 & 0 & s_{44}^E & 0 & 0 & 0 & d_{15} & 0 \\ 0 & 0 & 0 & 0 & s_{44}^E & 0 & d_{15} & 0 & 0 \\ 0 & 0 & 0 & 0 & 0 & s_{66}^E & 0 & 0 & 0 \\ \hline 0 & 0 & 0 & 0 & d_{15} & 0 & \chi_{11}^\sigma & 0 & 0 \\ 0 & 0 & 0 & d_{15} & 0 & 0 & 0 & \chi_{11}^\sigma & 0 \\ d_{31} & d_{31} & d_{33} & 0 & 0 & 0 & 0 & 0 & \chi_{33}^\sigma \end{array}\right] \begin{bmatrix} \sigma_1 \\ \sigma_2 \\ \sigma_3 \\ \sigma_4 \\ \sigma_5 \\ \sigma_6 \\ \hline E_1 \\ E_2 \\ E_3 \end{bmatrix} .$$

The converse and direct piezoelectric components can be written as

$$\begin{aligned} \varepsilon &= s^E \sigma + d^* E \\ P &= d\sigma + \chi^\sigma E \end{aligned} \tag{2.5}$$

where d^* denotes the transpose of d.

For the development of system models for transducer characterization, it is advantageous to formulate the stress as a linear function of strain with the Young's modulus $Y = s^{-1}$ as a proportionality constant, which is simply a manifestation of Hooke's law. Furthermore, it is advantageous to posit that for low drive regimes, strain is linearly proportional to P rather than E — while best motivated by experimental measurements, Ikeda notes that the linear P-ε assumption precedes the E-ε relation from a historical perspective [234]. In combination, this yields the constitutive relations

$$\begin{aligned} \sigma &= Y^P \varepsilon - a^* P \\ P &= d\sigma + \chi^\sigma E \end{aligned} \tag{2.6}$$

where a^* is a second piezoelectric coupling coefficient and Y^P denotes the Young's modulus measured at constant polarization.

Remark 2.2.1. *It will be demonstrated through the analysis in Section 2.2.2 that the linear constitutive relations (2.5) are thermodynamically consistent in the sense that the dependent variables (ε, P) are specified in terms of independent variables (σ, E). This is a natural formulation for systems in which (σ, E) constitute inputs. The relation (2.6) is not thermodynamically consistent since P is specified in terms of (σ, E) whereas (ε, P) provide the inputs for σ. It will demonstrated in (2.20) of Section 2.2.2 that instead, the relations*

$$\begin{aligned} \sigma &= Y^P \varepsilon - a^* P \\ E &= -a\varepsilon + \alpha^\varepsilon P \end{aligned} \tag{2.7}$$

consistently quantify the piezoelectric material behavior for (ε, P) inputs.

Although not thermodynamically consistent, the consideration of (2.6) motivates a framework employed in the hysteresis models developed in Sections 2.3 through 2.6. When characterizing hysteretic behavior in the context of thermodynamic analysis, one has two options: (i) employ irreversible thermodynamic tenets [101, 259, 331], or (ii) combine the reversible theory, which incorporates certain fundamental energy mechanisms, with phenomenological, mathematical, or stochastic homogenization techniques to obtain nonlinear constitutive relations. The formulation (2.6) reflects this latter philosophy. The nonlinear models in Sections 2.3 – 2.6 will have the general form

$$\begin{aligned} \sigma &= Y^P \varepsilon - a^* P \\ P &= \mathcal{F}(\sigma, E) \end{aligned} \tag{2.8}$$

where the map $\mathcal{F}$ characterizes the hysteretic dependence of P on (σ, E) for materials exhibiting a linear dependence of σ on P. The converse relation is extended to incorporate electrostrictive behavior through inclusion of a quadratic term $a_2 P^2$.

2.2.2 Energy Formulation

The constitutive relations can be posed in an energy framework through the definition of functionals which quantify the elastic, thermal, electrostatic and electromechanical energies. This facilitates thermodynamic analysis of the compounds and provides a framework which can be extended in a natural manner to incorporate nonlinear and hysteretic behavior as detailed in subsequent sections. Details regarding the classical formulation of energy relations for piezoelectric materials can be found in [73, 134, 234, 256, 325, 326, 365].

Thermodynamic Framework

Since piezoelectricity constitutes interactions between electrical and mechanical processes, and the related phenomenon pyroelectricity is due to thermal-electrical coupling, we consider the general set of electrical, mechanical and thermal processes depicted in Figure 2.13. The intensive variables (also termed generalized forces) electric field E, stress σ, and temperature T are represented at vertices of the outer triangle whereas the corresponding conjugate extensive variables polarization P, strain ε, and change in entropy δS are represented on the inner triangle. To quantify relations between these variables, we consider the energy for the system.

On a unit volume, the *first law of thermodynamics* (energy conservation) dictates that if an amount of work dW is done on the material by external forces or fields and an infinitesimal quantify of heat dQ enters the material, the change in the internal energy U is given by the exact or total differential

$$dU = dW + dQ.$$

It is shown on page 136 of [365] that the work per unit volume done by a stress σ_{ij} is

$$dW_\sigma = \sigma_{ij} d\varepsilon_{ij} \tag{2.9}$$

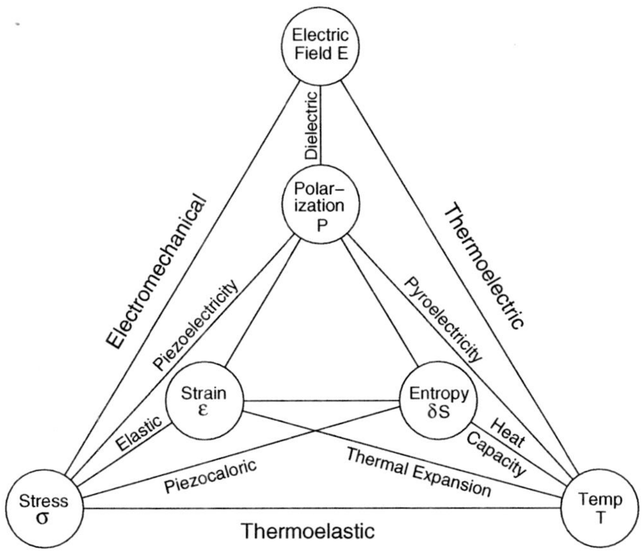

Figure 2.13. *Interaction between the electrical, mechanical and thermal processes with the intensive variables field E, stress σ and temperature T, and the extensive variables polarization P, strain ε and entropy S.*

whereas it is shown in Section 2.8 that the electrical work done on the material is

$$dW_E = E_n dP_n \quad \text{or} \quad dW_E = E_n dD_n \tag{2.10}$$

which we note are not exact differentials.[6] Hence the combined work is

$$dW = \sigma_{ij} d\varepsilon_{ij} + E_n dP_n.$$

From the second law of thermodynamics, the entropy S can be defined in differential form as

$$dS = \frac{dQ}{T} + dS_{irr}$$

where dS_{irr} denotes the entropy production due to irreversible processes — see page 416 in Appendix A for additional discussion. Thus dQ can be expressed as

$$dQ = TdS - TdS_{irr}$$

which yields the differential form

$$dU = \sigma_{ij} d\varepsilon_{ij} + E_n dP_n + TdS - TdS_{irr} \tag{2.11}$$

for the internal energy. We note that the term $-dS_{irr}$, which quantifies the irreversibility of transformations, is neglected in the thermodynamic analysis leading

[6]Details regarding the work relations and ramifications of the fact that dW, dW_σ, dW_E and dQ are not exact differentials whereas dU is an exact differential can be found in [3] and Appendix A.

to the construction of *reversible*, linear constitutive relations, but plays a fundamental role in the interpretation of thermodynamic potentials and determination of equilibrium states for the system.

The internal energy U resulting from (2.11) is a function of the independent variables ε_{ij}, P_n and S. However, for many applications, other choices of independent variables may be more reasonable. For example, in transducer design, it is often more natural to prescribe or control stresses σ_{ij} or electric fields E_n rather than strain or polarization values. The different choices of independent variables yield varying formulations for appropriate thermodynamic potentials whose minima quantify equilibrium states for the system.

To illustrate, consider the thermodynamic energy relation

$$G = U - \sigma_{ij}\varepsilon_{ij} - E_n P_n - TS \tag{2.12}$$

which, under the assumption of reversibility so that $dS_{irr} = 0$, has the differential

$$\begin{aligned} dG &= dU - \sigma_{ij} d\varepsilon_{ij} - \varepsilon_{ij} d\sigma_{ij} - E_n dP_n - P_n dE_n - TdS - SdT \\ &= -\varepsilon_{ij} d\sigma_{ij} - P_n dE_n - SdT. \end{aligned} \tag{2.13}$$

Hence G is a function of the independent variable set (σ_{ij}, E_n, T) so it follows from (2.13) that

$$\varepsilon_{ij} = -\left[\frac{\partial G}{\partial \sigma_{ij}}\right]_{E,T} \quad , \quad P_n = -\left[\frac{\partial G}{\partial E_n}\right]_{\sigma,T} \quad , \quad S = -\left[\frac{\partial G}{\partial T}\right]_{\sigma,E} \tag{2.14}$$

where subscripted quantities denote fixed process conditions. Equivalently, the thermodynamic constitutive relations (2.14) can be interpreted as providing necessary conditions for minimizing G under the condition that U is a function of ε_{ij}, P_i and S. The thermodynamic potential G given by (2.12) is the *Gibbs free energy* for the uncoupled electric, mechanical and thermal processes.

Continuing under the assumption that (σ_{ij}, E_n, T) constitutes the independent variable set and the conjugate variables $(\varepsilon_{ij}, P_n, S)$ are dependent, we can form the differentials

$$\begin{aligned}
d\varepsilon_{ij} &= \underbrace{\left[\frac{\partial \varepsilon_{ij}}{\partial \sigma_{k\ell}}\right]_{E,T} d\sigma_{k\ell}}_{\text{elasticity}} + \underbrace{\left[\frac{\partial \varepsilon_{ij}}{\partial E_n}\right]_{\sigma,T} dE_n}_{\text{converse piezoelectricity}} + \underbrace{\left[\frac{\partial \varepsilon_{ij}}{\partial T}\right]_{\sigma,E} dT}_{\text{thermal expansion}} \\
dP_n &= \underbrace{\left[\frac{\partial P_n}{\partial \sigma_{ij}}\right]_{E,T} d\sigma_{ij}}_{\text{direct piezoelectricity}} + \underbrace{\left[\frac{\partial P_n}{\partial E_m}\right]_{\sigma,T} dE_m}_{\text{permittivity}} + \underbrace{\left[\frac{\partial P_n}{\partial T}\right]_{\sigma,E} dT}_{\text{pyroelectricity}} \\
dS &= \underbrace{\left[\frac{\partial S}{\partial \sigma_{ij}}\right]_{E,T} d\sigma_{ij}}_{\text{piezocaloric effect}} + \underbrace{\left[\frac{\partial S}{\partial E_n}\right]_{\sigma,T} dE_n}_{\text{electrocaloric effect}} + \underbrace{\left[\frac{\partial S}{\partial T}\right]_{\sigma,E} dT}_{\text{heat capacity}}.
\end{aligned} \tag{2.15}$$

Furthermore, differentiation of the equations (2.14) yields the relations

$$
\begin{aligned}
-\left[\frac{\partial^2 G}{\partial \sigma_{ij}\partial E_n}\right]_T &= \left[\frac{\partial \varepsilon_{ij}}{\partial E_n}\right]_{\sigma,T} = \left[\frac{\partial P_n}{\partial \sigma_{ij}}\right]_{E,T} \equiv d^T_{nij} \\
-\left[\frac{\partial^2 G}{\partial \sigma_{ij}\partial T}\right]_E &= \left[\frac{\partial \varepsilon_{ij}}{\partial T}\right]_{\sigma,E} = \left[\frac{\partial S}{\partial \sigma_{ij}}\right]_{E,T} \equiv \kappa^E_{ij} \qquad (2.16) \\
-\left[\frac{\partial^2 G}{\partial E_n\partial T}\right]_\sigma &= \left[\frac{\partial P_n}{\partial T}\right]_{\sigma,E} = \left[\frac{\partial S}{\partial E_n}\right]_{\sigma,T} \equiv p^\sigma_n
\end{aligned}
$$

for the electromechanical and thermal constants. Note that the relations (2.16) are valid for both linear and nonlinear electromechanical processes as long as they are *reversible*.[7] Furthermore, they establish one of the fundamental principles resulting from the thermodynamic analysis; namely, the equivalence of coefficients in the system (2.15) exhibiting electric, mechanical, and thermal components. In this setting, the equivalence can be summarized as follows:

(i) The coefficients for the converse and direct piezoelectric effects are equivalent;

(ii) the coefficients for the piezocaloric effect are the same as those for thermal expansion;

(iii) the coefficients for the pyroelectric and electrocaloric effects are the same.

Exploiting these equivalences and the fact that $d\varepsilon$, dP and dS are exact differentials, the equations (2.15) can be integrated to yield the *linear* constitutive relations

$$
\begin{aligned}
\varepsilon_{ij} &= s^{E,T}_{ijk\ell}\sigma_{k\ell} + d^T_{nij}E_n + \kappa^E_{ij}\Delta T \\
P_n &= d^T_{nij}\sigma_{ij} + \chi^{\sigma,T}_{nm}E_m + p^\sigma_n\Delta T \qquad (2.17) \\
\Delta S &= \kappa^E_{ij}\sigma_{ij} + p^\sigma_i E_i + C^{\sigma,E}/T\Delta T
\end{aligned}
$$

which are appropriate for quantifying first-order effects.

At this point, it is instructive to note what the thermodynamic theory *does not* provide. It does not predict the occurrence of piezoelectric coupling or thermal effects — it simply provides a very effective framework for quantifying the relationships between these effects when they do occur. It also does not provide specific energy relations characterizing the effects of electromechanical or thermomechanical coupling. This must be accomplished using other physical criteria as detailed in the next section.

We note that the Gibbs functional G is but one of several thermodynamic potentials which can be formulated to accommodate other combinations of independent and dependent variable sets. We indicate several alternative choices in the following discussion focused on the construction of energy relations for coupled electromechanical effects.

[7] It is noted in [3] and Appendix A that these relations provide necessary and sufficient conditions ensuring that the differential dG is exact which guarantees the existence of a function $G(\sigma, E, T)$.

Energy Relations and Constitutive Relations for Piezoelectric Materials

The derivation of the linear constitutive relations (2.17) relied on the integration of the differential (2.15) whose form was dictated by the choice of independent variables. Whereas the general Gibbs relation (2.12) was used to establish the equivalence of coefficients, the relations (2.17) were not derived through the minimization of an energy functional. We consider here the choice of energy functionals which directly yield linear constitutive relations for a variety of independent variable sets. For simplicity, we consider fixed T, S conditions to focus the discussion on piezoelectric effects.

We consider first the choice of (ε, P) as an independent variable set. An appropriate energy functional which incorporates *linear* electromechanical coupling is the *Helmholtz relation*

$$\psi(\varepsilon, P) = \frac{1}{2} Y_{ijk\ell}^{P} \varepsilon_{ij} \varepsilon_{k\ell} - a_{nij} \varepsilon_{ij} P_n + \frac{1}{2} \alpha_{nm}^{\varepsilon} P_m P_n \tag{2.18}$$

where $Y_{ijk\ell}^{P}$ denotes the Young's modulus at constant polarization, a_{nij} is a piezoelectric coupling coefficient and $\alpha_{mn}^{\varepsilon}$ is the dielectric coefficient. The first and third terms on the right side of (2.18) respectively quantify the elastic and electrostatic energies resulting from the work relations (2.9) and (2.10). The second term incorporates linear contributions due to electromechanical coupling with the negative sign chosen to ensure that the coefficient has the same sign as the piezoelectric coefficient d.

The exact differential for this independent variable set is

$$d\psi = \sigma_{ij} d\varepsilon_{ij} + E_n dP_n$$

from which it follows that

$$\sigma_{ij} = \left[\frac{\partial \psi}{\partial \varepsilon_{ij}} \right]_P \quad , \quad E_n = \left[\frac{\partial \psi}{\partial P_n} \right]_\varepsilon . \tag{2.19}$$

The linear constitutive relations expressed in terms of the independent variables ε and P are thus

$$\begin{aligned} \sigma_{ij} &= Y_{ijk\ell}^{P} \varepsilon_{k\ell} - a_{nij} P_n \\ E_n &= -a_{nij} \varepsilon_{ij} + \alpha_{nm}^{\varepsilon} P_m . \end{aligned} \tag{2.20}$$

We note that (2.20) provides the thermodynamically consistent relation (2.7) in Remark 2.2.1.

For transducer design, it is often preferable to consider the stress σ_{ij} and field E_n as input variables rather than strain and polarization. To obtain an exact differential which is dependent on σ and E, one employs the relation

$$G = \psi - EP - \sigma\varepsilon \tag{2.21}$$

to get

$$dG = -P_n dE_n - \varepsilon_{ij} d\sigma_{ij}$$

as noted in (2.13). In this case, one employs the Gibbs relation

$$G(\sigma, E) = -\frac{1}{2}s^E_{ijk\ell}\sigma_{ij}\sigma_{k\ell} - d_{nij}\sigma_{ij}E_n - \frac{1}{2}\chi^\sigma_{nm}E_mE_n \tag{2.22}$$

where s^E and χ^σ respectively denote the compliance and scaled dielectric susceptibility at fixed field and stress.[8] From the relations

$$P_n = -\left[\frac{\partial G}{\partial E_n}\right]_\sigma \quad , \quad \varepsilon_{ij} = -\left[\frac{\partial G}{\partial \sigma_{ij}}\right]_E ,$$

one then obtains the linear constitutive equations

$$\begin{aligned} \varepsilon_{ij} &= s^E_{ijk\ell}\sigma_{k\ell} + d_{nij}E_n \\ P_n &= d_{nij}\sigma_{ij} + \chi^\sigma_{nm}E_m. \end{aligned}$$

These are the relations (2.5) which were obtained directly from linear electromechanical properties.

Additional choices for the independent variable sets, the relevant Legendre transforms and thermodynamic functions, and the resulting linear constitutive relations are summarized in Table 2.1. To simplify notation, we have dropped the tensor subscripts but retain superscripts to specify fixed quantities.

[8] As detailed in Appendix C.1 and Section 2.2.3, The Gibbs relation (2.22), formulated solely in terms of the independent variables (σ, E), is the negative Legendre transform of G with respect to (ε, P). Although a misnomer, the *negative* is often omitted in the literature in which case $G(\sigma, E)$ is termed the Legendre transform of $\psi(\varepsilon, P)$.

Independent variables	Constitutive relations	Thermodynamic function (Negative Legendre transform)
(ε, P)	$\begin{cases} \sigma = Y^P\varepsilon - aP \\ E = -a\varepsilon + \alpha^\varepsilon P \end{cases}$	Helmholtz free energy $\psi(\varepsilon, P) = \frac{1}{2}Y^P\varepsilon^2 - a\varepsilon P + \frac{1}{2}\alpha^\varepsilon P^2$
(σ, E)	$\begin{cases} \varepsilon = s^E\sigma + dE \\ P = d\sigma + \chi^\sigma E \end{cases}$	Gibbs free energy $G(\sigma, E) = -\frac{1}{2}s^E\sigma^2 - d\sigma E - \frac{1}{2}\chi^\sigma E^2$ $(G = \psi - \sigma\varepsilon - EP)$
(σ, P)	$\begin{cases} \varepsilon = s^P\sigma + bP \\ E = -b\sigma + \alpha^\sigma P \end{cases}$	Elastic Gibbs energy $G_1(\sigma, P) = -\frac{1}{2}s^P\sigma^2 - b\sigma P - \frac{1}{2}\alpha^\sigma P^2$ $(G_1 = \psi - \sigma\varepsilon)$
(ε, E)	$\begin{cases} \sigma = Y^E\varepsilon - eE \\ P = e\varepsilon + \chi^\varepsilon E \end{cases}$	Electric Gibbs energy $G_2(\varepsilon, E) = -\frac{1}{2}Y^E\varepsilon^2 - e\varepsilon E - \frac{1}{2}\chi^\varepsilon E^2$ $(G_2 = \psi - EP)$

Table 2.1. *Independent variable sets, thermodynamic functions and Legendre transforms, and linear piezoelectric relations formulated in terms of the polarization.*

Independent variables	Constitutive relations	Thermodynamic function (Negative Legendre transform)
(ε, D)	$\begin{cases} \sigma = Y^D\varepsilon - hD \\ E = -h\varepsilon + \beta^\varepsilon D \end{cases}$	Helmholtz free energy $\psi(\varepsilon, D) = \frac{1}{2}Y^D\varepsilon^2 - h\varepsilon D + \frac{1}{2}\beta^\varepsilon D^2$
(σ, E)	$\begin{cases} \varepsilon = s^E\sigma + dE \\ D = d\sigma + \epsilon^\sigma E \end{cases}$	Gibbs free energy $G(\sigma, E) = -\frac{1}{2}s^E\sigma^2 - d\sigma E - \frac{1}{2}\epsilon^\sigma E^2$ $(G = \psi - \sigma\varepsilon - ED)$
(σ, D)	$\begin{cases} \varepsilon = s^D\sigma + gD \\ E = -g\sigma + \beta^\sigma D \end{cases}$	Elastic Gibbs energy $G_1(\sigma, D) = -\frac{1}{2}s^D\sigma^2 - b\sigma D - \frac{1}{2}\beta^\sigma D^2$ $(G_1 = \psi - \sigma\varepsilon)$
(ε, E)	$\begin{cases} \sigma = Y^E\varepsilon - eE \\ D = e\varepsilon + \epsilon^\varepsilon E \end{cases}$	Electric Gibbs energy $G_2(\varepsilon, E) = -\frac{1}{2}Y^E\varepsilon^2 - e\varepsilon E - \frac{1}{2}\epsilon^\varepsilon E^2$ $(G_2 = \psi - ED)$

Table 2.2. *Independent variable sets, thermodynamic functions and Legendre transforms, and linear piezoelectric relations formulated in terms of the D field.*

As detailed in [73, 234, 365], the previous energy and thermodynamic analysis also applies with the electrostatic energy formulated in terms of the D field and the resulting independent variable sets, thermodynamic functions, and linear constitutive relations for this choice of electric variables are summarized in Table 2.2.

Remark 2.2.2. *The specification of which input condition is fixed is important since different conditions will in general yield different parameter values for ferroelectric materials — e.g., $Y^E \neq Y^P$ for general conditions. The specification is determined by the input variable pair.*

Remark 2.2.3. *The equivalence of parameters in the direct and converse piezoelectric equations, as dictated by (2.16), is dependent on reversible behavior and will not hold in hysteretic operating conditions where the parameters are nonlinear maps of the input. This latter property is easily observed in Figure 2.10(b) where the field dependence of d is illustrated.*

Thermodynamic Potentials

In addition to establishing reciprocity properties of piezoelectric or thermal coefficients, the thermodynamic relations can be employed to establish equilibrium states for the system subject to specified field and stress conditions. This relies on the fact that they constitute *thermodynamic potentials* which are defined as functionals whose minima yield equilibrium states of the system in the presence of constraints. As detailed in [234,284,298,365], the internal energy, Helmholtz energy, and various Gibbs energy relations are all thermodynamic potentials for different independent variable and constraint sets.

For subsequent model development, the Gibbs energy proves a widely applicable potential for specifying the polarization due to fixed electric field inputs or strains due to fixed stresses so we illustrate the concept in this context.

Consider the independent variable set (E, σ, T) and the Gibbs relation

$$G = \psi - EP - \sigma\varepsilon \tag{2.23}$$

where the Helmholtz energy

$$\psi = U - TS$$

is a function of P, ε, T — here P and ε are dependent variables. With the inclusion of irreversible entropy effects, the differential form of the internal energy, specified by (2.11), is

$$dU = \sigma d\varepsilon + EdP + TdS - TdS_{irr}$$

so that G has the differential form

$$dG = -PdE - \varepsilon d\sigma - SdT - TdS_{irr}. \tag{2.24}$$

For fixed field, stress and temperature regimes, (2.24) reduces to

$$dG = -TdS_{irr}.$$

Because the second law of thermodynamics dictates that dS_{irr} is necessarily nonnegative, G is nonincreasing with minima at equilibrium states. Hence for a specified field E, the polarization state is determined by the relation

$$\frac{\partial G}{\partial P} = 0 \Leftrightarrow E = \left[\frac{\partial \psi}{\partial P}\right]_{\varepsilon,T}. \tag{2.25}$$

Analogously, the equilibrium strains for a specified, fixed stress are determined by the condition

$$\frac{\partial G}{\partial \varepsilon} = 0 \Leftrightarrow \sigma = \left[\frac{\partial \psi}{\partial \varepsilon}\right]_{P,T}. \tag{2.26}$$

Not surprisingly, the relations (2.25) and (2.26) are equivalent to those in (2.19) derived through differential analysis. However, the use of Gibbs potentials to determine equilibrium states is very general and plays a fundamental role in the development of nonlinear hysteresis models for ferroelectric, ferromagnetic and ferroelastic materials.

Remark 2.2.4. *From a physical perspective, the Gibbs relation (2.23) can be interpreted as a combined measure of the internal energy and entropy quantified by ψ and the work due to external fields. The equilibrium conditions (2.25) and (2.26) quantify the manner through which the polarization and strain adjust so that the internal responses balance the external fields.*

2.2.3 Equivalence of Energy Formulations

For readers used to formulating the Gibbs energy G solely in terms of the conjugate variables E and σ — e.g., see (2.22) — the necessary condition (2.25) may cause alarm. However, under the assumption that dipole processes are *reversible* and in *equilibrium*, the analysis used to formulate G in terms of E and σ is equivalent to enforcement of the conditions $\frac{\partial G}{\partial P} = 0$ and $\frac{\partial G}{\partial \varepsilon} = 0$ for fixed fields and stresses.

To illustrate, consider the Helmholtz energy relation $\psi(P) = \frac{1}{2}\alpha P^2$ and the simplified Gibbs relation

$$G(E,P) = \psi(P) - EP \tag{2.27}$$

which is a function of the *independent* variable E and *dependent* variable P. To quantify the dependence of E on P, and hence formulate G solely in terms of E, we consider Legendre transform analysis as detailed in Appendix C.1. Because ψ is differentiable, the Legendre transform is given by

$$\begin{aligned}(L\psi)(E) &= \max_P[PE - \psi(P)] \\ &= P_E E - \psi(P_E)\end{aligned}$$

where P_E solves

$$E = \psi'(P_E). \tag{2.28}$$

Hence

$$E = \alpha P_E \Leftrightarrow P_E = \frac{1}{\alpha}E \tag{2.29}$$

which implies that

$$G(E) = -(L\psi)(E) = -\left[\frac{E^2}{\alpha} - \frac{1}{2}\alpha\left(\frac{E}{\alpha}\right)^2\right] = -\frac{1}{2}\chi E^2.$$

The property that (2.29) always has a solution follows from the fact that $E = \psi'(P_E) \Leftrightarrow P = G'(E_P)$ which is a direct consequence of the fact that the Legendre transform is self-dual — i.e., $L(L(f)) = f$; this requires *reversible* and hence monotone behavior but *does not require linearity*. The necessary condition (2.28) requires that dipole processes are in *equilibrium*.

For G given by (2.27), consideration of the Legendre transform indicates that the following operations are equivalent:

$$\begin{aligned}\text{specification of } G \text{ in terms of } E &\Leftrightarrow \text{solution of } E = \psi'(P_E) \\ &\Leftrightarrow \text{determination of minima } \tfrac{\partial G}{\partial P} = 0 \\ &\qquad \text{for fixed values of } E.\end{aligned}$$

The analysis of the electromechanical Gibbs relation (2.23) is analogous.

Hence the three strategies consisting of differential analysis, Legendre transform analysis and enforcement of the necessary condition $\frac{\partial G}{\partial P} = 0$ for fixed E yield identical equilibrium relations between P and E when dipole processes are reversible and in equilibrium. However, the third strategy is significantly more general and provides a natural framework for characterizing the irreversible processes associated with hysteresis. The physical interpretation associated with the minimization of G is summarized in Remark 2.2.4.

2.2.4 Electromechanical Coupling Factor

For design purposes, it is advantageous to define a metric for piezoelectric materials which incorporates the elastic, dielectric and piezoelectric coefficients while excluding external forces. A common choice is the *electromechanical coupling factor* which is defined as a ratio of the interaction energy to the product of the elastic and electrostatic energies [37, 114, 234].

For example, consider the Gibbs relation (2.22) which can be written as

$$\begin{aligned} G &= -\frac{1}{2}\left(s^E_{ijk\ell}\sigma_{ij}\sigma_{k\ell} + 2d_{nij}\sigma_{ij}E_n + \chi^\sigma_{nm}E_mE_n\right) \\ &\equiv G_e + 2G_{int} + D_d \end{aligned}$$

where G_e, G_{int} and G_d respectively denote the elastic, interaction or coupled, and dielectric contribution to the energy. The electromechanical coupling factor is then defined by

$$k^2 = \frac{G^2_{int}}{G_eG_d}. \tag{2.30}$$

To illustrate the form of k for a common actuator configuration, consider a 1-D piezoceramic patch with the electric field applied through the thickness. The Gibbs energy associated with motion in the x direction is then

$$G = -\frac{1}{2}\left(s^E_{11}\sigma_1^2 + 2d_{31}\sigma_1E_3 + \chi^\sigma_3E_3^2\right)$$

and the electromechanical coupling factor is

$$k_{31} = \frac{d_{31}}{\sqrt{s^E_{11}\chi^\sigma_3}}.$$

The definition (2.30) is valid only for *static* or *quasistatic* regimes and, as noted in [114], the electromechanical coupling factor exhibits a significant dependence on the stress and strain near resonant frequencies. Furthermore, the formulation is often too complicated in two and three-dimensional models to cancel stresses so that the factor involves external variables rather than specifying solely intrinsic material properties. Despite these limitations, coupling factors are widely employed to quantify the degree to which electrical energy is converted to mechanical energy or mechanical energy is converted to electrical. For example, a comparison of the values of k_{31} for PZT and PVDF in Table 1.1 on page 28 indicates the advantage provided by the former with regard to electromechanical energy conversion.

2.2.5 Model Summary and Attributes

The advantages and disadvantages of linear constitutive relations can be summarized in one word — *linear*. If through feedback design, low drive levels, amplifier design, or lucky alignment of the stars, one can employ linear models, do so! Otherwise, the nonlinear hysteresis models described in subsequent sections should be considered.

The constitutive relations for various input pairs are summarized in Tables 2.1 and 2.2. We illustrate here the extension of the (ε, E) expression to incorporate voltage inputs and Kelvin–Voigt damping for 1-D and 2-D geometries. The latter case provides constitutive relations for PZT plates and shells.

For applications in which the voltage V is measured or specified rather than the field E, one typically employs the approximate relation $V = Eh$ where h denotes the thickness of the transducer material. Furthermore, internal damping in materials such as PZT, aluminum or steel is commonly modeled through the Kelvin–Voigt hypothesis which posits that damping is proportional to strain rate. The proportionality constant, termed the Kelvin–Voigt damping parameter, is denoted by c. To simplify notation, we drop superscripts and simply note that fixed quantities are indicated by the input variable set.

We illustrate d_{31} electromechanical coupling since it will be the case for numerous unimorph, bimorph, beam, plate and shell designs in which transduction is coupled with transverse or out-of-plane motion — analogous constructs follow for d_{33} or d_{15} transduction.

1-D Constitutive Relations

The damped 1-D constitutive relations for the input variable set (ε, V) are

$$\begin{aligned} \sigma &= Y\varepsilon + c\dot{\varepsilon} - Y\frac{d_{31}}{h}V \\ P &= Yd_{31}\varepsilon + \chi\frac{V}{h} \end{aligned} \tag{2.31}$$

where temporal derivatives are denoted by $\dot{\varepsilon} \equiv \frac{d\varepsilon}{dt}$. The extensions to other combinations of electric variables follow in an analogous manner.

2-D Constitutive Relations

Let ε_x, σ_x and ε_y, σ_y denote the strains and stresses in the x and y directions and let ν denote the Poisson ratio. For this geometry, the converse relations for (σ, E) inputs are

$$\begin{aligned} \varepsilon_x &= s(\sigma_x - \nu\sigma_y) + d_{31}E \\ \varepsilon_y &= s(\sigma_y - \nu\sigma_x) + d_{31}E \end{aligned}$$

where s again denotes the compliance. The inclusion of Kelvin–Voigt damping and formulation in terms of V then yields

$$\begin{aligned} \sigma_x &= \frac{Y}{1-\nu^2}(\varepsilon_x + \nu\varepsilon_y) + \frac{c}{1-\nu^2}(\dot{\varepsilon}_x + \nu\dot{\varepsilon}_y) - \frac{Yd_{31}}{1-\nu}\frac{V}{h} \\ \sigma_y &= \frac{Y}{1-\nu^2}(\varepsilon_y + \nu\varepsilon_x) + \frac{c}{1-\nu^2}(\dot{\varepsilon}_y + \nu\dot{\varepsilon}_x) - \frac{Yd_{31}}{1-\nu}\frac{V}{h} \\ P &= Yd_{31}\varepsilon + \chi\frac{V}{h}. \end{aligned} \tag{2.32}$$

Analogous relations for cylindrical geometries in which x and θ delineate the longitudinal and circumferential coordinates can be obtained simply by replacing y by θ.

2.3 Higher-Order Energy Relations

In this section, we extend the energy analysis to incorporate ferroelectric hysteresis and phase transitions. To accomplish this, we first construct Helmholtz energy relations which quantify certain internal processes such as dipole switching or entropic effects in the absence of applied stresses or electromechanical coupling. We then incorporate reversible but potentially nonlinear electromechanical coupling mechanisms to construct Helmholtz energy relations ψ whose minima quantify polarization or strain distributions in the absence of applied fields. The effects of applied stresses or fields are incorporated through consideration of Gibbs energy relations

$$G(E, P, T) = \psi(P, T) - EP \tag{2.33}$$

or

$$G(\sigma, E, \varepsilon, P, T) = \psi(\varepsilon, P, T) - EP - \sigma\varepsilon \tag{2.34}$$

whose minima $\frac{\partial G}{\partial P} = 0$ and $\frac{\partial G}{\partial \varepsilon} = 0$ specify the manner through which the polarization and strains adjust so that the internal responses balance the applied fields.[9] As noted in (2.25) and (2.26), this yields the equilibrium conditions

$$\sigma = \left[\frac{\partial \psi}{\partial \varepsilon}\right]_{P,T} \;,\; E = \left[\frac{\partial \psi}{\partial P}\right]_{\varepsilon,T} \tag{2.35}$$

which provide constitutive relations.

We consider three techniques for specifying the Helmholtz energies ψ. (i) The first is phenomenological and is based on the truncation of power series with coefficients chosen to ensure observed physical phenomena. This forms a basis for theory proposed by Landau and Lifschitz [57, 284, 482] and Devonshire [134] and has the advantage of characterizing both first and second-order transition phenomena. (ii) The second is based on statistical mechanics principles in combination with mean field approximations. This yields an Ising relation for the polarization which includes the second-order phase transition of Devonshire as a special case. (iii) The third Helmholtz relation is based on a piecewise quadratic approximation to the globally defined models of (i) and (ii). This improves efficiency and eliminates stability issues associated with the higher-order polynomials employed in the first two constructs.

The resulting constitutive relations quantify instantaneous changes in polarization due to dipole switching in *homogeneous, single crystal* compounds and hence provide reasonably accurate approximations for behavior of the type depicted in Figure 2.8(a). However, they do not incorporate the mollified transitions depicted in Figure 2.8(b) which are due to material nonhomogeneities, relaxation mechanisms, nonuniform effective fields, polycrystallinity, or domain wall losses. The incorporation of these phenomena provides the basis for the domain wall and stochastically homogenized energy models described in Sections 2.5 and 2.6.

---

[9] Details regarding the dependence of ε and P on σ and E are provided in Section 2.2.3.

2.3.1 Polynomial Energy Relations

In the theory of Landau and Devonshire, it is assumed that the Helmholtz energy can be expressed as a truncated power series with coefficients chosen to ensure measured material properties. For example, the Helmholtz relation (2.18) incorporates linear electromechanical coupling and is truncated after quadratic strain and polarization terms since only linear effects are considered. Here we consider the retention of quadratic electromechanical coefficients and 4th and 6th order polarization terms to quantify the behavior of materials exhibiting first and second-order phase transitions.

The order of a phase transition quantifies a variety of properties concerning the transition behavior of materials including the release or absorption of energy in the form of latent heat, the continuity (or discontinuity) of the order parameter (P in ferroelectric materials) at the Curie point, and the number of stable equilibria present during the transition process. Materials exhibiting *first-order transitions* include the ferroelectric compounds $BaTiO_3$, $PbTiO_3$ and $PbZrO_3$ [256, pages 236 and 246] and shape memory alloys. The melting of ice also constitutes a familiar example of a first-order phase transition. *Second-order phase transitions* are observed in ferroelectric materials including Rochelle salts and KH_2PO_4 [134] and ferromagnetic compounds. We further elucidate the properties of first and second-order transitions in the context of developing appropriate free energy relations.

Second-Order Phase Transitions

For materials which exhibit second-order phase transitions, we consider the Helmholtz expression

$$\psi(\varepsilon, P, T) = \frac{1}{2}Y^P\varepsilon^2 - a_1\varepsilon P - a_2\varepsilon P^2 + \psi_0(T) + \alpha_1(T - T_c)P^2 + \alpha_2 P^4 \tag{2.36}$$

where Y^P denotes the Young's modulus at constant polarization, T_c is the Curie point, a_1 and a_2 are positive coupling coefficients, ψ_0 incorporates temperature effects independent of ε and P, and α_1, α_2 are positive constants. This relation extends the Helmholtz expression (2.18) employed for linear model construction through the incorporation of the quadratic coupling term, the inclusion of temperature-dependency and the addition of the quartic polarization contribution. The inclusion of the quadratic electromechanical coupling coefficient permits quantification of the nonlinear strain behavior depicted in Figure 2.10 and also observed in electrostrictive compounds. In the absence of strains, ψ exhibits single well behavior for $T > T_c$ and double well behavior for $T < T_c$ as depicted in Figure 2.14.

The Gibbs energy relations (2.33) and (2.34) respectively incorporate the work due to external fields and stresses. For $T < T_c$, the Gibbs energy (2.33) for $E = 0$ and $E = E_c$ is plotted in Figure 2.15(a). It is observed that the field distorts the energy landscape which results in the elimination of stable equilibria at the coercive field values E_c and $-E_c$. Hence this formulation for the Gibbs energy quantifies both the paraelectric to ferroelectric phase transition which occurs as T is decreased through the Curie point and the instantaneous polarization switch which results at $E = E_c$ and $E = -E_c$.

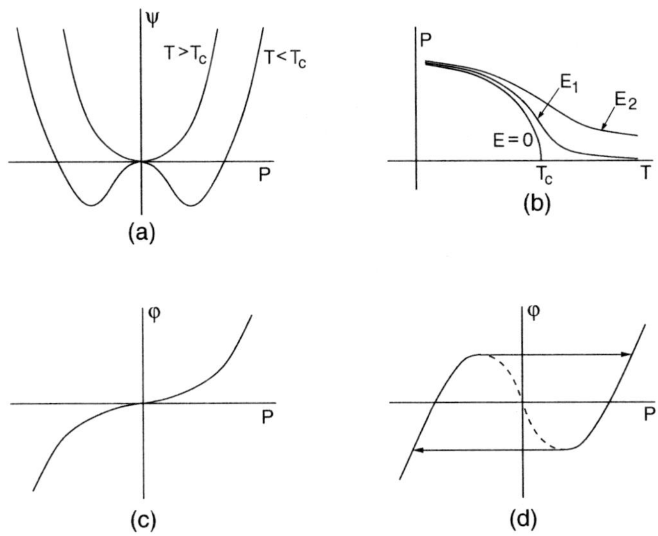

Figure 2.14. *Second-order phase transition. (a) Helmholtz energy ψ as a function of the spontaneous polarization P. (b) Continuous dependence of P on temperature in the absence and presence of a biasing field. (c) Response function $\varphi = \frac{\partial \psi}{\partial P}$ for $T > T_c$ and (d) for $T < T_c$ with unstable states indicated by dashes.*

Enforcement of the necessary conditions (2.35) yields the constitutive relations

$$\begin{aligned} \sigma &= Y^P \varepsilon - a_1 P - a_2 P^2 \\ E &= -a_1 \varepsilon - a_2 \varepsilon P + 2\alpha_1 (T - T_c) P + 4\alpha_2 P^3. \end{aligned} \tag{2.37}$$

The hysteron obtained by solving the second equation for P as a function of E is plotted in Figure 2.15(b) to illustrate the nonlinear and hysteretic behavior encapsulated in the formulation along with the instantaneous polarization switch which occurs at the coercive values $E = E_c$ and $E = -E_c$.

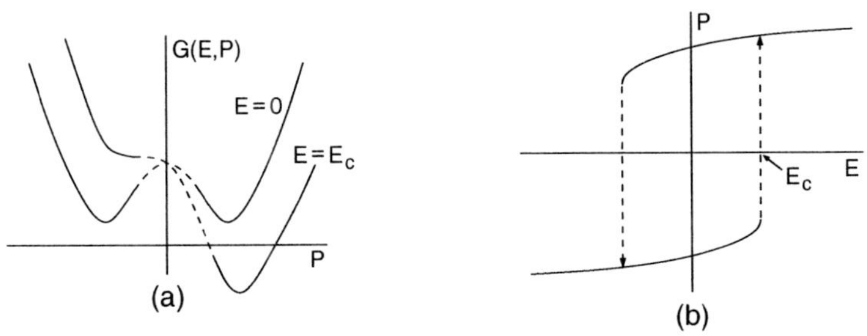

Figure 2.15. *Second-order phase transition with $\varepsilon = 0$ and fixed T. (a) Gibbs energy G for $E = 0$ and $E = E_c$ with unstable and hence inaccessible states denoted by dashes. (b) Relation between P and E given by (2.37).*

First-Order Phase Transitions

Materials exhibiting first-order phase transitions have more complicated behavior in neighborhoods of the transition temperature which must be incorporated in the energy functionals. To illustrate, consider the depiction in Figure 2.16 of $BaTiO_3$ data from Merz [338] for temperatures ranging from the Curie point $T_c = 107\ ^oC$ to $T = 128.3\ ^oC$. It is observed that the material develops a double hysteresis loop analogous to that of pseudoelastic SMA at $T = 114.4\ ^oC$ followed by a nonlinear anhysteretic curve at $T = 116.1\ ^oC$. Finally, approximately linear paraelectric behavior is observed at $T = 128.3\ ^oC$.

To quantify this behavior, one can employ the Helmholtz relation

$$\psi(\varepsilon, P, T) = \frac{1}{2}Y^P\varepsilon^2 - a_1\varepsilon P - a_2\varepsilon P^2 + \psi_0(T) + \alpha_1(T - T_0)P^2 - \alpha_2 P^4 + \alpha_3 P^6 \quad (2.38)$$

where $\alpha_1, \alpha_2, \alpha_3$ are positive constants. Here T_0 denotes the *Curie temperature* which, in materials exhibiting first-order phase transitions, is typically not equal to, and can be more than 10 ^oC lower than, the *transition temperature* or *Curie point* T_c [243].[10] The behavior of ψ for $\varepsilon = 0$ is plotted in Figure 2.17(a) to illustrate that it exhibits double well behavior below $T = T_0$ and metastable states for $T_0 < T < T_2$ where T_2 denotes the temperature at which the material becomes nonpolar (paraelectric), even in the presence of an external field.[11]

[10] The indiscriminate use of the terms *Curie temperature* and *Curie point* is a common mistake in the literature of ferroic materials undergoing first-order phase transformations. This is aggravated by the fact that the terms are often used synonymously for second-order materials where the two temperatures may be essentially identical. Finally, we note that the notation is not universal and is occasionally interchanged by authors.

[11] Metastable states are those associated with relative rather than absolute minima.

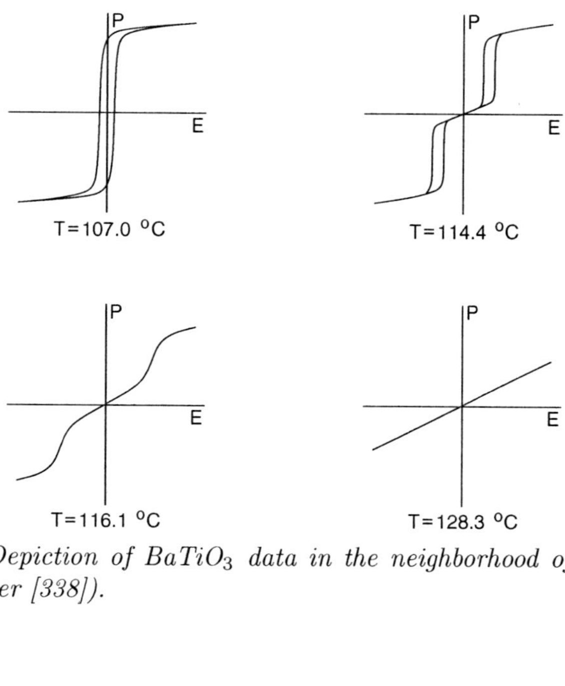

Figure 2.16. *Depiction of $BaTiO_3$ data in the neighborhood of the Curie point $T_c = 107\ ^oC$ (after [338]).*

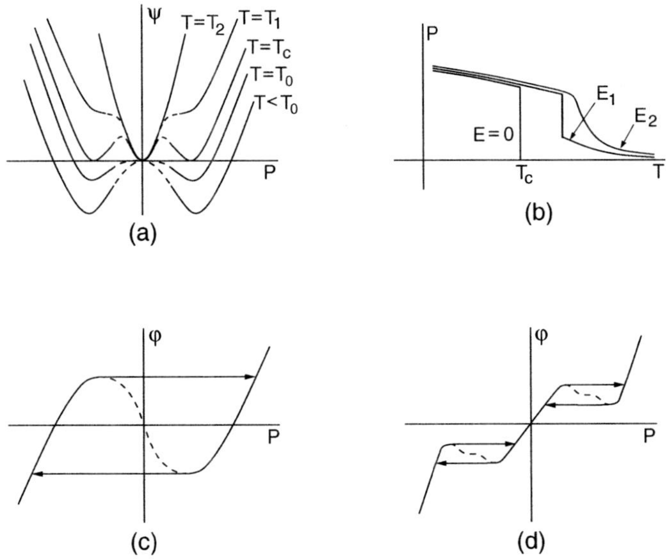

Figure 2.17. *First-order phase transition. (a) Helmholtz energy ψ as a function of P for $\varepsilon = 0$. (b) Dependence of P on temperature for increasing field values. (c) Response function $\varphi = \frac{\partial \psi}{\partial P}$ for $T = T_0$ and (d) $T_0 < T < T_2$ with unstable states designated by dashes.*

To determine the transition behavior of metastable states, it is necessary to incorporate the fields through the Gibbs relation (2.34). Application of the necessary conditions (2.35) yields the constitutive relations

$$\begin{aligned} \sigma &= Y^P \varepsilon - a_1 P - a_2 P^2 \\ E &= -a_1 \varepsilon - a_2 \varepsilon P + 2\alpha_1 (T - T_0) P - 4\alpha_2 P^3 + 6\alpha_3 P^6. \end{aligned} \tag{2.39}$$

We now focus on the behavior of the E-P relation with $\varepsilon = 0$ to illustrate the field-induced behavior in first-order phase transitions.

We first note that unlike the second-order behavior depicted in Figure 2.14(b), materials exhibiting first-order transitions retain a discontinuity in the polarization at temperatures $T > T_c$ for fields less than the critical field E_c. This is due to the metastability of the high temperature variants or twins in the presence of applied fields. The Gibbs relation and E-P behavior predicted by (2.39) are plotted in Figure 2.18 to demonstrate the onset of double hysteresis loop behavior for temperatures $T_1 < T < T_2$.

The first-order behavior can be summarized as follows. Materials exhibit a transition between absolutely stable states at the Curie point T_c with the polar ferroelectric phase being *absolutely stable* for $T < T_c$. However, the nonpolar phase can remain metastable down to $T_0 < T_c$ and, in the absence of a field, the polar phase is metastable up to $T_1 > T_c$ above which the twinned equilibria are eliminated as illustrated in Figure 2.17(a). For temperatures in the range $T_1 < T < T_2$, the

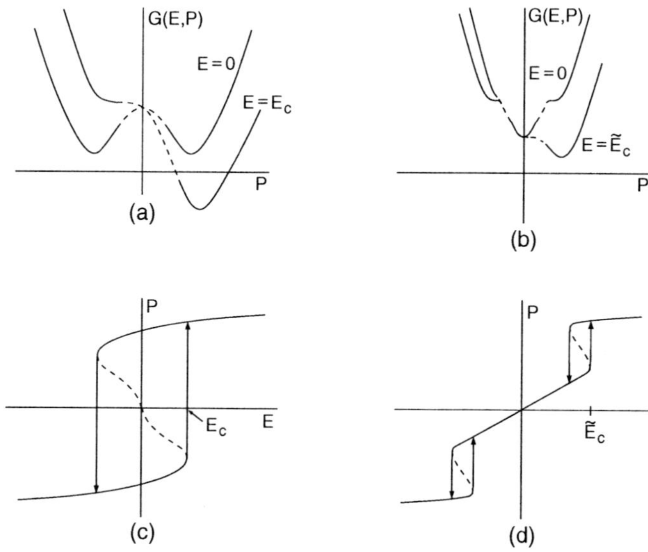

Figure 2.18. *First-order phase transition. (a) Gibbs energy for $T = T_0$ and input fields $E = 0$, $E = E_c$ and (b) for $T_1 < T < T_2$ with input fields $E = 0$, $E = \widetilde{E}_c$. (c) Polar behavior exhibited at $T = T_0$, and (d) for $T_1 < T < T_2$.*

ferroelectric phase can be induced by an applied field whereas for $T > T_2$, the material is always nonpolar. Note that the absolutely polar (ferroelectric) behavior for $T < T_0$ and the absolutely nonpolar (paraelectric) response for $T > T_2$ correspond to the double and single potential wells associated with the ion positions depicted in Figure 2.1. Finally, we note that this provides only a summary of first-order transition behavior, and the reader is referred to [57, 134, 256, 298, 482] for details regarding the phenomenology and analysis of this topic.

2.3.2 Boltzmann Energy Relations

The Helmholtz relations (2.36) and (2.38) were both constructed by truncating even power series for the polarization and determining coefficients which reflect observed physical behavior. Hence they are solely *phenomenological* in nature. An alternative technique for constructing ψ is to employ statistical mechanics tenets to quantify the internal energy and entropy. This provides a Helmholtz relation which corroborates properties of (2.36) and (2.38) and motivates relations employed in Section 2.6 as well as Chapter 4 where analogous hysteresis models for ferromagnetic materials are developed.

We first consider the construction of a Helmholtz relation in the absence of strains $\varepsilon = 0$. We assume a homogeneous material and consider a uniform lattice of volume V and mass ν having N cells of the form depicted in Figure 2.2(b). Each cell is assumed to have dipole moment p_0 and two possible dipole orientations $s_i = \pm 1$ so that $N = N_- + N_+$ where N_- and N_+ respectively denote the number of negatively and positively oriented dipoles.

Under these assumptions, the polarization for the lattice is

$$\begin{aligned} P &= \frac{p_0}{V}\sum_{i=1}^{N} s_i \\ &= \frac{P_s}{N}(N_+ - N_-) \end{aligned} \tag{2.40}$$

where $P_s = Np_0/V$ denotes the *saturation polarization* which occurs when all dipoles are positively aligned. It follows immediately that

$$N_+ = \frac{N}{2}\left(1+\frac{P}{P_s}\right) \quad , \quad N_- = \frac{N}{2}\left(1-\frac{P}{P_s}\right). \tag{2.41}$$

To quantify the energy required to reorient dipoles, we employ the *mean field approximation* of Bragg and Williams [52, 53, 511] and make the assumption that the *average exchange energy* Φ is proportional to P/P_s; that is,

$$\Phi = \Phi_0 \frac{P}{P_s} \tag{2.42}$$

where Φ_0 denotes the energy required to reorient a single dipole if the lattice is completely ordered ($P = P_s$). For the case of a homogeneous lattice, Φ_0 is considered to be constant. For nonhomogeneous and polycrystalline materials, Φ_0 will be considered as a manifestation of an underlying statistical distribution as discussed in Section 2.6. For further details regarding the mean field assumptions leading to (2.42), see [52,53,191,198,376,511].

We now consider the decrease in internal energy due to a change from N_+ to $N_+ + dN_+$. From the mean field approximation (2.42), each switch requires $\Phi_0\frac{P}{P_s}$ in energy so the change in internal energy for a unit volume is

$$\begin{aligned} dU &= -\frac{\Phi_0 P}{VP_s}dN_+ \\ &= -\frac{\Phi_0 P}{VP_s}\cdot\frac{N}{2Ps}dP \end{aligned} \tag{2.43}$$

where the second equality follows from (2.41). Thus the internal energy is

$$U = \frac{\Phi_0 N}{4V}\left(1-\frac{P^2}{P_s^2}\right) + U_0.$$

Since we are interested in relative rather than absolute measures of energy, we take $U_0 = 0$ which specifies that the completely ordered state has an internal energy of zero.

As detailed in [198], the entropy S for the system is given by

$$S = \frac{k}{V}\ln W \tag{2.44}$$

where $k = 1.38 \times 10^{-23}$ J/K is Boltzmann's constant and W quantifies the number of ways dipoles can be arranged in the lattice to yield the magnetization P. By noting that this is equivalent to arranging N_+ dipoles in N sites, it follows that

$$W = \begin{pmatrix} N \\ N_+ \end{pmatrix} = \frac{N!}{N_-!\,N_+!}.$$

By employing Stirling's formula

$$\ln x! \approx x \ln x - x, \tag{2.45}$$

in combination with the relations (2.41) for N_- and N_+, the entropy can be formulated as

$$\begin{aligned} S &= \frac{kN}{V}\left[\ln 2 - \frac{1+P/P_s}{2}\ln\left(1+\frac{P}{P_s}\right) - \frac{1-P/P_s}{2}\ln\left(1-\frac{P}{P_s}\right)\right] \\ &= \frac{-kN}{2VP_s}\left[P\ln\left(\frac{P+P_s}{P_s-P}\right) + P_s\ln\left(1-\left(\frac{P}{P_s}\right)^2\right)\right] + S_0 \end{aligned} \tag{2.46}$$

where $S_0 = \frac{kN}{V}\ln 2$. As with U_0, we neglect S_0 in the final relation for the Helmholtz energy since we are interested in a relative, rather than absolute, measure of energy.

The Helmholtz energy for the lattice is then

$$\begin{aligned} \psi &= U - ST \\ &= \frac{E_h P_s}{2}\left[1-(P/P_s)^2\right] + \frac{E_h T}{2T_c}\left[P\ln\left(\frac{P+P_s}{P_s-P}\right) + P_s\ln(1-(P/P_s)^2)\right] \end{aligned} \tag{2.47}$$

where $E_h = \frac{N\Phi_0}{2VP_s}$ is a bias field and $T_c = \frac{\Phi_0}{2k}$ denotes the Curie point for the material. The initial assumption that the exchange energy Φ_0 is constant implies that E_h will also be constant for homogeneous materials. This assumption will be relaxed in Section 2.6 to include statistically distributed values of E_h for modeling nonhomogeneous and polycrystalline materials.

As illustrated in Figure 2.19, the Helmholtz energy, in the absence of stresses or strains, exhibits double well behavior for $T < T_c$ and single well behavior for $T > T_c$. This is consistent with the ferroelectric and paraelectric behavior depicted in Figure 2.2.

To incorporate elastic behavior, we make the assumption that stresses produce reversible changes in the polarization but are not sufficiently large to produce the ferroelastic switching depicted in Figure 2.11. This assumption is reasonable for hard piezoelectric materials and soft compounds subjected to moderate stresses. Under this assumption, the Helmholtz relation is

$$\begin{aligned} \psi(\varepsilon, P, T) &= \frac{1}{2}Y^P\varepsilon^2 - a_1\varepsilon P - a_2\varepsilon P^2 + \frac{E_h P_s}{2}\left[1-(P/P_s)^2\right] \\ &+ \frac{E_h T}{2T_c}\left[P\ln\left(\frac{P+P_s}{P_s-P}\right) + P_s\ln\left(1-(P/P_s)^2\right)\right] \end{aligned} \tag{2.48}$$

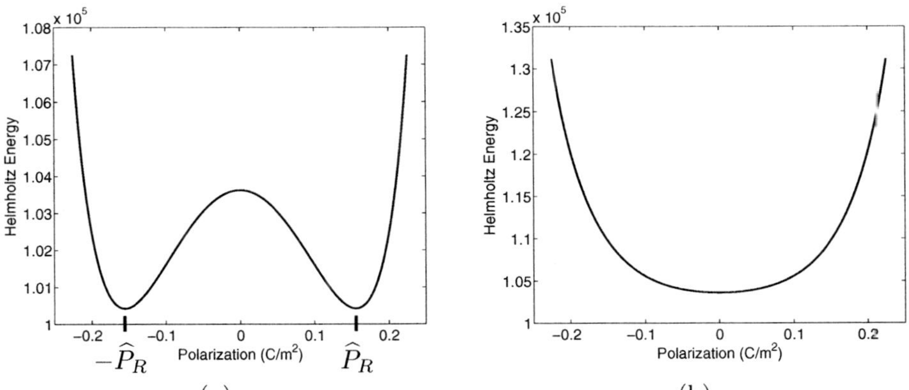

Figure 2.19. *Helmholtz energy specified by (2.47) for (a) $T < T_c$ and (b) $T > T_c$.*

which incorporates both linear and quadratic electromechanical coupling and ferroelectric dipole switching. A corresponding Gibbs energy relation is then given by (2.34).

Equilibrium states are again established by the conditions (2.35) which yields the constitutive relations

$$\begin{aligned} \sigma &= Y^P\varepsilon - a_1P - a_2P^2 \\ E &= -a_1\varepsilon - 2a_2\varepsilon P - \frac{E_h}{P_s}P + \frac{E_hT}{T_c}\text{arctanh}(P/P_s). \end{aligned} \tag{2.49}$$

To compare with the polynomial field-polarization relations in Section 2.3.1, the hyperbolic arctangent is represented by its Taylor series to yield

$$\begin{aligned} E &= -a_1\varepsilon - 2a_2\varepsilon P - \frac{E_h}{P_s}P + \frac{E_hT}{T_c}\left[\left(\frac{P}{P_s}\right) + \frac{1}{3}\left(\frac{P}{P_s}\right)^3 + \frac{1}{5}\left(\frac{P}{P_s}\right)^5 + \cdots\right] \\ &= -a_1\varepsilon - 2a_2\varepsilon P + \frac{E_h}{T_cP_s}(T - T_c)P + \frac{E_hT}{3T_cP_s^3}P^3 + \frac{E_hT}{5T_cP_s^5}P^5 + \cdots. \end{aligned}$$

Because $\frac{E_hT}{12T_cP_s^3} > 0$, we observe that (2.49) quantifies *second-order phase transition* behavior. Moreover, the truncated Helmholtz relation (2.36) is a special case of (2.49) in which the parameters α_1 and α_2 are given by

$$\alpha_1 = \frac{E_h}{2T_cP_s} \quad , \quad \alpha_2(T) = \frac{E_hT}{12T_cP_s^3}.$$

We note that while the energy expression (2.48) and constitutive relations (2.49) will not quantify the material behavior observed at T_0 for first-order transitions, they are applicable in the ferroelectric regime $T < T_0$ where piezoelectric transducers are required to operate.

2.3.3 Piecewise Polynomial Relations

The polarization components in the Helmholtz expressions (2.36), (2.38) and (2.48) share the property that they are globally defined throughout the polarization range of interest. This provides the advantage of compact representations but also induces certain disadvantages. The logarithmic components of (2.48) and hyperbolic arctangent relation in (2.49) add complexity to resulting macroscopic models which reduces the efficiency of algorithms required for real-time implementation. Furthermore, the high-order polynomials required in (2.36) and (2.38) can induce oscillatory and unstable behavior through slight modifications or uncertainties in the coefficients. This motivates the construction of a simplified piecewise quadratic Helmholtz relation which retains the physical behavior encapsulated in (2.36), (2.38) and (2.48) but facilitates implementation.

To motivate the form of this relation, we consider the Taylor expansion

$$\psi(P,T) = \psi(P_0,T) + (P-P_0)\frac{\partial \psi}{\partial P}(P_0,T) + \frac{(P-P_0)^2}{2}\frac{\partial^2 \psi}{\partial P^2}(P_0,T) + \mathcal{O}(P^3) \quad (2.50)$$

for fixed $T < T_c$, where P_0 is taken to be an equilibrium, and

$$\frac{\partial \psi}{\partial P} = \frac{-E_h}{P_s}P + \frac{E_h T}{T_c}\text{arctanh}(P/P_s),$$

$$\frac{\partial^2 \psi}{\partial P^2} = \frac{-E_h}{P_s} + \frac{E_h T}{T_c P_s\left[1-(P/P_s)^2\right]}.$$

From the necessary condition $\frac{\partial \psi}{\partial P}(P_0,T) = 0$, the equilibria are determined to be the two stable solutions to

$$P = P_s \tanh\left(\frac{\alpha P}{a(T)}\right),$$

designated $-\widehat{P}_R$ and $\widehat{P}_R$ as depicted in Figure 2.19(a), in addition to the unstable solution $P = 0$. The parameters α and $a(T)$ are given by

$$\alpha = \frac{E_h}{P_s} \quad , \quad a(T) = \frac{E_h T}{T_c}.$$

It can be directly established that the quadratic approximations to (2.50) in neighborhoods of the equilibria $P_0 = 0, -\widehat{P}_R$ and $\widehat{P}_R$ are

$$\psi(P,T) = \begin{cases} c_1 - k_1^2(T)P^2 & , \quad P_0 = 0 \\ c_2(T) + k_2^2(T)(P+\widehat{P}_R)^2 & , \quad P_0 = -\widehat{P}_R \\ c_2(T) + k_2^2(T)(P-\widehat{P}_R)^2 & , \quad P_0 = \widehat{P}_R \end{cases}$$

where

$$c_1 = \frac{E_h P_s}{2} \quad , \quad k_1^2(T) = \frac{E_h}{2P_s T_c}(T_c - T).$$

Analogous, but more complicated, expressions follow for $c_2(T)$ and $k_2(T)$.

For fixed temperature regimes, this motivates consideration of the piecewise quadratic definition

$$\psi(P) = \begin{cases} \frac{1}{2}\eta(P+P_R)^2 & , \; P \le -P_I \\ \frac{1}{2}\eta(P-P_R)^2 & , \; P \ge P_I \\ \frac{1}{2}\eta(P_I-P_R)\left(\frac{P^2}{P_I}-P_R\right) & , \; |P| < P_I . \end{cases} \tag{2.51}$$

As illustrated in Figure 2.20, P_I and P_R respectively denote the inflection point and polarization at which the minimum of ψ occurs.

Under the assumption that stresses have magnitude less than the ferroelastic coercive value σ_c for the material, the coupled electromechanical Helmholtz relation is taken to be

$$\psi(\varepsilon, P) = \psi(P) + \frac{1}{2}Y^P\varepsilon^2 - a_1\varepsilon P - a_2\varepsilon P^2. \tag{2.52}$$

Whereas the quadratic Helmholtz definition (2.52) is conceptually simpler than the statistical mechanics formulation (2.48), the piecewise nature must be accommodated in the manner detailed in Section 2.6 when specifying initial conditions and switching between branches.

The behavior of the Gibbs energy relation (2.34) and the hysteron $\overline{P}(E)$ resulting from the necessary conditions (2.35) are plotted in Figure 2.20. Details concerning the construction of $\overline{P}$ are provided in Section 2.6.

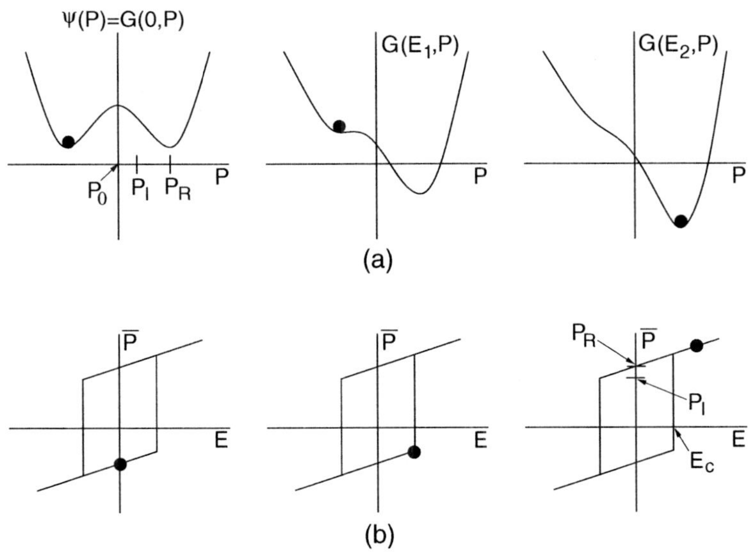

Figure 2.20. *(a) Helmholtz energy ψ and Gibbs energy G for increasing field E ($E_2 > E_1 > 0$). (b) Dependence of the local average polarization $\overline{P}$ on the field E.*

2.4 Preisach Hysteresis Models

The Preisach method is the earliest of the unified frameworks which we present for quantifying hysteresis and constitutive nonlinearities in ferroelectric, ferromagnetic and ferroelastic materials. It was developed by Preisach in 1935 in the context of magnetic materials [382] and the majority of research has focused in this regime. For this reason, we postpone much of the discussion on Preisach models to Chapter 4 where it is considered for ferromagnetic compounds.

One of the early investigations of Preisach models for ferroelectric compounds was by Sreeran, Salvady and Naganathan [399] in 1993, and Hughes and Wen in 1997 discussed both the unified framework provided by Preisach models for PZT, magnetic compounds and SMA, and the construction of inverse filters based on Preisach models to compensate for hysteresis inherent to ferroelectric materials [227]. Subsequent investigations (e.g., [179,180,394]) have focused on extensions of the theory for characterization and control design in system employing piezoceramic transducers. However, several issues persist and the extension of generalized Preisach techniques to ferroelectric materials remains an open research area.

2.4.1 Model Attributes

It is established in Chapter 4 that the classical Preisach model is completely characterized by two criteria: (i) the *congruency* of minor loops, and (ii) the *deletion* or *wiping out* property for minor loops. The first states that all minor loops generated by inputs having the same extrema must be congruent, whereas the deletion property states that inputs having a large extremum will delete the effects of previous inputs having smaller extrema. By establishing that these two criteria are necessary and sufficient conditions for classical Preisach models, Mayergoyz [334,335] provided mathematical criteria which can be easily checked experimentally to establish the suitability of classical Preisach models for various hysteretic phenomena.

These mathematical conditions hint at the greatest strength of the Preisach framework — namely, its highly structured and mathematical nature. Moreover, because Preisach models can be specified solely in terms of mathematical criteria, they can be applied in regimes where the underlying physics is unknown or poorly understood. For these reasons, Preisach techniques are widely employed for characterizing hysteresis for a broad range of phenomena and are growing in popularity for piezoelectric material characterization and model-based control design.

However, the general mathematical nature of the classical Preisach model also contributes to a number of its limitations. (i) For systems in which Preisach models are uncorrelated with physical principles, it is difficult to incorporate known physics when identifying model parameters or updating models to accommodate changing environmental conditions such as temperature-dependent dynamics. (ii) Many materials exhibit nonclosure of minor loops due to relaxation phenomena or noncongruent minor loop behavior, neither of which can be accommodated by classical formulations. (iii) Frequency or rate-dependence in the hysteresis will not be quantified by classical Preisach models which are typically based on the assumption that hysteresis is rate-independent. (iv) Similarly, the classical Preisach framework does not accommodate temperature or stress-dependence in the polarization or magnetization relations thus requiring extensions to vector-valued formulations or lookup tables. (v) Classical formulations do not provide the mechanisms required to represent the reversible components of hysteresis nonlinearities.

All of these limitations have been addressed for magnetic hysteresis models through the development of *extended Preisach formulations.* While the extension of these extended models to ferroelectric materials is less mature, the mathematical nature of the framework will facilitate this development. Furthermore, it is demonstrated in [447] and illustrated in Sections 2.6 and 4.7.10 that the statistically homogenized energy models provide an energy basis for Preisach models with the following important difference — the energy-based formulation automatically addresses the limitations (i)–(v). It is thus anticipated that the symbiotic investigation of energy-based and extended Preisach models will strongly contribute to the growth of unified characterization frameworks with the former providing strengths inherent to the energy basis and the latter providing a mature mathematical framework for model analysis and identification.

2.4.2 Model Formulation

Preisach models can be defined as a weighted superposition of kernels $[k_s(v,\xi)](t)$ where v and ξ respectively denote inputs and initial states and $s = (s_1, s_2)$ is considered in the closed Preisach plane

$$S = \{(s_1, s_2) \,|\, s_1 \le s_2\}$$

or the compact subset

$$S_\triangle = \{(s_1, s_2) \,|\, \underline{s}_1 \le s_1 \le s_2 \le \overline{s}_2\} \tag{2.53}$$

depicted in Figure 2.21. If the weights are denoted by ν, then the general model can be formulated as

$$[F_\nu(v,\xi)](t) = \iint_S [k_s(v,\xi)](t)\nu(s)\, ds \tag{2.54}$$

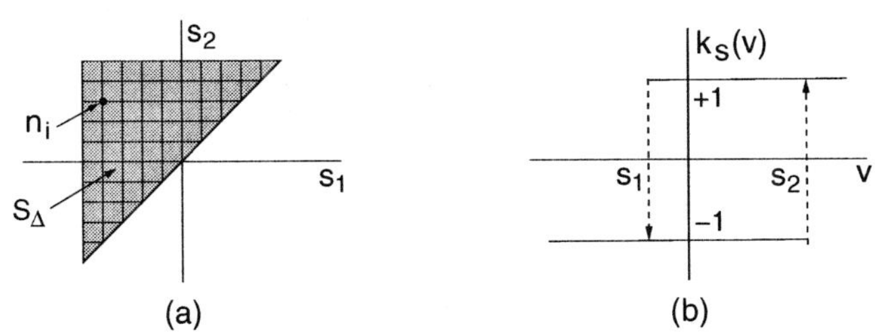

Figure 2.21. *(a) Compact Preisach plane $S_\triangle$ given by (2.53) and (b) classical Preisach relay with threshold values s_1 and s_2.*

with an analogous definition on $S_\triangle$ when decay in ν is invoked to truncate the domain. For piezoelectric applications, the input v consists of the electric field E or voltage V and the operator F_ν can represent the resulting hysteretic and nonlinear polarization or strains. Hence the relation (2.54) can be employed to quantify the hysteretic behavior depicted in Figures 1.1 and 1.25 for electrostrictive polymers or the hysteretic AFM data plotted in Figure 1.10.

The formulation (2.54) is commonly employed to characterize hysteresis in ferroelectric and other materials. However, it is advantageous when establishing convergence properties for approximation methods or providing a framework which includes related models, such as that of Prantl, to consider the more general formulation

$$[F_\mu(v,\xi)](t) = \iint\limits_S [k_s(v,\xi(s))](t)\, d\mu(s) \tag{2.55}$$

where $\mu \in \mathbb{M}$ is a signed Borel measure in the set $\mathbb{M}$ of all finite, signed Borel measures on S. Details regarding the continuity and convergence properties associated with (2.55) can be found in [28, 29, 57, 312].

The construction of the models (2.54) or (2.55) requires the specification of an appropriate basis or kernel k_s, approximation of the integrals, and identification of the weights ν or measure μ given a set of data. We summarize two choices for k_s to illustrate the classical Preisach and Krasnosel'skiĭ–Pokrovskiĭ kernels. Additional choices and a more extensive discussion of the Preisach framework are provided in Chapter 4.

Classical Preisach Model

The classical Preisach model employs the multivalued relay operator

$$[k_s(v,\xi)](t) = \begin{cases} [k_s(v,\xi)](0) & \text{if} \quad \tau(t) = \emptyset \\ -1 & \text{if} \quad \tau(t) \neq \emptyset \text{ and } v(\max \tau(t)) = s_1 \\ +1 & \text{if} \quad \tau(t) \neq \emptyset \text{ and } v(\max \tau(t)) = s_2 \end{cases} \tag{2.56}$$

where the crossing times τ are defined by

$$\tau(t) = \{t_s \in (0,t] \,|\, v(t_s) = s_1 \text{ or } v(t_s) = s_2\}. \tag{2.57}$$

The starting value

$$[k_s(v,\xi)](0) = \begin{cases} -1 & \text{if} \quad v(0) \leq s_1 \\ \xi & \text{if} \quad s_1 < v(0) < s_2 \\ +1 & \text{if} \quad v(0) \geq s_2 \end{cases} \tag{2.58}$$

defines the initial state of the kernel in terms of the parameter $\xi \in \{-1, 1\}$.

As illustrated in Figure 2.21(b), the parameters s_1 and s_2 provide threshold values which dictate the switching properties of the kernel. From a physical perspective, the kernel can be roughly interpreted as quantifying the dipole switch which occurs in the lattice cell when a Ti ion achieves sufficient energy to overcome an unstable equilibrium — e.g., the center of the double well potential depicted in Figure 2.2. For a single crystal, the homogeneous nature of lattice cells will produce a uniform integration of the kernels producing a reasonable characterization of single behavior of the type illustrated for $BaTiO_3$ in Figure 2.8(a). Moreover, due to the generality of the weights or measures, the representation can characterize fairly general hysteresis phenomena, including biased minor loops, as long as the behavior satisfies the congruency and deletion properties established by Mayergoyz as necessary and sufficient conditions for the classical formulation.

Several of the model limitations noted in Section 2.4.1 are readily apparent in the formations (2.54) or (2.55) using the kernel (2.56). For materials which exhibit noncongruent minor loops or nonclosure due to relaxation mechanisms, the formulation is overly stringent. The nonphysical nature of the weights ν or measures μ precludes direct incorporation of physical properties of the data for model development, and the horizontal arms of k_s prohibit the quantification of reversible effects. Finally, the method must be extended to incorporate rate, stress or temperature-dependent dynamics as well as certain continuity properties.

Krasnosel'skiĭ–Pokrovskiĭ (K–P) Kernel

The second choice of kernel illustrates the construction of a general Krasnosel'skiĭ–Pokrovskiĭ (K–P) operator which provides improved continuity, approximation, and convergence properties. To define the K–P kernel, we consider translates $r(v-s_1)$ and $r(v-s_2)$ of a Lipschitz continuous ridge function $r(v)$ as depicted in Figure 2.22. For time intervals $[t_{k-1}, t_k]$ where the input v is monotone, a monotone operator is recursively defined by

$$[\mathcal{R}(v, \mathcal{R}_{k-1})](t) = \begin{cases} \max\{\mathcal{R}_{k-1}, r(v(t) - s_2)\} & \text{if } v \text{ is non-decreasing} \\ \min\{\mathcal{R}_{k-1}, r(v(t) - s_1)\} & \text{if } v \text{ is non-increasing} \end{cases}$$

where

$$\mathcal{R}_k = \begin{cases} \mathcal{R}(v, \mathcal{R}_{k-1})(t_k) \ , \ k = 2, \ldots, j \\ \mathcal{R}_0 = \xi \ , \quad k = 1, \xi \in \{-1, 1\} \end{cases}$$

defines the values of $\mathcal{R}$ at times t_k. The K–P kernel is then defined recursively on each subinterval by

$$[k_s(v, \xi)](t) = [\mathcal{R}(v, \mathcal{R}_{k-1})](t) \quad , \quad t \in [t_{k-1}, t_k] \,. \tag{2.59}$$

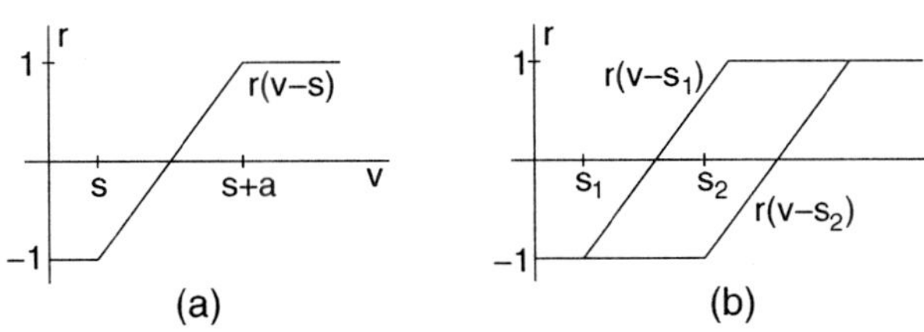

Figure 2.22. *(a) Ridge function $r(v-s)$ and (b) hysteresis envelop provided by the translates $r(v-s_1)$ and $r(v-s_2)$ of the ridge function $r(v)$.*

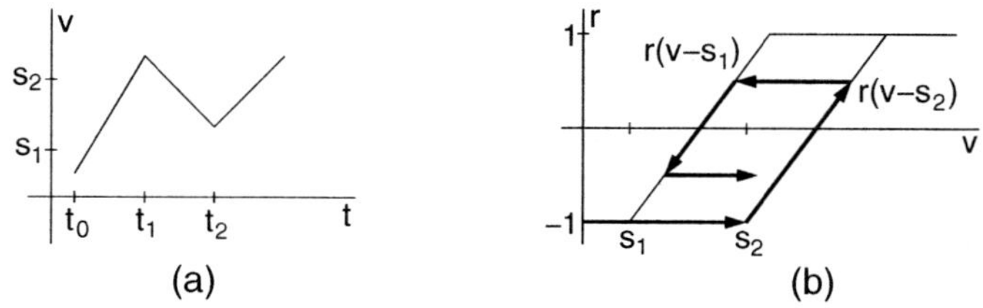

Figure 2.23. *(a) Piecewise monotone input $v(t)$ and (b) output from the K–P kernel in response to the input.*

A typical path for k_s is depicted in Figure 2.23. Details concerning the properties of the kernel are provided in [28, 29, 268].

Approximation and Identification

To numerically implement the model (2.55) with the K–P kernel (2.59), the measure $\mu(s)$ is approximated by the expansion

$$\mu_m = \sum_{i=1}^{m} \alpha_{n_i} \delta_{n_i}$$

where, as depicted in Figure 2.21(a), n_i are nodes in the restricted Preisach plane (2.53), α_{n_i} are weights, and δ_{n_i} denotes the Dirac measure having an atom at n_i [148, 287]. The K–P operator F_μ is then approximated by F_m given by

$$[F_m(v,\xi)](t) = \sum_{i=1}^{m} [k_{n_i}(v,\xi_{n_i})](t)\alpha_{n_i}\,.$$

As detailed in [28, 29, 174], the continuity and convergence properties of the K–P operator lead to well-posedness of the corresponding identification problem and provide a framework which is amenable to approximation and implementation. Additional discussion regarding classical and extended Preisach formulations are provided in Chapter 4.

2.5 Domain Wall Model

The domain wall models incorporate both switching mechanisms and the energy required to translate domain walls to provide a characterization framework which quantifies both the burst region and gradual transition behavior depicted in Figure 2.8 for $BaTiO_3$. This theory was preceded and motivated by the framework developed by Jiles and Atherton for ferromagnetic materials [251] and constitutes the second general framework considered for quantifying hysteresis and constitutive nonlinearities in ferroic compounds.

To motivate the domain processes incorporated in the models, we briefly summarize historical aspects of domain theory for ferroelectric materials. A more detailed historical summary can be found in [438] whereas additional details regarding the theory of domain processes is provided in [298].

Historical Development

Initial investigations focusing on nucleation and growth of domains in $BaTiO_3$ were performed by Caspari and Merz [81], Merz [339] and Little [300] with subsequent domain mechanisms provided and validated by Drougard [137] and Miller and Weirich [342, 344]. In concert, these investigations served to establish that both nucleation and domain wall motion contribute to polarization reversal in $BaTiO_3$. The role of impurities or inclusions on domain processes was elucidated by Miller and Savage [343] who observed that domain wall motion is dependent upon the impurity levels in the crystals. Based in part on experimental observations from [165, 182], which illustrated that domain walls exhibit an "effective viscosity" dependent on the concentration of material defects, Laikhtman in 1973 proposed a model quantifying the bending of domain walls pinned at material inclusions [280]. This model was developed under the premise that the application of an external field produces a force which acts as a pressure on the wall and causes it to bend in a manner analogous to a pinned membrane. Hence this model provides a precursor to the reversible theory discussed here. The irreversible component of the domain wall model is motivated in part by investigations of Fedosov and Sidorkin [160] who concluded that for low displacements, domain walls move as a whole whereas for higher levels, pinning effects become significant and local bending adjacent to pinning sites should be considered. The irreversible polarization changes due to domain wall translation are further corroborated by the analysis of Chen and Wang for relaxor ferroelectric materials [83]. These authors note that at low field levels with high frequencies, the dielectric behavior of the material can be primarily attributed to the bending of domain walls pinned at inclusions. For large applied or internal fields, the local energy barriers associated with pinning sites are overcome and the domain walls move for extended distances.

In combination, the loss mechanisms due to dipole switching and domain wall motion provide the basis for the theory proposed in [438, 442, 443] for characterizing hysteresis in ferroelectric materials. The development is considered in three steps: (i) quantification of dipole switching through energy formulations, (ii) quantification of irreversible polarization changes due to domain wall translations, and (iii) characterization of reversible polarization changes due to domain wall bending.

2.5.1 Polarization Switching

The first component of the model quantifies irreversible polarization changes due to dipole switching. In the original model formulation, this component was referred to as the anhysteretic (hysteresis-free) polarization P_{an} through analogy with the corresponding anhysteretic magnetization M_{an}. However, this is a misnomer for ferroelectric materials since the switching processes are irreversible and hysteretic. We retain the notation P_{an} to ensure uniformity in the unified framework of Chapter 6 but refer to the processes as *equilibrium* rather than *anhysteretic*.

The equilibrium component P_{an} can be constructed through either the minimization of the energy relations detailed in Section 2.3 or the probabilistic quantification of achieving various dipole orientations. We summarize both approaches to provide a basis for various ferroic domain wall models.

Energy Formulation

The construction of P_{an} from energy analysis follows directly from the minimization of the Gibbs energy (2.34) with ψ given by (2.48). The resulting constitutive relations formulated as a function of strain are given in (2.49). For transducer models, it is advantageous to reformulate the direct equation in terms of stresses rather than strains. Substitution of the elastic converse relation into the second expression, neglect of quadratic and cubic coupling terms, and solution for P in terms of E yields

$$P_{an}(\sigma, E) = P_s \tanh\left(\frac{E_e(\sigma, E)}{a(T)}\right) \tag{2.60}$$

where the effective field is given by

$$E_e(\sigma, E) = E + \alpha_1 \sigma + (\alpha + \alpha_2 \sigma)P. \tag{2.61}$$

The parameters

$$\alpha_1 = s^P a_1 \, , \quad \alpha_2 = s^P a_2 \, , \quad \alpha = \frac{E_h}{P_s} + s^P a_1^2 \, , \quad a(T) = \frac{E_h T}{T_c} \tag{2.62}$$

must be identified through either a least squares fit to data or online adaptive estimation techniques for a given material or transducer design. We note that the temperature-dependence encapsulated in the parameter a should be interpreted as *qualitative* rather than *quantitative*, and more sophisticated techniques for characterizing temperature effects are discussed in Chapter 3 in the context of relaxor ferroelectric materials. Finally, we point out that the inclusion of stress and polarization dependencies in the effective field relation (2.61) is consistent with effective field theory discussed in [12, 345] for ferroelectric materials and serves as a prelude to analogous ferromagnetic and ferroelastic mechanisms discussed in the models of Chapters 4 and 5.

Quantification of Dipole Orientations — Boltzmann Probabilities

A second technique for specifying P_{an} is based on the use of Boltzmann principles to quantify the probability that dipoles occupy certain energy states in the presence of a field. Given that the Helmholtz energy relation (2.48) was derived through the application of Boltzmann's law, it is not surprising that the two techniques provide equivalent results under certain conditions. However, the quantification of dipole orientations provides additional insights and can be used to derive the Langevin relation widely employed in magnetic models.

Two Assumed Dipole Orientations

The different models resulting from this approach are due to differing assumptions regarding possible dipole orientations. We consider first the assumptions employed in Section 2.3.2 which posited that dipoles have two possible orientations: in the direction of the external field E or diametrically opposite to it. We again let N_+ and N_- denote the number of dipoles that are oriented with the field or opposite to it. Unlike the analysis of Section 2.3.2, however, we do not presently incorporate dipole or cell interactions.

For a dipole $\mathbf{p}_0$ in an electric field $\mathbf{E}$, it follows from the work expression (2.143) that the potential energy is

$$U = -\mathbf{p}_0 \cdot \mathbf{E} = -p_0 E \cos\theta \tag{2.63}$$

where $p_0 = |\mathbf{p}_0|, E = |\mathbf{E}|$. For noninteracting dipoles, Boltzmann statistics can be used to express the probability of dipoles occupying certain energy states. Letting k again denote Boltzmann's constant, the thermal energy is kT and the probability that a dipole occupies the energy state U is

$$\mu(U) = Ce^{-U/kT}. \tag{2.64}$$

The constant C is specified to ensure that summation over all possible dipole configurations yields the total number N of dipoles per unit volume.

The application of Boltzmann's relation then yields the relations

$$N_+ = Ce^{p_0 E/kT} \quad , \quad N_- = Ce^{-p_0 E/kT}$$

quantifying the distribution of positive and negative dipoles. We then employ the fact that $N = N_+ + N_-$ to obtain

$$N = 2C\cosh\left(\frac{p_0 E}{kT}\right).$$

Furthermore, by noting that the polarization is given by

$$\begin{aligned} P &= p_0 N_+ - p_0 N_- \\ &= 2p_0 C\sinh\left(\frac{p_0 E}{kT}\right), \end{aligned}$$

we can eliminate C to obtain the Ising relation

$$P = p_0 N\tanh\left(\frac{p_0 E}{kT}\right).$$

To obtain a macroscopic version, we assume that the polarization saturates to the value P_s and has the initial permittivity $\epsilon = \frac{P_s E_h T}{E T_c}$ where E_h and T_c again denote a scaling field and the Curie point. This yields the switching expression

$$P_{an} = P_s\tanh\left(\frac{E}{a(T)}\right) \tag{2.65}$$

where $a(T)$ is specified by (2.62).

A comparison of (2.60) and (2.65) derived from energy principles reveals that the latter does not incorporate stresses or interaction field effects due, in part, to the hypothesis that dipoles are noninteracting. From a rigorous perspective, the Ising relation (2.65) is thus limited to *paraelectric* operating regimes which are devoid of dipole interactions.

However, in practice the relation is often extended to ferroelectric (or ferromagnetic) regimes by positing that the effective field has the form specified in (2.61) which yields the relation

$$P_{an}(\sigma, E) = P_s \tanh\left(\frac{E_e(\sigma, E)}{a(T)}\right) \tag{2.66}$$

quantifying the equilibrium polarization. Because the effective field follows directly from energy considerations in the previous derivation, it may appear counterproductive to consider this second method which requires a purely phenomenological extension to ferroelectric regimes. However, it is more easily extended to the less restrictive regime in which dipoles are assumed capable of orienting in any direction.

Continuum of Dipole Orientations

We now make the assumption that dipoles can achieve *any orientation* in response to an applied field which is equivalent to the situation for a dielectric fluid [161]. The number of dipoles between θ and $\theta + d\theta$ is subsequently proportional to the surface area $2\pi \sin\theta d\theta$ on a unit sphere as depicted in Figure 2.24. From (2.63) it follows that the number of moments in this configuration can be expressed as

$$dN = 2\pi \sin\theta d\theta C e^{p_0 E \cos\theta / kT} .$$

To evaluate C, we note that the integration of dN over all possible configurations must equal the total number of moments per unit volume N; hence

$$\begin{aligned} C &= \frac{N}{2\pi \int_0^\pi e^{p_0 E \cos\theta / kT} \sin\theta \, d\theta} \\ &= \frac{N}{\frac{4\pi kT}{p_0 E} \sinh\left(\frac{p_0 E}{kT}\right)} . \end{aligned} \tag{2.67}$$

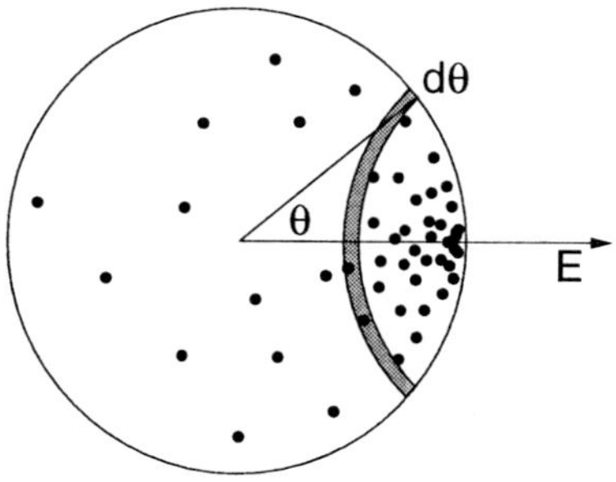

Figure 2.24. *Angular distribution of dipoles in an applied field E.*

Because each cell contributes $p_0 \cos\theta$ to the polarization, it follows that

$$\begin{aligned} P &= 2\pi \int_0^\pi p_0 C e^{p_0 E \cos\theta / kT} \cos\theta \sin\theta \, d\theta \\ &= 2\pi p_0 C \left[\frac{kT}{p_0 E} \cosh\left(\frac{p_0 E}{kT}\right) - \left(\frac{kT}{p_0 E}\right)^2 \sinh\left(\frac{p_0 E}{kT}\right) \right]. \end{aligned}$$

Using (2.67) to eliminate C, we obtain the Langevin relation

$$P = p_0 N \left[\coth\left(\frac{p_0 E}{kT}\right) - \frac{kT}{p_0 E} \right] \tag{2.68}$$

relating the field to polarization.

To obtain a macroscopic model, we invoke the assumptions made for the Ising relation to obtain

$$P_{an}(\sigma, E) = P_s \left[\coth\left(\frac{E_e(\sigma, E)}{a}\right) - \frac{a}{E_e(\sigma, E)} \right] \tag{2.69}$$

where E_e and a are again defined by (2.61) and (2.62). Like the Ising relation, the Langevin expression (2.69) rigorously applies to noninteracting paraelectric regimes and is extended to ferroelectric conditions solely through phenomenological arguments.

A comparison of the two models for ferroelectric and paraelectric regimes is provided in Figure 2.25 where it is observed that both quantify saturation behavior and have nearly equivalent low-field behavior. The latter observation is due to the fact that the relations (2.66) and (2.69) agree through first-order terms. Finally, it is noted that the Langevin relation saturates less quickly than the Ising model since dipoles in the former have more allowable degrees of freedom. Both relations have been employed in the construction of domain wall models.

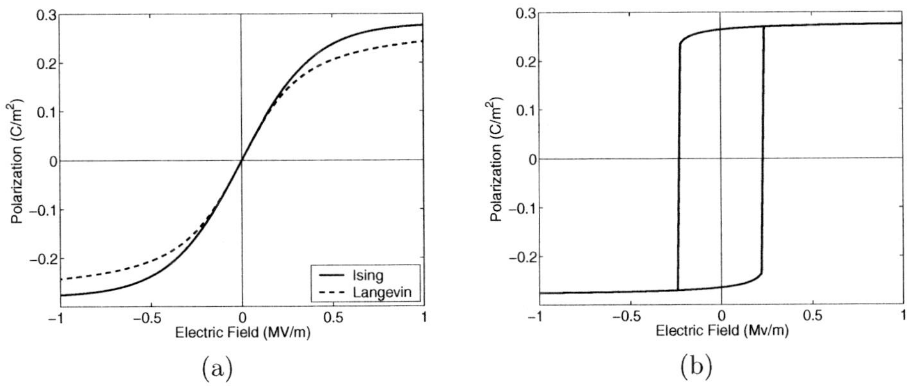

Figure 2.25. *(a) Ising and Langevin models for the equilibrium polarization P_{an} for paraelectric regimes ($T > T_c$). (b) Langevin model for P_{an} in ferroelectric regimes ($T < T_c$).*

2.5.2 Irreversible, Reversible and Total Polarization

It is observed that while the Ising and Langevin relations characterize the qualitative single crystal $BaTiO_3$ behavior depicted in Figure 2.8(a), there are certain mechanisms which they do not incorporate. For example, both relations predict a discontinuous (instantaneous) change in the polarization at coercivity whereas the single crystal data exhibits mollified coercive behavior with a gradual transition to the saturation region. The discrepancy between these switching models and the polycrystalline data depicted in Figure 2.8(b) is significantly more pronounced due to the gradual transition through remanence and coercivity exhibited by the materials. The mollified E-P behavior in both cases is due in part to the impediment of domain wall motion by material inclusions, and the quantification of these effects forms the basis for the domain wall modeling framework.

As noted previously, it is energetically favorable for domain walls to form at inclusions due to material nonhomogeneities, defects, or stress concentrations [298]. We quantify first the energy required to translate domain walls and the resulting irreversible changes in the polarization.

Irreversible Polarization

Consider the dipole configuration depicted in Figure 2.26. Following the theory established by Jiles and Atherton for ferromagnetic materials [251], we hypothesize that the energy required to overcome pinning sites and hence translate the domain wall is proportional to the change in energy required to align dipoles with the field. Employing (2.63), this implies that

$$\mathcal{E}_{pin} = -pE_e c_1(1 - \cos\theta)$$

where $p = |\mathbf{p}|$ is the dipole strength. To eliminate c_1, we let $\mathcal{E}_\pi$ denote the energy for 180° domain walls so that $\mathcal{E}_\pi = -2pE_e c_1$. This implies that

$$\mathcal{E}_{pin} = \frac{1}{2}\mathcal{E}_\pi(1 - \cos\theta). \tag{2.70}$$

The relation (2.70) quantifies the energy required to break a single pinning site. To obtain a corresponding macroscopic relation which characterizes the bulk

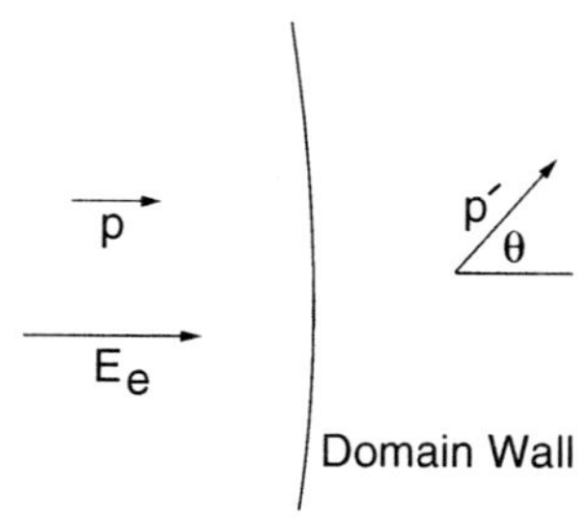

Figure 2.26. *Domain wall and reorientation of dipoles from* $\mathbf{p}'$ *to* $\mathbf{p}$ *due to an effective field* $\mathbf{E}_e$.

material behavior, it is necessary to *homogenize* over cells to formulate losses in terms of an effective material parameter. Employing the approach of Jiles and Atherton, we let n denote the average density of pinning sites and $\langle \mathcal{E}_\pi \rangle$ denote the average energy for 180° walls. The energy required to break pinning sites if a domain wall of area A moves a distance dx is then

$$d\mathcal{E}_{pin} = \frac{n\,\langle \mathcal{E}_\pi \rangle}{2}(1 - \cos\theta)A\,dx. \tag{2.71}$$

Moreover, the irreversible polarization change due to the reorientation of dipoles is

$$dP_{irr} = p(1 - \cos\theta)Adx\,. \tag{2.72}$$

With the definition $k_p = \frac{n\langle \mathcal{E}_\pi \rangle}{2p}$, (2.71) and (2.72) can be combined to yield the expression

$$d\mathcal{E}_{pin} = k_p dP_{irr} \tag{2.73}$$

quantifying the energy lost by moving the domain wall.

To specify the irreversible polarization, it is necessary to quantify the work performed in the polarization process. As detailed in Section 2.8, the energy supplied by the effective field $D_e = \epsilon_0(E + \alpha P)$ over a full cycle is

$$\begin{aligned} w &= \oint \mathbf{E}\cdot d\mathbf{P} \\ &= -\frac{1}{\epsilon_0}\oint \mathbf{P}\cdot d\mathbf{D}_E \end{aligned}$$

so we *postulate* that the work over a partial cycle can be expressed as

$$w = -\frac{1}{\epsilon_0}\int_0^{\mathbf{D}_e} \mathbf{P}\cdot d\mathbf{D}_E.$$

The total energy supplied to the material can then be expressed as that producing switching plus energy losses due to domain wall movement which gives

$$\frac{1}{\epsilon_0}\int_0^{D_e} P_{irr}(E_e)\,dD_e = \frac{1}{\epsilon_0}\int_0^{D_e} P_{an}(E_e)\,dD_e - \int_0^{D_e} k_p \frac{dP_{irr}}{dD_e}\,dD_e.$$

Differentiation yields the local relation

$$P_{irr} = P_{an} - \delta\epsilon_0 k_p \frac{dP_{irr}}{dD_e} \tag{2.74}$$

where the parameter $\delta = \text{sign}(dE)$ ensures that the energy required to break pinning sites always opposes changes in the polarization. To obtain a differential equation which facilitates implementation, we formulate (2.74) as

$$P_{an} - P_{irr} = \delta k_p \frac{dP_{irr}}{dE}\left(\frac{1}{1 + \alpha\frac{dP_{irr}}{dE}}\right)$$

from which it follows that

$$\frac{dP_{irr}}{dE} = \frac{P_{an} - P_{irr}}{\delta k_p - \alpha(P_{an} - P_{irr})}. \tag{2.75}$$

The differential equation (2.75) dictates that P_{irr} evolves toward P_{an}. This provides correct physical behavior for most operating regimes but leads to nonphysical predictions in the following instances: (i) negative differential susceptibility $\frac{\partial P}{\partial E}$ following field reversal, and (ii) nonclosure of biased minor loops. The first discrepancy is easily addressed by enforcing the condition $\frac{\partial P_{irr}}{\partial E} = 0$ until the polarization crosses the switching curve P_{an}. This produces the physically more realistic expression

$$\frac{dP_{irr}}{dE} = \widetilde{\delta}\frac{P_{an} - P_{irr}}{k_p\delta - \alpha\,(P_{an} - P_{irr})} \tag{2.76}$$

where

$$\widetilde{\delta} = \begin{cases} 1\,, & \{dE < 0 \text{ and } P > P_{an}\} \text{ or } \{dE > 0 \text{ and } P < P_{an}\} \\ 0\,, & \text{otherwise}\,. \end{cases} \tag{2.77}$$

From a physical perspective, this simply ensures reversible polarization behavior immediately following field reversal.

The second issue is more fundamental and poses a limitation to the model for feedback control design. As will be illustrated in Section 2.5.6, biased minor loop closure can be guaranteed if one has *a priori* knowledge of turning points. However, this is not the case in typical state feedback algorithms which limits the utility of the model for such applications.

Reversible Polarization

The second component of the polarization is the reversible polarization P_{rev} which quantifies domain wall bending observed before the onset of irreversibility. Details regarding this component of the model can be found in [438] and we provide only a summary of the development.

By employing the energy relation (2.63), one can establish that input of a field E induces a pressure

$$\mathcal{P} = k_1(P - P_{an})$$

on the wall where k_1 is a material constant. To quantify the reversible polarization which results when the domain wall bows in response to this pressure, one approximates the displaced volume to obtain

$$P_{rev} = c(P_{an} - P_{irr}). \tag{2.78}$$

Here c denotes an effective parameter which quantifies the degree of reversibility observed for the material.

Total Polarization

The total polarization is given by the sum

$$P = P_{rev} + P_{irr} \tag{2.79}$$

where P_{irr} and P_{rev} are respectively defined in (2.76) and (2.78). The algorithm employed for implementation is summarized in Section 2.5.6 with additional details provided in [438]. The model in this form is employed for material or device characterization.

For control applications and the development of inverse filters, it is advantageous to reformulate the model as a single ordinary differential equation specifying the polarization as a function of input field. For the Ising model (2.66) with effective field $E_e = E + \alpha P$, this yields the differential equation

$$\begin{aligned} \frac{dP}{dE} &= \mathcal{G}(E,P) \\ P(E_0) &= P_0 \end{aligned} \tag{2.80}$$

where

$$\mathcal{G}(E,P) = \frac{1}{1 - \frac{cP_s\alpha}{a}\mathrm{sech}^2\left(\frac{E+\alpha P}{a}\right)} \cdot \left\{ \frac{\widetilde{\delta}\left[P_s \tanh\left(\frac{E+\alpha P}{a}\right) - P\right]}{k_p\delta - \tilde{\alpha}\left[P_s \tanh\left(\frac{E+\alpha P}{a}\right) - P\right]} + \frac{cP_s}{a}\mathrm{sech}^2\left(\frac{E+\alpha P}{a}\right)\right\}. \tag{2.81}$$

As detailed in [438], one obtains an analogous differential equation if the Langevin relation (2.68) is employed for the switching model

Parameters and Initial Conditions

Both of the formulations (2.79) and (2.80) require the specification of initial values (E_0, P_0) and determination of appropriate material parameters $P_s, k_p, c, \alpha, a, a_1, a_2$. For material characterization, one typically employs the value $(E_0, P_0) = (0,0)$ when initializing the model. For model-based control implementation, it is often necessary to specify initial conditions for the experimental setup which is much harder since the depolarized state is typically unavailable. Potential choices in this case include positive and negative saturation or remanence values.

As detailed in [443], k_p and c satisfy the asymptotic relations

$$k_p \approx E_c \quad , \quad c \approx \frac{\chi_m^-}{\chi_m^+} \tag{2.82}$$

where χ_m^+ and χ_m^- denote the differential susceptibilities $\frac{\partial P}{\partial E}$ before and after field reversal at the loop tip. This provides initial estimates for k_p and c using easily measured properties of the data. Furthermore, an initial estimate for P_s can be obtained from polarization data measured at high input field levels. Final values for these and the remaining parameters are typically obtained through a least squares fit to data.

It has been noted that the hysteretic properties of ferroelectric materials are temperature-dependent which necessitates the updating of model parameters if the material heats or cools significantly during operation. This is reflected in part by the temperature-dependence of $a(T)$ in (2.62). For slowly varying temperature changes, this can be accomplished through adaptive algorithms designed to accommodate the nonlinear parameterization inherent to these models. The development of such algorithms for the analogous ferromagnetic hysteresis model is illustrated in [357].

2.5.3 Inverse Model

The model in the form (2.80) is amenable to inversion through consideration of the complementary differential equation

$$\begin{aligned} \frac{dP^{-1}}{dE} &= \frac{1}{\mathcal{G}(P^{-1},E)} \\ P^{-1}(E_0) &= \widehat{P}_0. \end{aligned} \tag{2.83}$$

As detailed in [431], an efficient model inverse can be employed to construct an inverse filter which, at least approximately, compensates for hysteresis and hence permits linear control design for linear plants. Important properties of such inverse filters are sufficient efficiency to permit real-time implementation and robustness with regard to noise or model uncertainties. As illustrated in Chapter 4 for the analogous ferromagnetic model and its inverse, the formulation (2.83) is efficient to experimentally implement when low-order finite difference techniques are used to discretize the differential equation. The major issue concerning robustness is the identification or initialization of the initial condition $\widehat{P}_0$. For applications in which $\widehat{P}_0$ can be accurately determined, approximate inverse filters based on (2.83) prove to be effective. However, sufficient uncertainty in $\widehat{P}_0$ can produce phase delays which can degrade control performance or even lead to instabilities. For such cases, the incorporation of adaptive mechanisms or state estimation techniques may prove beneficial.

2.5.4 Electromechanical Constitutive Relations

The expressions (2.79) and (2.80) quantify P as a function of E. For transducer design, it is necessary to combine these expressions with relations characterizing the elastic and electromechanical properties of the actuator or sensor material.

To construct the converse relation, we retain the elastic and electromechanical terms in the Helmholtz expression (2.36) or (2.38) to construct the elastic Gibbs relation

$$G = \frac{1}{2}Y^P\varepsilon^2 - a_1\varepsilon P - a_2\varepsilon P^2 - \varepsilon\sigma. \tag{2.84}$$

Enforcement of the necessary condition $\frac{\partial G}{\partial \varepsilon} = 0$ yields the constitutive relation

$$\sigma = Y^P \varepsilon - a_1 P - a_2 P^2 \tag{2.85}$$

which can be used to quantify the effect of polarization on either strains or stresses.

Conversely, stresses have two effects on polarization. The first is to modify the equilibrium component P_{an} due to the stress-dependence of the effective field — e.g., see (2.61). Secondly, the application of stresses tends to drive the polarization toward P_{an} in a manner analogous to that described in [247,249,381] and Chapter 4 for ferromagnetic materials. For low to moderate stress levels, the latter mechanism is often neglected. In this case, the constitutive behavior is quantified by (2.85) with the polarization modeled by (2.79) and (2.80) using the stress-dependent relation (2.61) for E_e.

2.5.5 Material Characterization

To illustrate attributes of the model, we consider its performance when characterizing the hysteretic behavior of PZT5A and PVDF at multiple drive levels. Additional examples illustrating its accuracy for the softer material PZT5H and the hard compound PZT4 can be found in [442] whereas the incorporation of certain relaxation mechanisms is addressed in [444]. All of these cases focus on material characterization in the context of symmetric major hysteresis loops and inherent limitations regarding the closure of biased minor loops are addressed in Section 2.5.6.

PZT5A Characterization

We consider first the characterization of hysteresis exhibited by a circular PZT wafer having a diameter of 2.54 cm and thickness of 0.0254 cm (10 mils). Input voltages ranging from 600 V to 1600 V were input to the wafer yielding the E-P relations plotted in Figure 2.27. Note that the correspondence between E and V is given by the approximate relation $E = V/h$ where $h = 0.0254$ cm is the thickness of the wafer.

The model employing the Langevin switching expression (2.69) was constructed using attributes of the 1600 V data and then used to predict the lower input behavior. Specifically, the 1600 V coercive field $E_c = 1.2 \times 10^6$ V/m, slope $\chi_m^-/\chi_m^+ = 0.18$ at field reversal, and maximum measured polarization value $P = 0.46$ C/m^2 were used as initial estimates for the parameters k_p, c and P_s. A least squares fit to the 1600 V data subsequently yielded the final parameter set $k_p = 1.8 \times 10^6$ V/m, $c = 0.30$, $P_s = 0.49$ C/m^2, $\alpha = 3.6 \times 10^6$ Vm/C and $a = 4.2 \times 10^5$ V/m. The resulting fit and predictions at 600 V, 800 V and 1000 V are compared with the data in Figure 2.27. For this operating regime, the model accurately characterizes both the hysteresis and nonlinear saturation behavior of the material.

PVDF Characterization

To illustrate the model performance for a piezoelectric polymer, we consider the characterization of polyvinylidene fluoride (PVDF) driven at an input frequency

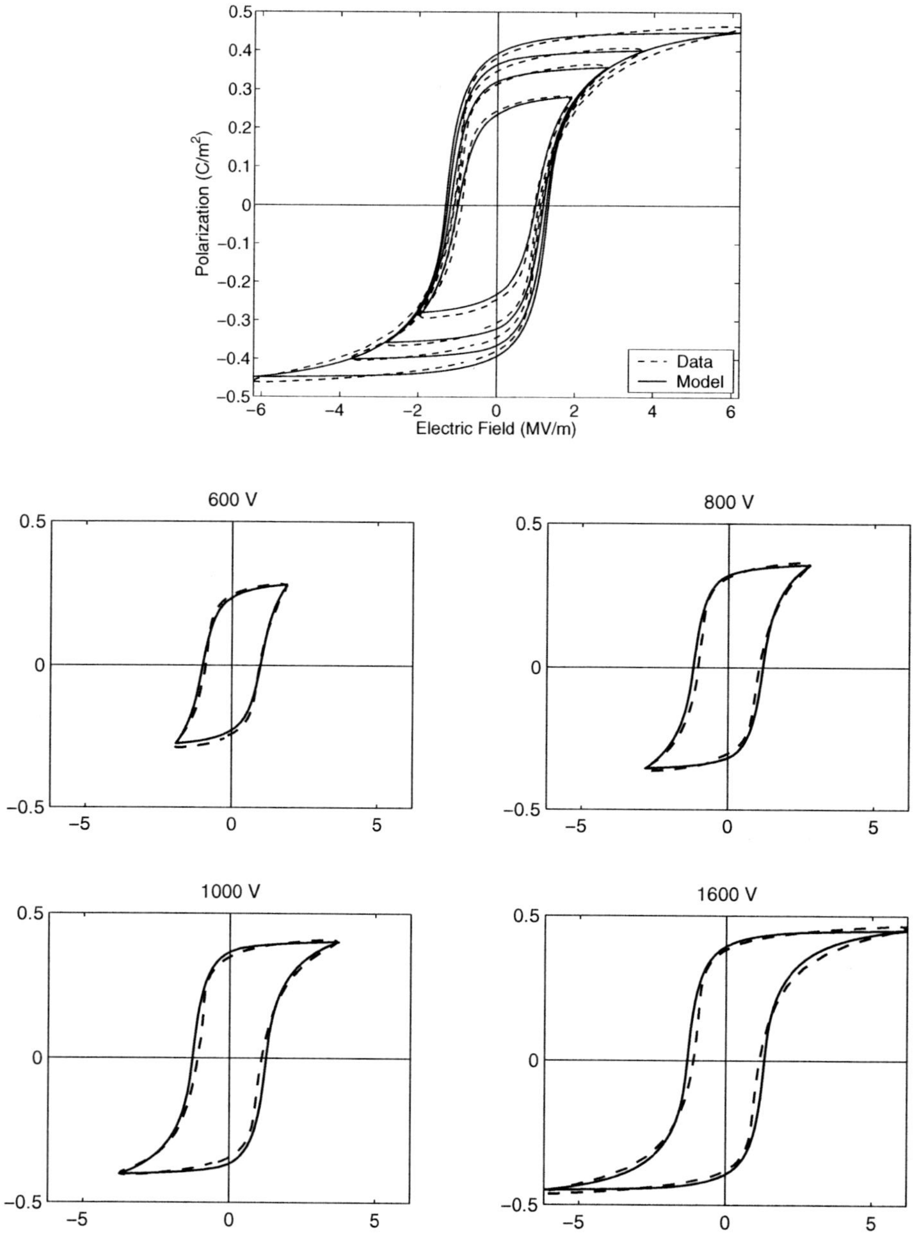

Figure 2.27. *Model fit to 1600 V data for the circular PZT5A patch and model predictions at 600 V, 800 V and 1000 V (from [442]).*

of 20 mHz. As summarized in Section 1.5.1, PVDF has lower strain constants than PZT but higher stress constants. Hence it is primarily considered for sensor applications or actuator designs which exploit its flexibility and high field capabilities.

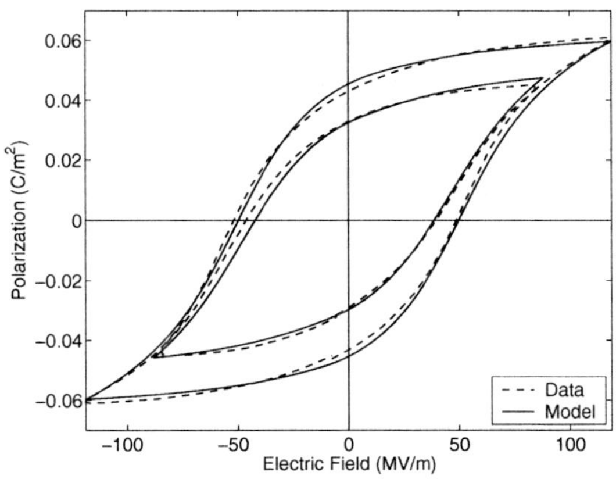

Figure 2.28. *Model fit to 120 MV/m PVDF data and prediction at 85 MV/m.*

A fit to PVDF data having peak input fields of 120×10^6 V/m yields the parameter values $k_p = 85 \times 10^6$ V/m, $c = 0.35$, $P_s = 0.097$ C/m^2, $\alpha = 8.2 \times 10^8$ Vm/C and $a = 2.9 \times 10^7$ V/m. The resulting model fit and the prediction for 85 MV/m inputs are compared with the data in Figure 2.28. For this quasistatic, major loop drive regime, the model very accurately quantifies the significant hysteresis exhibited by the material. Additional examples illustrating the incorporation of low frequency relaxation effects can be found in [443].

2.5.6 Model Construction and Attributes

The algorithm for constructing the domain wall model is summarized in Algorithm 2.5.1. The examples of Section 2.5.5 illustrate that the domain wall theory can accurately predict the *unbiased E-P* behavior of both piezoceramic compounds and piezoelectric polymers. However, care must be taken if operating regimes include biased minor loops of the form depicted in Figure 2.29. For quasistatic input

Algorithm 2.5.1.

(i) $E_e = E + \alpha_1\sigma + (\alpha + \alpha_2\sigma)P$

(iia) $P_{an} = P_s\left[\coth\left(E_e/a\right) - \left(a/E_e\right)\right]$ *(Langevin Model)*

(iib) $P_{an} = P_s\tanh\left(E_e/a\right)$ *(Ising Model)*

(iii) $$\frac{dP_{irr}}{dE} = \frac{\tilde{\delta}\left[P_{an} - P_{irr}\right]}{k_p\delta - \alpha[P_{an} - P_{irr}]}$$

(iv) $P_{rev} = c[P_{an} - P_{irr}]$

(v) $P = P_{rev} + P_{irr}$

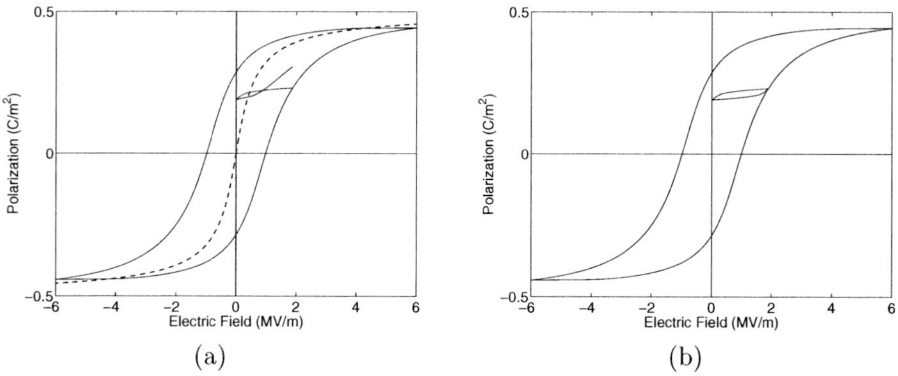

Figure 2.29. *(a) Nonclosure of the domain wall model for biased minor hysteresis loops due to the evolution of P (——) toward P_{an} (- - -). (b) Closure obtained using prior knowledge of turning points.*

fields, the materials exhibit closed minor loops whereas the model may predict nonclosure as illustrated in Figure 2.29(a). The reason for this discrepancy is the formulation (2.76) for P_{irr} which produces evolution toward P_{an}.

Jiles addressed this in the magnetization model [248] by enforcing closure through *a priori* knowledge of both turning points as demonstrated in Figure 2.29(b). While this can be argued from the physical perspective of maintaining volume fractions, it is problematic for state feedback control designs since one often does not have advanced knowledge regarding the state of the system. For regimes in which inaccuracies due to nonclosure prove too great, alternative models such as the Preisach techniques described in Section 2.4 or the homogenized energy model developed in Section 2.6 should be considered since they guarantee biased minor loop closure for quasistatic inputs.

2.6 Homogenized Energy Model

The Gibbs energy relations developed in Section 2.3 quantify instantaneous dipole switching in the absence of material nonhomogeneities and hence provide reasonable accuracy when characterizing hysteresis in certain single crystal compounds — e.g., the $BaTiO_3$ behavior depicted in Figure 2.8(a). However, they do not provide mechanisms for incorporating thermal relaxation, material nonhomogeneities, variable effective fields or the effects of polycrystallinity, a subset of which can produce the behavior depicted in Figure 2.8(b). The homogenized energy framework incorporates these effects by employing the energy relations developed in Section 2.3 as a kernel at the lattice level whose parameters are considered manifestations of material-dependent distributions. *Homogenization* or *averaging* in this manner provides highly flexible and efficient macroscopic models which guarantee minor loop closure in quasistatic drive regimes, accommodate reversible polarization changes, incorporate certain rate, stress and temperature-dependencies, and retain physical interpretations for a number of the model parameters. Moreover, this technique provides a unified framework for quantifying hysteresis and constitutive nonlinearities in ferroelectric, ferromagnetic and ferroelastic compounds.

The homogenized energy framework is the most recent of the unified frameworks discussed for characterizing hysteresis in ferroic compounds. The method has its genesis in the Müller-Achenbach-Seelecke theory for SMA [1,2,409–413] and has recently been extended to ferroelectric and ferromagnetic materials [434, 447, 450] and SMA thin films [324]. The models can be inverted to provide inverse filters for linear control design and, as detailed in Section 4.7.10, the method provides an energy basis for certain extended Preisach models.

This section is long due not to the complexity of the framework but rather to its highly comprehensive nature. Hence we provide significant detail both to illustrate the development for ferroelectric materials and to provide the theory used to establish analogous ferromagnetic, ferroelastic and ferroic models in Chapters 5–7. The following outline summarizes the topics in the remainder of the section. For implementation purposes, the reader can focus on those sections covering operating conditions of interest and skip sections delineated with an asterisk which address theoretical issues. Finally, the primary attributes of the framework are summarized in Section 2.6.11 to provide an omega for the discussion.

Section 2.6.1: Local Polarization — Negligible Thermal Relaxation. If thermal relaxation mechanisms are negligible, the local polarization $\overline{P}$ follows directly from the minimization of Gibbs energy relations from Section 2.3.

Section 2.6.2: Local Polarization — Thermal Activation. The effects of thermal activation and relaxation are incorporated through Boltzmann principles which balance the Gibbs energy G and relative thermal energy kT/V.

Section 2.6.3:* Local Polarization — Limiting Behavior. We establish here the rigorous convergence of the second formulation to the first as thermal effects become negligible.

Section 2.6.4: Macroscopic Polarization Model. Stochastic homogenization techniques are used to incorporate material nonhomogeneities, polycrystallinity, and variable interaction field effects by positing that parameters in the kernels from Sections 2.6.1 and 2.6.2 are manifestations of underlying distributions rather than constants. Two choices are considered: (i) general nonparametric densities based solely on a fit to data, and (ii) *a priori* choices of lognormal and normal distributions which facilitate identification but can reduce accuracy.

Section 2.6.5: Polarization Model Implementation. Highly efficient algorithms are constructed to facilitate implementation of both nonlinearly and linearly parameterized models.

Section 2.6.6: Parameter Estimation. Following a discussion of the infinite-dimensional parameter estimation problem to establish the ill-posedess of the inverse problem, finite-dimensional algorithms for identifying parameters are discussed.

Section 2.6.7: Model inversion. Algorithms are constructed to provide model inverses for use in linear control design.

Section 2.6.8: Thermal Evolution. This section summarizes the evolution equations used to quantify thermal processes in ferroelectric materials.

Section 2.6.9: Electromechanical Constitutive Relations. The polarization model developed in Section 2.6.4 is combined with electromechanical Helmholtz and Gibbs energy relations developed in Section 2.3 to provide coupled constitutive relations for ferroelectric compounds.

Section 2.6.10: Material Characterization. The accuracy of the framework and its capacity to accommodate biased minor loop closure, and reversibility are illustrated through comparison with PZT5H data and prediction of various physical mechanisms.

Section 2.6.11: Model Attributes. A summary of the model construction is provided. Details regarding attributes of the model and the manner in which it relates to the domain wall and Preisach frameworks can be found in Section 4.7.10. Associated MATLAB software can be found at the website `http://www.siam.org/books/fr32`.

2.6.1 Local Polarization — Negligible Thermal Relaxation

For operating regimes in which thermal relaxation mechanisms are considered negligible, as quantified in Section 2.6.3, the local average polarization $\overline{P}$ is determined directly through the minimization of the Gibbs energy

$$G(E,P,T) = \psi(P,T) - EP \tag{2.86}$$

where the Helmholtz energy is taken to be

$$\psi(P,T) = \frac{E_h P_s}{2}\left[1-(P/P_s)^2\right] + \frac{E_h T}{2T_c}\left[P\ln\left(\frac{P+P_s}{P_s-P}\right) + P_s\ln(1-(P/P_s)^2)\right]$$

or

$$\psi(P) = \begin{cases} \frac{1}{2}\eta(P+P_R)^2 & , \; P \le -P_I \\ \frac{1}{2}\eta(P-P_R)^2 & , \; P \ge P_I \\ \frac{1}{2}\eta(P_I-P_R)\left(\frac{P^2}{P_I}-P_R\right) & , \; |P| < P_I \end{cases} \tag{2.87}$$

as originally defined in (2.47) and (2.51).

For the first case, it is illustrated in (2.60) that the necessary condition $\frac{\partial G}{\partial P} = 0$ yields

$$\overline{P}(E) = P_s \tanh\left(\frac{E+\alpha P}{a(T)}\right) \tag{2.88}$$

where $\alpha = \frac{E_h}{P_s}$ and $a(T) = \frac{E_h T}{T_c}$.

The kernel resulting from the relation (2.87) has the general form

$$\overline{P}(E) = \frac{1}{\eta}E + P_R\delta \tag{2.89}$$

where $\delta = -1$ for negatively oriented dipoles and $\delta = 1$ for those with positive orientation. To specify δ and hence $\overline{P}$ more specifically in terms of the initial dipole orientations and previous switches, we employ the Preisach notation (2.56) and take

$$[\overline{P}(E; E_c, \xi)](t) = \begin{cases} [\overline{P}(E; E_c, \xi)](0) & , \tau(t) = \emptyset \\ \frac{E}{\eta} - P_R & , \tau(t) \neq \emptyset \text{ and } E(\max \tau(t)) = -E_c \\ \frac{E}{\eta} + P_R & , \tau(t) \neq \emptyset \text{ and } E(\max \tau(t)) = E_c. \end{cases} \tag{2.90}$$

Here $\emptyset$ denotes the empty set and

$$[\overline{P}(E; E_c, \xi)](0) = \begin{cases} \frac{E}{\eta} - P_R & , E(0) \leq -E_c \\ \xi & , -E_c < E(0) < E_c \\ \frac{E}{\eta} + P_R & , E(0) \geq E_c \end{cases}$$

defines initial kernel values in terms of the parameter $\xi = \frac{E_0}{\eta} \pm P_R$ — e.g., see Figure 2.34 on page 113. The local coercive field

$$E_c = \eta(P_R - P_I) \tag{2.91}$$

quantifies the field at which the negative well ceases to exist and hence a dipole switch occurs. Finally, the set of switching times is given by

$$\tau(t) = \{t_s \in (0, t] \mid E(t_s) = -E_c \text{ or } E(t_s) = E_c\}. \tag{2.92}$$

To illustrate the manner through which (2.90) quantifies switching, consider the condition $\tau \neq \emptyset$ and $E(\max \tau(t)) = E_c$. This designates that switching has occurred and the last switch was at E_c, thus yielding $[\overline{P}(E; E_c, \xi)](t) = \frac{E(t)}{\eta} + P_R$.

The kernels provided by (2.88) and (2.90) are compared in Figure 2.30. While both characterize the hysteretic behavior of $\overline{P}$ due to dipole switching, the relation (2.90) predicts linear behavior with slope $\frac{1}{\eta}$ after switching whereas (2.88) exhibits a local saturation value of P_s and decreasing slope for increasing fields. While both can be employed as kernels for macroscopic models, we typically consider the piecewise linear relation (2.89) or (2.90) due to its flexibility and efficiency.

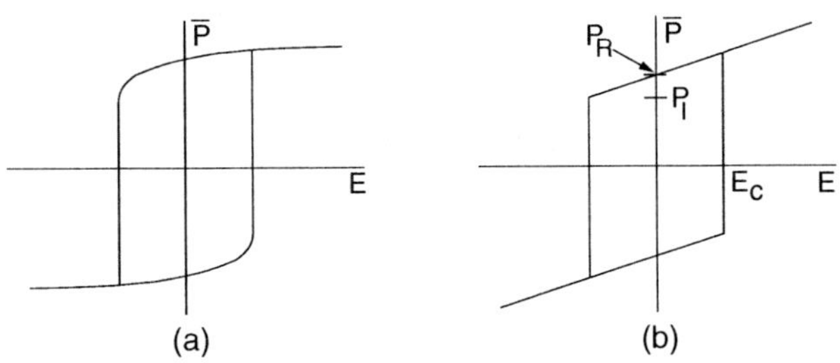

Figure 2.30. *(a) Kernel provided by (2.88) and (b) kernel provided by (2.90).*

2.6.2 Local Polarization — Thermal Relaxation

For more general operating regimes which include the possibility of thermal relaxation mechanisms, it is necessary to include the effects of thermal activation when quantifying $\overline{P}$. This can be accomplished formally through the use of the Boltzmann relation

$$\mu(G) = Ce^{-GV/kT} \tag{2.93}$$

which balances the Gibbs energy and relative thermal energy kT/V, or more fundamentally through the minimization of an energy functional which incorporates internal kinetic and entropy components. We consider the formal approach first to illustrate properties of the model. We then provide the energy derivation at the end of the section to motivate the origin of (2.93) and illustrate fundamental properties of the model.

Behavior of the Boltzmann Density

As in Section 2.3.2, we consider a uniform lattice of volume V having N cells of the form depicted in Figure 2.2. Based on the assumption that dipoles are restricted to two configurations $s_i = \pm 1$, we can designate the number of negatively and positively oriented dipoles as N_- and N_+. The corresponding dipole fractions are designated by

$$x_- = \frac{N_-}{N} \quad , \quad x_+ = \frac{N_+}{N}$$

where the conservation relation

$$x_- + x_+ = 1 \tag{2.94}$$

follows immediately from the fact that $N_- + N_+ = N$. The average polarization values associated with negative and positive dipoles are denoted by $\langle P_- \rangle$ and $\langle P_+ \rangle$, and the likelihoods of switching from negative to positive and conversely are respectively denoted by p_{-+} and p_{+-}; we avoid the terminology probabilities since p_{-+} and p_{+-} can be greater than one.

Throughout the development we consider the Gibbs energy (2.86) with the piecewise quadratic Helmholtz energy (2.87). To illustrate the behavior of (2.93) we consider the specific energy profile depicted in Figure 2.31 for which it is assumed that $E > 0$ and $G(P_{min}^+) < G(P_{min}^-) \leq G(P_0)$ where P_0 denotes the unstable equilibrium of G. The relative minima

$$\begin{aligned} P_{min}^-(E) &= \frac{E}{\eta} - P_R\,, \\ P_{min}^+(E) &= \frac{E}{\eta} + P_R \end{aligned} \tag{2.95}$$

result from the necessary condition $\frac{\partial G}{\partial P} = 0$ used to construct the limiting model (2.89) or (2.90).

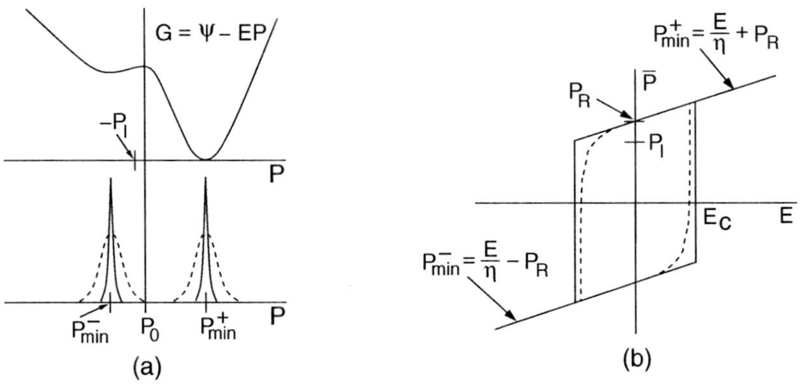

Figure 2.31. *(a) Gibbs energy profile and Boltzmann densities $\mu(G) = Ce^{-GV/kT}$ with high (- - -) and low levels (——) of thermal activation. (b) Local polarization $\overline{P}$ given by equation (2.99) with high thermal activation (- - -) and limiting polarization $\overline{P}$ specified by (2.90) in the absence of thermal activation (——).*

From (2.87), it follows that for $P < -P_I$, the Boltzmann probability density function can be formulated as

$$\begin{aligned} \mu(G(E,P)) &= C(T)e^{-G(E,P)V/kT} \\ &= \frac{e^{-\left[\frac{1}{2}\eta(P+P_R)^2 - EP\right]V/kT}}{\displaystyle\int_{-\infty}^{-P_I} e^{-\left[\frac{1}{2}\eta(P+P_R)^2 - EP\right]V/kT} dP} \\ &= \frac{e^{-(P-P_{min}^-)^2\eta V/2kT}}{\displaystyle\int_{-\infty}^{-P_I} e^{-(P-P_{min}^-)^2\eta V/2kT} dP} \\ &= \overline{C}(T,\beta)e^{-(P-P_{min}^-)^2/2\beta^2} \end{aligned} \tag{2.96}$$

where

$$\begin{aligned} \beta &= \sqrt{\frac{kT}{\eta V}}, \\ \overline{C}(T,\beta) &= \left[\int_{-\infty}^{-P_I} e^{-(P-P_{min}^-)^2/2\beta^2(T)} dP\right]^{-1}. \end{aligned} \tag{2.97}$$

Similarly, for $P > P_I$, the probability density is defined by

$$\mu(G(E,P)) = \frac{e^{-(P-P_{min}^+)^2\eta V/2kT}}{\displaystyle\int_{P_I}^{\infty} e^{-(P-P_{min}^+)^2\eta V/2kT} dP}. \tag{2.98}$$

Relations (2.96) and (2.98) illustrate the Gaussian behavior of the Boltzmann density for the piecewise quadratic Helmholtz function ψ while (2.97) illustrates that the variance β^2 is proportional to the relative thermal energy kT/V. From a physical perspective, low relative thermal energy implies that fewer dipoles achieve the energy required to overcome energy barriers thus producing steep transitions in the local relation between E and $\overline{P}$ as depicted in Figure 2.31(b).

Transition Likelihoods and Average Polarization

For a uniform lattice, the local average polarization is

$$\overline{P} = x_+ \langle P_+ \rangle + x_- \langle P_- \rangle . \tag{2.99}$$

The evolution of dipole fractions is quantified by the differential equations

$$\begin{aligned} \dot{x}_+ &= -p_{+-}x_+ + p_{-+}x_- \\ \dot{x}_- &= -p_{-+}x_- + p_{+-}x_+ \end{aligned} \tag{2.100}$$

which can be simplified to

$$\dot{x}_+ = -p_{+-}x_+ + p_{-+}(1 - x_+) \tag{2.101}$$

through the identity (2.94). The expected polarizations are quantified by the general relations

$$\langle P_+ \rangle = \int_{P_0(T)}^{\infty} P\mu(G)\, dP \quad , \quad \langle P_- \rangle = \int_{-\infty}^{P_0(T)} P\mu(G)\, dP.$$

For implementation, the unstable equilibrium P_0 is often replaced by the positive and negative inflection points to yield

$$\langle P_+ \rangle = \frac{\displaystyle\int_{P_I}^{\infty} Pe^{-G(E,P)V/kT} dP}{\displaystyle\int_{P_I}^{\infty} e^{-G(E,P)V/kT} dP} \quad , \quad \langle P_- \rangle = \frac{\displaystyle\int_{-\infty}^{-P_I} Pe^{-G(E,P)V/kT} dP}{\displaystyle\int_{-\infty}^{-P_I} e^{-G(E,P)V/kT} dP} \tag{2.102}$$

where the denominator results from the evaluation of C in (2.93) to guarantee integration to unity when evaluated over all admissible states. The use of P_I and $-P_I$ simplifies the evaluation of the integrals by restricting the integrand to one component of the piecewise energy definitions. This is motivated by the observation that if one considers the forces $\frac{\partial G}{\partial P}$ due to the applied field, maximal restoring forces occur at P_I and $-P_I$ as illustrated in Figure 2.32 and detailed on pages 332–333 of [111].

The transition likelihoods are given by

$$p_{+-} = \frac{1}{\mathcal{T}(T)} \frac{\displaystyle\int_{P_I-\epsilon}^{P_I} e^{-G(E,P)V/kT} dP}{\displaystyle\int_{P_I-\epsilon}^{\infty} e^{-G(E,P)V/kT} dP} \; , \; p_{-+} = \frac{1}{\mathcal{T}(T)} \frac{\displaystyle\int_{-P_I}^{-P_I+\epsilon} e^{-G(E,P)V/kT} dP}{\displaystyle\int_{-\infty}^{-P_I+\epsilon} e^{-G(E,P)V/kT} dP} \tag{2.103}$$

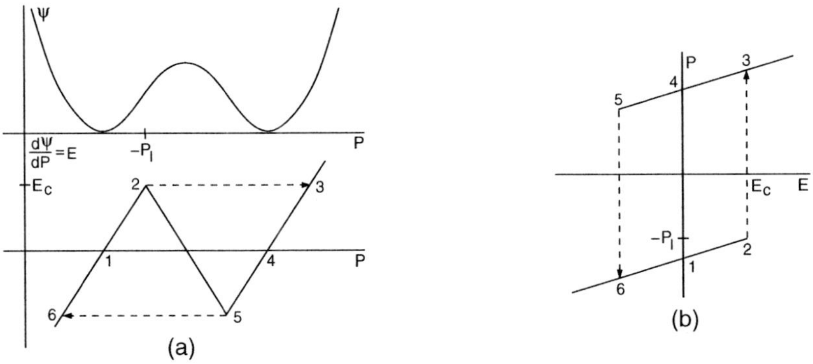

Figure 2.32. *(a) Helmholtz energy ψ given by (2.87) and derivative $\frac{d\psi}{dP}$. The necessary condition $\frac{\partial G}{\partial P} = 0 \Leftrightarrow E = \frac{d\psi}{dP}$ produces an irreversible jump at $-P_I$ as E is increased through E_c. (b) Corresponding E-P relation.*

where ϵ is taken to be a small positive constant. The quotient of integrals is a probability and hence is unitless. The relaxation time $\mathcal{T}$ is the reciprocal of the frequency at which dipoles attempt to switch so $\frac{1}{\mathcal{T}}$ has units of $\frac{1}{s}$. This yields the correct units in the differential equations (2.100) and (2.101). Moreover, we note that $\mathcal{T}^2$ is considered to be inversely proportional to the relative thermal energy so that $\mathcal{T}(T) = \mathcal{T}_1\sqrt{V/kT}$; hence increased temperatures lead to increased thermal relaxation behavior. For materials having a single relaxation time, $\mathcal{T}_1$ is constant whereas for the relaxor ferroelectric compounds discussed in Chapter 3, it is distributed which requires identification techniques analogous to those discussed in Section 2.6.6 to estimate densities for the coercive and interaction fields E_c and E_I.

In Section 2.6.3, we rigorously demonstrate that the local average polarization $\overline{P}$ given by (2.99) converges to the piecewise linear relation (2.89) or (2.90) in the limit $kT/V \to 0$ of decreasing relative thermal energy.

Remark 2.6.1. *We note that in practice, one often employs the likelihood relations*

$$p_{+-} = \frac{1}{\mathcal{T}(T)} \frac{e^{-G(E,P_I)V/kT}}{\displaystyle\int_{P_I}^{\infty} e^{-G(E,P)V/kT}dP} \quad , \quad p_{-+} = \frac{1}{\mathcal{T}(T)} \frac{e^{-G(E,-P_I)V/kT}}{\displaystyle\int_{-\infty}^{-P_I} e^{-G(E,P)V/kT}dP} \qquad (2.104)$$

which can be obtained using left or right endpoint approximates to the integrals in the numerator of (2.103) and incorporating the factor of ϵ in $\mathcal{T}(T)$. Interpretation in this manner simplifies implementation and avoids the discrepancy associated with point evaluation of a continuous density.

Alternative Derivation Based on Energy Minimization

The relations for the average polarizations and transition likelihoods are fundamentally dependent on the nature of the Boltzmann relation (2.93). Here we derive this relation through the minimization of an appropriate energy functional to provide insights regarding dipole switching in the presence of thermal processes.

We make the following assumptions regarding the transition process. First, we assume that the times required to achieve thermal and mechanical equilibrium are short compared with the process of dipole equilibration as quantified by the differential equations (2.100) or (2.101). To specify equilibrium behavior for positively and negatively oriented dipoles, we consider the minimization for each orientation separately. We illustrate the process for the potential well associated with negative dipole orientations and note that analogous results hold for positive dipoles.

We again let N_- denote the number of negatively oriented dipoles and let P_- denote the polarization due to these dipoles. We also generalize the assumption from Section 2.3.2 of fixed dipole values p_0 to include the possibility of either a finite number of dipole states or a continuum of values. We let S_- denote the set of negative dipole states so that $S_- = (-\infty, P_0]$ for the continuum.

If we sum over the index set of negative dipoles in the lattice, the resulting polarization is

$$P_- = \frac{1}{V}\sum_{i=1}^{N_-} N_i p_i$$

— compare to (2.40) for the case of fixed dipole strength p_0. However, for the purpose of energy formulation and minimization, it is advantageous to sum over the set of *possible states* rather than the *index set* (see Chapter 1 of [213] for discussion of this concept for general energy states). For a finite set of polarization states, this yields the relations

$$N_- = \sum_{p\in S_-} N_p \quad , \quad P_- = \frac{1}{V}\sum_{p\in S_-} pN_p \tag{2.105}$$

where N_p is a distribution quantifying the number of dipoles having strength p. For a continuum of dipole values, N_- and P_- are given by

$$N_- = \int_{p\in S_-} N(p)dp \quad , \quad P_- = \frac{1}{V}\int_{p\in S_-} pN(p)dp. \tag{2.106}$$

We consider first the case when S_- is finite.

The Helmholtz energy associated with negatively oriented dipoles is

$$\psi = \phi + K - ST$$

where ϕ and K respectively denote internal and kinetic energies. The internal energy is taken to be

$$\phi = \sum_{p\in S_-} \phi(p)N_p$$

where $\phi(p)$ quantifies the internal energy for each dipole of strength p. Motivated by the work expression (2.144), we employ the quadratic relation

$$\phi(p) = \frac{1}{2}\eta p^2 + \eta_1 \tag{2.107}$$

for this latter quantify. To quantify the kinetic energy, we extend the microscopic relation $K = \frac{1}{2}kT$ to the mesoscopic lattice by considering the formulation

$$K = \frac{1}{2}\frac{kTN_-}{V}.$$

From (2.44), Boltzmann's law yields the entropy relation

$$S = \frac{k}{V}\ln W$$

where, in this case,

$$W = \frac{N_-!}{\prod_{p\in S_-} N_p!}$$

quantifies the number of ways to arrange N_- dipoles so that their states p have the distribution N_p — e.g., see [213, page 12]. Application of Stirling's formula (2.45) then yields the entropy relation

$$S = \frac{k}{V}N_-\ln N_- - \frac{k}{V}\sum_{p\in S_-} N_p \ln N_p$$

and subsequent Helmholtz energy formulation

$$\psi = \sum_{p\in S_-}\left(\phi(p)N_p + \frac{kT}{V}N_p\ln N_p\right) + N_-\frac{kT}{V}\left(\frac{1}{2} - \ln N_n\right) \tag{2.108}$$

with ϕ given by (2.107). We note that the relation (2.108) differs from that in (2.47) due to differing forms of ϕ and W and the inclusion of the kinetic energy K.

The dipole distribution in the absence of an applied field is determined through the minimization of (2.108) subject to the constraints (2.105). To reformulate this as an unconstrained optimization problem, we let α and γ denote Lagrange multipliers and consider the augmented functional

$$\begin{aligned}
\psi^* &= \psi + \alpha\left(N_- - \sum_{p\in S_-} N_p\right) + \gamma\left(P - \frac{1}{V}\sum_{p\in S_-} pN_p\right) \\
&= \sum_{p\in S_-}\left(\phi(p)N_p + \frac{kT}{V}N_p\ln N_p\right) + N_-\frac{kT}{V}\left(\frac{1}{2} - \ln N_-\right) \\
&\quad + \alpha\left(N_- - \sum_{p\in S_-} N_p\right) + \gamma\left(P - \frac{1}{V}\sum_{p\in S_-} pN_p\right).
\end{aligned}$$

For a fixed set of dipole states, the distribution N_p is determined by the equilibrium condition

$$\delta\psi^* = 0$$

where the δ-notation denotes the first variation with respect to N_p. Under the assumption that for a given state p, N_p is sufficiently large to permit consideration of the derivative $\frac{\partial \psi^*}{\partial N_p}$, this yields

$$N_p = \exp\left(\frac{\alpha V}{kt} - 1\right) \exp\left[-\left(\frac{\phi(p) - \gamma p/V}{kT/V}\right)\right].$$

From (2.105), it follows that

$$\frac{N_p}{N_-} = \frac{\exp\left[-\left(\frac{\phi(p)-\gamma p/V}{kT/V}\right)\right]}{\sum_{p\in S_-} \exp\left[-\left(\frac{\phi(p)-\gamma p/V}{kT/V}\right)\right]}.$$

To specify γ, we now fix N_p and consider variations in p. This yields

$$\gamma = \frac{\eta P_- V^2}{N_-}$$

so that

$$\frac{N_p}{N_-} = \frac{\exp\left[-\frac{\eta}{2}(p - p_0)^2 V/kT\right]}{\sum_{p\in S_-} \exp\left[-\frac{\eta}{2}(p - p_0)^2 V/kT\right]}. \tag{2.109}$$

Here

$$p_0 = \frac{P_- V}{N_-}$$

represents the average dipole strength since $P_- = \frac{N_- p_0}{V}$. We point out that (2.109) specifies the probability of finding a dipole of strength p_0.

The analysis when $S_- = (-\infty, P_0]$ includes a continuum of dipole values is analogous and yields the dipole fraction

$$\frac{N_p}{N_-} = \frac{\exp\left[-\frac{\eta}{2}(p - p_0)^2 V/kT\right]}{\int_{p\in S_-} \exp\left[-\frac{\eta}{2}(p - p_0)^2 V/kT\right] dp}. \tag{2.110}$$

The relation (2.110) defines a continuous probability density function which can be physically interpreted as specifying the probability of finding a negative dipole having a value between p_0 and $p_0 + dp$; that is

$$d\mu(p_0) = \frac{\exp\left[-\frac{\eta}{2}(p - p_0)^2 V/kT\right] dp}{\int_{p\in S_-} \exp\left[-\frac{\eta}{2}(p - p_0)^2 V/kT\right] dp}.$$

Hence it defines a local version of the global density (2.96) which was obtained directly from the Boltzmann relation (2.94). This indicates the manner through which the familiar expression (2.94) follows from the constrained minimization of an appropriate Helmholtz energy relation.

2.6.3 * Local Polarization Model — Limiting Behavior

The local model (2.99) incorporates thermal relaxation mechanisms by employing the Boltzmann relation (2.94) to balance the relative thermal energy kT/V and Gibbs energy $G = \psi - EP$ whereas the local model (2.89) or (2.90) was derived in the absence of thermal relaxation mechanisms simply by minimizing the Gibbs energy. We establish here the convergence of (2.99) to (2.90) in the limit $kT/V \to 0$ of increasing control volumes and hence diminishing relative thermal energies

To clarify the discussion, we consider the representative energy landscape depicted in Figure 2.31(a) — however, the analysis techniques encompass general energy configurations.

Boltzmann Distribution

Relations (2.96) and (2.97) illustrate that for quadratic internal energy relations, the Boltzmann distribution is comprised of a normal distribution with variance β^2 proportional to the relative thermal energy kT/V. To illustrate the Dirac nature of $\mu(G)$ in (2.96) as kT/V decreases, let $j = 1/\beta$ and define the sequence

$$\phi_j(P - P_{min}^-) = \begin{cases} \overline{C}(T,j)e^{-(P-P_{min}^-)^2 j^2/2} & , \quad P \leq -P_I \\ 0 & , \quad P > -P_I \, . \end{cases} \tag{2.111}$$

The sequence $\{\phi_j\}$ satisfies properties (i)-(iii) of Theorem B.1 of Appendix B.1 and hence constitutes a Dirac family. It follows immediately that

$$\begin{aligned} \lim_{kT/V \to 0} \mu(G(P)) &= \lim_{j\to\infty} \phi_j(P) \\ &= \delta(P - P_{min}^-). \end{aligned}$$

Analogous behavior is exhibited at P_{min}^+ as depicted in Figure 2.31.

Expected Polarization and Transition Likelihoods

We consider first the convergence of the expected polarization relations (2.102). For negative dipoles, we consider the Dirac sequence $\{\phi_j\}$ defined in (2.111), define the function $f(P) = P$, and consider the interval $[a, b] = [P_{min}^-, P_0]$ where P_{min}^- is defined in (2.95). Hence f is continuous on $\mathbb{R}$ and satisfies the decay property (iv) in Theorem B.1, Appendix B.1. It then follows that

$$\begin{aligned} \lim_{kT/V \to 0} \langle P_- \rangle &= \lim_{j\to\infty} \int_{-\infty}^{\infty} P\phi_j(P - P_{min}^-)dP \\ &= P_{min}^-. \end{aligned}$$

Analogous arguments can be used to demonstrate that $\langle P_+ \rangle \to P_{min}^+$ as $kT/V \to 0$.

To illustrate the convergence of the transition likelihoods for a *fixed* relaxation time $\mathcal{T}(T)$, we modify the sequence $\{\phi_j\}$ defined in (2.111) for the interval $(-\infty, -P_I + \epsilon]$. The function f is specified to be

$$f(P) = \begin{cases} 0 & , \quad P < -P_I \\ 1 & , \quad P \geq -P_I \end{cases}$$

and the interval $[a, b]$ is taken to be $[-2P_{min}^-, -P_I]$ or $[-P_I, P_0]$. Since f again satisfies (iv) in Theorem B.1, we obtain the convergence

$$\begin{aligned}\lim_{kT/V \to 0} p_{-+} &= \lim_{j \to \infty} \frac{1}{\mathcal{T}(T)} \int_{-\infty}^{\infty} f(P)\phi_j(P - P_{min}^-)\,dP \\ &= \frac{1}{\mathcal{T}(T)} \begin{cases} 0 & , \quad E < E_c \\ 1 & , \quad E \geq E_c \end{cases}\end{aligned}$$

for E_c defined by (2.91). Similar analysis for positively oriented dipoles in the considered energy landscape yields

$$\lim_{kT/V \to 0} p_{+-} = 0.$$

We let $\zeta_+ = x_+(0)$ and $\zeta_- = x_-(0)$, $\zeta_+ + \zeta_- = 1$, denote the initial dipole fractions and consider the behavior of the differential equation (2.101) governing the evolution of x_+. Under the assumption that E, which is parameterized with respect to time, is increasing, we let $t = t_c$ denote the time at which $E(t) = E_c$. The solution to (2.101) in the limit $kT/V \to 0$ is then

$$x_+(t) = \begin{cases} \zeta_+ & , \quad t < t_c \\ 1 - (1 - \zeta_+)e^{-(t-t_c)/\mathcal{T}} & , \quad t \geq t_c \end{cases}$$

and the limiting local polarization is

$$[\overline{P}(E)](t) = \begin{cases} \frac{E(t)}{\eta} + (2\zeta_+ - 1)P_R & , \quad t < t_c \\ \frac{E(t)}{\eta} + \left[1 - 2(1 - \zeta_+)e^{-(t-t_c)/\mathcal{T}}\right] P_R & , \quad t \geq t_c. \end{cases} \tag{2.112}$$

For large t or small $\mathcal{T}$, $\overline{P}(E)$ limits to

$$\begin{aligned}[\overline{P}(E)](t) &= \begin{cases} \frac{E(t)}{\eta} + (2\zeta_+ - 1)P_R & , \quad t < t_c \\ \frac{E(t)}{\eta} + P_R & , \quad t \geq t_c \end{cases} \\ &= \begin{cases} \zeta_-[P_{min}^-(E)](t) + \zeta_+[P_{min}^+(E)](t) & , \quad t < t_c \\ [P_{min}^+(E)](t) & , \quad t \geq t_c \end{cases}\end{aligned}$$

which is precisely (2.90).

Remark 2.6.2. *To summarize, the linear kernel (2.89) or (2.90) can be accurately employed when thermal effects are negligible (kT/V is small) and relaxation times $\mathcal{T}$ are small compared with drive frequencies. Otherwise, one should employ the kernel (2.99) or an asymptotic relation of the form (2.112) to accommodate thermal activation or long relaxation times.*

2.6.4 Homogenized Macroscopic Polarization Model

The mesoscopic hysteresis models developed in Sections 2.6.1 and 2.6.2 were derived under the assumption of *homogeneous* and *isotropic* material properties throughout the representative lattice volume V. Under this assumption, the parameters α and a in the polarization kernel (2.88) and E_c, η, P_R in the piecewise linear kernel (2.89) or (2.90) are constant and uniform throughout V. Analogously, the parameters P_I, P_0 and $\mathcal{T}$ in the relaxation model (2.99) will be spatially invariant.

For isotropic, homogeneous, single crystal compounds, these local relations can be extended throughout the material to provide bulk or macroscopic constitutive relations. To illustrate, consider the macroscopic models which result from the kernels (2.89) or (2.90) depicted in Figure 2.30. Both will adequately characterize the single crystal $BaTiO_3$ behavior depicted in Figure 2.8(a) and reported in [351].

However, these local relations do not adequately quantify the gradual transitions and pre-remanent switching typical for polycrystalline compounds as illustrated in Figure 2.8(b). This mollification is due to a wide range of physical mechanisms including material, stress and field nonhomogeneities, nonuniform lattice variations across grain boundaries, and stress and crystalline anisotropies. These effects can in theory be incorporated directly into energy relations through micromechanical analysis [86, 87, 166, 303, 395] as summarized in Section 2.7. However, this yields models whose complexity precludes fast material characterization, transducer design, or model-based control implementation. Alternatively, one can employ the local models as kernels from which macroscopic models are derived either through *homogenization techniques* or the determination of *bulk effective parameters* through stochastic or empirical means. We accomplish this by assuming that certain parameters are manifestations of underlying distributions rather than constants as posited for homogeneous compounds. *Stochastic homogenization in this manner produces macroscopic models which retain energy characteristics but are sufficiently low-order to permit implementation.*

Macroscopic Model — General Densities

We consider first the incorporation of lattice variations due to material or stress nonhomogeneities, impurities, grain boundaries or polycrystallinity. As illustrated in Figure 2.33, this produces variations in the Helmholtz or Gibbs energies which in turn yield a distribution of hysterons or hysteresis kernels. To incorporate this variability, we first consider the local coercive field $E_c = \eta(P_R - P_I)$ specified in (2.91) to be the manifestation of an underlying distribution rather than a constant. We denote the associated density by ν_1.

Secondly, we consider variations in the effective field at the lattice level. As noted in [12, 187, 345, 438, 443], the applied field E is augmented by an interaction field E_I due to neighboring dipoles as well as certain electromechanical interactions. While microelectric energy analysis can quantify some of these mechanisms, the required complexity of subsequent models precludes transducer design or real-time control implementation. Alternatively, we make the assumption that effective fields $E_e = E + E_I$ are distributed about the applied field E with an underlying density ν_2

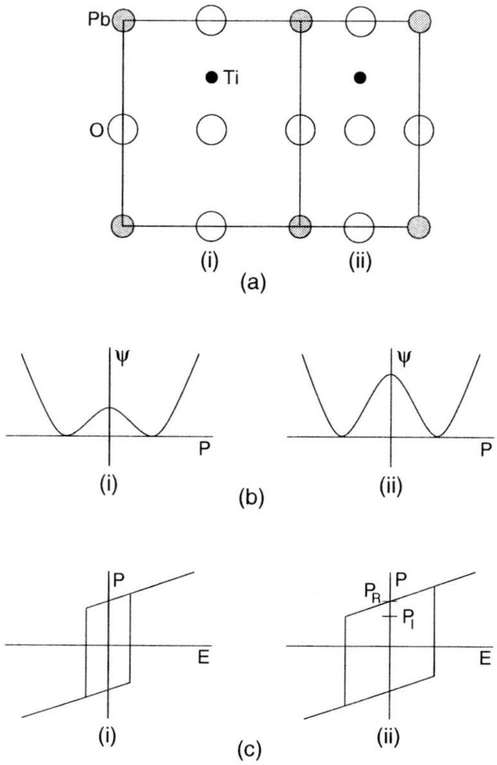

Figure 2.33. *(a) Nonuniform PZT lattice due to polycrystallinity or material nonhomogeneities. (b) Helmholtz energies associated with lattice structures (i) and (ii). (c) Variations in hysteresis kernels due to differing energy profiles.*

characteristic of the ferroelectric material under consideration. Note that this is equivalent to considering a density ν_2 for the interaction field E_I.

To accommodate physical criteria, we assume that $\nu_1 \geq 0$ and $\nu_2 \geq 0$ satisfy the conditions

$$
\begin{aligned}
&\text{(i)} \quad \nu_1(x) \text{ defined for } x > 0, \\
&\text{(ii)} \quad \nu_2(-x) = \nu_2(x), \\
&\text{(iii)} \quad |\nu_1(x)| \leq c_1 e^{-a_1 x}, \\
&\qquad\quad |\nu_2(x)| \leq c_2 e^{-a_2 |x|}
\end{aligned}
\tag{2.113}
$$

for positive c_1, a_1, c_2, a_2. The restricted domain in (i) reflects the fact that the coercive field E_c is positive whereas the symmetry enforced in the interaction field through (ii) yields the symmetry observed in low-field Rayleigh loops. Hypothesis (iii) incorporates the physical observation that the coercive and interaction fields decay as a function of distance and guarantees that integration against the piecewise linear kernel yields finite polarization values. Typical behavior exemplified by ν_1 and ν_2 is illustrated in Figure 2.34.

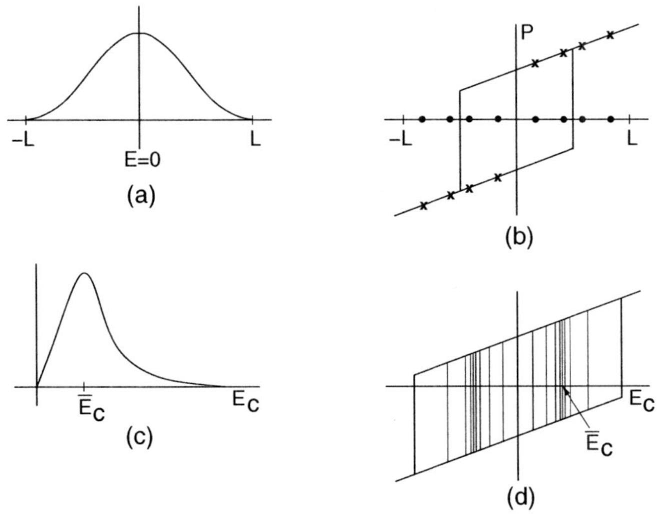

Figure 2.34. *(a) Symmetry and decay exhibited by ν_2 and truncated domain $[-L, L]$. (b) Gaussian quadrature points • and initial local polarization values ξ_j (indicated by* **x***) for $N_j = 8$. (c) Density ν_1 having mean $\overline{E}_c$ and (d) distribution of hysteresis kernels having coercive fields E_c.*

The general polarization model can then be formulated as

$$\begin{aligned}[P(E)](t) &= \int_0^\infty \int_{-\infty}^\infty \nu_1(E_c)\nu_2(E_I)[\overline{P}(E + E_I; E_c, \xi)](t)\, dE_I\, dE_c \\ &= \int_0^\infty \int_{-\infty}^\infty \nu(E_c, E_I)[\overline{P}(E + E_I; E_c, \xi)](t)\, dE_I\, dE_c\end{aligned} \tag{2.114}$$

where $\overline{P}$ is specified in (2.89), (2.90) or (2.99) and ξ denotes the initial distribution of dipoles.[12] Formulation in terms of the joint density ν is more general whereas retention of the components ν_1 and ν_2 can facilitate subsequent implementation.

Remark 2.6.3. *The densities ν_1, ν_2 and ν are unscaled in the sense that*

$$\int_0^\infty \int_{-\infty}^\infty \nu(E_c, E_I)\, dE_I dE_c = C$$

where C is a scaling constant. We retain the convention of employing unscaled densities to remain consistent with previous Preisach notation. This serves as a prelude to Section 4.7.10 in which we illustrate that this framework provides an energy basis for certain extended Preisach models.

[12]The notation $\overline{P}(E + E_I; E_c, \xi)](t)$ and $[P(E)](t)$ facilitates analysis of the operators and is consistent with certain Preisach notation. This can be interpreted as $\overline{P}(E(t) + E_I; E_c, \xi)$ and $P(E(t))$ to emphasize that the time-dependence occurs in E rather than in the structure of the kernel.

Discretized Macroscopic Model — General Densities

To implement the model (2.114), it is necessary to approximate the integrals. This can be accomplished either by employing Gaussian quadrature routines constructed for infinite or semi-infinite domains or by exploiting the decay dictated by Assumption (iii) of (2.113) to truncate the domains and employ Gauss-Legendre quadrature rules in the manner illustrated in Figure 2.34. Both strategies yield discrete models of the form

$$\begin{aligned}[P(E)](t) &= \sum_{i=1}^{N_i}\sum_{j=1}^{N_j} \nu_1(E_{c_i})\nu_2(E_{I_j})[\overline{P}(E_{I_j}+E;E_{c_i},\xi_j)](t)v_i w_j \\ &= \sum_{i=1}^{N_i}\sum_{j=1}^{N_j} \nu(E_{c_i},E_{I_j})[\overline{P}(E_{I_j}+E;E_{c_i},\xi_j)](t)v_i w_j\end{aligned} \tag{2.115}$$

where E_{I_j}, E_{c_i} denote the abscissas associated with respective quadrature formulae and v_i, w_j are the respective weights — e.g., see Section 8.1. A representative distribution of initial dipole orientations $\{\xi_j\}$ is depicted in Figure 2.34(b).

We illustrate here a composite 4-point Gauss-Legendre quadrature rule for the truncated domains. Consider first the approximation of the interaction field integral. For a given threshold ϵ and index N_q, we let $[-L, L]$ denote the interval where $|\nu_2(E_I)| \geq \epsilon$ and consider the partition $h_q = -L + qh, h = 2L/N_q$. On each subinterval $[h_{q-1}, h_q]$, the quadrature points and weights are defined to be

$$\begin{aligned}
E_{I_{q1}} &= h_{q-1} + h\left[\tfrac{1}{2} - \tfrac{\sqrt{15+2\sqrt{30}}}{2\sqrt{35}}\right] \quad , \quad & w_{q1} &= \tfrac{49h}{12(18+\sqrt{30})} \\
E_{I_{q2}} &= h_{q-1} + h\left[\tfrac{1}{2} - \tfrac{\sqrt{15-2\sqrt{30}}}{2\sqrt{35}}\right] \quad , \quad & w_{q2} &= \tfrac{49h}{12(18-\sqrt{30})} \\
E_{I_{q3}} &= h_{q-1} + h\left[\tfrac{1}{2} + \tfrac{\sqrt{15-2\sqrt{30}}}{2\sqrt{35}}\right] \quad , \quad & w_{q3} &= \tfrac{49h}{12(18-\sqrt{30})} \\
E_{I_{q4}} &= h_{q-1} + h\left[\tfrac{1}{2} + \tfrac{\sqrt{15+2\sqrt{30}}}{2\sqrt{35}}\right] \quad , \quad & w_{q4} &= \tfrac{49h}{12(18+\sqrt{30})}.
\end{aligned}$$

Similar analysis yields the quadrature points $E_{c_{pk}}$ and weights v_{pk} for the approximation of the coercive field integral on a grid with N_p subintervals. In this case, $N_i = 4N_p$, $N_j = 4N_q$, and (2.115) has the form

$$[P(E)](t) = \sum_{p=1}^{N_p}\sum_{k=1}^{4}\left(\sum_{q=1}^{N_q}\sum_{\ell=1}^{4}[\overline{P}(E_{I_{q\ell}}+E;E_{c_{pk}},\xi_{q\ell})](t)\nu_2(E_{I_{q\ell}})w_{q\ell}\right)\nu_1(E_{c_{pk}})v_{pk}.$$

The construction of the discretized model (2.115) in terms of ν_1 and ν_2 requires identification of the $N_i + N_j + 1$ parameters $\{q_\ell\}$, where

$$q = [\eta, \nu_1(E_{c_1}), \ldots, \nu_1(E_{c_{N_i}}), \nu_2(E_{I_1}), \ldots, \nu_2(E_{I_{N_j}})]. \tag{2.116}$$

If one employs the joint density ν, discretization yields $N_i \cdot N_j + 1$ parameters since one must identify all of the values $\nu(E_{c_i}, E_{I_j})$ in this formulation. Since one often requires N_i and N_j on the order of 20 to 80 to achieve convergence, the identification of the general densities using either the isolated densities ν_1 and ν_2, or the joint density ν, requires highly efficient optimization and parameter estimation techniques. This will be discussed in Section 2.6.6.

Macroscopic Model — Lognormal and Normal Densities

An alternative to identifying the general densities ν_1 and ν_2, or the joint density $\nu = \nu_1 \cdot \nu_2$, is to make *a priori* choices for ν_1 and ν_2 which satisfy the assumptions in (2.113). Motivated by densities employed in Preisach models for magnetic compounds, one such choice is

$$\begin{aligned} \nu_1(E_c) &= c_1 e^{-[\ln(E_c/\overline{E}_c)/2c]^2} \\ \nu_2(E_I) &= c_2 e^{-E_I^2/2b^2}. \end{aligned} \tag{2.117}$$

As noted in Remark 2.6.3, the densities are unscaled in the sense that $c_1 \cdot c_2 = C$.

It is illustrated in [126] that if c is small compared with $\overline{E}_c$, the norm and variance for the lognormal distribution have the approximate values

$$\langle E_c \rangle \approx \overline{E}_c \quad , \quad \sigma \approx 2\overline{E}_c c. \tag{2.118}$$

These relations can be employed to obtain initial parameter estimates which reflect properties of the measured data. The coercive field E_c for the data provides an initial estimate for $\overline{E}_c$. The magnitude of c quantifies the variability observed at coercivity. Materials such as single crystal $BaTiO_3$ which exhibit steep transitions — as illustrated in Figure 2.8(a) — have small variability in ν_1 and hence smaller values of c than the polycrystalline behavior depicted in Figure 2.8(b) which is characterized by a gradual transition at coercivity.

The magnitude of the variance b^2 in the normal interaction field distribution dictates the degree to which switching occurs prior to the remanence polarization. Large values of b^2 produce models with significant switching as the applied field E is reduced to zero with correspondingly large changes in the slope of the hysteresis curve whereas small values of b are employed when modeling the hysteretic response of materials exhibiting nearly linear E-P behavior at remanence. Hence the model for the single crystal $BaTiO_3$ data depicted in Figure 2.8(a) would have a smaller value of b^2 than that employed when characterizing the polycrystalline behavior in Figure 2.8(b).

Macroscopic Model — Comparison of General and A Priori Density Choices

(i) The formulation employing the *a priori* choices (2.117) requires identification of the 5 parameters $\eta, \overline{E}_c, c, b$ and $C = c_1 \cdot c_2$, all of which have at least qualitative physical interpretations. The identification of general parameters ν_1 and ν_2 requires estimation of the $N_i + N_j + 1$ coefficients $\{q_\ell\}$ specified in (2.116) whereas estimation of ν requires $\nu_1 \cdot \nu_2 + 1$ coefficients — which, with the exception of η are *nonphysical*. Hence the model employing the functions (2.117) is significantly more efficient to construct and update than that formulated in terms of general densities.

(ii) The general formulation (2.115) exhibits a *linear* dependence on parameters whereas use of the lognormal densities yields a *nonlinear* parameterization. This has significant ramifications for adaptive estimation and control design where the overwhelming majority of theory and algorithms require *linear* parameterizations.

(iii) As illustrated in Section 2.6.10, the *a priori* assumption (2.117) can limit the model's accuracy for certain models and drive regimes. The general model is highly accurate if ν_1 and ν_2 are estimated using a comprehensive set of major and minor loop data.

(iv) Whereas the general model is more expensive to construct, it is illustrated in Section 2.6.5 that once parameters have been identified, the two formulations require identical overhead to implement since the quadrature limits dictating matrix dimensions are the same. *Hence they will be equally efficient for model-based control design.*

2.6.5 Polarization Model Implementation

The implementation of the discretized polarization model (2.115) using the piecewise linear kernel $\overline{P}$ requires highly efficient evaluation of the conditional definition (2.90) specifying the orientation of dipoles. While it is algorithmically straightforward to implement these conditions using the transition times $\tau(t)$ defined in (2.92), this must be done for all quadrature points E_{I_j} and E_{c_i} for each input field value. Implementation in this manner significantly diminishes the speed of the model evaluation and would likely prohibit the use of the model for real-time control design and implementation.

Implementation Algorithm — Nonlinear Parameterization

An alternative is to employ the relation (2.89),

$$\overline{P}(E) = \frac{1}{\eta}E + P_R\delta(E; E_c, E_I),$$

where $\delta = 1$ if dipoles are positively oriented and $\delta = -1$ if they are negative. The crux of the algorithm focuses on the efficient construction of a corresponding matrix formulation which specifies this dependence in terms of the quadrature points E_{c_i} and E_{I_j}.

Intuitively, the local polarization values associated with each interaction field value E_{I_j} will jump when they reach a coercive field E_{c_i}. Because both the interaction and coercive field values are distributed in the manner depicted in Figure 2.34, this leads to $N_i \times N_j$ relations which must be checked for each input field value E. Hence for the distributed algorithm, Δ is an $N_i \times N_j$ matrix in which the ijth element specifies whether the jth interaction field value E_{I_j} has crossed the ith coercive value E_{c_i} to determine whether the associated polarization value is on the upper or lower branch of the hysteron. The high efficiency of the algorithm is achieved by employing algebraic matrix operations to construct Δ rather than conditional statements implemented through `if-then` constructs.

We first define the matrices

$$\Delta_{init} = \begin{bmatrix} -1 & \cdots & -1 & 1 & \cdots & 1 \\ \vdots & & \vdots & \vdots & & \vdots \\ -1 & \cdots & -1 & 1 & \cdots & 1 \end{bmatrix}_{N_i \times N_j}, \quad \mathcal{E}_c = \begin{bmatrix} E_{c_1} & \cdots & E_{c_1} \\ \vdots & & \vdots \\ E_{c_{N_i}} & \cdots & E_{c_{N_i}} \end{bmatrix}_{N_i \times N_j}$$

$$\mathcal{E}_k = \begin{bmatrix} E_k + E_{I_1} & \cdots & E_k + E_{I_{N_j}} \\ \vdots & & \vdots \\ E_k + E_{I_1} & \cdots & E_k + E_{I_{N_j}} \end{bmatrix}_{N_i \times N_j} \tag{2.119}$$

and weight vectors

$$V^T = \left[v_1 \nu_1(E_{c_1}), \; \cdots, v_{N_i} \nu_1(E_{c_{N_i}}) \right]_{1 \times N_i}$$

$$W^T = \left[w_1 \nu_2(E_{I_1}), \; \cdots, w_{N_j} \nu_2(E_{I_{N_j}}) \right]_{1 \times N_j}$$

where $E_k = E(t_k)$ is the kth value of the input field. The polarization $P_k \approx P(E_k)$ is specified by Algorithm 2.6.4.

Algorithm 2.6.4.

$\Delta_{prev} = \Delta_{init}$

for $k = 1 : N_k$

$\quad \Delta = \text{sgn}(\mathcal{E}_k + \mathcal{E}_c .* \Delta_{prev})$

$\quad \overline{P} = \frac{1}{\eta}\mathcal{E}_k + P_R \Delta$

$\quad P_k = V^T \overline{P} W$

$\quad \Delta_{prev} = \Delta$

end

In this algorithm, $.*$ indicates componentwise matrix multiplication and sgn denotes the signum function. The first step in the `for`-loop updates Δ by incorporating the status of the previous coercive field switch.

Depending on the methods used for programming, the use of Algorithm 2.6.4 rather than utilizing conditional `if-then` constructs can reduce runtimes by factors in excess of 100 so that full multiloop model simulations run in the order of seconds on a workstation. This level of efficiency is necessary to achieve real-time implementation of control algorithms utilizing the model.

Implementation Algorithm — Linear Parameterization

The formulation (2.115) with $\overline{P}$ given by (2.90) and separate coercive and interaction field densities ν_1 and ν_2 is highly efficient when implemented using Algorithm 2.6.4 but depends *nonlinearly* on the $N_i + N_j + 1$ parameters $\{q_\ell\}$ given by (2.116). This limits the utility of the formulation for certain parameter estimation techniques, including adaptive identification, and linear adaptive control designs. Alternatively, formulation in terms of the joint density ν yields a system which depends *linearly* on the $N_i \cdot N_j$ parameters $\{\nu(E_{c_i}, E_{I_j})\}$. This permits the implementation of linear least squares, adaptive identification, and control techniques but comes at the cost of significantly increased dimensionality — thus illustrating a manifestation of the *Law of Conservation of Difficulty.*

Consider the discretized problem

$$[P(E)](t) = \sum_{i=1}^{N_i} \sum_{j=1}^{N_j} \nu(E_{c_i}, E_{I_j})[\overline{P}(E + E_{I_j}; E_{c_i}, \xi)](t) v_i w_j \tag{2.120}$$

where $\nu : \mathbb{R}^{N_i \cdot N_j} \to \mathbb{R}$. To formulate (2.120) as a linear system, we define the $N_i \times N_j$ matrices $A(E)$ and Φ to have components

$$[A(E)]_{ij} = \left[\frac{E + E_{I_j}}{\eta} + P_R \delta(E; E_{c_i}, E_{I_j}) \right] v_i w_j$$

$$[Q]_{ij} = \nu(E_{c_i}, E_{I_j}).$$

For $N = N_i \cdot N_j$, we define the $N \times 1$ vector q and $1 \times N$ vector $a(E)$ by

$$q = \text{vec}(Q) \quad , \quad a(E) = [\text{vec}(A(E))]^T$$

where '`vec`' denotes the vector concatenation of the respective matrices. The discretized polarization model (2.120) can then be formulated as the *linear* system

$$P(E) = a(E)q. \tag{2.121}$$

We note that η is considered known and fixed in this formulation and is incorporated in $a(E)$. We also point out that for implementation purposes, $A(E)$ is constructed in the manner used to construct the matrix $\overline{P}$ in Algorithm 2.6.4.

2.6.6 Parameter Estimation

Infinite-Dimensional Problem

Consider the polarization model (2.114) formulated in terms of the joint density $\nu = \nu_1 \cdot \nu_2$. Assumption (iii) of (2.113) dictates exponential decay criteria in accordance with physical observations and guarantees that integration against the piecewise linear kernel yields finite polarization values. To define the parameter estimation problem on a compact domain, we define the region

$$\Omega_2 = \{(E_c, E_I) \in \mathbb{R}_+ \times \mathbb{R} \mid \nu(E_c, E_I) \geq \epsilon\} \tag{2.122}$$

and consider parameters $q = \nu$ in the parameter space

$$\mathcal{Q} = L^2(\Omega_2).$$

We consider polarization data $\widehat{P}$ corresponding to inputs $\widehat{E} \in L^2(E_{min}, E_{max})$ so the observation operator $\mathcal{C}$ is defined to be

$$\mathcal{C}P = P(\widehat{E})$$

on the observation space

$$\mathcal{Y} = L^2(P_{min}, P_{max}).$$

The parameter-to-observation map $\mathcal{K} : \mathcal{Q} \to \mathcal{Y}$ is then defined to be

$$\mathcal{K}q = \mathcal{C} \iint_{\Omega_2} \overline{P}(\cdot + E_I, E_c) q(E_c, E_I) \, dE_I dE_c.$$

In this abstract framework, the parameter estimation problem can be formulated as follows: find $q \in \mathcal{Q}$ so that

$$\mathcal{K}q = \widehat{P}. \tag{2.123}$$

It is proven in Appendix B.2 that $\mathcal{K}$ is a compact operator having infinite-dimensional range which in turn implies that the Moore–Penrose inverse $\mathcal{K}^\dagger$ is discontinuous. This has significant ramifications regarding the well-posedness of (2.123).

We first note that (2.123) has a classical solution if and only if $\widehat{P} \in \mathcal{R}(\mathcal{K})$, where $\mathcal{R}(\mathcal{K})$ denotes the range of $\mathcal{K}$, which, in general, will not be true. Instead it is more reasonable to consider the least squares problem

$$\min_{q \in \mathcal{Q}} T(q) \; , \; T(q) = \frac{1}{2} \|\mathcal{K}q - \widehat{P}\|^2_{\mathcal{Y}}. \tag{2.124}$$

The following theorem from [196] provides equivalent characterizations of least squares solutions to general operator relations of the form (2.123).

Theorem 2.6.5. *Let Π denote the orthogonal projection of $\mathcal{Y}$ onto $\overline{\mathcal{R}(\mathcal{K})}$. Then the following conditions are equivalent:*

(i) q is a least squares solution to $\mathcal{K}q = \widehat{P}$;

(ii) $\mathcal{K}q = \Pi \widehat{P}$;

(iii) $\mathcal{K}^*\mathcal{K}q = \mathcal{K}^*\widehat{P}$.

From Theorem 2.6.5, it follows that a least squares solution to (2.123) exists if and only if $\Pi\widehat{P} \in \mathcal{R}(\mathcal{K})$. With the definition

$$\mathcal{W} = \{\widehat{P} \in \mathcal{Y} \mid \Pi\widehat{P} \in \mathcal{R}(\mathcal{K})\},$$

the Moore–Penrose generalized inverse $\mathcal{K}^\dagger : \mathcal{W} \to \mathcal{Q}$ assigns to each $\widehat{P} \in \mathcal{W}$ the least squares solution to (2.123) — that is,

$$q = \mathcal{K}^\dagger \widehat{P} \text{ minimizes } \frac{1}{2}\|\mathcal{K}q - \widehat{P}\|_{\mathcal{Y}}^2.$$

This result is analogous to that for finite-dimensional systems [89, page 188].

To quantify the ill-posedness induced by discontinuous inverses $\mathcal{K}^\dagger$, we employ the following definition and theorem from [196].

Definition 2.6.6. *Let $\mathcal{Q}$ and $\mathcal{Y}$ be Banach spaces with $\mathcal{K} : \mathcal{Q} \to \mathcal{Y}$. The problem $\mathcal{K}q = \widehat{P}$ is well-posed in the sense of Hadamard if the following conditions hold:*

(i) for all $\widehat{P} \in \mathcal{Y}$, there exists a solution $q \in \mathcal{Q}$ such that $\mathcal{K}q = \widehat{P}$;

(ii) the solution q is unique;

(iii) the solution q depends continuously on the data.

The problem is said to be ill-posed if it is not well-posed.

Theorem 2.6.7. *Let $\mathcal{K} : \mathcal{Q} \to \mathcal{Y}$ be a compact operator. Then $\mathcal{K}^\dagger$ is continuous if and only if $dim\mathcal{R}(\mathcal{K}) < \infty$.*

Consider now the well-posedness of (2.123). We first note that there will be no solution if $\widehat{P} \notin \mathcal{R}(\mathcal{K})$. Of even more concern is the fact that the discontinuous behavior of $\mathcal{K}^\dagger$ can augment small perturbations in the data $\widehat{P}$ into arbitrarily large perturbations in solutions $q \in \mathcal{Q}$; in this case, even the least squares formulation (2.124) will be ill-posed. This motivates consideration of *regularization* techniques to accurately approximate $q = \nu$.

Consider the augmented functional

$$T_\alpha(q) = \frac{1}{2}\|\mathcal{K}q - \widehat{P}\|_{\mathcal{Y}}^2 + \alpha\mathcal{J}(q) \tag{2.125}$$

and the regularized least squares minimization problem

$$\min_{q\in\mathcal{Q}} T_\alpha(q). \tag{2.126}$$

The regularization parameter $\alpha > 0$ controls the tradeoff between goodness of fit to the data and stability whereas the penalty functional $\mathcal{J}$ provides stability and allows the inclusion of *a priori* information regarding the parameter q. One choice for $\mathcal{J}$ is the Tikhonov functional which we illustrate in the context of the discretized problem.

As detailed in [150, 415, 499], the assumption that $\mathcal{K}$ is weakly continuous, in combination with the specification that $\mathcal{J} : \mathcal{Q} \to \mathbb{R}$ is weakly semicontinuous and convex, guarantees that there exists a solution q_α to (2.126). Hence the regularized least squares formulation is well-posed.

Finite-Dimensional Problem

The minimization problems (2.124) and (2.126) involve infinite-dimensional parameter spaces which precludes numerical implementation. To approximate the density $q = \nu$, we first consider the least squares problem corresponding to the discretized model (2.120) formulated in terms of the joint density ν. We do not consider the convergence of approximate parameters as discretization levels are increased but instead let the previous infinite-dimensional analysis motivate potential sources of ill-posedness in the discrete least squares formulations.

Linear Parameterization

To estimate the discretized parameter $\nu : \mathbb{R}^{N_i \cdot N_j} \to \mathbb{R}$, we consider a least squares fit to data $\{(\widehat{E}_k, \widehat{P}_k)\}, k = 1, \ldots, N_d$. The accuracy of resulting models will be improved if the data is chosen to include the drive regimes under consideration and, in general, inclusion of a highly varied set of drive regimes will provide more comprehensive characterization of the densities. For example, identification of the densities solely based on symmetric major loop data will provide a model which has moderate accuracy when predicting biased minor loops whereas identification using data that includes some biased drive levels will provide a model with improved accuracy in these regimes.

To formulate the least squares functional, we modify the linearly parameterized system (2.121) to reflect measured data. We define the $N_i \times N_j$ matrices

$$[A_k]_{ij} = \left[\frac{\widehat{E}_k + E_{I_j}}{\eta} + P_R \delta(\widehat{E}_k; E_{c_i}, E_{I_j}) \right] v_i w_j$$

$$[Q]_{ij} = \nu(E_{c_i}, E_{I_j})$$

and vector concatenations

$$q = \text{vec}(Q) \quad , \quad a_k = [\text{vec}(A_k)]^T$$

so that q and a_k are respectively $1 \times N$ and $N \times 1$ where $N = N_i \cdot N_j$. Additionally, the $N_d \times 1$ vectors $\mathcal{P}$ and $\widehat{\mathcal{P}}$ are defined componentwise by

$$[\mathcal{P}]_k = P(\widehat{E}_k; q) \quad , \quad [\widehat{\mathcal{P}}]_k = \widehat{P}_k \tag{2.127}$$

and the $N_d \times N$ matrix A is defined row-wise by

$$[A]_k = a_k.$$

The polarization model (2.120) can then be formulated as the linearly parameterized system

$$\mathcal{P}(\widehat{E}_k) = Aq.$$

The least squares problem used to estimate $q = \nu \in Q = \mathbb{R}^{N_i \cdot N_j}$ given measurements $\{(\widehat{E}_k, \widehat{P}_k)\}, k = 1, \ldots, N_d$ is the following:

$$\begin{aligned} &\min_{q \in Q} T(q) \ , \ T(q) = \frac{1}{2} \|Aq - \widehat{\mathcal{P}}\|^2 \\ &\text{subject to } q_\ell \geq 0, \ \ell = 1, \ldots, N. \end{aligned} \tag{2.128}$$

Here $\|\cdot\|$ denotes the Euclidean norm in $\mathbb{R}^N$. A solution technique for (2.128) based on reformulating the problem as a quadratic programming problem in terms of a singular value decomposition can be found in [435, 437].

It was proven in the last section that the integral operator is compact which can produce a discontinuous dependence on data in the least squares formulation of the inverse problem. It is illustrated in the validation examples of Section 2.6.10 that this ill-posedness is also manifested in the finite dimensional least squares formulations (2.128) as the discretization limits $N = N_i \cdot N_j$ are increased. This necessitates the consideration of regularized least squares formulations analogous to (2.126).

Consider the augmented functional

$$T_\alpha(q) = \frac{1}{2}\|Aq - \widehat{\mathcal{P}}\|^2 + \alpha J(q)$$

where the penalty functional J is chosen to incorporate physical or regularity properties of q. One common choice for J is the Tikhonov functional

$$J(q) = \frac{1}{2}\|q\|^2$$

which enforces stability by penalizing oscillatory parameter behavior. As detailed in [499], the Tikhonov functional can be interpreted as shifting the spectrum of $A^T A$ to $A^T A + \alpha I$ to avoid the deleterious effects of small singular values. The constrained minimization problem in this case is

$$\begin{aligned} &\min_{q \in Q} T_\alpha(q)\ ,\ T_\alpha(q) = \frac{1}{2}\|Aq - \widehat{\mathcal{P}}_k\|^2 + \frac{\alpha}{2}\|q\|^2 \\ &\text{subject to } q_\ell \geq 0,\ \ell = 1, \ldots, N. \end{aligned} \tag{2.129}$$

Techniques for choosing α to avoid oversmoothing solutions as well as a solution algorithm for (2.129) can be found in Vogel [499]. The reader is also referred to [499] for details regarding the construction of alternative functionals such as the Total Variation (TV) functional.

Nonlinear Parameterization

The regularized least squares problem (2.129) has the advantage of a linear parameterization but with that comes the disadvantage of requiring $N = N_i \cdot N_j$ parameters. For discretization limits of $N_i = N_j = 48$ to 80, this can lead to prohibitively large optimization problems. Alternatively, one can retain the nonlinear parameterization (2.115) formulated in terms of the constituent densities ν_1 and ν_2 which reduces the dimensionality of the minization problem to $N_i + N_j + 1$.

In this case, we use the definition (2.116) for q, so that $Q = \mathbb{R}^{N_i+N_j+1}$, and let $P(\widehat{E}_k; q)$ denote the parameter-dependent solutions to (2.115) as computed using Algorithm 2.6.4. With the definitions (2.127), the constrained nonlinear least squares formulation is

$$\min_{q \in Q} T(q)\ ,\ T(q) = \frac{1}{2}\|\mathcal{P} - \widehat{\mathcal{P}}\|^2 \qquad (2.130)$$

$$\text{subject to } q_\ell \geq 0$$

and the regularized minimization problem can be formulated as

$$\min_{q \in Q} T_\alpha(q)\ ,\ T(q) = \frac{1}{2}\|\mathcal{P} - \widehat{\mathcal{P}}\|^2 + \frac{\alpha}{2}\|q\|^2 \qquad (2.131)$$

$$\text{subject to } q_\ell \geq 0.$$

For moderate values of N_i and N_j — e.g., $N_i = N_j = 80$ — minimization can be performed using the Matlab routine `fmincon.m` or other algorithms which enforce the positivity constraint. While these formulations significantly reduce the problem dimensionality, they come at the cost of nonlinear parameterizations.

Finally, the *a priori* function choices

$$\nu_1(E_c) = c_1 e^{-[\ln(E_c/\overline{E}_c)/2c]^2}$$

$$\nu_2(E_I) = c_2 e^{-E_I^2/2b^2},$$

given by (2.117), *significantly* simplify the minimization problem. In this case,

$$q = (\eta, \overline{E}_c, c, b, C)$$

where properties of the measured data provide initial values for $\eta, \overline{E}_c$ and C and qualitative information regarding b and c. This greatly facilitates model construction and updating but can lead to reduced accuracy for certain materials and operating regimes. These properties are illustrated in Section 2.6.10.

2.6.7 Inverse Model

One strategy for model-based control design in systems with hysteretic or nonlinear transducers is to construct an inverse which is employed as a filter before the actuator in the manner depicted in Figure 2.35. This linearizes the transducer input to the plant and reduces the degree to which control algorithms must attenuate unmodeled or nonlinear transducer dynamics.

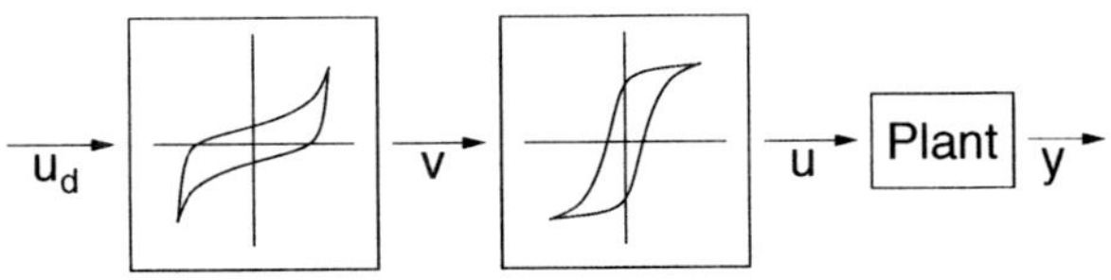

Figure 2.35. *Use of an inverse filter to linearize the response of a nonlinear and hysteretic actuator.*

For the discretized polarization model (2.115), this entails the construction of an inverse relation which specifies E as a function of P and hence gives $P^{-1}(E)$. Due to the highly nonlinear nature of both the limiting expression (2.90) for $\overline{P}$ as well as (2.99) which incorporates frequency and thermal relaxation mechanisms, direct inversion is not feasible. Instead, we rely on the monotonicity inherent to the E-P relation in combination with highly efficient forward algorithms to construct robust inverse algorithms feasible for real-time implementation.

Consider first the piecewise linear kernel $\overline{P}$ given by (2.90) and Algorithm 2.6.4 used to implement the forward model. The manner through which a field value E is determined for a given polarization P is outlined in Algorithm 2.6.8 and detailed in Algorithm D.1.1 of Appendix D. For given values of P_{k-1} and P_k, forward Algorithm 2.6.4 is used to increment the field until the predicted polarization P_{tmp} has advanced beyond the specified value P_k. At this point, the final field value E_k is determined through linear interpolation between the final two predicted field values. As detailed in Appendix D, the stepsize ΔE is adaptively updated when implementing the algorithm to ensure that the efficiency of the inverse algorithm is close to that of the forward algorithm.

Algorithm 2.6.8.

for $k = 2 : N_k$

  Specify ΔE

  $dP = P_k - P_{k-1}$

  $E_{tmp} = E_{k-1}\,,\; P_{tmp} = P_{k-1}$

  while $\text{sgn}(dP) \cdot (P_k - P_{tmp}) >= 0$

    $E_{tmp} = E_{tmp} + \Delta E$

    P_{tmp} given by Algorithm 2.6.4

  end

  E_k given by linear interpolation

end

The flexibility and robustness provided by the inverse Algorithm 2.6.8 or Algorithm D.1.1 and forward Algorithm 2.6.4 are illustrated in Figure 2.36. The polarization plotted in Figure 2.36(c) is employed as input to Algorithm 2.6.8 to yield the P-E relation plotted in Figure 2.36(a). At each time step, the resulting field value is then employed as input to Algorithm 2.6.4 to yield the E-P curve shown in Figure 2.36(b). These output polarization values P_{out} are compared with inputs P_{in} in Figure 2.36(c) and the absolute errors $|P_{in} - P_{out}|$ are plotted in (d).

From these results, a number of conclusions can be drawn. (i) We first note that the model and its inverse provide the capability for characterizing a wide range of symmetric and biased minor loop behavior. This will be further illustrated in Section 2.6.10 through comparison with PZT5H data. (ii) The composition of the inverse and model in the manner depicted in Figure 2.35 can effectively linearize

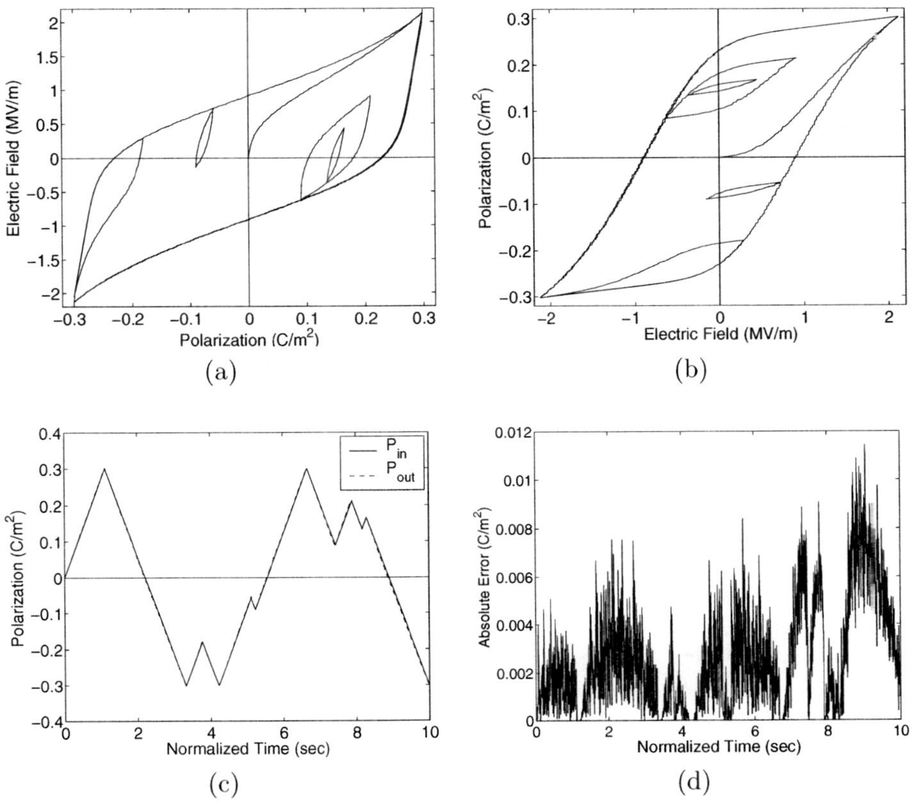

Figure 2.36. *(a) Inverse relation P_{in}-E_{out} given by Algorithm 2.6.8. (b) Forward relation E_{out}-P_{out} from Algorithm 2.6.4. (c) Comparison between P_{in} and P_{out}. (d) Absolute error $|P_{in} - P_{out}|$ for complete inversion process depicted in Figure 2.35.*

the nonlinear transducer behavior with the numerical accuracy $|P_{in} - P_{out}|$ limited only by dP. While the accuracy in a physical system will be diminished due to modeling error, linearization in this manner can significantly improve control authority since less control effort is focused on unmodeled or nonlinear dynamics. This forms the crux of various linear control designs [358,359]. (iii) Although faster implementation algorithms can be constructed for the inversion process, the algorithm described here is highly robust and avoids the potential for losing track of the memory incorporated in the model. Furthermore, the use of adaptive stepsizes ΔE ensures that Algorithm 2.6.8 is approximately a factor of two slower than forward Algorithm 2.6.4 which is reasonable for physical implementation. (iv) Whereas Algorithm 2.6.8 employs the limiting piecewise linear kernel $\overline{P}$ given by (2.90), analogous algorithms can be employed for the more general kernel (2.99) which incorporate thermal relaxation and additional dynamic effects.

2.6.8 Thermal Evolution

To quantify changes in temperature due to convection and conduction to surrounding media, Joule heating, and dipole switching, we employ a balance of energy to obtain the evolution equation

$$\bar{c}\mathcal{M}\dot{T}(t) = -h_c\Omega[T - T_E(t)] - \frac{\lambda\Omega}{\ell}[T - T_E(t)] + J(t) - [h_+\dot{x}_+ + h_-\dot{x}_-]. \quad (2.132)$$

Here $\bar{c}, \mathcal{M}, h_c, \Omega, \lambda$ and ℓ respectively denote the specific heat for the material, the mass of the actuator, a heat transfer coefficient, the surface area of the PZT, the thermal conductivity of the surrounding medium, and the interval over which conduction occurs [224]. The first term on the right hand side of (2.132) quantifies heat exchange due to convection whereas the second term incorporates potential heat loss or sources due to conduction. The term $J(t)$ characterizes temperature changes due to Joule heating. For certain operating regimes, this can be quantified by the relation

$$J(t) = \bar{\rho}^e h \frac{I^2(t)}{\mathcal{A}} \quad (2.133)$$

where $\bar{\rho}^e, h$ and $\mathcal{A}$ respectively denote the average electrical resistivity, thickness, and cross-sectional area of the actuator [417]. To incorporate additional geometric, electromagnetic, or frequency-dependent effects, one can quantify $J(t)$ either through experimental measurements or more comprehensive models. We note that the incorporation of Joule heating mechanisms becomes increasingly important in applications requiring high drive frequencies.

The final component of (2.132) quantifies heat transduction due to dipole switching so that h_+ and h_- are analogous to the specific enthalpies in the corresponding SMA relation. This relation also becomes increasingly significant for high frequency transduction. In general, validation experiments will be required to establish regimes when this latter contribution should be retained as well as operating conditions where it can be considered negligible. Finally, we point out that $\bar{c}$ and $\bar{\rho}_e$ are considered as averages and the specification of phase-dependent components to these coefficients may be required when quantifying the temperature changes which occur during phase transitions.

2.6.9 Electromechanical Constitutive Relations

The relations (2.114) and (2.115) characterize the hysteresis and nonlinear E-P behavior exhibited by a number of ferroelectric compounds for fixed stress conditions. To provide electromechanical constitutive relations, it is necessary to incorporate elastic and electromechanical energy terms into the Helmholtz and Gibbs energy relations. As in Section 2.3, we make the assumption that stresses have amplitude less than the coercive stress σ_c so that no ferroelastic switching occurs.

From (2.52) and (2.34), the electromechanical Helmholtz and Gibbs energy relations are taken to be

$$\psi(\varepsilon, P) = \psi(P) + \frac{1}{2}Y^P\varepsilon^2 - a_1\varepsilon P - a_2\varepsilon P^2$$

and

$$G(\sigma, E, \varepsilon, P) = \psi(\varepsilon, P) - EP - \sigma\varepsilon \qquad (2.134)$$

where $\psi(P)$ is given by (2.87). As before, Y^P denotes the Young's modulus for the material and a_1, a_2 are electromechanical coupling coefficients.

For regimes in which thermal activation is significant, the local polarization $\overline{P}$ is specified by (2.99) with the Gibbs energy relation (2.134) employed in the integrals (2.102) and (2.103). For the limiting case of negligible thermal activation, which is determined through solution of $\frac{\partial G}{\partial P} = 0$, $\overline{P}$ is given by

$$[\overline{P}(E, \varepsilon; E_c, \xi)](t) = \begin{cases} [\overline{P}(E, \varepsilon; E_c, \xi)](0) & , \ \tau(t) = \emptyset \\ \frac{E}{\eta - 2a_2\varepsilon} - \frac{P_R\eta - a_1\varepsilon}{\eta - 2a_2\varepsilon} & , \ \tau(t) \neq \emptyset \text{ and } E(\max \tau(t)) = -E_c \\ \frac{E}{\eta - 2a_2\varepsilon} + \frac{P_R\eta + a_1\varepsilon}{\eta - 2a_2\varepsilon} & , \ \tau(t) \neq \emptyset \text{ and } E(\max \tau(t)) = E_c \end{cases}$$

where $E_c = \eta(P_R - P_I)$, τ denotes the set of switching times (2.92), and

$$[\overline{P}(E, \varepsilon; E_c, \xi)](0) = \begin{cases} \frac{E}{\eta - 2a_2\varepsilon} - \frac{P_R\eta - a_1\varepsilon}{\eta - 2a_2\varepsilon} & , \ E(0) \leq -E_c \\ \xi & , \ -E_c < E(0) < E_c \\ \frac{E}{\eta - 2a_2\varepsilon} + \frac{P_R\eta + a_1\varepsilon}{\eta - 2a_2\varepsilon} & , \ E(0) \geq E_c \, . \end{cases}$$

Enforcement of the necessary condition $\frac{\partial G}{\partial \varepsilon} = 0$ and combination with the polarization relation (2.114) yields the constitutive relations

$$\begin{aligned} \sigma &= Y^P\varepsilon - a_1 P - a_2 P^2 \\ [P(E)](t) &= \int_0^\infty \int_{-\infty}^\infty \nu_1(E_c)\nu_2(E_I)[\overline{P}(E + E_I; E_c, \xi)](t) \, dE_I \, dE_c. \end{aligned} \qquad (2.135)$$

We note that the elastic constitutive relation incorporates both the linear piezoelectric effect and quadratic electrostrictive relation. Hence it is appropriate for a broad range of ferroelectric and relaxor ferroelectric materials operating below the threshold σ_c where ferroelastic switching begins to occur — see [24] for extensions to the theory which incorporate ferroelastic switching. We also point out that the constitutive relations (2.135) embody the philosophy imbued in Remark 2.2.1 of Section 2.2. A thermodynamic framework is employed to the extent possible to quantify fundamental energy relations for reversible regimes. This is subsequently augmented with theory of thermal processes and stochastic homogenization techniques to characterize the hysteretic E-P relation embodied in the general map $\mathcal{F}$ of (2.8).

2.6.10 Material Characterization

The constitutive relations (2.135) provide a means of quantifying a broad range of nonlinear and hysteretic behavior. To demonstrate attributes of the model, we consider the characterization of PZT5H in both symmetric and biased regimes with data collected at 0.2 Hz to minimize frequency effects. Additional examples illustrating properties of the polarization model for characterizing general PZT compounds can be found in [437, 450]. The use of (2.135) to characterize AFM displacements generated by the stage depicted in Figure 1.9(a) is illustrated in Section 7.3.3.

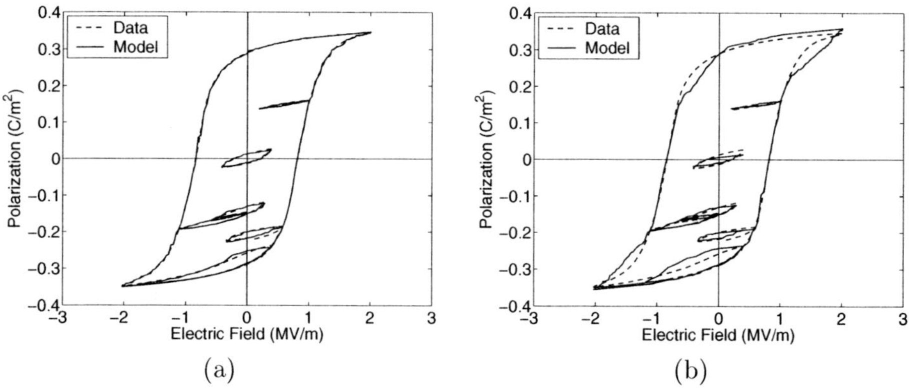

Figure 2.37. *PZT5H data and model fit with general joint density ν estimated using the unregularized functional (2.128) with data from all 7 loops: (a) $N_i = N_j = 24$ ($N = 576$) and (b) $N_i = N_j = 48$ ($N = 2304$).*

General Densities

We consider first the formulation (2.115) in terms of the joint density ν which yields a linearly parameterized system. To estimate the $N = N_i \cdot N_j$ parameters $\{\nu(E_{c_i}, E_{I_j})\}$, we employ both the unregularized least squares functional (2.128) and the regularized Tikhonov functional (2.129). The unregularized model fits obtained with $N_i = N_j = 24$ and $N_i = N_j = 48$ using data from all seven hysteresis loops are plotted in Figure 2.37 whereas those obtained using the same quadrature limits in the regularized functional are given in Figure 2.38. Without regularization, the ill-posedness associated with inversion of the compact integral operator, as quantified in Section 2.6.6, yields increasingly inaccurate model predictions as discretization limits are increased. Regularization through the inclusion of the penalty term $\frac{\alpha}{2}\|q\|^2$

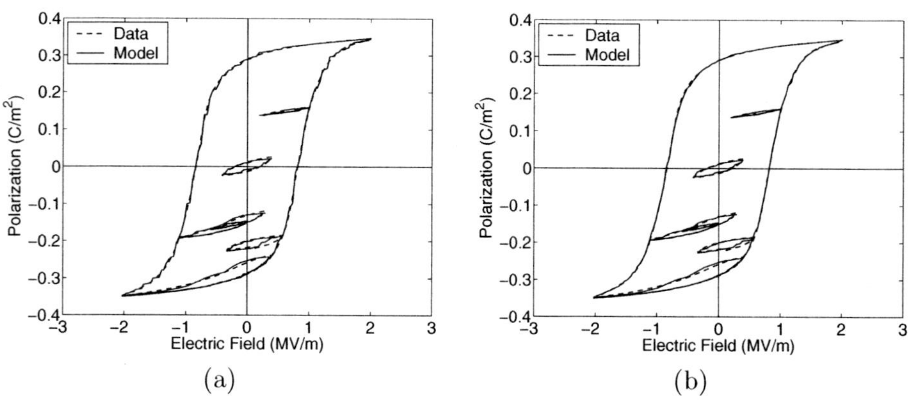

Figure 2.38. *PZT5H data and model fit with general joint density ν estimated using the regularized Tikhonov functional (2.129) with data from all 7 loops: (a) $N_i = N_j = 24$ ($N = 576$) and (b) $N_i = N_j = 48$ ($N = 2304$).*

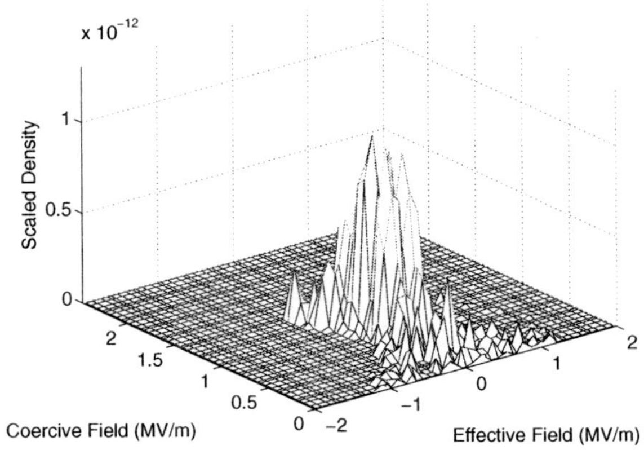

Figure 2.39. *General joint density ν obtained using Tikhonov regularization with $N_i = N_j = 48$.*

stabilizes the pseudoinverse by shifting singular values away from the origin thus yielding the highly accurate fit observed in Figure 2.38. The corresponding densities shown in Figure 2.39 demonstrate the validity of the exponential decay assumption (iii) in (2.113) and restriction of the solution operators to the compact domain Ω_2 specified in (2.122).

This formulation has the advantage of a linear parameterization but comes at the cost of high dimensionality. Alternatively, one can employ the nonlinear least squares formulations (2.130) and (2.131) to estimate the $N_i + N_j + 1$ parameters q specified in (2.116). The unregularized fit obtained using data from all seven loops is plotted in Figure 2.40(a) whereas that obtained using only symmetric and

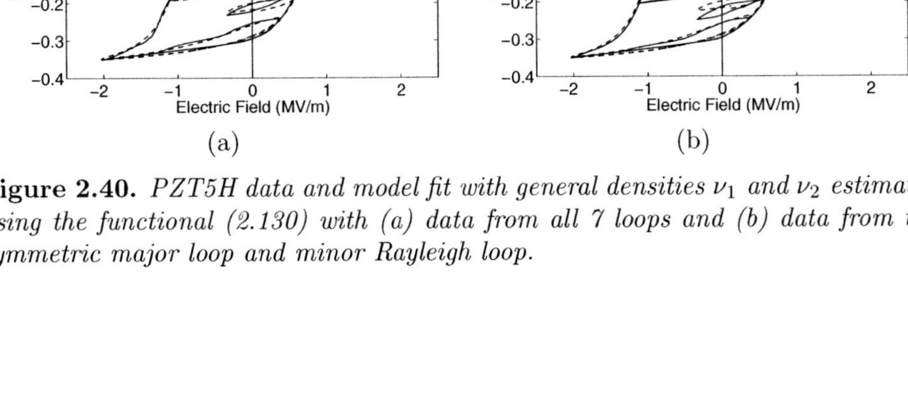

Figure 2.40. *PZT5H data and model fit with general densities ν_1 and ν_2 estimated using the functional (2.130) with (a) data from all 7 loops and (b) data from the symmetric major loop and minor Rayleigh loop.*

Rayleigh loop data is plotted in Figure 2.40(b). The fit in the first case is as accurate as that obtained with the joint density formulation but requires a significantly reduced number of parameters. Figure 2.40(b) illustrates the necessity of employing a comprehensive data set in the identification process. While the fits to the symmetric loops are highly accurate, the predictions of biased minor loop behavior are significantly less accurate than when the full data set is employed for identification.

Lognormal and Normal Densities

An alternative to the identification of general coercive and interaction field densities is the *a priori* choice of functions $\nu_1(E_c) = c_1 e^{-[\ln(E_c/\overline{E}_c)/2c]^2}$ and $\nu_2(E_I) = c_2 e^{-E_I^2/2b^2}$ which satisfy the criteria (2.113). This reduces the dimensionality of the optimization problem to $N = 5$ and permits the association of measured data attributes with properties of the parameters. In this case, the major symmetric loop coercive field $E_c = 8.15 \times 10^5 V/m$ and reciprocal slope $\frac{dE}{dP} = 2.1 \times 10^7$ after switching provide initial estimates for $\overline{E}_c$ and η. Solution of (2.130) using data from all seven loops yields the final parameter values $\overline{E}_c = 7.6 \times 10^5$ V/m, $\eta = 8.9 \times 10^6$, $c = 0.237$ V^2/m^2, $b = 1.26 \times 10^5$ V^2/m^2, $C = c_1 \cdot c_2 = 1.4 \times 10^{-12}$. The resulting model fit and densities are plotted in Figures 2.41 and 2.42. Whereas the fit to the major loop is quite accurate, a comparison with the general density fits in Figures 2.38 and 2.40(a) illustrates that this *a priori* specification of ν_1 and ν_2 leads to reduced accuracy for minor loop characterization.

Hence the efficiency gained by *a priori* density choices must be balanced against the loss of accuracy when choosing the manner through which ν_1 and ν_2, or $\nu = \nu_1 \cdot \nu_2$, are constructed for a given application.

2.6.11 Model Construction and Attributes

In the following relations, the Boltzmann probability $\mu(G)$ is specified by (2.93) and domains of integration are defined by $\chi_+ = (P_I, \infty)$, $\chi_- = (-\infty, -P_I)$, $\widehat{\chi}_+ = (P_I - \epsilon, P_I)$ and $\widehat{\chi}_- = (-P_I, -P_I + \epsilon)$. If quantifying only the E-P relation for fixed-

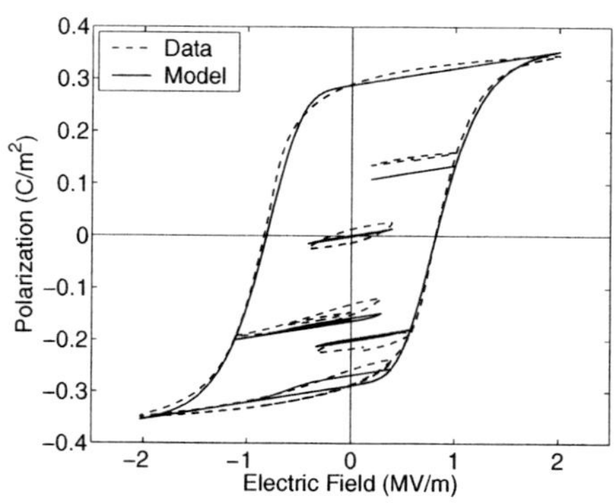

Figure 2.41. *PZT5H data and model fit with lognormal density ν_1 and normal density ν_2 estimated with data from all 7 loops.*

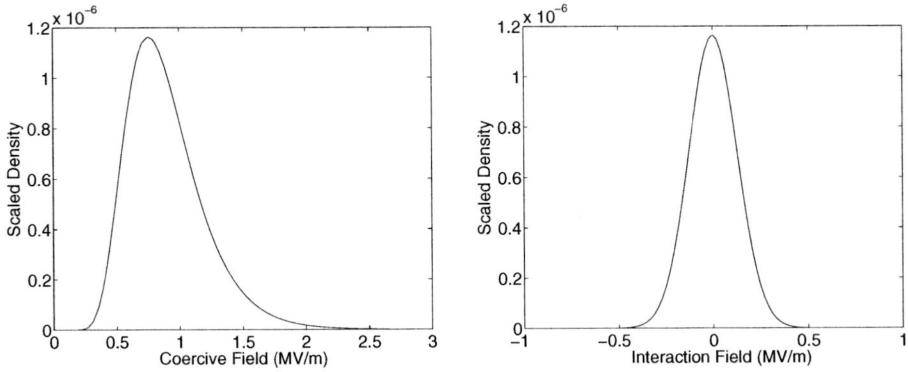

Figure 2.42. *Lognormal coercive and normal interaction field densities estimated through a fit to the full data set.*

stress conditions, one can employ $\varepsilon = 0$ in the polarization expressions. Finally, the inclusion of the strain rate term $c\dot{\varepsilon}$ in the converse constitutive relation incorporates a Kelvin–Voigt form of internal damping.

Helmholtz and Gibbs Energy Relations: [Sections 2.6.1, 2.6.9]

$$\begin{aligned} G(\sigma, E, \varepsilon, P, T) &= \psi(P, \varepsilon, T) - EP - \sigma\varepsilon \\ \psi(\varepsilon, P, T) &= \psi(P, T) + \frac{1}{2} Y^P \varepsilon^2 - a_1 \varepsilon P - a_2 \varepsilon P^2 \end{aligned}$$

where

$$\psi(P, T) = \frac{E_h P_s}{2}\left[1 - (P/P_s)^2\right] + \frac{E_h T}{2T_c}\left[P \ln\left(\frac{P + P_s}{P_s - P}\right) + P_s \ln(1 - (P/P_s)^2)\right]$$

or

$$\psi(P) = \begin{cases} \frac{1}{2}\eta(P + P_R)^2 & , \ P \le -P_I \\ \frac{1}{2}\eta(P - P_R)^2 & , \ P \ge P_I \\ \frac{1}{2}\eta(P_I - P_R)\left(\frac{P^2}{P_I} - P_R\right) & , \ |P| < P_I \end{cases}$$

Mesoscopic Model — No Relaxation Mechanisms: [Section 2.6.1]

$$[\overline{P}(E, \varepsilon; E_c, \xi)](t) = \begin{cases} [\overline{P}(E, \varepsilon; E_c, \xi)](0) & , \ \tau(t) = \emptyset \\ \frac{E}{\eta - 2a_2\varepsilon} - \frac{P_R\eta - a_1\varepsilon}{\eta - 2a_2\varepsilon} & , \ \tau(t) \neq \emptyset \text{ and } E(\max \tau(t)) = -E_c \\ \frac{E}{\eta - 2a_2\varepsilon} + \frac{P_R\eta + a_1\varepsilon}{\eta - 2a_2\varepsilon} & , \ \tau(t) \neq \emptyset \text{ and } E(\max \tau(t)) = E_c \end{cases}$$

$$[\overline{P}(E,\varepsilon;E_c,\xi)](0) = \begin{cases} \frac{E}{\eta-2a_2\varepsilon} - \frac{P_R\eta-a_1\varepsilon}{\eta-2a_2\varepsilon} & , \; E(0) \le -E_c \\ \xi & , \; -E_c < E(0) < E_c \\ \frac{E}{\eta-2a_2\varepsilon} + \frac{P_R\eta+a_1\varepsilon}{\eta-2a_2\varepsilon} & , \; E(0) \ge E_c \end{cases}$$

$$\tau(t) = \{t_s \in (0,t] \mid E(t_s) = -E_c \text{ or } E(t_s) = E_c\}$$

Mesoscopic Model — With Relaxation Mechanisms: [Section 2.6.2]

$$\begin{aligned} \overline{P} &= x_+ \langle P_+ \rangle + x_- \langle P_- \rangle \\ \langle P_\pm \rangle &= \int_{\chi_\pm} P\mu(G(E,P,T)dP \\ \dot{x}_+ &= -p_{+-}x_+ + p_{-+}(1-x_+) \\ p_\pm &= \frac{1}{\mathcal{T}(T)} \int_{\widehat{\chi}_\pm} \mu(G(E,P,T)dP \end{aligned}$$

Thermal Evolution: [Section 2.6.8]

$$\bar{c}\mathcal{M}\dot{T}(t) = -\Omega[h_c + \lambda/\ell][T - T_E(t)] + J(t) - \sum_\alpha h_\alpha \dot{x}_\alpha$$

Macroscopic Constitutive Relations: [Sections 2.6.4 and 2.6.9]

$$\begin{aligned} \sigma &= Y^P\varepsilon + c\dot{\varepsilon} - a_1 P - a_2 P^2 \\ [P(E)](t) &= \int_0^\infty \int_{-\infty}^\infty \nu_1(E_c)\nu_2(E_I)[\overline{P}(E+E_I;E_c,\xi)](t)\, dE_I\, dE_c. \end{aligned}$$

Model Attributes and Comparison with Previous Models

It was noted at the beginning of the section that the homogenized energy framework has also been employed for ferromagnetic materials (Chapter 4) and ferroelastic compounds (Chapter 5) thus providing a unified modeling framework for ferroic compounds (Chapter 6). We refer the reader to Section 4.7.9 where a detailed summary of the model attributes and its relation to previous models is provided in the context of ferromagnetic compounds.

Software

MATLAB m-files for implementing the homogenized energy model for ferroelectric compounds can be found at the website `http://www.siam.org/books/fr32`.

2.7 Ginzburg–Landau Relations

In Section 2.3.1, polynomial energy relations for ferroelectric materials were constructed by truncating series expansions for the elastic, electrostatic and coupling energies and choosing coefficients to ensure measured material behavior. For example, the Helmholtz energy relation (2.36) for materials exhibiting second-order phase transitions retains quadratic and quartic polarization terms to characterize paraelectric-ferroelectric phase transitions, and linear and quadratic electromechanical coupling terms. As illustrated in Figure 2.15(b), this provides a macroscopic model capable of characterizing hysteresis in certain single crystal compounds. However, it does not provide the capability for quantifying more complex polycrystalline behavior or ferroelastic switching mechanisms leading to the stress-dependent behavior depicted in Figures 2.11 and 2.12. These macroscopic relations also do not provide a way to quantify mesoscopic or microscopic phenomena associated with domain nucleation or growth in the presence of input fields or stresses.

We summarize here micromechanical theory which extends the polynomial energy relations from Section 2.3.1 to incorporate field and stress effects at the domain level. The resulting time-dependent Ginzburg–Landau (TDGL) relations provide a framework which can be used for fundamental material characterization and analysis but have a level of complexity which presently precludes model-based control design.

A fundamental difference between the Ginzburg–Landau relations and the Landau expressions discussed in Section 2.3.1 is the inclusion of gradient terms of the form $\frac{\partial P}{\partial x}$. This has a number of physical and mathematical ramifications. From a physical perspective, these derivative relations incorporate local behavior or gradients such as those occurring in the domain walls which delineate transition regions between domains. This observation is corroborated by the fact that the gradient energy approximates dipole-dipole interactions [298]. From a mathematical perspective, the inclusion of gradient components serves to *regularize* or *stabilize* models in the manner of Tikhonov functionals in least squares formulations — e.g., see (2.125) or Vogel [499].

The Ginzburg–Landau theory for micromechanical analysis is vast and we summarize here only certain facets to illustrate issues and capabilities provided by the framework. To simplify the discussion, we consider 2-D relations for second-order phase transitions so that polarization relations are truncated after quartic terms. We consider field $\mathbf{E}$ and stress $\boldsymbol{\sigma} = (\sigma_x, \sigma_y)$ inputs so that the resulting Helmholtz energy is formulated in terms of the polarization $\mathbf{P} = (P_x, P_y) = (P_1, P_2)$ and strains $\boldsymbol{\varepsilon} = (\varepsilon_x, \varepsilon_y)$.

Additional details regarding general micromechanical theory for ferroelectric materials can be found in [226, 285, 286, 298] whereas models and implementation techniques for specific compounds are provided in [6, 76, 102, 211, 296, 329, 330, 355]. Finally, mathematical ramifications of Ginzburg–Landau theory for ferroic compounds are discussed in [57, 502].

Free Energy Relations

We begin by considering the total free energy

$$\psi(\varepsilon, \mathbf{P}) = \int [f_p + f_e + f_c + f_g]\, d\mathbf{r} \tag{2.136}$$

where f_p, f_e, f_c and f_g respectively denote the Landau-Devonshire, elastic, coupling and gradient energy densities. The 2-D Landau-Devonshire free energy density for materials exhibiting second-order phase transitions is

$$f_p(\mathbf{P}) = \alpha_1(T - T_c)(P_x^2 + P_y^2) + \alpha_{11}(P_x^4 + P_y^4) + \alpha_{12}P_x^2 P_y^2 \tag{2.137}$$

where T_c denotes the Curie point and $\alpha_1, \alpha_{11}, \alpha_{12} > 0$. The elastic free energy density is

$$f_e(\varepsilon) = \frac{Y_{11}}{2}(\varepsilon_x^2 + \varepsilon_y^2) + Y_{12}\varepsilon_x\varepsilon_y + \frac{Y_{44}}{2}\varepsilon_{xy}^2 \tag{2.138}$$

where Y_{11}, Y_{12} and Y_{44} are elastic coefficients and ε_{xy} is the shear strain. In the usual manner, strains can be related to displacements $\mathbf{u} = (u, v)$ through the relations

$$\varepsilon_x = \frac{\partial u}{\partial x} \quad , \quad \varepsilon_y = \frac{\partial v}{\partial y} \quad , \quad \varepsilon_{xy} = \frac{\partial u}{\partial y} + \frac{\partial v}{\partial x}.$$

The electromechanical coupling energy density is taken to be

$$\begin{aligned} f_c(\varepsilon, \mathbf{P}) &= -a_{11}(\varepsilon_x P_x + \varepsilon_y P_y) - a_{12}(\varepsilon_x P_y + \varepsilon_y P_x) \\ &\quad -q_{11}(\varepsilon_x P_x^2 + \varepsilon_y P_y^2) - q_{12}(\varepsilon_x P_y^2 + \varepsilon_y P_x^2) - q_{44}\varepsilon_{xy}P_x P_y \end{aligned} \tag{2.139}$$

where a_{ij} and q_{ij} respectively denote piezoelectric and electrostrictive coefficients. Finally, the gradient energy density is expressed as

$$\begin{aligned} f_c\left(\frac{\partial P_i}{\partial x_j}\right) &= \frac{g_1}{2}\left[\left(\frac{\partial P_x}{\partial x}\right)^2 + \left(\frac{\partial P_y}{\partial y}\right)^2\right] + \frac{g_2}{2}\left[\left(\frac{\partial P_x}{\partial y}\right)^2 + \left(\frac{\partial P_y}{\partial x}\right)^2\right] \\ &\quad + g_3\left(\frac{\partial P_x}{\partial x}\right)\left(\frac{\partial P_y}{\partial y}\right). \end{aligned} \tag{2.140}$$

A comparison between the energy relations (2.137)–(2.139) and the Helmholtz energy expression (2.36) in Section 2.3.1 indicates analogous constructs with the former incorporating 2-D polarization and strain effects. The primary difference between the Helmholtz relations (2.136) and (2.36) is due to the inclusion of the gradient energy (2.140) which incorporates local interactions necessary to characterize domain and domain wall attributes.

The work due to an applied field $\mathbf{E}$ can be incorporated by employing a Gibbs total energy relation

$$G(\mathbf{E}, \varepsilon, \mathbf{P}) = \psi(\varepsilon, \mathbf{P}) + \int f_{ext}(\mathbf{E}, \mathbf{P})\, d\mathbf{r} \tag{2.141}$$

where

$$f_{ext}(\mathbf{E},\mathbf{P}) = -\mathbf{E}\cdot\mathbf{P}.$$

This is consistent with the 1-D Gibbs relation (2.34).

Time-Dependent Ginzburg–Landau Equations

Time-dependent Ginzburg–Landau equations quantifying domain properties of ferroelectric materials are derived under the assumption that elastic dynamics are fast compared with changes in the polarization and hence the elastic field equilibrates quickly to a given dipole or polarization configuration. The polarization dynamics are quantified by the nonlinear TDGL equations

$$\frac{\partial P_i}{\partial t} = -\Gamma\frac{\delta\psi}{\delta P_i} + \xi_i \ , \ i = 1,2, \tag{2.142}$$

where ψ is given by (2.136), Γ is a kinetic coefficient, and ξ_i represents Gaussian random noise. Details regarding the numerical implementation and simulation capabilities provided by (2.142) can be found in the previously cited references.

2.8 Work in the Polarization Process

We summarize here the analysis used to establish various work relations employed in the development of energy relations for ferroelectric materials. A similar analysis for ferromagnetic materials can be found in Brown [60] and additional details regarding the magnetic model are given by Iyer and Krishnaprasad [240, 496]. Note that throughout this discussion, it is assumed that field and polarization changes are sufficiently slow to permit the use of electrostatic field relations. In many applications, involving the characterization of materials, this is a reasonable assumption.

To quantify the effective field contributions, it is advantageous to consider the work required to polarize the body rather than the work necessary to achieve a specified field level. Let $\rho = \frac{dQ}{dV}$ denote a charge density in the region V as depicted in Figure 2.43. As detailed in [417, page 72], this charge configuration then produces the electric field

$$\mathbf{E} = \int_V \frac{\rho\hat{\mathbf{r}}}{4\pi\epsilon_0 r^2}\, dV$$

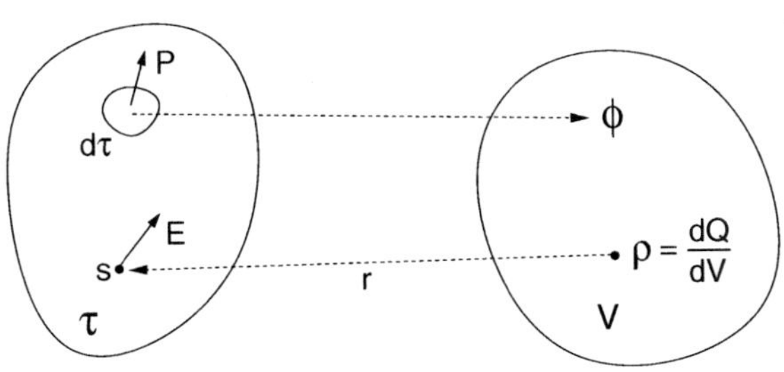

Figure 2.43. *Orientation of the body V with charge ρ. This generates a field $\mathbf{E}$ in the dielectric τ which in turn produces a potential ϕ in the charged body.*

at the point s in the dielectric material τ. Note that $\hat{\mathbf{r}} \equiv \frac{\mathbf{r}}{|\mathbf{r}|}$ and $r \equiv |\mathbf{r}|$. Furthermore, the electric potential due to the dielectric is

$$\phi = \frac{1}{4\pi\epsilon_0} \int_\tau \frac{\mathbf{P} \cdot \hat{\mathbf{r}}}{r^2} \, d\tau$$

(see [417, page 260]). Finally, the time rate of work due to the changing polarization is

$$\frac{dW}{dt} = \int_V \rho \frac{d\phi}{dt} \, dV.$$

It then follows that

$$\begin{aligned} \frac{dW}{dt} &= \int_V \rho(V) \left[\frac{1}{4\pi\epsilon_0} \frac{d}{dt} \int_\tau \frac{\mathbf{P} \cdot \hat{\mathbf{r}}}{r^2} d\tau \right] dV \\ &= \int_V \rho(V) \left[\int_\tau \frac{1}{\rho(V)} \frac{d\mathbf{P}}{\partial t} \cdot \frac{\rho(V)\hat{\mathbf{r}}}{4\pi\epsilon_0 r^2} \, d\tau \right] dV \\ &= \int_\tau \frac{d\mathbf{P}}{dt} \cdot \left[\int_V \frac{\rho(V)\hat{\mathbf{r}}}{4\pi\epsilon_0 r^2} \, dV \right] d\tau \\ &= \int_\tau \mathbf{E} \cdot \frac{d\mathbf{P}}{dt} \, d\tau \end{aligned}$$

so that

$$\delta W = \int_\tau \mathbf{E} \cdot \delta\mathbf{P} \, d\tau \, .$$

This implies that the work required to polarize a unit volume of the ferroelectric material is

$$\delta W = \mathbf{E} \cdot \delta\mathbf{P}.$$

The work required to change the polarization of a unit volume from $P = P_0$ to $P = P_1$ is subsequently

$$W = \int_{P_0}^{P_1} \mathbf{E} \cdot d\mathbf{P} \tag{2.143}$$

which yields the area of the shaded region in Figure 2.44(a) when a depolarized material is polarized to the saturation level $P = P_s$. The energy due to this work is partially stored as potential energy with the remainder dissipated as heat. Over a full cycle, the potential energy returns to its original value which implies that the energy loss is given by

$$W = \oint \mathbf{E} \cdot d\mathbf{P}. \tag{2.144}$$

As illustrated in Figure 2.44(b), *the energy dissipated by a hysteretic ferroelectric material is thus equal to the area of the hysteresis loop.*

Thus far, the field $\mathbf{E}$ has been considered to be generated by the charge configuration ρ or external source. We now consider the effective field $\mathbf{E}_e = \mathbf{E} + \mathbf{E}_1$

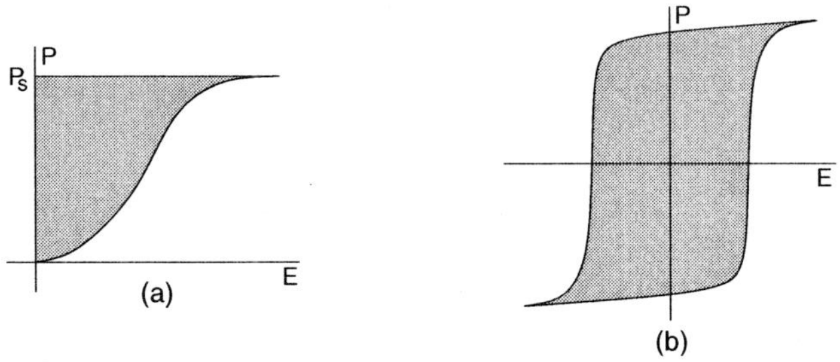

Figure 2.44. *(a) Work required to polarize to saturation a unit volume of ferroelectric material. (b) Energy dissipated during one cycle of the hysteresis loop.*

where $\mathbf{E}_1 = \alpha\mathbf{P}$ quantifies the field contributions due to the polarized dielectric body. Expansion of (2.143) yields

$$W_e = W + W_1$$

where

$$W_1 = \int_0^{\mathbf{P}} \mathbf{E}_1 \cdot d\mathbf{P}$$

denotes the electrostatic self-energy. This quantity characterizes the shape-dependent component to the energy which does not contribute to the total irreversible hysteresis losses since

$$\oint W_1 = \oint \alpha\mathbf{P} \cdot d\mathbf{P} = 0\,. \tag{2.145}$$

Hence we have some flexibility when specifying an appropriate energy functional which quantifies energy losses due to hysteresis.

From (2.145), the energy due to the effective field over a full polarization cycle can be expressed as

$$\begin{aligned} W_e &= \oint \mathbf{E} \cdot d\mathbf{P} \\ &= -\oint \mathbf{P} \cdot d\mathbf{E} \\ &= -\oint \mathbf{P} \cdot d\mathbf{E} - \alpha \oint \mathbf{P} \cdot d\mathbf{P} \\ &= -\frac{1}{\epsilon_0} \oint \mathbf{P} \cdot d\mathbf{D}_e \end{aligned}$$

for arbitrary α. These arguments can be extended to partial cycles by employing the fact that $\int_{c_1} \mathbf{P} \cdot d\mathbf{P} = 0$ for any closed curve c_1. This motivates the use of the functional

$$W = -\frac{1}{\epsilon_0} \int_0^{\mathbf{D}_e} \mathbf{P} \cdot d\mathbf{D}_e \tag{2.146}$$

when quantifying irreversible components to the polarization. We point out that (2.146) is analogous to the expressions employed when constructing domain wall models for ferromagnetic and ferroelastic materials as detailed in Sections 4.6 and 5.4.

Chapter 3

Model Development for Relaxor Ferroelectric Compounds

Relaxor ferroelectric materials operating below their freezing temperature exhibit ferroelectric behavior similar to that of $BaTiO_3$ and PZT — hence in these low temperature regimes the hysteresis models and theory developed in Chapter 2 are applicable. As temperatures are increased within their diffuse transition region, however, the materials exhibit negligible hysteresis as illustrated in Figure 3.1. When combined with their capacity to generate large strains, this makes them advantageous in applications ranging from sonar transduction to precision positioning in the Hubble

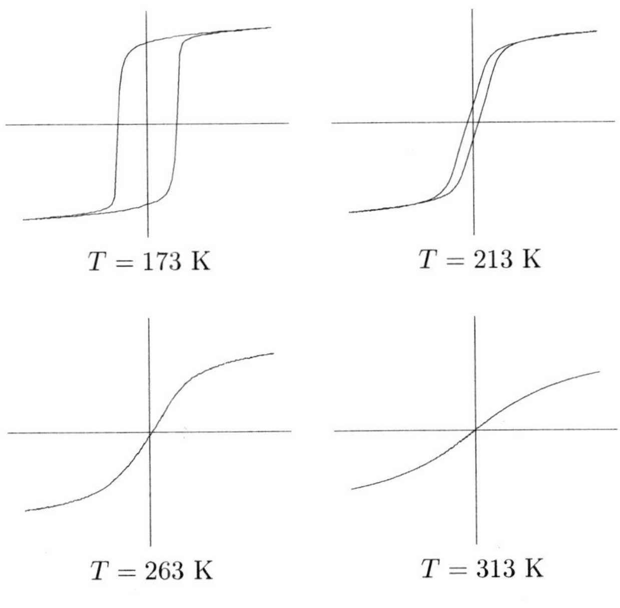

Figure 3.1. *Decreasing hysteresis and saturation polarization of PMN as a function of increasing temperature (from Glazounov, Bell and Tagantsev [186]). Abscissas: electric field (MV/m), ordinates: polarization (C/m²).*

telescope. In this chapter, we discuss the development of nonlinear models which characterize the temperature-dependent behavior of relaxor ferroelectric materials. Because the materials are ferroelectric at low temperatures, these models are equivalent to certain models developed in Chapter 2 at fixed temperatures below the freezing temperature.

As noted in Section 1.2, the low hysteresis, high strain capabilities of lead magnesium niobate derivatives (PMN-PT), have led to their inception as drivers for underwater sonar transducers where they has been demonstrated to produce at least 6 dB increases in acoustic source levels over PZT [221, 379]. In addition, the compounds' high set point accuracies make them leading candidates for a number of micro- and nanopositioning actuator designs. In the field of adaptive optics, the use of multi-layered PMN arrays for deformable mirrors has increased sensitivity from 0.3 nm/V to 12.0 nm/V relative to monolithic PZT [146, 218, 219].

Initial models, proposed by Hom and Shankar [220] and experimentally validated by Brown et al. [58], quantified the electrostrictive strains and saturation behavior of relaxor compounds for isothermal and anhysteretic (hysteresis-free) operating regimes. This theory was later extended by Glazounov, Bell and Tagantsev [186] and Hom and Shankar [222] to provide a quasistatic, temperature-dependent model for PMN which quantifies the nonlinear behavior of the material but does not characterize the hysteresis present at temperatures below the freezing temperature. We summarize the fundamental aspects of this anhysteretic theory in Section 3.2.

The hysteretic behavior present in the low-temperature ferroelectric regimes was subsequently modeled by Smith and Hom [439, 440] by extending the domain wall theory discussed in Section 2.5 to incorporate the decrease in hysteresis and saturation polarization illustrated in Figure 3.1 as a function of increasing temperature. As detailed in Section 3.3, this was accomplished by employing statistical and thermodynamic tenets to incorporate thermal dependence in material parameters quantifying the saturation polarization, density of polar regions, and energy required to translate domain walls.

Initial relaxor ferroelectric models based on the homogenized free energy theory developed in Section 2.6 are summarized in Section 3.4. The development and experimental validation of this theory for general operating regimes is under investigation and a number of open research issues remain.

3.1 Physical Properties of Relaxor Compounds

The behavior of relaxor ferroelectric materials differs from that of regular ferroelectric compounds in at least three aspects. The first is the nature of the phase transition from the paraelectric to ferroelectric state. Ferroelectric materials, such as PZT, exhibit relatively sharp transitions — even in the case of materials exhibiting first-order phase transitions — and are employed well below the transition temperature. In contrast, relaxor ferroelectrics have a diffuse transition region over a broad temperature range [109, 186, 452].

Since relaxors are typically employed at temperatures within this transition region, the material's behavior strongly depends on the temperature and frequency

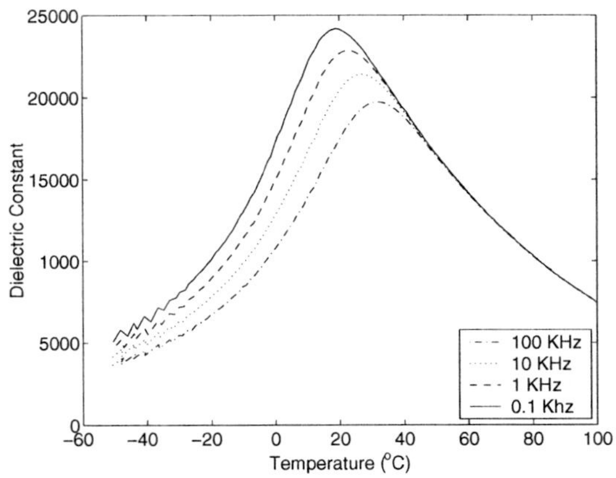

Figure 3.2. *Dielectric response of a PMN-PT-BT (10% PT, 3% BT) at frequencies ranging from* 0.1 *kHz to* 100 *Hz.*

of the applied electric load as illustrated in Figure 3.2. The transition temperature is often tailored to a specific application by mixing PMN, with lead titanate (PT), and additional minor dopants such as barium titanate (BT) and strontium titanate (ST).

Secondly, the dielectric hysteresis exhibited by relaxor ferroelectric compounds is strongly temperature-dependent as illustrated in Figure 3.1. At low temperatures, the materials exhibit significant hysteresis whereas the response is essentially anhysteretic for temperatures within the transition region. It is due to the lack of hysteresis in these latter regimes that relaxor compounds are employed in applications, such as deformable mirrors for adaptive optics, which require high accuracy.

Accommodation of the highly temperature-dependent dielectric behavior is crucial for transducer design as has been illustrated for sonar transducers. When driven at high fields and high frequency, PMN will heat due to dielectric hysteresis. Since PMN's hysteresis decreases with temperature, actuators reach an equilibrium temperature when heat dissipation due to conduction, convection and radiation balances the internal heat generated by the ceramic. The final equilibrium temperature depends on the configuration of the actuator, its electric drive cycle, and its surrounding environment. Shankar and Hom [423] predicted a 40 ^{o}C rise for a flextensional sonar transducer submersed in water using a simple heat generation model for PMN. These predictions agreed quantitatively with experimental measurements made on an actual transducer.

The third difference noted by Cross is the lack of a macroscopic phase change below the maximum Curie point T_c [109]. In combination, these properties produce transition behavior which is significantly more complex than the first and second-order ferroelectric phase transitions described in Section 2.3.1.

The behavior of these materials is typically attributed to their heterogeneous nature. To illustrate, we consider the structure of the prototypical compound lead magnesium niobate, $Pb(Mg_{1/3},Nb_{2/3})O_3$ (PMN), which is one of the most commonly employed relaxors. As illustrated in Figure 3.3, PMN has an $A(B',B'')O_3$

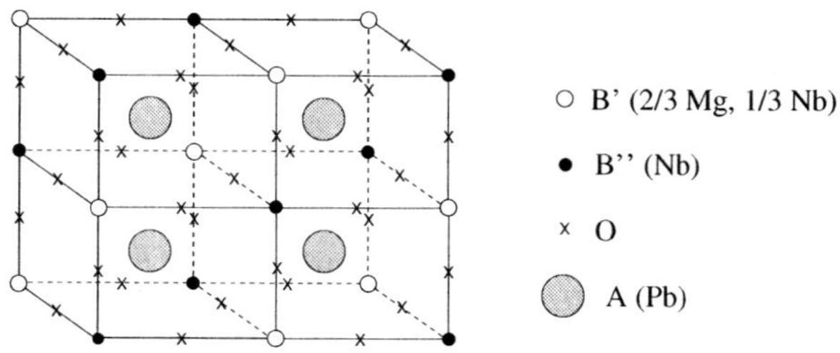

Figure 3.3. *Atomic structure of a perovskite relaxor ferroelectric A(B',B")O_3. The B' and B" sites indicate the cation ordering for the random layer model.*

perovskite structure comprised of Pb anions at A sites and a mixture of Nb^{5+} and Mg^{2+} at B sites. The transition behavior of relaxor compounds is attributed to the manner through which B' and B" sites are populated by Nb^{5+} and Mg^{2+} cations.

Based on experimental studies for $Pb(Sc_{1/2}Ta_{1/2})O_3$ (PST) reported in [416, 463], Cross in [109] made the following observations regarding that compound. Case (i): Thermally annealed specimens having a *highly ordered* population of B' and B" sites by Sc and Ta cations exhibited the following *ferroelectric* properties: sharp first-order phase transition behavior at T_c, stable remanent polarization, and no frequency-dependence at radio frequencies in the ferroelectric state. Case (ii): Quenched specimens having statistical fluctuations in the Sc, Ta population of B sites yield *superparaelectric* behavior: diffuse transition region, no stable remanence, and strong frequency dependence. The general conclusion for PMN compounds is the following: the diffuse transition is attributed to the random population of B' sites by Nb^{5+} and Mg^{2+} cations which produces local regions having different *local* Curie points.

The mechanisms proposed by Cross to explain the transition behavior of relaxor compounds are summarized in Figure 3.4. At high temperatures, the materials exhibit normal paraelectric behavior and zero macroscopic polarization. As the materials are cooled, regions in which cation distributions produce high local Curie points begin to coalesce into nanoscale polar regions (2–3 nm) having local dipole moments. Because they have low energy barriers, these regions are randomly oriented due to thermal energy so that the materials still exhibit zero macroscopic polarization and behave in a manner analogous to superparamagnetic compounds. Upon further cooling, the thermal fluctuations decrease and interaction between micropolar regions freezes the polarization in a manner similar to magnetic spin glasses. As observed by Viehland et al. [497, 498], the materials begin to deviate from the Curie–Weiss law, possess the frequency-dependence indicated in Figure 3.2, and exhibit non-Debye relaxation. As temperatures are cooled through the freezing temperature T_f, the micropolar regions coalesce into domains and a remanent polarization is observed. Further cooling produces normal ferroelectric behavior including the manifestation of hysteresis as illustrated by the PMN data in Figure 3.1.

Because the ordering of B sites plays a fundamental role in defining the transition behavior of relaxor compounds, several investigations have focused on this aspect of material characterization and have led to the proposal of two mecha-

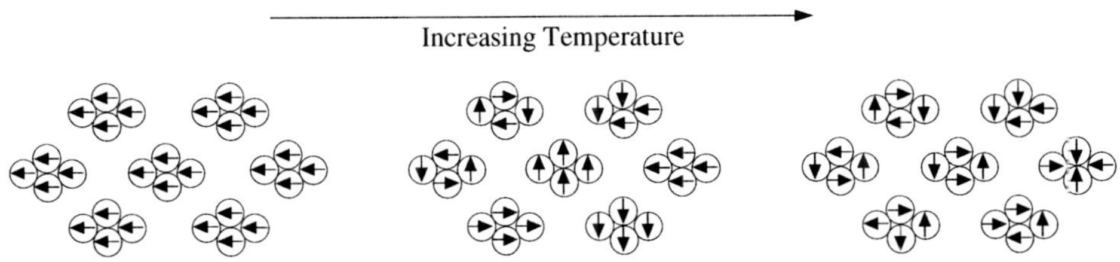

Ferroelectric: Micropolar clusters interact and material exhibits a remanent polarization.

Superparaelectric: Micropolar regions develop but have random orientation due to thermal excitation.

Paraelectric: Regions are nonpolar and material exhibits no macroscopic polarization.

Figure 3.4. *Polar mechanisms in relaxor ferroelectric materials as postulated by Cross [109]. In the superparaelectric phase, the materials exhibit a nonzero RMS dipole moment but have zero macroscopic polarization.*

nisms describing B site population. In the *space charge* model for PMN, it is assumed that ordered regions consisting of Mg exclusively occupy the B' sites whereas Nb exclusively occupies the B" sites [216, 497]. The ordered regions thus carry a net negative charge which should inhibit domain growth. In the *random layer* model [9, 84], Mg and Nb in a 2-to-1 ratio randomly populate the B' sites, while Nb exclusively occupies the B" sites. In this model, the ordered regions are charge-balanced so domains should coarsen during heat treatment. Experiments by Akbas and Davies [9] have shown that domain growth does occur during annealing in $Pb(Mg_{1/3}\,Ta_{2/3})O_3$ (PMT), which supports the *random-layer* model of the B-site cation distribution even though this mechanism has not been observed in PMN. We employ this *random-layer* mechanism in the equilibrium model summarized in Section 3.2 and the hysteresis model described in Section 3.3.

3.2 Temperature-Dependent Equilibrium Model

The PMN data plotted in Figure 3.1 illustrates that relaxor materials exhibit negligible hysteresis above the freezing temperature T_f and have a temperature-dependent saturation polarization $P_s(T)$ which decreases as temperatures are increased. The development of modeling anhysteretic relations is motivated by the results of Glazounov, Bell and Tagantsev [186] and follows the theory developed by Hom and Shankar [222] as detailed in [440]. Following a summary of the isothermal energy relations developed in Section 2.3, the equilibrium polarization model is developed in two steps. In the first, we consider the thermodynamics of the micropolar regions in order to predict the saturation polarization and density of Mg cations as a function of temperature. Secondly, we incorporate these effects into macroscopic energy relations to obtain constitutive relations quantifying the temperature-dependent anhysteretic behavior of relaxor materials.

3.2.1 Isothermal Energy Relations and Equilibrium Polarization

In Section 2.3.2, it was demonstrated that application of mean field principles to a homogeneous lattice of volume V having N cells, each with dipole moment p_0 and two possible orientations $s_i = \pm 1$, yields the Helmholtz energy relation

$$\psi = \frac{N\Phi_0}{4V}\left[1-(P/P_s)^2\right] + \frac{TkN}{2VP_s}\left[P\ln\left(\frac{P+P_s}{P_s-P}\right) + P_s\ln(1-(P/P_s)^2)\right]. \quad (3.1)$$

Here k is Boltzmann's constant, Φ_0 denotes the energy required to reorient a single dipole in a completely ordered lattice, and the saturation polarization is

$$P_s = \frac{Np_0}{V}. \quad (3.2)$$

Reformulation in terms of a bias field E_h and the *global* Curie point T_c, where

$$E_h = \frac{N\Phi_0}{2VP_s} \quad , \quad T_c = \frac{\Phi_0}{2k}, \quad (3.3)$$

and inclusion of elastic and electromechanical energy terms yields the elastic Helmholtz relation

$$\begin{aligned}\psi(\varepsilon, P, T) = & \frac{1}{2}Y^P\varepsilon^2 - a_2\varepsilon P^2 + \frac{E_h P_s}{2}\left[1-(P/P_s)^2\right] \\ & + \frac{E_h T}{2T_c}\left[P\ln\left(\frac{P+P_s}{P_s-P}\right) + P_s\ln(1-(P/P_s)^2)\right].\end{aligned} \quad (3.4)$$

A comparison between (3.4) and (2.48) reveals that we have retained only the quadratic electromechanical coupling term since relaxor materials are unpoled and exhibit quadratic, or primarily electrostrictive, P-ε relations.

As in (2.34), we incorporate the work due to applied fields and stresses through consideration of the Gibbs energy relation

$$G = \psi - EP - \sigma\varepsilon.$$

Finally, enforcement of the necessary conditions

$$\frac{\partial G}{\partial \varepsilon} = 0 \quad , \quad \frac{\partial G}{\partial P} = 0$$

yields the constitutive relations

$$\begin{aligned}\sigma &= Y^P\varepsilon - a_2 P^2 \\ E &= -2a_2\varepsilon P - \alpha P + a(T)\operatorname{arctanh}(P/P_s)\end{aligned} \quad (3.5)$$

where

$$\alpha = \frac{E_h}{P_s} \quad , \quad a(T) = \frac{E_h T}{T_c}.$$

It was illustrated in Section 2.5.1 that inversion of the direct relation yields the equilibrium polarization relation

$$P_{an}(\sigma, E) = P_s \tanh\left(\frac{E_e(E,\sigma)}{a(T)}\right).$$

The effective field is given by

$$E_e(\sigma, E) = E + (\alpha + \alpha_2\sigma)P$$

where $\alpha_2 = s^P a_2$ and s^P denotes the compliance.

As illustrated in Figure 2.19, the temperature-dependence encapsulated in this formulation yields the transition from double to single-well energy associated with a second-order ferroelectric to paraelectric phase transformation. However, the formulation does not incorporate the temperature-dependent behavior of P_s illustrated in Figure 3.1 since, as quantified in (3.2), P_s is assumed constant. To incorporate this effect, we consider statistical and thermodynamic mechanisms which lead to the formation of micropolar regions under the assumption that the distribution of Mg and Nb in B' sites is characterized by the *random layer* model.

3.2.2 Thermodynamics of Micropolar Regions

It was noted in Section 3.1 that the random layer model employs the assumption that Nb exclusively occupies B" sites whereas Mg and Nb randomly occupy B' sites while maintaining a 2-1 ratio. Since the uncertainty lies in the population of the B' sites, we focus on quantifying the distribution of Mg in these locations which provides sufficient information to determine the density of micropolar regions, the distribution of local Curie points, and the temperature-dependent behavior of P_s.

We consider a reference volume V containing N_0 Mg cations, each assumed to have dipole moment p_0 and associated dipole orientation $s_i = \pm 1$. We consider the material to be comprised of N_R micropolar regions as depicted in Figure 3.5 with the composition of regions assumed to satisfy the following properties.

(i) Each region contains N Mg cations so that $N_0 = N \cdot N_R$. The number of positively and negatively oriented dipoles within a region are denoted by N_+ and N_- so that $N = N_+ + N_-$.

(ii) The density of Mg cations X is uniform throughout each region but varies between regions. As a result of Assumption (i), the volume V_i in each region is inversely proportional to X_i.

(iii) The strength of interactions between regions is directly proportional to the number of adjoining nearest neighbors. We also assume that dipoles interact only with nearest neighbors.

(iv) Below its local transition temperature, each region acts as a single dipole and exhibits a remanent polarization.

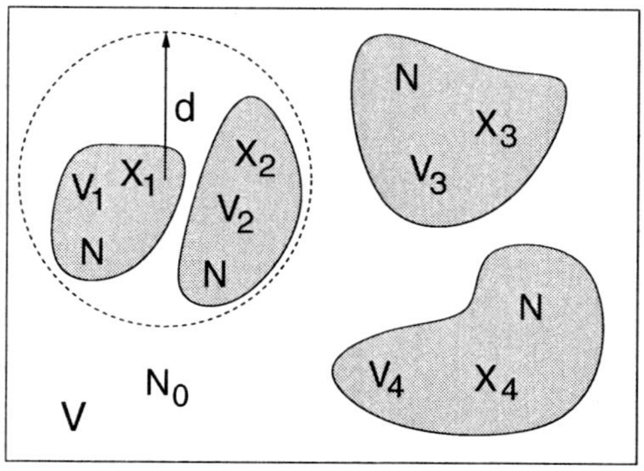

Figure 3.5. *Relaxor ferroelectric material consisting of an aggregate of micropolar regions which satisfy Assumptions (i) and (ii). In this depiction, there are $N_R = 4$ regions and a total of $N_0 = N_R N$ Mg cations in the material.*

We now define the correlation between cells comprised of Mg cations as the average

$$R = \frac{1}{N}\sum_{i=1}^{N} s_i.$$

Since $N = N_+ + N_-$ within a micropolar region, the relations

$$N_+ = \frac{N}{2}(1+R) \quad , \quad N_- = \frac{N}{2}(1-R)$$

can be used to quantify the number of positively and negatively oriented dipoles in terms of R.

The density X is defined as the number of neighboring B' sites that contain a Mg cation. As indicated by Figure 3.3, X can range from 0 to 12. Furthermore X is assumed to be uniform throughout the region in accordance with Assumption (ii).

As noted in Section 3.1, the microregions have a distribution of Curie points due to differing cation densities in the regions. To determine these densities, we note that B' sites have a ratio of 2/3 Mg cations to 1/3 Nb ions. For a Mg populated B' site, each of the 12 adjacent B' sites has a 65.4% probability of being populated by a Mg cation and 34.6% probability of being populated by Nb. The probability that the site is surrounded by X Mg cations is computed by assuming a binomial distribution

$$B(X) = \frac{12!}{X!(12-X)!}(0.654)^X(0.346)^{12-X}.$$

To obtain a continuous distribution appropriate for subsequent integration, the binomial distribution is fit with the normal distribution

$$\Pi(x) = \frac{1}{b\sqrt{2\pi}}e^{-(x-a)^2/2b^2} \tag{3.6}$$

having a mean $a = 0.654 \times 12 = 7.85$ and standard deviation $b = (0.654 \times 12 \times 0.346)^{1/2} = 1.65$. Finally, the distribution is normalized to the interval $[0, 1]$ through the introduction of the variables

$$\zeta = \frac{x}{12} \quad , \quad \bar{\zeta} = \frac{a}{12}$$

to obtain

$$\Pi(\zeta) = \frac{1}{b\sqrt{2\pi}} e^{-144(\zeta - \bar{\zeta})^2/2b^2}. \tag{3.7}$$

The distributions (3.6) and (3.7) are employed in subsequent discussion to compute ferroelectric properties of the materials which depend upon cation densities in the micropolar regions.

We now compute the change in internal energy due to dipole reorientation among the Mg cations. As in Section 2.3.2, we let Φ_0 denote the energy required to reorient dipoles if the lattice is completely ordered ($R = \pm 1$). From Assumption (iii), it follows that the energies required to switch from positive to negative, and conversely, are

$$\Phi_{+-} = \frac{x}{12} \frac{N_+}{N} \Phi_0 \quad , \quad \Phi_{-+} = \frac{x}{12} \frac{N_-}{N} \Phi_0 .$$

The change in internal energy due to dipole reorientation is computed in the same manner as (2.43) which yields the relation

$$U = \frac{\Phi_0 N \zeta}{4V} (1 - R^2) + U_0 .$$

As in Section 2.3.2, we take $U_0 = 0$ since we are interested in relative rather than absolute measures of energy.

The Helmholtz energy for the microregion is

$$\psi = U - ST$$

where, again, the entropy is

$$S = \frac{k}{V} \ln W$$

and

$$W = \begin{pmatrix} N \\ N_+ \end{pmatrix} = \frac{N!}{N_+! N_-!}$$

quantifies the number of ways dipoles can be oriented to yield the correlation R. Simplification in the manner detailed in (2.46) yields the entropy relation

$$S = \frac{kN}{V} \left[\ln(2) - \frac{1+R}{2} \ln(1+R) - \frac{1-R}{2} \ln(1-R) \right]$$

and the Helmholtz energy relation

$$\psi = \frac{\Phi_0 N \zeta}{2V} (1 - R^2) + \frac{kNT}{4V} \left[R \ln\left(\frac{1+R}{1-R} \right) - \ln(1 - R^2) \right] \tag{3.8}$$

for the region. We note that the first term in (3.8) incorporates the internal energy due to dipole ordering whereas the second incorporates disorder due to thermal excitation. Further details concerning the construction of energy relations which quantify such order-disorder phenomena can be found in [198, Section 9.6].

The local electric field E_{loc} per unit volume in the micropolar region is determined by

$$E_{loc} = \frac{\partial \psi}{\partial R} = -\frac{\zeta \Phi_0 R N}{2V} + \frac{TkN}{V}\text{arctanh}(R)\,.$$

In the absence of a local field, the equation can be expressed as

$$R = \frac{T}{\overline{T}_c}\text{arctanh}(R)$$

where

$$\overline{T}_c = \frac{\zeta \Phi_0}{2k} \tag{3.9}$$

denotes the *local Curie point* for that region — see (3.3). To motivate the definition (3.9) and this terminology, it is noted that (3.9) has a single solution for $T > \overline{T}_c$ and three solutions comprised of two stable and one unstable solution for temperatures $T < \overline{T}_c$. Hence a pitchfork bifurcation occurs at $T = \overline{T}_c$ which is consistent with a second-order phase transition.

The expression (3.9) quantifies the Curie point for a single micropolar region. The macroscopic Curie point for the material is specified as the average Curie point for the regions or the Curie point of the mean,

$$T_c = \frac{\bar{\zeta}\Phi_0}{2k}\,.$$

The behavior of each polar region can then be expressed as

$$E_{loc} = \frac{\Phi_0 N}{2V}\left[-\zeta R + \frac{\bar{\zeta} T}{T_c}\text{arctanh}(R)\right] \tag{3.10}$$

in the absence of interactions between regions.

3.2.3 Temperature-Dependent Energy Relations and Equilibrium Polarization

The Helmholtz energy expression (3.1) quantifies the balance between internal energy and entropic effects for a region having a *uniform* cation density, constant Curie point T_c specified by (3.3), and fixed saturation polarization P_s given by (3.2). In this section, we employ the relations derived in Section 3.2.2 to incorporate the heterogeneity due to random B' site occupation which in turn yields constitutive relations characterizing the temperature-dependent anhysteretic behavior indicated in Figure 3.1. Details regarding this development can be found in [222, 440].

To quantify interactions between the micropolar regions depicted in Figure 3.5, we employ Assumption (iii) and assume that strength of interactions is proportional to the number N_d of regions within an interaction distance d of a cluster. The assumption that microregions are uniformly distributed yields

$$N_d = \frac{4}{3}\pi d^3 \cdot \frac{N_R}{V}$$

where N_R/V denotes the number of regions per unit volume. We also let Φ designate the relative measure of interaction energy between regions in a manner analogous to Φ_0 which quantified interaction energy between cells or dipoles.

Under these assumptions, the Helmholtz energy quantifying the aggregate material behavior is taken to be

$$\psi = \frac{N_d\Phi}{4V}\left[1-(P/P_s)^2\right] + \frac{TkN_0}{2VP_s}\left[P\ln\left(\frac{P+P_s}{P_s-P}\right) + P_s\ln(1-(P/P_s)^2)\right] \tag{3.11}$$

using *formal* analysis similar to that employed at the lattice level to obtain (3.1). However, the balance between the internal energy and entropic effects leading to (3.11) occurs on much larger length scales then is typical for mean field approximations thus imbuing it with phenomenological properties. Recall that $N_0 = N_R N$ designates the total number of cations in the reference volume V.

In the absence of applied stresses, minimization of the Gibbs energy

$$G = \psi - EP$$

yields the constitutive relation

$$E(T) = \frac{2\pi d^3\Phi}{3V^2P_s}\left[\frac{-PN_R(T)}{P_s} + \frac{TN_R(T_f)}{T_f}\operatorname{arctanh}\left(\frac{P}{P_s}\right)\right] \tag{3.12}$$

where

$$T_f = \frac{2\pi d^3\Phi N_R(T_f)}{3kVN_0}$$

denotes the *freezing temperature*. We now employ the theory of Section 3.2.2 to quantify the temperature-dependent behavior of P_s and N_R.

The saturation polarization occurs when the dipoles in all the microregions are aligned with the field. The degree of alignment for microregions is quantified by the correlation $R(T,x)$ specified by (3.10) whereas the magnitude is dependent upon the number $n(x)$ of Mg cations having x nearest neighbors. Recalling that p_0 denotes the dipole strength of a single cell, the saturation polarization can be expressed as

$$P_s(T) = \frac{1}{V}\int_0^{12} |R(T,x)|\, p_0 n(x)dx\,.$$

Since N_0 denotes the total number of Mg cations in the volume V, we can express $n(x) = N_0\Pi(x)$. Normalization to the unit interval then yields the expression

$$P_s(T) = \widetilde{P}_s\int_0^1 |R(T,\zeta)|\,\Pi(\zeta)d\zeta\,. \tag{3.13}$$

The distribution Π is computed using (3.7) and the parameter $\widetilde{P}_s = \frac{N_0 p_0}{V}$ is estimated either directly from the data or through a least squares fit to the data. The correlation R is computed through iterative solution of the relation

$$\beta E_{loc} = -\zeta R + \frac{\bar{\zeta} T}{T_c} \text{arctanh}(R) \tag{3.14}$$

where $\beta = \overline{\frac{2V}{\Phi_0 N}}$ denotes the average degree to which local fields turn the polarization and hence change the correlation. We note that at low drive levels, this contribution is small and one can employ $\beta = 0$. For high drive levels, however, P_s is more accurately approximated by treating β as a parameter to be estimated and solving (3.14) with $E_{loc} = E_{max}$.

The number of micropolar regions $N_R(T)$, at a fixed temperature T, is computed by determining those regions with nonzero correlation R, or from (3.14), those regions in which $T < \overline{T}_c$. Since the number of regions with density X is given by $n(X)/N = N_0 \Pi(X)/N$, the total number of regions in the volume is

$$\frac{N_R(T)}{V} = \frac{N_0}{VN} \int_{\bar{\zeta} T/T_c}^{1} 12\Pi(\zeta) d\zeta \,. \tag{3.15}$$

The use of (3.13) and (3.15) in (3.12) yields the temperature-dependent constitutive relation

$$E(T) = \frac{-2\pi d^3 \Phi N_0}{3V^2 N} \frac{I_1(T)}{P_s^2(T)} P + \frac{2\pi d^3 \Phi N_0}{3V^2 N} \frac{I_2}{T_f P_s(T)} \text{arctanh}\left(\frac{P}{P_s(T)}\right) \tag{3.16}$$

where

$$I_1(T) = \int_{\bar{\zeta} T/T_c}^{1} 12\Pi(\zeta)\, d\zeta$$

$$I_2 = \int_{\bar{\zeta} T_f/T_c}^{1} 12\Pi(\zeta)\, d\zeta$$

$$P_s(T) = \widetilde{P}_s \int_0^1 12|R(T,\zeta)|\, \Pi(\zeta)\, d\zeta \,.$$

For implementation purposes, we formulate (3.16) as

$$E = \alpha(T) P + a(T) \text{arctanh}\left(\frac{P}{P_s(T)}\right)$$

with the temperature-dependent parameters

$$\alpha(T) = \widetilde{\alpha} \frac{I_1(T)}{P_s^2(T)} \,, \quad a(T) = \widetilde{a} \frac{I_2}{T_f P_s(T)} \,, \quad P_s(T) = \widetilde{P}_s \int_0^1 12|R(T,\zeta)| \Pi(\zeta)\, d\zeta .$$

Here $\widetilde{P}_s = \frac{p_0 N_0}{V}$, $\widetilde{\alpha}$ and $\widetilde{a}$ are parameters which are estimated for a given material through a least squares fit to data.

The resulting model for the equilibrium polarization is

$$P_{an}(T) = P_s(T)\tanh\left[\frac{E + \alpha(T)P_{an}(T)}{a(T)}\right]. \tag{3.17}$$

Inclusion of elastic and electromechanical energy contributions in a manner analogous to that summarized in Section 3.2.1 yields the constitutive relations

$$\begin{aligned} \sigma &= Y^P\varepsilon - a_2P^2 \\ E &= -2a_2\varepsilon P - \alpha(T)P + a(T)\text{arctanh}(P/P_s(T)). \end{aligned} \tag{3.18}$$

Comparison with (3.5) reveals that the relations (3.18) have the same general form but include temperature-dependence through the coefficients $\alpha(T), a(T)$ and $P_s(T)$.

3.3 Temperature-Dependent Domain Wall Model

The domain wall theory developed in Section 2.5 and summarized in Algorithm 2.5.1 of Section 2.5.6 quantifies hysteresis in ferroelectric materials in two steps: (i) characterization of switching mechanisms using Langevin or Ising relations, and (ii) incorporation of irreversible and reversible energy losses due to domain wall movement. The anhysteretic relation (3.17) incorporates the temperature-dependent saturation behavior of relaxor compounds in the Ising component of the domain wall model. We summarize here the modification of the irreversible polarization component (2.76) to incorporate the increase in hysteresis illustrated in Figure 3.1 as temperatures are cooled below the freezing temperature.

The increase in hysteresis observed for decreasing temperatures is manifested in the coercive field E_c which, as noted in (2.82), is asymptotically related to the pinning constant k_p; that is,

$$k_p \approx E_c.$$

As detailed in [440], the temperature-dependence of E_c can be quantified using energy analysis drawn from the theory of dislocation plasticity [264], which is similar to the approach employed by Chen and Wang [83], or simply from phenomenological arguments. Both approaches yield relations of the form

$$E_c(T) = \widetilde{k}_p(T_f - T)^p \tag{3.19}$$

where T_f is the freezing temperature at which remanence disappears, $\widetilde{k}_p$ is a material-dependent parameter, and p is a real-valued constant.

The temperature-dependent parameter $k_p(T)$ is thus specified as

$$k_p(T) = \begin{cases} \widetilde{k}_p(T_f - T)^p & , T < T_f \\ 0 & , T \geq T_f. \end{cases}$$

Note that for $T \geq T_f$, the material exhibits no hysteresis so $k_p(T) = 0$ and the irreversible component (2.76) reduces to the anhysteretic model (3.17).

3.3.1 Model Summary

The construction of the temperature-dependent domain wall hysteresis model is summarized in Algorithm 3.3.1. The performance of the model is illustrated in Section 3.3.2.

Algorithm 3.3.1.

(A) Determine Temperature-Dependent Parameters

(1) At a specified temperature, estimate $\widetilde{\alpha}, \widetilde{a}, \widetilde{k}_p, \beta, c, p$ and $\widetilde{P}_s = p_0 N_0/V$ using the asymptotic relations from [442] or a least squares fit to data. Note that for small drive level applications and moderate temperature ranges, one can often employ $\beta = 0$ and $p = 1$.

(2) Compute the temperature-dependent parameters

$$\alpha(T) = \widetilde{\alpha}\frac{I_1(T)}{P_s^2(T)} \quad , \quad P_s(T) = \widetilde{P}_s \int_0^1 12|R(T,\zeta)|\,\Pi(\zeta)\,d\zeta$$

$$a(T) = \widetilde{a}\frac{I_2}{T_f P_s(T)} \quad , \quad k_p(T) = \widetilde{k}_p(T_f - T)^p \tag{3.20}$$

where

$$I_1(T) = \int_{\bar{\zeta}T/T_c}^1 12\Pi(\zeta)\,d\zeta \quad , \quad I_2 = \int_{\bar{\zeta}T_f/T_c}^1 12\Pi(\zeta)\,d\zeta$$

$$\Pi(\zeta) = \frac{1}{b\sqrt{2\pi}}e^{-144(\zeta-\bar{\zeta})^2/2b^2} \quad , \quad \bar{\zeta} = a/12\ ,\ a = 7.85, b = 1.65$$

$$\beta E_{max} = -\zeta R + \frac{\bar{\zeta}T}{T_c}\mathrm{arctanh}(R)$$

(B) Temperature-Dependent Polarization

(1) Equilibrium Polarization

$$P_{an}(T) = P_s(T)\tanh\left[\frac{E + \alpha(T)P_{irr}(T)}{a(T)}\right]$$

(2) Irreversible Polarization

$$\frac{dP_{irr}}{dE}(T) = \widetilde{\delta}\frac{P_{an}(T) - P_{irr}(T)}{k_p\delta - \alpha\left[P_{an}(T) - P_{irr}(T)\right]}$$

(3) Reversible Polarization

$$P_{rev}(T) = c[P_{an}(T) - P_{irr}(T)]$$

(4) Total Polarization

$$P(T) = P_{irr}(T) + P_{rev}(T)$$

3.3.2 Material Characterization

To illustrate the performance of the model, we consider the characterization of PMN-PT-BT and PMN measured in quasistatic regimes. For both data sets, a single set of parameters was employed throughout the full temperature range and the temperature and electric field provided the sole inputs to the model. Additional details regarding these examples can be found in [440].

PMN-PT-BT Characterization

We illustrate first the characterization of the PMN-PT-BT data plotted in Figure 3.6 for temperatures ranging from $T = 263$ K to $T = 313$ K. The data was collected from a stress-free sample ($\sigma = 0$) at 1 Hz to minimize frequency effects.

The freezing temperature was specified as $T_f = 288$ K, the Curie point was taken to be $T_c = 313$ K, and the parameter values $\widetilde{\alpha} = 1.5 \times 10^7$ Vm/C, $\widetilde{a} = 7.65 \times 10^7$ C/m, $\widetilde{k}_p = 4.0 \times 10^3$ V/m, $\widetilde{P}_s = 0.53$ C/m^2, $\beta = 7 \times 10^{-9}, c = .3$ and $p = 1$ were estimated through a least squares fit to the data. The model fit provided by these parameter values, with temperature-dependence incorporated through the

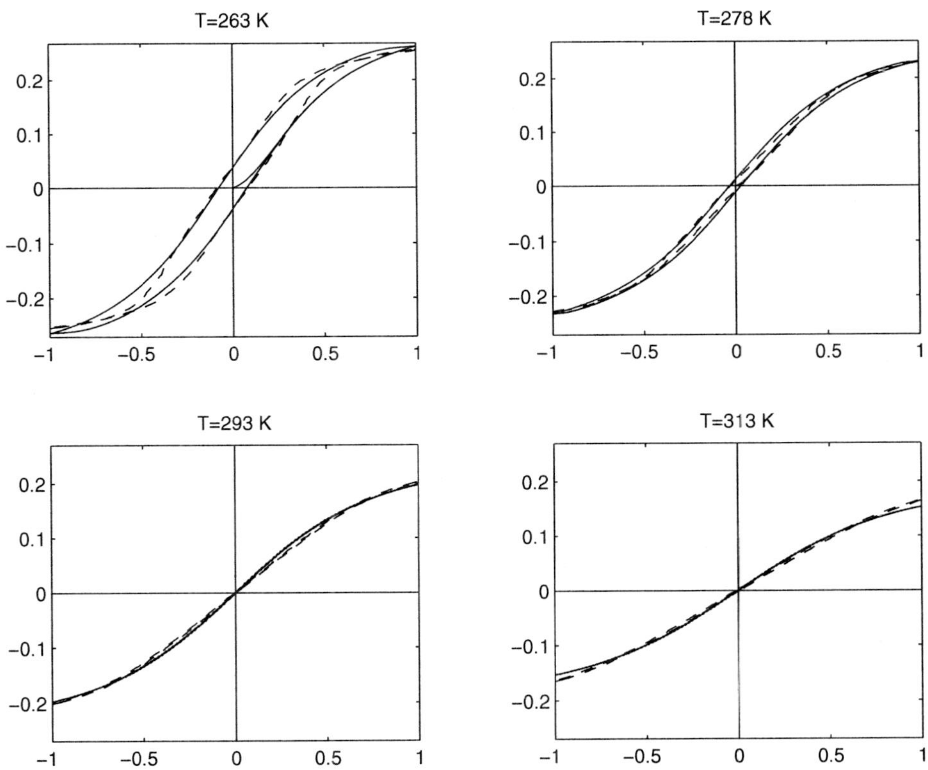

Figure 3.6. *PMN-PT-BT data (- - -) from [440] and the domain wall model (——). Abscissas: electric field (MV/m), ordinates: polarization (C/m^2).*

parameters $\alpha(T), a(T), k(T)$ and $P_s(T)$ defined in (3.20), is compared with the measured data in Figure 3.6. It is observed that the model accurately quantifies the increase in hysteresis and saturation values that occur as temperatures are decreased.

PMN Characterization

To provide a second illustration of the model attributes over a more extensive temperature range, we consider its capability for predicting the dielectric behavior of 0.1 Hz PMN data collected by Glazounov et al. [186].

For this material, we employed $T_f = 242$ K, $T_c = 398$ K, and $p = 2$ with the parameters $\widetilde{\alpha} = 1.8\times 10^8$ Vm/C, $\widetilde{a} = 2.5\times 10^8$ C/m, $\widetilde{k}_p = 190$ V/m, $\widetilde{P}_s = 0.38$ C/m^2, $\beta = 1.42 \times 10^{-8}$ and $c = 0.1$ obtained through a fit to the data. The data and model are plotted in Figure 3.7. It is observed that the model quantifies both the hysteretic behavior in the low temperature ferroelectric regime and the decaying anhysteretic response above the freezing temperature. We note that whereas the model fits at any given temperature can be improved by optimizing the parameters for that temperature, the specified parameter set provides a good fit throughout the 180° temperature range with thermal dependence in $\alpha(T), a(T), k_p(T)$ and $P_s(T)$ incorporated solely through the expressions (3.20).

3.4 Temperature-Dependent Homogenized Energy Model

Investigations focused on extending the homogenized energy framework discussed in Section 2.6 to incorporate the temperature-dependent mechanisms inherent to relaxor ferroelectric compounds have only recently been initiated and this discussion should be viewed primarily as a summary of present research and preview of directions to be pursued.

3.4.1 Temperature-Dependent Model with Negligible Relaxation

As noted in Section 3.1, two properties which differentiate relaxor ferroelectric compounds from conventional ferroelectric materials are the diffuse transition region and the strongly temperature-dependent decay in both hysteresis and saturation polarization as temperatures are increased through this transition region. Within the transition region, the materials exhibit the frequency-dependence dielectric behavior illustrated in Figure 3.2 — this results in dielectric relaxation which gives the materials their name.

Initial extensions to the homogenized energy theory have focused on the incorporation of mechanisms to characterize the reduction in saturation polarization, decrease in hysteresis, and transition to nonlinear anhysteretic E-P relations as temperatures are increased through the freezing temperature T_f. The extensions summarized here do not accommodate the frequency-dependent behavior illustrated in Figure 3.2 and should be employed only in quasistatic regimes.

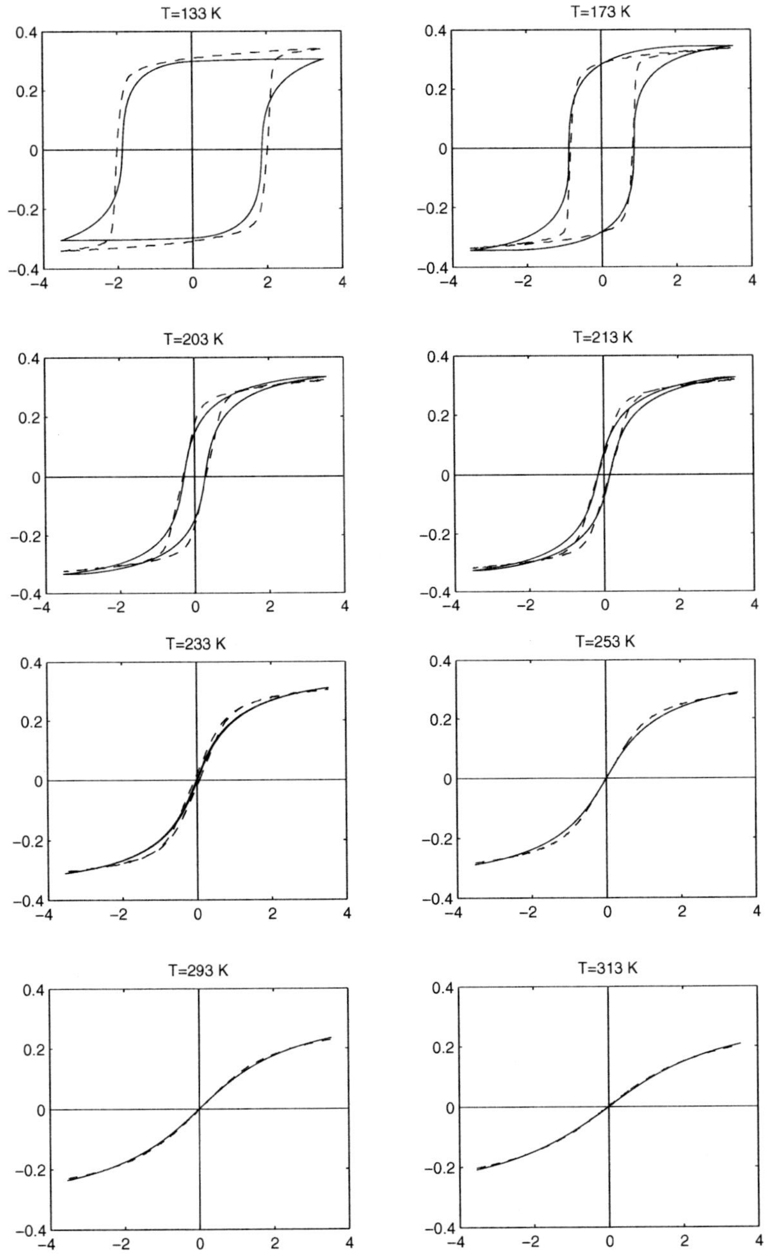

Figure 3.7. *PMN data (– – –) from Glazounov et al. [186] and domain wall model (——). Abscissas: electric field (MV/m), ordinates: polarization (C/m^2).*

To motivate mechanisms for incorporating temperature-dependence, we start with the Helmholtz energy relation (2.47) derived through statistical mechanics tenets and employed as one of the kernels in the homogenized energy model. As illustrated in Figure 2.19, this energy functional quantifies a second-order ferroelectric to paraelectric phase transition at the Curie point but, in general, is too restrictive to quantify the extensive temperature-dependence exhibited by relaxor compounds. Instead we use (2.47) to motivate phenomenological modifications to the piecewise quadratic Helmholtz definition (2.51) which yields piecewise linear kernels in the final hysteresis model.

As noted in Sections 2.3.2 and 3.2.1, minimization of the Gibbs energy

$$G = \psi - EP,$$

with ψ given by (2.47) or (3.1), yields

$$E = -\frac{E_h}{P_s}P + \frac{E_h T}{T_c}\text{arctanh}(P/P_s)$$

where E_h is a bias field and T_c is the Curie point. Enforcement of the condition $\frac{\partial^2 G}{\partial P^2} = 0$ yields the inflection points

$$\pm P_I(T) = \pm P_s\sqrt{1 - T/T_c}$$

and coercive field values

$$\begin{aligned} E_c(T) &= \frac{E_h}{P_s}P_I(T) - \frac{E_h T}{T_c}\text{arctanh}(P_I(T)/P_s) \\ &= E_h\sqrt{1 - T/T_c} - \frac{E_h T}{T_c}\text{arctanh}\sqrt{1 - T/T_c} \\ &= E_h\sqrt{1 - T/T_c} - \frac{E_h T}{T_c}\left[\sqrt{1 - T/T_c} + \frac{1}{3}(1 - T/T_c)^{3/2} + \cdots\right] \\ &= E_h(1 - T/T_c)^{3/2}\left[1 - \frac{1}{3}\frac{T}{T_c} + \cdots\right]. \end{aligned} \tag{3.21}$$

Moreover, through the relation (2.91) the local remanence polarization

$$\begin{aligned} P_R(T) &= \frac{E_c(T)}{\eta} + P_I(T) \\ &= \frac{E_h}{\eta}(1 - T/T_c)^{3/2}\left[1 - \frac{1}{3}\frac{T}{T_c} + \cdots\right] + P_s\sqrt{1 - T/T_c} \end{aligned} \tag{3.22}$$

inherits temperature-dependence from both E_c and P_I. The relations (3.21) and (3.22) motivate phenomenological expressions to incorporate temperature effects in the homogenized energy model employing piecewise linear kernels.

To retain generality, we employ the expression

$$\overline{E}_c(T) = \widetilde{E}_c(1 - T/T_f)^p \tag{3.23}$$

for the mean in the lognormal density ν_1 used to characterize coercive field variations — see (2.117). We note that (3.23) is equivalent to the relation (3.19) used to characterize $E_c(T)$ and hence $k_p(T)$ in the domain wall model. Similarly, (3.22) motivates the expression

$$P_R(T) = \widetilde{P}_R(1 - T/T_f)^q \tag{3.24}$$

where q and $\widetilde{P}_R$ are real-valued constants which must be identified for a given material.

3.4.2 PMN-PT-BT Characterization

To illustrate the performance of the model, we revisit the characterization of PMN-PT-BT discussed in the first example of Section 3.3.2 in the context of the domain wall model. Data collected at 1 Hz to minimize frequency effects is plotted in Figure 3.8.

The construction of the model requires identification of the parameters $T_f, \widetilde{E}_c, \widetilde{P}_R, \eta, b, c, C, p$ and q. It is noted in Section 2.6.11, $\widetilde{P}_R$ and C can be combined to yield a single scaling parameter. As detailed in [388], p was taken to be $p = 3/2$ in accordance with (3.21) and the relation (3.22) was used to characterize the temperature-dependence of P_R. The remaining parameter values $T_f = 321.6$ K, $E_h = 8.8 \times 10^5$ V/m, $\eta = 1.0 \times 10^8$, $b = 2.9 \times 10^{11}$ and $c = 0.17$ were obtained through a least squares fit to data collected at 263 K, 283 K, 293 K and 313 K. The resulting

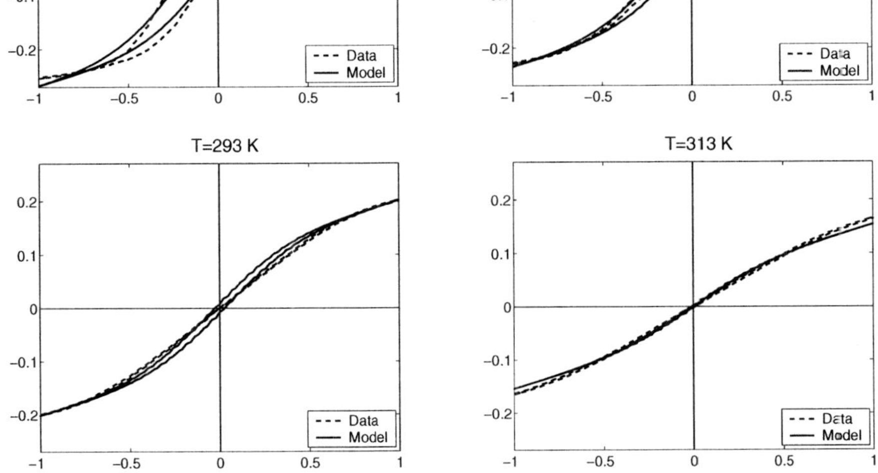

Figure 3.8. *PMN-PT-BT data and homogenized energy model. Abscissas: electric field (MV/m), ordinates: polarization (C/m^2).*

model fits are compared with data in Figure 3.8. It is observed that through the temperature-dependent relations (3.23) and (3.24), the model incorporates the onset and increase in hysteresis as well as the increasing polarization values exhibited by PMN-PT-BT as temperatures are decreased through the diffuse transition region.

Chapter 4

Model Development for Ferromagnetic Compounds

At temperatures below the Curie point T_c, *ferromagnetic materials* exhibit a *domain structure* and *spontaneous magnetization* which imbue them with fundamental properties that are both crucial to technology and challenging to characterize. The term ferromagnetic preceded and motivated the designations *ferroelectric* and *ferroelastic* for related electric and elastic domain phenomena as well as the unified expression *ferroic* for the combined class of compounds. At the mesoscopic and macroscopic levels, the domain properties of the three compounds are sufficiently analogous to permit construction of the unified models described in Chapter 6. At the microscopic level, however, the phenomena differ significantly with domain formation in ferromagnetic materials attributed to the alignment of electron spins. We summarize certain microscopic mechanisms to motivate mesoscopic domain properties but focus primarily on the latter due to its utility for model development in the context of unified frameworks for device characterization and model-based control design.

Properties of magnetic materials have been known for millenia with the discovery of lodestones, Fe_3O_4, attributed to Greeks on the island of Magnesia roughly 3500 years ago. Whereas subsequent technologies exploiting magnetic materials and properties are too numerous to delineate, three areas in which ferromagnetic materials have played fundamental roles are (i) data storage and processing, (ii) electric power generation, and (iii) telecommunications. More recent investigations have focused on the development and utilization of ferromagnetic compounds for use as transducers in smart systems, and it is on this regime that we will focus.

As was the case for ferroelectric materials, hysteresis is a fundamental property of all ferromagnetic materials due to the inherent domain structure. However, for many applications and drive regimes, hysteresis and constitutive nonlinearities can be minimized thus permitting the development of linear *piezomagnetic* constitutive relations analogous to the piezoelectric equations developed in Section 2.2 for ferroelectric materials. For general operating conditions, however, constitutive nonlinearities and hysteresis must be accommodated to achieve optimal material or device performance, and development of nonlinear hysteresis models occupies the majority of this chapter.

The fact that this chapter is shorter than Chapter 2 should in no sense be construed as an indictment that ferromagnetism is fundamentally simpler than ferroelectricity — in certain aspects, the opposite is true and a number of ferromagnetic phenomena are still poorly understood. Instead, the focus on modeling frameworks which encompass ferroic compounds permits the direct application of principles detailed in Chapter 2 to aspects of model development for ferromagnetic compounds.

Chapter Organization

- Section 4.1 – Summary of ferromagnetic properties which contribute to the nonlinear and hysteretic material behavior exhibited by the compounds.
- Section 4.2 – Construction of fundamental energy relations incorporating exchange, anisotropic, magnetostatic and magnetoelastic components.
- Section 4.3 – Development of linear constitutive relations appropriate for low to moderate drive regimes.
- Section 4.4 – Higher-order energy relations constructed to provide a basis for subsequent hysteresis models.
- Section 4.5 – Preisach models: Classical and extended Preisach theories provide a rigorous mathematical framework for characterizing hysteresis and material nonlinearities.
- Section 4.6 – Domain wall theory: Provides an efficient framework for characterizing symmetric hysteresis loops but *a priori* knowledge of turning points is required to ensure closure of biased minor loops. This precludes its use for model-based control design for certain applications exhibiting transient dynamics.
- Section 4.7 – Homogenized energy theory: Provides a framework which characterizes both major and biased minor loop behavior for a broad range of operating regimes including those which exhibit certain rate-dependencies. Theory provides mechanisms for incorporating temperature and stress-dependencies and provides an energy basis for certain Preisach formulations.

4.1 Physical Properties of Ferromagnetic Materials

The complexity of ferromagnetic material behavior precludes a complete treatment so instead we focus on physical properties pertinent to model development for smart material applications. The reader is referred to [4,40,88,111,247,492] for a comprehensive treatment regarding the physics of ferromagnetism. The classical text by Bozorth [51] contains extensive data illustrating a number of ferromagnetic mechanisms whereas [501] provides a comprehensive bibliography summarizing both the Eastern and Western literature on ferromagnetism at the time the book was published. The categorical citations in Bertotti [40] may prove useful to readers seeking an overview of topics related to hysteresis in ferromagnetic materials.

4.1.1 Crystallographic Notation

To facilitate discussion regarding crystallographic properties of ferromagnetic materials, we summarize first vector conventions in which directional indices are represented by square brackets and planes are denoted by round brackets. We recall that the latter are defined in terms of their normal vectors. As illustrated in Figure 4.1, the vertices of a unit cube are thus specified by $[100], [110], [010], [000], [101], [111], [011], [001]$ whereas the faces are denoted by $(100), (010), (001), (\bar{1}00), (0\bar{1}0), (00\bar{1})$. In both cases, $\bar{1}$ indicates -1. Finally, pointed brackets are used to summarize an entire set of indices. Hence $\langle 111 \rangle$ represents all eight vertices of the unit cube. The reader is referred to Cullity [111] for additional details regarding this notation.

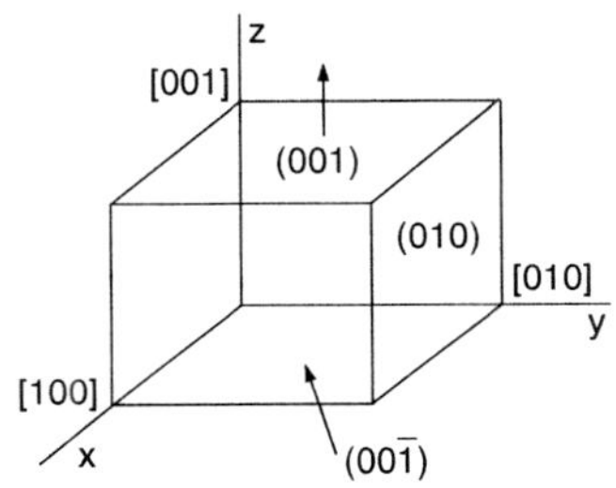

Figure 4.1. *Vector convention specifying faces and indices of a unit cube.*

4.1.2 Ferromagnetic Domains

In two papers appearing in 1906 and 1907, Pierre Weiss posited a theory of ferromagnetism based on two hypotheses: (i) a ferromagnetic material is comprised of a domain structure consisting of regions (domains) in which magnetic moments are aligned and hence exhibit a spontaneous magnetization, and (ii) the moment alignment within domains is due to a mean or molecular field $H_I = \alpha M$ which is proportional to the magnetization [506,507]. In the demagnetized state, the random ordering of domains yields zero net magnetization whereas the application of a field aligns domains to produce a bulk magnetization.

This theory was indirectly validated in 1919 by Barkhausen [35] who measured discrete changes in the magnetic induction B which were attributed to irreversible changes in the domain structure. Direct validation of the domain structure was provided in 1931 by Bitter [43] through visual techniques in which carrier liquids containing fine magnetic powers were spread on the surface of the ferromagnetic material. In 1935, Landau and Lifshitz established that domain formation produces a reduction in magnetostatic energy and hence domain structures minimize the material's internal energy [282]. This is analogous to the minimization of electrostatic energy which produces the ferroelectric domains depicted in Figure 2.3.

To illustrate, consider the domain structures for a fully magnetized and demagnetized material as depicted in Figure 4.2 with a more comprehensive depiction in Figure 4.3. In a single crystal, the fully magnetized specimen consists of a single

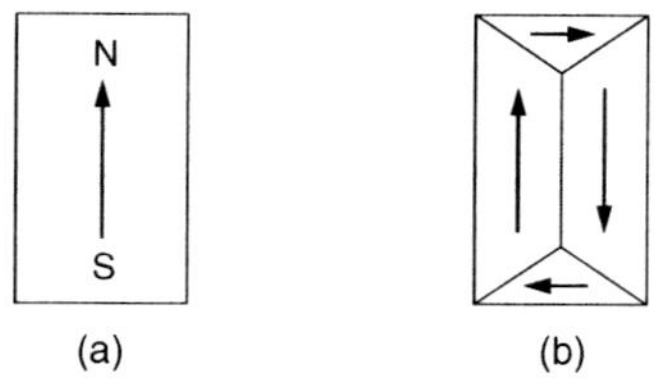

Figure 4.2. *(a) Single domain for a fully magnetized material and (b)* 180° *domains and* 90° *closure domains which minimize magnetostatic energy in the demagnetized material.*

domain and exhibits spontaneous magnetization M_0. As the material is demagnetized, 180° domains separated by transition regions termed *domain walls* nucleate to reduce the magnetostatic energy. In cubic materials, they are accompanied by 90° closure domains which complete the magnetic flux paths.[13] As in ferroelectric materials, it is the 90° domains which respond to stresses in the field direction. We point out that for the demagnetized material depicted in Figure 4.2(b), each domain exhibits a spontaneous magnetization M_0 but the bulk magnetization is $M = 0$.

We note that analogies between ferromagnetic and ferroelectric domains provide a valuable basis for formulating unified mesoscopic and macroscopic models. However, care must be taken not to overextend these analogies and at least three fundamental criteria differentiate ferromagnetic and ferroelectric domain processes. (i) The incorporation of stresses and the effect of coupled elastic forces on 90° domain formation is more significant in ferroelectric materials since electrostrictive strains are one to two orders of magnitude larger than typical magnetostrictive strains — e.g., strains in iron Fe or nickel Ni. This is not true for giant magnetostrictive materials such as Terfenol-D and for such compounds, magnetoelastic interactions play a prominent role. (ii) Ferroelectric materials do not exhibit a mechanism equivalent to the quantum exchange interaction which produces long range ordering in ferromagnetic materials. In summary, ferromagnetism results

[13]Closure domains can occur in hexagonal materials but are less energetically favorable since they are orthogonal to easy axes.

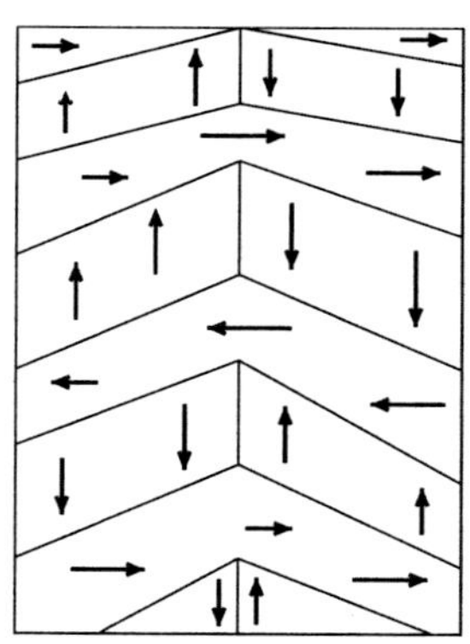

Figure 4.3. *Depiction of* 90° *and* 180° *domains in a cubic ferromagnetic compound.*

from interactions between magnetic dipoles whereas ferroelectricity is due to the ionic structure of the material. The balance between the exchange and anisotropy energy, as detailed in the next section, also causes magnetic domain walls to be significantly thicker than their ferroelectric counterparts as illustrated in Figure 4.4. To quantify this difference, we note that the domain wall thickness for Fe is approximately 138 lattice parameters (40 nm) [247, page 134], whereas $BaTiO_3$ walls are estimated to have thicknesses on the order of 1 lattice constant. (iii) Ferroelectric materials have electrons which freely carry charge whereas an analogous construct (e.g., a magnetic monopole) is absent for magnetic materials. Details regarding the ramifications of these phenomena can be found in [502].

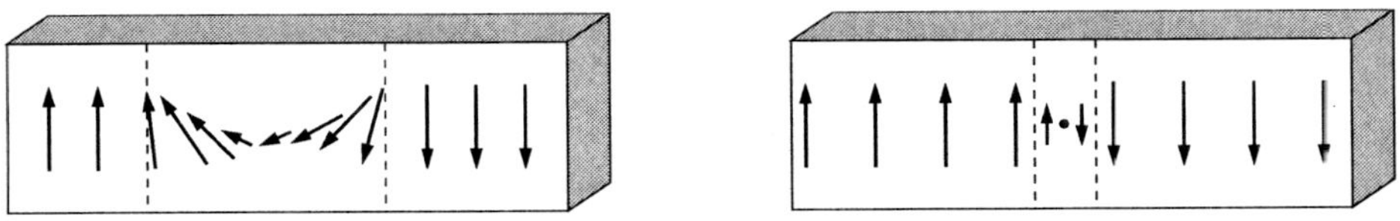

Figure 4.4. *Domain walls between* 180° *domains for (a) ferromagnetic and (b) ferroelectric materials.*

4.1.3 Material Anisotropies

All materials exhibit varying degrees of *anisotropy* (orientation-dependence) due to their crystalline structure, stress orientations, shape and configuration. This includes *magnetocrystalline* or *crystal anisotropies*, which are characterized by a change in the internal energy as the direction of magnetization measurement is varied, and *stress anisotropies* which are due to magnetoelastic coupling. We note that the crystalline anisotropy is an inherent property of the material whereas stress and shape anisotropies are externally induced properties. For single crystal compounds such as nickel, these effects are relatively minimal and the behavior can be considered as approximately *isotropic*. For other compounds such as iron or Terfenol-D, anisotropies are more significant and must be accommodated in models for various operating regimes. For polycrystalline materials, the effects of magnetic anisotropies may average thus producing isotropic bulk behavior. However, this is not always the case and the degree to which material anisotropies must be accommodated in models is often dependent on specific operating conditions.

As detailed in Chikazumi [88] and Jiles [247], crystal anisotropy is due to interactions between spins and crystal axes which produce energetically preferable spin configurations termed *easy axes*. For materials such as cobalt and various rare-earth compounds, there are two preferred orientations and the materials exhibit *hexagonal* or *uniaxial* anisotropies as illustrated in Figure 4.5(a). Other materials such as iron exhibit a crystallographic symmetry which yields easy axes in the $\langle 100 \rangle$ direction of cubic face centers as depicted in Figure 4.5(b). Alternatively, materials like nickel have easy axes in the $\langle 111 \rangle$ direction of cube corners and thus exhibit another form of cubic anisotropy.

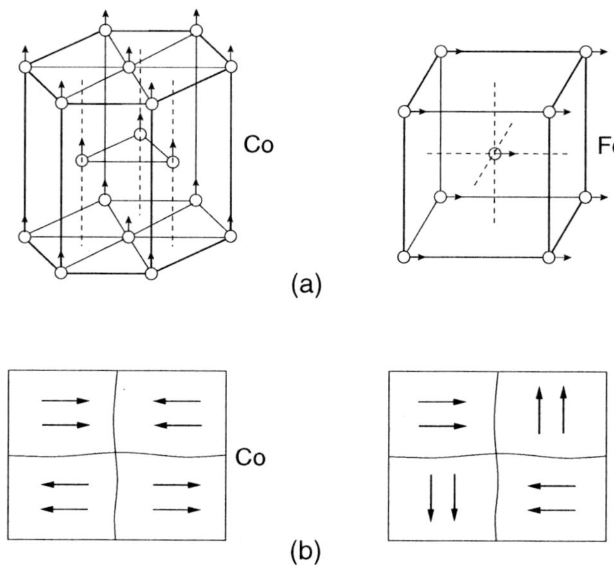

Figure 4.5. *(a) Crystallographic orientation of magnetic moments in cobalt and iron (after [247]). (b) Corresponding domain structure which reflects the uniaxial or hexagonal anisotropy of cobalt and the cubic anisotropy of iron.*

The *anisotropy energy* quantifies the increase in energy required to rotate away from the easy axis. For a uniaxial anisotropy, the energy can be represented by the relation

$$U_A = K \sin^2 \phi \tag{4.1}$$

where ϕ is the angle between the magnetization and crystallographic axis. The anisotropy energy for cubic materials is represented by

$$\begin{aligned} U_A &= K\left(\cos^2\theta_1 \cos^2\theta_2 + \cos^2\theta_2 \cos^2\theta_3 + \cos^2\theta_3 \cos^2\theta_1\right) \\ &= K\left(\alpha_1^2\alpha_2^2 + \alpha_2^2\alpha_3^2 + \alpha_3^2\alpha_1^2\right). \end{aligned} \tag{4.2}$$

Here $\theta_1, \theta_2, \theta_3$ are the angles between the magnetization and three crystallographic axes and $\alpha_1, \alpha_2, \alpha_3$ are the corresponding direction cosines. Details regarding the magnitude of the anisotropy constants K and examples illustrating the effect of uniaxial and cubic anisotropies are provided in Chikazumi [88] and Jiles [247].

Crystalline anisotropies have a variety of effects on domain and domain wall characteristics. As illustrated in Figure 4.5, 90° closure domains are more prevalent in cubic materials than those with uniaxial anisotropies since they occupy an easy axis in the former case. As detailed in Section 7.1.2 of Jiles [247], domain wall energies, which quantify domain wall thicknesses, are determined through a balance between the exchange energy and anisotropy energy. The former produces thicker walls since aligned moments are energetically favored whereas the latter decreases wall thickness since the anisotropy energy is lowest when moments are aligned with easy axes.

Crystalline anisotropies and magnetomechanical coupling are intrinsically coupled and strain generation in response to an applied field is due in part to deformations in the crystalline lattice to minimize internal energy. The interactions will be detailed in the context of magnetoelastic interactions.

4.1.4 Magnetic Materials

The purpose here is not to catalogue magnetic materials or their attributes, but rather to summarize properties of representative compounds which subsequently prove important when describing and characterizing hysteresis and constitutive nonlinearities, stress and temperature dependencies, and material properties important for transducer design. The reader is referred to Bozorth [51], Chikazumi [88], Cullity [111] or Jiles [247] for details about the various materials.

As noted previously, iron and nickel exhibit cubic anisotropies whereas the ionic structure of cobalt produces a hexagonal or uniaxial anisotropy. Several figures of merit for these materials are summarized in Table 4.1. The saturation magnetostriction λ_s quantifies strains due to magnetomechanical coupling and provides a metric for characterizing the transduction capabilities of the material. The saturation magnetization M_s or induction B_s quantifies the extent to which moments can reorient in response to an applied field and hence this quantity should also be large for strong transduction. The Curie point T_c delineates the temperature at which a ferromagnetic to paramagnetic phase transition occurs. Since significant moment rotation and subsequent changes in magnetization occur only in the ferromagnetic state, materials must be chosen or designed to operate well below T_c. The linear coupling coefficient k and elastic modulus Y prove important for transducer design. As discussed in Section 2.2.4 in the context of ferroelectric materials, k is defined as the sqare root of the ratio of the mechanical energy stored to the total energy stored and hence large values are advantageous. The elastic modulus quantifies the material stiffness which should be larger than that of coupled components to permit effective transduction. Details elaborating on a number of these material properties will be provided in later sections.

Material	$\frac{3}{2}\lambda_s(\times 10^{-6})$	B_s (T)	T_c (^{o}C)	Y (GPa)	k
Fe	-14	2.15	770	285	
Ni	-50	0.61	358	210	0.31
Co	-93	1.79	1120	210	
Tb	3000 (-196 ^{o}C)		-48	55.7	
Dy	6000 (-196 ^{o}C)		-184	61.4	
Terfenol-D	1620	1.0	380	110	0.77
Metglas 2605SC	60	1.65	370	25-200	0.92

Table 4.1. *Magnetoelastic material properties from [117]. All measurements are at room temperature except where specified.*

The ordering in Table 4.1 reflects a historical perspective ranging from iron to Terfenol-D (terbium: Ter, iron: Fe, Naval Ordnance Laboratory: NOL, dysprosium: D) and metglas. As was noted in Section 1.3, the demonstration that terbium and dysprosium have 'giant' magnetostrictive capabilities at cryogenic temperatures and that Terfenol-D can produce strains up to 1600 μL/L at room temperature has greatly expanded the role of magnetostrictive transducers for smart material applications.

In the present manufacturing process, Terfenol-D crystals are grown in dendrite sheets oriented in the $[11\bar{2}]$ direction as depicted in Figure 4.6. At room temperature, the easy axes lie approximately in the $\langle 111 \rangle$ set of directions. To optimize transducer outputs and to maintain the Terfenol-D rod in a state of compression, prestresses are applied using mechanisms such as the compression bolt and spring washers in the transducer depicted in Figure 1.11. For sufficiently large prestresses — e.g., 1–2 ksi for the design depicted in Figure 1.11 — the preferred orientation of domains is shifted from the original eight $\langle 111 \rangle$ magnetic easy axes to the two axes $[111]$ and $[\overline{111}]$ perpendicular to the $[11\bar{2}]$ direction of the applied stress. Significant strains are then generated when the magnetization rotates from $[111]$ to $[11\bar{1}]$ or from $[\overline{111}]$ to $[\overline{11}1]$ in response to an applied field.

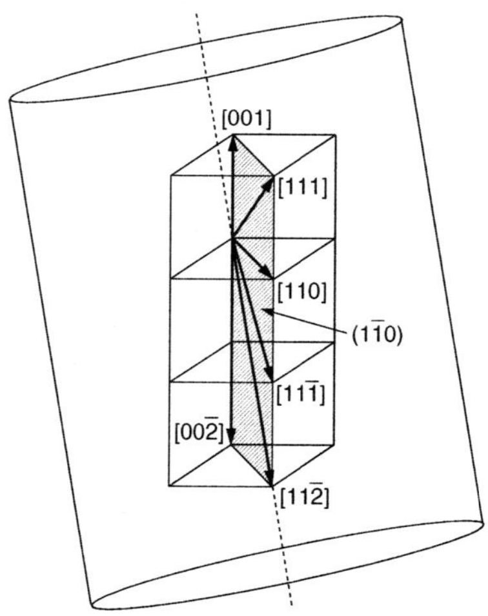

Figure 4.6. *Orientation of Terfenol-D crystals.*

4.1.5 Magnetic Processes and Hysteresis

Students taking elementary electromagnetism courses are typically taught that the magnetic field H, magnetism M, and magnetic induction B are related by the expressions

$$B = \mu H \tag{4.3}$$

and

$$M = \chi H \tag{4.4}$$

where μ and χ respectively denote the magnetic permeability and susceptibility. For certain paramagnetic and diamagnetic materials (4.3) and (4.4) are reasonably accurate since μ and χ are approximately constant over a broad range of field inputs. The accuracy range is significantly diminished for ferromagnetic materials, however, due to hysteresis and constitutive nonlinearities associated with inherent domain processes. As illustrated in Figure 4.7(a), hysteresis causes μ, and correspondingly χ, to be not only nonconstant but also multivalued. Hence the linear approximations (4.3) and (4.4) must be used with caution and for low drive regimes when considering ferromagnetic materials.

A quantity that is applicable for ferromagnetic materials — indeed all magnetic materials — is the relation

$$B = \mu_0(H + M) \tag{4.5}$$

which expresses the induction in terms of the components due to the field and magnetization. Here $\mu_0 = 4\pi \times 10^{-7}$ H/m in Sommerfeld (SI) units — the reader is referred to Jiles [247] for discussion about the Gaussian (CGS) and two MKS (Sommerfeld and Kennelly) unit systems employed in the magnetics literature. In the physics and materials science literature, one typically finds hysteresis discussed in the context of the magnetization whereas induction is more commonly employed in engineering literature. Due to (4.5), both are equivalent and we consider both throughout this discussion.

The magnetization process and resulting hysteretic H-M relation can be broadly categorized in terms of *domain wall movement* and *domain rotation*. As detailed in Jiles [247], it is energetically favorable for domain walls to form at *pinning sites* comprised of inclusions, impurities, and stress nonhomogeneities in the material, and domain growth is due to reversible bowing or irreversible displacements of domain walls pinned at these sites. Quantification of these effects forms the basis of the Jiles–Atherton domain wall models discussed in Section 4.6. Rotational effects are due to reversible rotation of a domain within a neighborhood of an easy axis or irreversible rotation from one easy axis to another. Characteri-

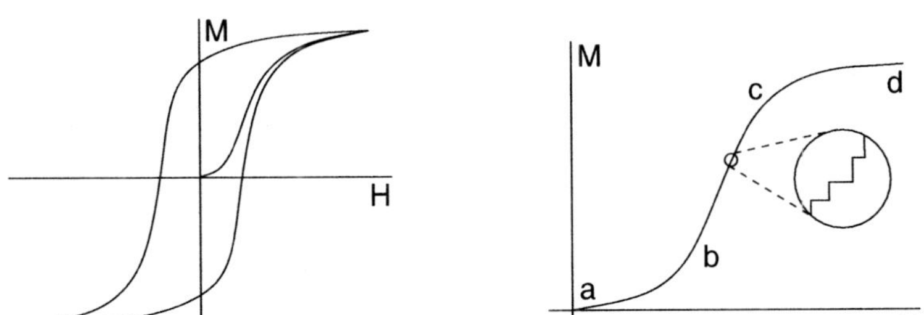

(a) (b)

Figure 4.7. *(a) Hysteresis in the H-M relation for a typical ferromagnetic material. (b) Initial magnetization curve for demagnetized Terfenol-D including the Barkhausen effect which is attributed primarily to discontinuous domain wall motion across pinning sites in the material.*

zation of these mechanisms requires consideration of the anisotropy energies (4.1) or (4.2) and forms the basis for the Stoner–Wohlfarth theory [464], the anisotropy models of Jiles and Thoelke [252] and Armstrong [13], as well as micromagnetic frameworks [4, 5, 40, 59–61, 270].

We illustrate the magnetization process for monolythic Terfenol-D having the crystalline configuration depicted in Figure 4.6. The initial magnetization curve is depicted in Figure 4.7(b) and associated domain processes are illustrated in Figure 4.8. (a) The demagnetized state is comprised of domains, each having nonzero spontaneous magnetization M_0, but yielding zero bulk magnetization M due to their random configuration. (b) At low field levels, changes in magnetization are due primarily to reversible domain wall movement and moment rotation. (c) As field levels are increased, two irreversible mechanisms emerge; domain wall displacement favors the growth of domains having components in the field direction and moments rotate to $[11\bar{1}]$ which is the easy axis closest to the field direction. This produces an irreversible burst region in the H-M or H-ε curve in which small changes in field produce large changes in the magnetization or strain. In the final state depicted in Figure 4.8(d), the material acts as a single domain as moments rotate coherently from the easy axis into the direction of the applied field. This produces the saturation behavior exhibited by the material.

A reversal of the field following saturation produces a region of reversible moment rotation and domain wall movement followed by irreversible reorientation and domain wall displacement. This yields the characteristic ferromagnetic hysteresis curve depicted in Figure 4.7(a).

Whereas both domain wall movement and domain rotation contribute irreversible processes, the former is predominant for most materials. Hence the Barkhausen discontinuities measured throughout the magnetization process are at-

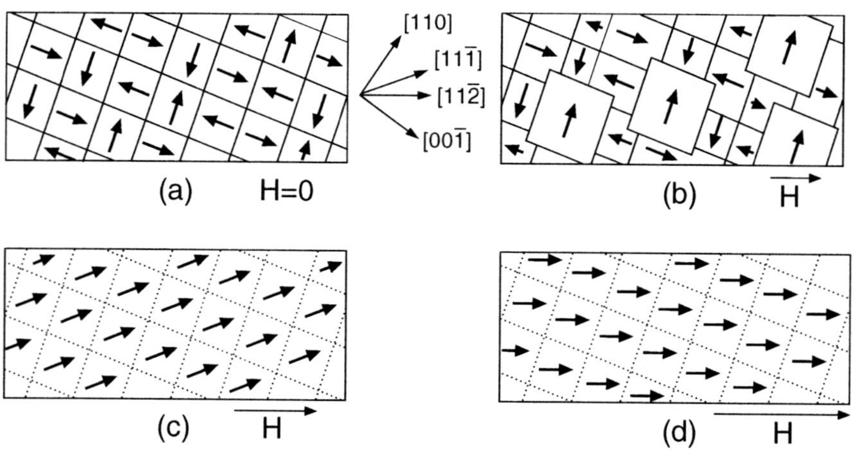

Figure 4.8. *Magnetization process in the* $(1\bar{1}0)$ *plane of single crystal Terfenol-D due to an applied field H in the* $[11\bar{2}]$ *direction. (a) Demagnetized state, and (b) growth of domains due to domain wall motion. (c) Rotation of moments to the easy* $[11\bar{1}]$ *axis and (d) rotation of moments to align with the applied field.*

tributed primarily to pinning–unpinning of domain walls to accommodate growth of favorably oriented domains. Two regimes in which domain rotation dominate domain wall movement are fine magnetic particles for magnet design which are too small to permit domain formation and thin magnetic films (~1000 Å) constructed so that flux changes are due to domain rotation [189].

Moreover, it is observed that whereas domain wall movement and domain rotation both involve the rotation of magnetic moments, the scales and cooperative interactions vary significantly for the two mechanisms. In the former case, changes in flux are due to a *sequential, local* rotation of moments in the region of the domain wall as it moves through a region. This is in contrast to domain rotation where all the moments within a domain rotate *in unison*. The property that domain rotation is inherently faster than domain wall movement should be considered for applications requiring fast flux reversal — e.g., high speed memory access.

The situation is more complicated in polycrystalline materials due to the random orientation of grains. As with ferroelectric materials, this necessitates consideration of homogenization techniques, such as those employed in the Preisach models of Section 4.5, domain wall theory in Section 4.6 and homogenized energy model developed in Section 4.7 to obtain macroscopic *effective parameters* which incorporate variability due to random grain orientations and material nonhomogeneities.

Hence hysteresis and constitutive nonlinearities are inherent properties of ferromagnetic compounds which must be accommodated in models to achieve full performance capabilities but also imbue materials with unique transduction capabilities — e.g., energy dissipation in tuned vibration absorbers or memory capabilities for magnetic recording media.

Hysteresis Losses

In Section 2.8, we illustrated that the work required to change the polarization from P_0 to P_1 in a unit volume of material is

$$W = \int_{P_0}^{P_1} E \cdot dP$$

so that the integral over an entire hysteresis loop represents the total energy converted from electrical energy to potential energy and heat. Magnetic materials exhibit similar behavior. For a magnetic moment $\mathbf{m}$, the energy per unit volume in the presence of an internal field $\mathbf{H}$ is

$$U_M = -\mu_0 \mathbf{m} \cdot \mathbf{H}.$$

The work required to change the magnetization from M_0 to M_1 is then

$$W = \mu_0 \int_{M_0}^{M_1} \mathbf{H} \cdot d\mathbf{M}$$

and the energy absorbed by the material or dissipated as heat during a full hysteresis loop is

$$W = \mu_0 \oint \mathbf{H} \cdot d\mathbf{M}.$$

Because

$$\oint \mathbf{H} \cdot d\mathbf{H} = 0,$$

(4.5) can be employed to provide the equivalent relation

$$W = \oint \mathbf{H} \cdot d\mathbf{B}$$

quantifying the work required to make one cycle.

Second-Order Phase Transitions

Unlike ferroelectric materials which exhibit both first and second-order phase transition behavior, ferromagnetic materials are primarily characterized by a second-order phase transition. At temperatures above the Curie point T_c, atomic moments are *disordered* and the material exhibits *paramagnetic behavior*. For $T < T_c$, the interaction of moments produces the domain structure which characterizes the *ferromagnetic* properties of the materials. Since most transducer applications require operation in the ferromagnetic phase, materials should be chosen to ensure that T_c is well above potential operating temperatures — see Table 4.1.

4.1.6 Anhysteretic Magnetization

Figure 4.7 illustrates the initial magnetization curve and hysteretic H-M relation for a typical ferromagnetic compound. A third magnetization curve which is important from both theoretical and practical perspectives is the anhysteretic (hysteresis-free) magnetization M_{an}. Experimentally, M_{an} can be obtained at a DC field value by superimposing a decaying AC field centered at that DC field value. The locus of magnetization values obtained for a range of DC fields constitutes M_{an} and has the single-valued sigmoid form depicted in Figure 4.9. As discussed in the context of the magnetomechanical effect, the application of stress will also drive the magnetization to the anhysteretic magnetization. In both cases, the external stimuli can be interpreted as providing the energy necessary to break pinning sites.

From a theoretical perspective, the anhysteretic magnetization represents the *global* equilibrium configuration of the magnetization for a specified field level. The pinning sites and easy axes provide *local* minima in underlying energy relations

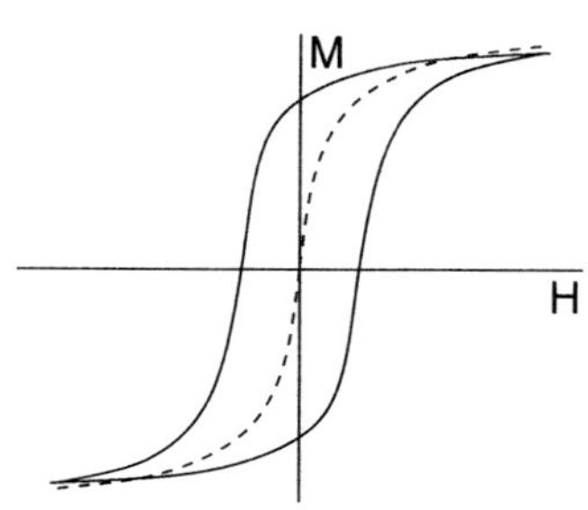

Figure 4.9. *Anhysteretic and hysteretic magnetization curves.*

which determine the magnetization unless sufficient energy is provided to overcome the local barriers and achieve the global minimum quantified by M_{an}.

The anhysteretic magnetization is analogous to the switching polarization P_{an} discussed in Section 2.5.1. Moreover, the fact that M_{an} quantifies the equilibrium magnetization in the absence of pinning sites provides a basis for the Jiles–Atherton model discussed in Section 4.6.

4.1.7 Magnetic After-Effects and Accommodation

Two phenomena which produce nonclosure of minor loops are magnetic after-effects and accommodation. Since both can arise in smart material applications, we summarize the fundamental physical mechanisms as a prelude to model development in Sections 4.5 and 4.7.

Magnetic After-Effects

Magnetic after-effects are attributed to either atomic diffusion (termed diffusion after-effects) or thermally-induced switching in advance of the local coercive field, which is termed thermal after-effects [88, 126]. As depicted in Figure 4.10, both yield a drift or creep in the magnetization if input fields and stresses are held fixed. This can cause nonclosure of minor loops thus negating the deletion property associated with Preisach models.

The time constants associated with such relaxation processes are typically quantified by Arrhenius relations of the form

$$\mathcal{T}(T) = \mathcal{T}_0 e^{U/kT} \tag{4.6}$$

where $\mathcal{T}_0$ is a scaling constant, U is the energy required to overcome barriers in the internal energy, and k is Boltzmann's constant. It is noted that (4.6) is closely related to the Boltzmann relations (2.93) and (4.77) used when constructing Helmholtz energy relations for ferroelectric and ferromagnetic materials.

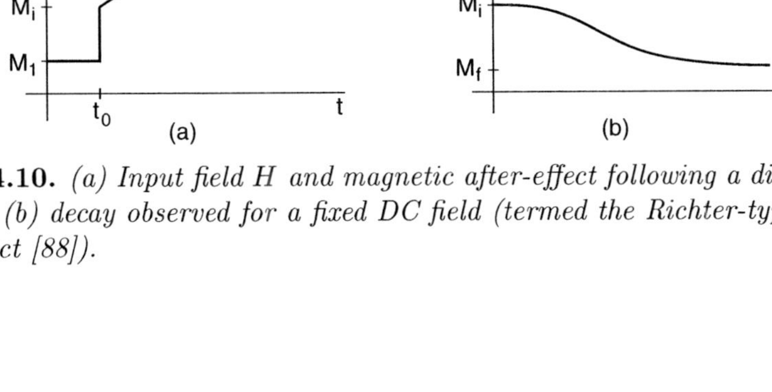

Figure 4.10. *(a) Input field H and magnetic after-effect following a discontinuous step and (b) decay observed for a fixed DC field (termed the Richter-type magnetic after-effect [88]).*

Accommodation (Reptation)

A second property of magnetic materials which can produce nonclosure of minor loops is the *accommodation* or *reptation* process depicted in Figure 4.11. In this case, application of a periodic field produces a shifting family of minor loop responses which converge to a final stable orbit. From a physical perspective, this is often referred to as a *training* or *breaking in* period for the material, and final transducer configurations are often employed only after the material has stabilized to produce repeatable behavior. Accommodation in ferromagnetic materials is phenomenologically similar to the plastic deformations and material training discussed on pages 249–251 for SMA.

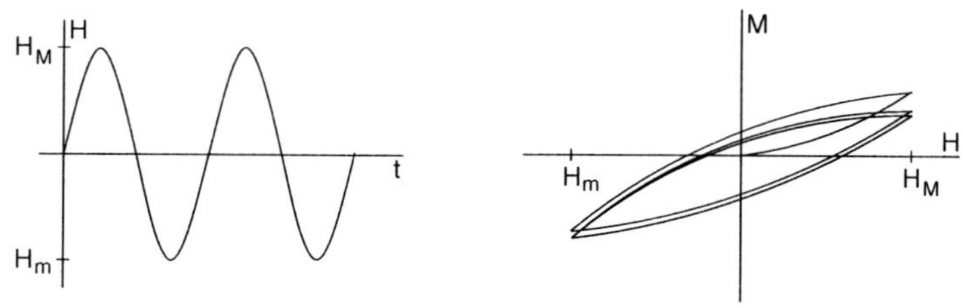

Figure 4.11. *Behavior of a minor hysteresis loop as it stabilizes or accommodates to an equilibrium orbit.*

4.1.8 Magnetomechanical Effects

The previously described magnetization processes encapsulate several fundamental attributes of ferromagnetic materials including inherent hysteresis and constitutive nonlinearities due to the cooperative domain structure of the materials. However, it is the magnetomechanical properties of the materials which provide them with actuator and sensor capabilities for smart system design. In general, these effects are highly complex and contain a number of facets which remain open research questions. We outline those mechanisms which are pertinent to transducer design and refer the reader to Chikazumi [88] for details regarding this phenomenon.

Very simply, the magnetomechanical effect is comprised of two coupled mechanisms: (i) applied stresses cause magnetic moments to rotate thus changing the magnetization, and (ii) the rotation of moments to align with an applied field generates strains in the material. These mechanisms respectively provide the materials with sensor and actuator capabilities. We consider the latter mechanisms first.

Magnetostriction

The general term *magnetostriction* refers to strains generated during the paramagnetic–ferromagnetic phase transition or in response to an applied field which constitutes magnetic-to-elastic coupling. Three phenomena of interest are the spontaneous magnetostriction λ_0, saturation magnetostriction λ_s, and Joule magnetostriction λ below saturation.

Spontaneous Magnetostriction

The spontaneous magnetostriction quantifies strains generated as materials are cooled through the Curie point T_c and domains form in the material. As detailed in Jiles [247] and depicted in Figure 4.12, this is commonly modeled by representing the *isotropic disordered* paramagnetic regions by spheres and the still isotropic but *ordered* ferromagnetic domains by ellipsoids. If the total spontaneous strain is denoted by ε, the change in length due to the transition from a sphere to an ellipsoid with major axis at θ is

$$\varepsilon(\theta) = \varepsilon \cos^2 \theta \tag{4.7}$$

and the spontaneous magnetostriction is

$$\lambda_0 = \int_{-\pi/2}^{\pi/2} \varepsilon \cos^2 \theta \sin \theta d\theta = \frac{\varepsilon}{3}. \tag{4.8}$$

Upon application of a field, domains in materials exhibiting magnetostriction will rotate in the field direction to produce the total strain ε. From the relation

$$\lambda_s = \varepsilon - \lambda_0 = \frac{2}{3}\varepsilon, \tag{4.9}$$

one can deduce values for λ_0 given the saturation magnetostriction λ_s.

Saturation Magnetostriction

The behavior of the saturation magnetostriction is dependent on both the anisotropy of the material and the angle θ at which strains are measured relative to the applied field. Consider first the case of isotropic materials — e.g., this produces a reasonable approximation for nickel and certain Terfenol-D compositions. Combination of (4.7) and (4.9) in this case yields

$$\lambda_s(\theta) = \frac{3}{2}\lambda_s \left(\cos^2 \theta - \frac{1}{3} \right) \tag{4.10}$$

quantifying the saturation magnetostriction at an angle θ from the field.

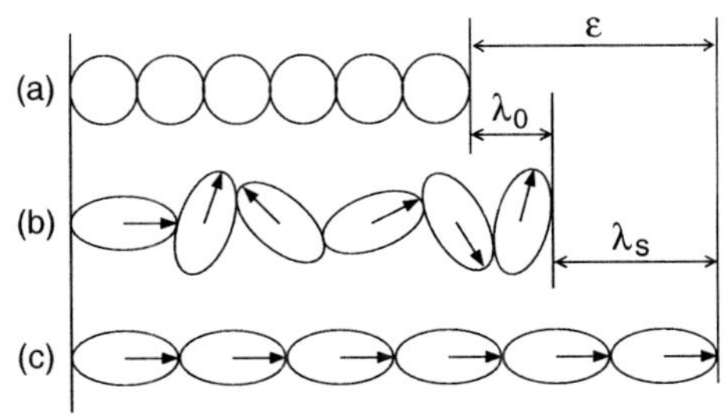

Figure 4.12. *(a) Spheres used to model isotropic, disordered material behavior in the paramagnetic phase and (b) ellipsoids representing the order encapsulated by domains for $T < T_c$. (c) Total strain ε and saturation magnetostriction λ_s due to orientation of domains in the field direction.*

In material characterization, the reported saturation magnetostriction is typically taken to be the difference between values measured parallel ($\lambda_{s\|}$) and perpendicular ($\lambda_{s\perp}$) to the field. Evaluation at $\theta = 0^o$ and $\theta = 90^o$ in (4.10) yields

$$\lambda_{s\|} - \lambda_{s\perp} = \frac{3}{2}\lambda_s \tag{4.11}$$

as a model for the measured saturation magnetostriction. For example, the measured value $\lambda_{s\|} - \lambda_{s\perp} \approx 1620 \times 10^{-6}$ for the nearly isotropic Terfenol composition $Tb_{0.3}Dy_{0.7}Fe_2$ yields the value $\frac{3}{2}\lambda_s = 1620 \times 10^{-6}$ reported in Table 4.1.

As detailed in Chikazumi [88], magnetoelastic effects are inherently coupled with the crystalline anisotropy and the isotropic relation (4.10) must be generalized for materials exhibiting hexagonal or cubic anisotropies. In the latter case, the saturation magnetostriction can be expressed as

$$\begin{aligned}\lambda_s &= \frac{3}{2}\lambda_{100}\left(\alpha_1^2\beta_1^2 + \alpha_2^2\beta_2^2 + \alpha_3^2\beta_3^2 - \frac{1}{3}\right) \\ &+ 3\lambda_{111}\left(\alpha_1\alpha_2\beta_1\beta_2 + \alpha_2\alpha_3\beta_2\beta_3 + \alpha_3\alpha_1\beta_3\beta_1\right)\end{aligned} \tag{4.12}$$

where λ_{100} and λ_{111} denote the saturation values in the $\langle 100\rangle$ and $\langle 111\rangle$ directions. The direction cosines α_i and β_i respectively denote the orientation of the domain magnetization and saturation magnetostriction, both considered relative to the field. When coincident ($\alpha_i = \beta_i$), (4.12) simplifies to

$$\lambda_s = \lambda_{100} + 3(\lambda_{111} - \lambda_{100})\left(\alpha_1^2\alpha_2^2 + \alpha_2^2\alpha_3^2 + \alpha_3^2\alpha_1^2\right). \tag{4.13}$$

The saturation magnetostriction relations (4.10), (4.12) and (4.13) are for single domain materials and the situation is significantly more complex for polycrystalline compounds having complex orientations and domain structures. In the absence of texture, the saturation effects in (4.13) can be averaged to obtain

$$\overline{\lambda_s} = \frac{2}{5}\lambda_{100} + \frac{3}{5}\lambda_{111}. \tag{4.14}$$

Joule Magnetostriction

The relations (4.10)–(4.14) characterize the *maximal strain capabilities* obtainable for various crystalline configurations and hence are important for design to ensure that materials provide requisite strains. For operation, however, it is also necessary to quantify intermediate magnetostrictions due to nonsaturating fields and, in general, this is much more difficult since these intermediate strains are highly dependent on magnetization states and the underlying crystalline structure. One case that can be directly quantified is that of strains due to 90° rotations which is typically a result of either crystalline or stress anisotropy. An example of the former is the moment rotation in a uniaxial compound in response to a field applied perpendicular to the easy axes. The second case is important for transducer designs in which prestress mechanisms — e.g., tensile stresses in nickel or compressive stresses in Terfenol-D — are needed to overcome crystalline anisotropies and orient moments perpendicular to the applied field. This serves to optimize transducer outputs and provides a regime in which models for strains due to 90° moment rotation yield reasonable accuracy.

To illustrate, consider the Terfenol-D transducer depicted in Figure 1.11 constructed from monolythic Terfenol crystals having the orientation shown in Figure 4.6 — specifically, the longitudinal rod axis is oriented in the $[11\overline{2}]$ direction. As noted previously, sufficiently large prestresses (e.g., 1.0 – 2.0 ksi) shift moment orientations from the eight $\langle 111\rangle$ easy axes to the axes $[111]$ and $[\overline{111}]$ perpendicular to the $[11\overline{2}]$ rod axis. Application of a field H in the $[11\overline{2}]$ direction via the surrounding solenoid causes 90° moment rotations which generates strains in the manner depicted in Figure 4.13.

To model strains due to rotation, the relations (4.7) and (4.9) are combined to yield

$$\lambda(\theta) = \frac{3}{2}\lambda_s \cos^2\theta.$$

Furthermore, it follows from the relation $M = M_s \cos\theta$ that

$$\lambda(M) = \frac{3}{2}\frac{\lambda_s}{M_s}M^2. \tag{4.15}$$

Hence the strains exhibit a quadratic dependence on magnetization in a manner analogous to the quadratic polarization dependence exhibited by electrostrictive materials and piezoelectric compounds in high drive regimes. The relation (4.15) will be employed in later sections when constructing nonlinear constitutive relations for ferromagnetic materials.

To illustrate the strain behavior exhibited by Terfenol-D under prestress conditions, H-M, M-ε and H-ε data collected under a prestress of 1 ksi (6.9 MPa) at 1 Hz is illustrated in Figure 4.14. The M-ε relation exhibits an even dependence on M and no saturation effects in accordance with the quadratic modeling relation (4.15). The hysteretic H-ε "butterfly" relation illustrates the behavior which must be characterized for actuator design and model-based control algorithms. A more detailed analysis of the data and magnetomechanical behavior of Terfenol-D can be found in [119].

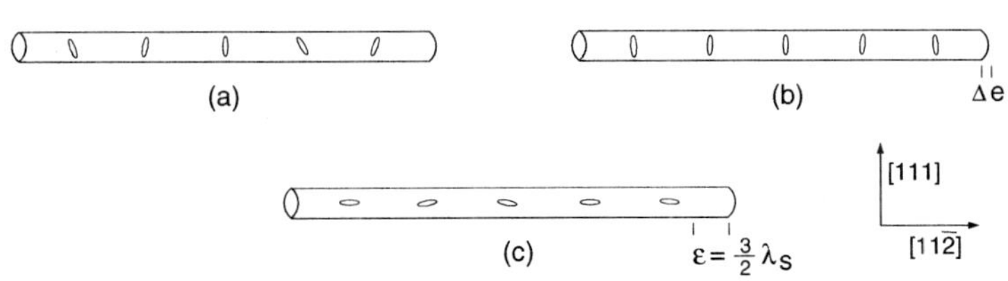

Figure 4.13. *Magnetic domains in a Terfenol-D rod: (a) orientation of unstressed rod in absence of applied magnetic field, (b) orientation of pre-stressed rod with no applied field, and (c) orientation of pre-stressed rod when field is applied in direction of longitudinal rod axis.*

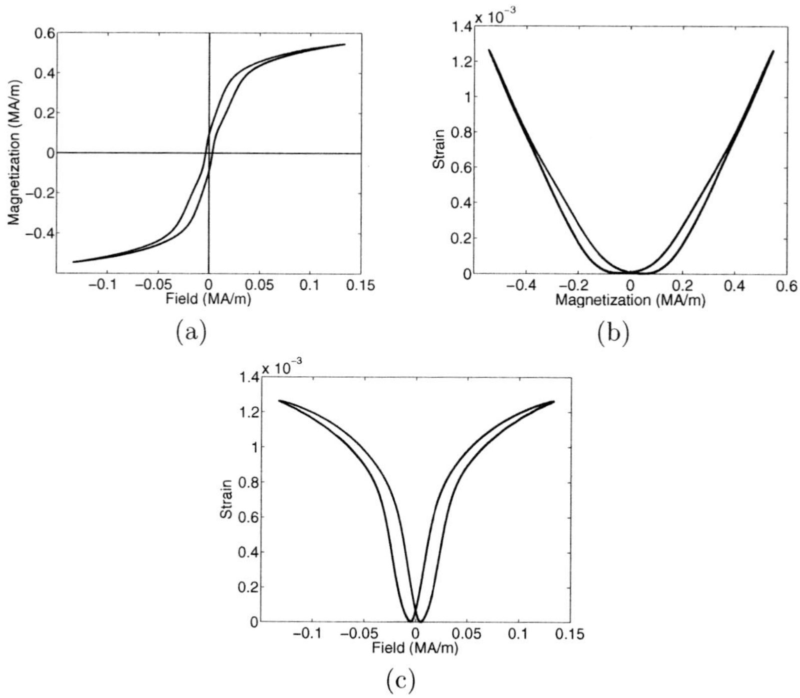

Figure 4.14. *Terfenol-D data from [119] collected at 1.0 Hz with a prestress of 1 ksi (6.9 MPa): (a) H-M behavior, (b) M-ε relation, and (c) H-ε dependence.*

Villari and Direct Magnetomechanical Effects

Even the most optimistic reader has probably concluded that the converse magnetomechanical effect comprised of magnetic-to-elastic interactions is pretty complicated — and with good reason, because it is! The good news for researchers but bad news for readers looking for an easy answer is that the direct magnetomechanical effect which provides ferromagnetic materials with sensor capabilities is even more complex! In general, the Villari effect, which constitutes changes in magnetization due to applied stresses, is due to a number of mechanisms, all of which are coupled. We summarize here mechanisms pertinent to sensor design and indicate issues which must be addressed in both present and future models.

The direct magnetomechanical effect can be broadly categorized as reflecting two cooperative phenomena; (i) the magnetization is driven to the anhysteretic magnetization M_{an}, and (ii) the behavior of M_{an} changes due to a number of mechanisms including stress-dependent behavior of local coercive and effective fields. Aspects of these phenomena are illustrated in Figures 4.15–4.17.

Approach to M_{an}

Figure 4.15 illustrates the manner through which application of an applied compressive stress drives the magnetization near positive and negative remanence

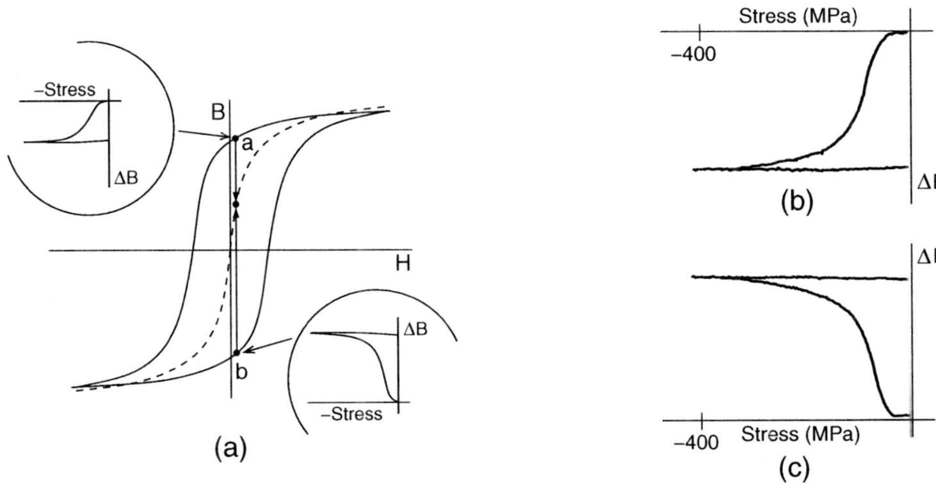

Figure 4.15. *(a) Manner through which the magnetization near positive remanence is driven to the anhysteretic curve through application of compressive stresses; (b) and (c) data from Pitman [381] quantifying the σ-B behavior for steel near positive and negative remanence.*

($H = 80$ A/m) to M_{an}. As detailed in Pitman [381], a steel specimen was driven to both positive and negative saturation and then held at a constant field value of 80 A/m while compressive stresses were applied and subsequently released. A comparison of the data plotted in Figures 4.15(a) and (b) illustrates that in both cases, the magnetization was driven to M_{an} by input stresses of approximately $\sigma = -300$ MPa. This reflects the fact that the stresses are sufficiently large to eliminate *local minima* associated with pinning sites so that the magnetization equilibrates to the *global minimum* associated with M_{an}. In other words, *local coercive fields* have been reduced to zero. Close examination of the σ-M relations upon stress release reveals that they are not constant and hence the global energy minimum associated with M_{an} is also stress-dependent.

Stress-Dependence of M_{an}

The stress-dependence of M_{an} and M is further illustrated by the H-M data from Pitman [381] which is plotted in Figure 4.16. It is observed that as stresses are decreased from +100 MPa to −400 MPa, M_{an} transitions from almost constant behavior at $\pm M_s$ to a highly mollified curve with decreased maximal values. This indicates that the variability in local effective fields increases as compressive stresses are increased. Similar behavior is shown for 68 permalloy in [189] where the variability in local fields is attributed to stress-induced formation of magnetic easy axes.

An additional hallmark of stress-dependence on the magnetization is the asymmetry in σ-M behavior for tensile and compressive stresses as illustrated in Figure 4.17 with steel data from Craik and Wood [106]. This data illustrates both the approach to the anhysteretic and the stress-dependence of M_{an} thus indicating the

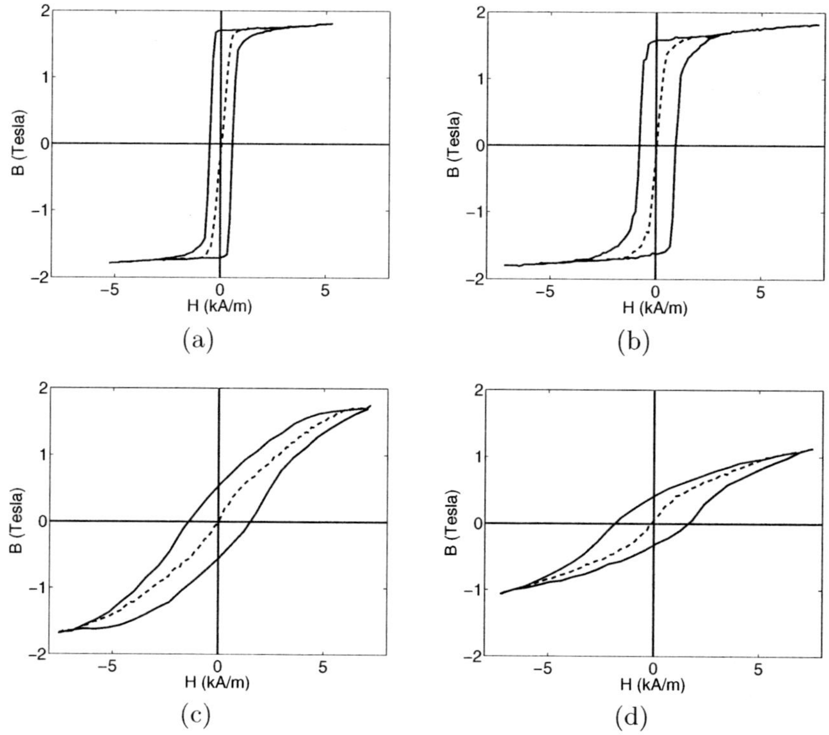

Figure 4.16. *Hysteretic (—) and anhysteretic (- - -) H-B behavior of steel data from Pitman [381] for differing input stress levels: (a) 100 MPa, (b) -100 MPa, (c) -200 MPa, and (d) -400 MPa.*

complexity of the direct magnetomechanical effect. Further experimental results illustrating the magnetomechanical effect can be found in [42] and macroscopic theory focused on aspects of the phenomenon are provided by Jiles [247].

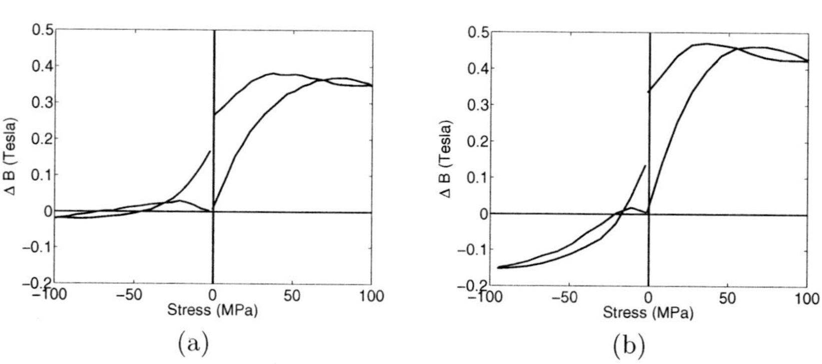

Figure 4.17. *σ-B behavior of steel data from Craik and Wood [106] at field levels of (a) 80 A/m and (b) 132 A/m.*

4.2 Fundamental Energy Relations

The microscopic and macroscopic behavior of ferromagnetic materials is defined by the exchange, anisotropy, magnetostatic and magnetomechanical energies, and hence a complete energy-based theory quantifying hysteresis and magnetic properties of these materials must include, or at least, accommodate these contributions. The magnetostatic energy quantifies long range interactions between magnetic moments and applied fields and is significant in both ferromagnetic and paramagnetic materials. Additionally, ferromagnetic materials exhibit exchange interactions between neighboring atomic spins which serves to align moments. This effect is short range and typically influences nearest, or next-nearest, neighbors. The exchange interactions and associated exchange energy also differentiate ferromagnetic and paramagnetic compounds. The anisotropy energy quantifies changes in the internal energy of materials due to changes in the direction of the magnetization. While magnetic anisotropy can be due to a number of factors, we will focus primarily on crystalline and stress anisotropies. Finally, the magnetoelastic energy quantifies magnetomechanical coupling inherent to the materials.

The first statistical mechanics model for the exchange interactions was posed by Ising in 1925 and was based on the assumption of a linear lattice of magnetic dipoles in which only neighbors interact [237]. This was later extended by Heisenberg in 1928 to include quantum effects and complete correlation between neighboring electrons having overlapping wave functions [214]. In the Heisenberg model, the interaction energy between spins $\mathbf{S}_i$ and $\mathbf{S}_j$ is

$$U_E = -2\mathcal{J}\mathbf{S}_i \cdot \mathbf{S}_j$$

where $\mathcal{J}$ is an exchange integral with $\mathcal{J} > 0$ for ferromagnetic materials and $\mathcal{J} < 0$ for antiferromagnetic compounds. The exchange energy for the system is obtained by summing over all magnetic moments which yields

$$U_E = -2\sum_{ij} \mathcal{J}_{ij}\mathbf{S}_i \cdot \mathbf{S}_j. \tag{4.16}$$

The Ising model can be obtained from that of Heisenberg by truncating the interaction energy for the lattice. For example, if quantization is assumed to take place only in the z-direction (e.g., the direction of an applied field $\mathbf{H}$), the full inner product $\mathbf{S}_i \cdot \mathbf{S}_j = S_{ix}S_{jx} + S_{iy}S_{jy} + S_{iz}S_{jz}$ is replaced in the Ising model by the z-component $S_{iz}S_{jz}$. If the restricted spins are denoted by $s_i = \pm 1$, the Ising relation for the exchange energy can be formulated as

$$U = -2\sum_{\langle ij\rangle} \mathcal{J}_{ij}s_i s_j \tag{4.17}$$

where the notation $\langle ij\rangle$ indicates summation over nearest neighbors. In addition to the simplification provided by reduced dimensionality, the Ising model admits a classical treatment of the spins, due in part to the fact that spins commute in the truncated expansion and hence can be treated as variables, whereas spins must be treated as quantum mechanical operators in the Heisenberg model. While the Ising model proves sufficiently accurate when characterizing the exchange energy in a number of applications, it is necessary to include the quantum effects incorporated in the Heisenberg model to quantify mean field properties from first principles.

As detailed in [4,461], the Heisenberg energy relation (4.16) is exactly solvable for only a few cases, one of which is the Ising model, and is completely isotropic. The inclusion of the anisotropy and magnetostatic energy relation in the Heisenberg Hamiltonian or energy formulation yields a model that is prohibitively expensive to approximate due to its generality. This necessitates the consideration of simplified theories motivated by the quantum energy relations but tractable for implementation.

One technique in this vein is the theory of micromagnetics which originated with the work by Landau and Lifshitz in 1935 on the analysis of domain walls [282]. Significant contributions to the theory were made by Brown [59–61] and subsequent researchers. The basic tenet in this approach is to employ classical theory in combination with quantum principles to predict the distribution of spins through the minimization of general energy relations of the form

$$U = U_E + U_A + U_\lambda + U_M \tag{4.18}$$

where U_E, U_A, U_λ and U_M respectively denote the exchange, anisotropy, magnetoelastic and magnetostatic energies. The capability this theory provides for predicting domain structures in ferromagnetic materials is illustrated in [512] where a micromagnetic energy relation of the form (4.18), with specific energy components

$$U_E = -2\sum_x \sum_y \sum_z \mathcal{J} \mathbf{S}_i \cdot \mathbf{S}_j$$

$$U_A = -2\sum_x \sum_y \sum_z \left[K_1\left(\alpha_1^2\alpha_2^2 + \alpha_2^2\alpha_3^2 + \alpha_3^2\alpha_1^2\right) + K_2\alpha_1^2\alpha_2^2\alpha_3^2\right]$$

$$U_\lambda = -\frac{3}{2}\sum_x \sum_y \sum_z \lambda\sigma\cos^2\phi$$

$$U_M = -\frac{1}{2}\sum_x \sum_y \sum_z \mathbf{m}_i \cdot \mathbf{H}_e$$

was minimized on a Cray X-MP/22 for a cubic grid consisting of $22 \times 22 \times 22$ exchange coupled spins. Here K_1 and K_2 are anisotropy constants and α_1, α_2 and α_3 are direction cosines of a moment $\mathbf{m}_i$ at the center of the cube. Furthermore, λ, σ and ϕ respectively denote the magnetostriction constant, magnetoelastic stress and angle between the magnetization and stress. Finally, $\mathbf{H}_e$ denotes the effective field at $\mathbf{m}_i$ due to all moments.

In addition to the capabilities that this theory provides for predicting the domain structure and dynamics of ferromagnetic materials, it is established in [4] that this approach provides a set of differential equations for which the Stoner–Wohlfarth model is an eigenfunction or mode.

Due to their complexity, however, models derived through micromagnetic principles presently preclude real-time implementation. The predictive capabilities provided by the models include more detail than is necessary when quantifying material behavior for subsequent system and control design. In subsequent sections, we develop macroscopic models which incorporate components of the microscopic energy relations but are sufficiently efficient to permit transducer design and model-based control implementation.

4.3 Linear Models

The H-M, H-B and H-ε data plotted in Figures 4.14 and 4.16 illustrate that hysteresis and constitutive nonlinearities are inherent properties of ferromagnetic materials due to their underlying domain structures. For biased, low drive regimes, however, linear approximations can be employed with reasonable accuracy which greatly facilitates model development, transducer characterization, and model-based control design. In this section, we summarize linear models which prove adequate for a number of smart material designs. However, a note of caution comes with these models since they neglect the hysteresis and nonlinearities which can be dominant features of ferromagnetic materials in moderate to high drive regimes.

4.3.1 Linear Constitutive Relations

The development of linear constitutive relations is analogous to that detailed in Section 2.2 for ferroelectric compounds and details regarding the general procedure can be found there. This similarity is encapsulated in the designation *piezomagnetic* constitutive relations which is motivated by the similar, linear *piezoelectric* relations.

For fixed temperatures, restriction of (4.3) to linear regimes yields

$$B_n = \mu^\sigma_{nm} H_m$$

whereas stress-induced changes in $\mathbf{B}$ are quantified by the relation

$$B_n = d_{nij}\sigma_{ij}.$$

The superscript σ designates that the permeability is evaluated at fixed stress and $d_{nij} \equiv \frac{dB_n}{d\sigma_{ij}}|_H$ denotes a piezomagnetic coefficient. Hooke's law is invoked to quantify elastic interactions, and linear magnetostrictive effects are modeled by the relation

$$\varepsilon_{ij} = d^*_{nij} H_n$$

where $d^*_{nij} \equiv \frac{d\varepsilon_{ij}}{dH_n}|_\sigma$ denotes a second piezomagnetic coefficient. Combination of these relations yields the linear piezomagnetic equations

$$\begin{aligned} \varepsilon_{ij} &= s^H_{ijk\ell}\sigma_{k\ell} + d^*_{nij} H_n \\ B_n &= d_{nij}\sigma_{ij} + \mu^\sigma_{nm} H_m. \end{aligned}$$

A comparison with (2.3) and (2.4) demonstrates the similarity with their ferroelectric counterparts. Moreover, one can invoke elastic and magnetic symmetries to formulate the corresponding matrix relations

$$\begin{aligned} \varepsilon &= s^H \sigma + d^* H \\ B &= d\sigma + \mu^\sigma H \end{aligned} \tag{4.19}$$

which are analogous to (2.5).

For applications in which formulation in terms of the magnetization is preferable to the flux, the direct magnetoelastic relation can be reformulated using (4.4) to yield

$$\begin{aligned} \varepsilon &= s^H \sigma + d^* H \\ M &= d\sigma + \chi^\sigma H. \end{aligned} \tag{4.20}$$

In both (4.19) and (4.20), d^* denotes the transpose of d.

4.3.2 Energy Formulation

The construction of underlying energy relations and an encompassing thermodynamic framework parallels the theory detailed in Section 2.2 so we provide here only a summary of the ferromagnetic relations. As before, we focus on low drive regimes for which linear, reversible, ferromagnetic and magnetomechanical relations provide sufficient accuracy.

The choice of thermodynamic potentials or energy functionals depends on the choice of inputs or independent variables. Two possibilities relevant for transducer design are the Helmholtz and Gibbs energies.

The Helmholtz energy employs (ε, M) as an independent variable set, and under the assumption of linear magnetomechanical coupling, is given by

$$\psi(\varepsilon, M) = \frac{1}{2} Y^M_{ijk\ell} \varepsilon_{ij} \varepsilon_{k\ell} - a_{nij} \varepsilon_{ij} M_n + \frac{1}{2} \alpha^\varepsilon_{nm} M_m M_n.$$

The independent variables dictate the exact differential

$$d\psi = \sigma_{ij} d\varepsilon_{ij} + H_n dM_n$$

which yields the necessary conditions

$$\sigma_{ij} = \left[\frac{\partial \psi}{\partial \varepsilon_{ij}} \right]_M \quad , \quad H_n = \left[\frac{\partial \psi}{\partial M_n} \right]_\varepsilon$$

and the corresponding linear constitutive relations

$$\begin{aligned} \sigma_{ij} &= Y^M_{ijk\ell} \varepsilon_{k\ell} - a_{nij} M_n \\ H_n &= -a_{nij} \varepsilon_{ij} + \alpha^\varepsilon_{nm} M_m. \end{aligned} \tag{4.21}$$

If one considers the analogy between H and E, M and P, the relation (4.21) is equivalent to the ferroelectric relation (2.20).

The Gibbs energy is formulated in terms of (σ, H) and can be obtained from the Helmholtz energy through the negative Legendre transform

$$G = \psi - \mu_0 HM - \sigma\varepsilon.$$

From the Gibbs relation

$$G(\sigma, H) = -\frac{1}{2}s^H_{ijk\ell}\sigma_{ij}\sigma_{k\ell} - \mu_0 d_{nij}\sigma_{ij}H_n - \frac{1}{2}\mu_0\chi^\sigma_{nm}H_m H_n, \tag{4.22}$$

one obtains the necessary conditions

$$M_n = -\frac{1}{\mu_0}\left[\frac{\partial G}{\partial H_n}\right]_\sigma \quad , \quad \varepsilon_{ij} = -\left[\frac{\partial G}{\partial \sigma_{ij}}\right]_H,$$

and the constitutive relations

$$\begin{aligned} \varepsilon_{ij} &= s^H_{ijk\ell}\sigma_{k\ell} + d_{nij}H_n \\ M_n &= d_{nij}\sigma_{ij} + \chi^\sigma_{nm}H_m. \end{aligned} \tag{4.23}$$

We make two observations regarding the constitutive relations (4.23). The first is that they are identical to (4.20) which is not surprising — the constitutive behavior encompassed in (4.20) is employed when constructing the energy functional. Secondly, the functional (4.22) and constitutive relations (4.23) are appropriate for low drive actuator characterization since they are formulated in terms of H and σ which are physically relevant input variables.

Different choices of independent variable sets yield elastic and magnetic Gibbs relations and linear constitutive equations analogous to those summarized in Table 2.1 for ferroelectric materials. Similar relations can be formulated in terms of the flux B.

4.3.3 Model Attributes

The linear constitutive relations have been incorporated in a number of finite element codes, including NASTRAN, ABAQUS, FEMLAB and the Magsoft package ATILA, and hence provide a highly flexible framework for characterizing linear ferromagnetic behavior in a wide range of devices having highly varied geometries. However, we reiterate the warning that because they neglect the inherent hysteresis and quadratic M-ε behavior illustrated in Figure 4.14, they should be employed only in operating regimes where linear approximations have been demonstrated to provide reasonable accuracy.

4.4 Higher–Order Energy Models

The Helmholtz and Gibbs energy relations, summarized in the last section, provide constitutive relations which characterize linear elastic, magnetic and magnetoelastic interactions but neglect hysteresis and nonlinearities such as the quadratic M-ε

relation shown in Figure 4.14. In this section, we develop higher-order energy relations which incorporate some of these phenomena and provide a basis for hysteresis models developed in later sections. The development is analogous to that provided in Section 2.3 for ferroelectric materials and details regarding the general approach can be found there.

We first construct Helmholtz relations ψ through three techniques: (i) mean field theory, (ii) high-order polynomials, and (iii) piecewise quadratic polynomials. The Gibbs relations

$$G(H, M, T) = \psi(M, T) - \mu_0 HM \tag{4.24}$$

or

$$G(\sigma, H, \varepsilon, M, T) = \psi(M, T) - \mu_0 HM - \sigma\varepsilon \tag{4.25}$$

provide functionals whose minima $\frac{\partial G}{\partial \varepsilon} = 0$ and $\frac{\partial G}{\partial M} = 0$ determine the manner through which strains and magnetization adjust to minimize energy.[14] Equivalently, the resulting necessary conditions

$$\sigma = \left[\frac{\partial \psi}{\partial \varepsilon}\right]_{M,T} , \quad H = \frac{1}{\mu_0}\left[\frac{\partial \psi}{\partial M}\right]_{\varepsilon,T} \tag{4.26}$$

can be interpreted as specifying how internal energies adjust to balance external stresses and magnetic fields. The relations (4.26) also determine constitutive relations appropriate for certain operating regimes.

4.4.1 Boltzmann Energy Relations

We consider first the development of a Helmholtz energy relation ψ based on statistical mechanics principles. This theory is based on a simplification of the Ising relation (4.17) for the exchange energy which has the requisite assumption that spins, or magnetic moments, are restricted to two possible orientations $s_i = \pm 1$.

The assumption that spins have two preferred orientations appears at first to be highly restrictive but can in fact be physically motivated for a number of regimes. From a classical perspective, this will be at least approximately true for materials having uniaxial crystalline anisotropies or systems in which uniaxial stresses dominate the crystalline structure. The first case includes materials such as cobalt and some rare earth metals and alloys (e.g., terbium single crystals) whereas the second encompasses transducer designs in which compressive or tensile prestresses are employed to optimize outputs — e.g., Terfenol-D transducers having the configuration depicted in Figures 1.11 and 4.6 and the magnetomechanical mechanisms described in Section 4.1.8. From a quantum perspective, this assumption can be motivated by noting that spins cannot be uniformly oriented and one permissible orientation is parallel and opposite to an applied field. This interpretation helps explain the accuracy of the theory for quantifying the behavior of materials such as iron which exhibit cubic anisotropies.

[14]See Section 2.2.3 for details motivating the formulation of G in terms of both the order parameters ε, M and conjugate fields σ, H.

The theory is based on an approximation to the Ising model first proposed by Gorsky in the analysis of order-disorder transitions in binary alloys [191]. In this context, it was later extended by Bragg and Williams to include the concept of long range order [52, 53, 511]. An underlying tenet in the Bragg–Williams theory, which simplifies subsequent computations, is the assumption that the energy of an individual atom is determined by the average order of the system rather than the fluctuating states of adjacent atoms. For this reason, the model is often termed the mean field or molecular field approximation to the Ising model. To construct a ferromagnetic model, we make the same assumption regarding magnetic moments or spins. Further details about this approach, including some discussion concerning its application to ferromagnetic materials, can be found in [198, 376].

As in Section 2.3.2, we consider an arbitrary, homogeneous lattice of volume V comprised of $N = N_+ + N_-$ cells, each of which is assumed to contain one spin or magnetic moment. In accordance with the Ising assumptions, spin orientations are constrained to be $s_i = \pm 1$ so that N_+ and N_- respectively denote the number of positive and negative spins in the lattice. If each dipole has moment m_0, the magnetization for the lattice is

$$M = \frac{m_0}{V} \sum_{i=1}^{N} s_i.$$

The development at this point is analogous to that provided in Section 2.3.2 for ferroelectric materials so we summarize only the results pertinent to ferromagnetic materials. To quantify the energy required to reorient spins or moments, we employ the assumption that the average exchange energy Φ is given by

$$\Phi = \frac{\Phi_0 M}{M_s}$$

where

$$M_s = \frac{N m_0}{V}$$

is the technical saturation magnetization which occurs when moments are fully aligned. The energy Φ_0 required to reorient a single moment in a fully ordered lattice is related to the exchange integral $\mathcal{J}$ employed in (4.17) by the expression

$$\Phi_0 = 2\xi\mathcal{J}$$

where ξ denotes the number of neighbors adjacent to a site. Hence $\xi = 2$ for a 1-D lattice chain, $\xi = 4$ for a 2-D rectangular lattice, and $\xi = 6, 8$ or 12, respectively, for 3-D cubic, body-centered cubic, or face-centered cubic lattices.

Quantification of changes in internal energy in a manner analogous to (2.43) yields the internal energy relation

$$U_E = \frac{\Phi_0 N}{2V}\left(1 - \frac{M^2}{M_s^2}\right) + U_0.$$

As before, we take $U_0 = 0$ since we are interested in relative rather than absolute measures of energy.

Consideration of an entropy relation analogous to (2.46), derived using Boltzmann's relation, yields the Helmholtz energy expression

$$\begin{aligned}\psi(M,T) &= \frac{H_h M_s}{2}\left[1-(M/M_s)^2\right]\\ &+\frac{H_h T}{2T_c}\left[M\ln\left(\frac{M+M_s}{M_s-M}\right)+M_s\ln\left(1-(M/M_s)^2\right)\right]\end{aligned} \tag{4.27}$$

for the lattice. Here $T_c = \frac{\Phi_0}{2k}$ and $H_h = \frac{N\Phi_0}{2VM_s}$ respectively denote the Curie point and a bias field. The behavior of ψ is plotted in Figure 4.18 to illustrate that for $T < T_c$, it exhibits double-well behavior characteristic of the ferromagnetic phase whereas single-well, paramagnetic behavior results for $T > T_c$.

For stresses below a coercive stress σ_c where ferroelastic switching commences, the coupled magnetoelastic Helmholtz energy is taken to be

$$\begin{aligned}\psi(\varepsilon,M,T) &= \frac{1}{2}Y^M\varepsilon^2 - a_1\varepsilon M - a_2\varepsilon M^2 + \frac{H_h M_s}{2}\left[1-(M/M_s)^2\right]\\ &+\frac{H_h T}{2T_c}\left[M\ln\left(\frac{M+M_s}{M_s-M}\right)+M_s\ln(1-(M/M_s)^2)\right]\end{aligned} \tag{4.28}$$

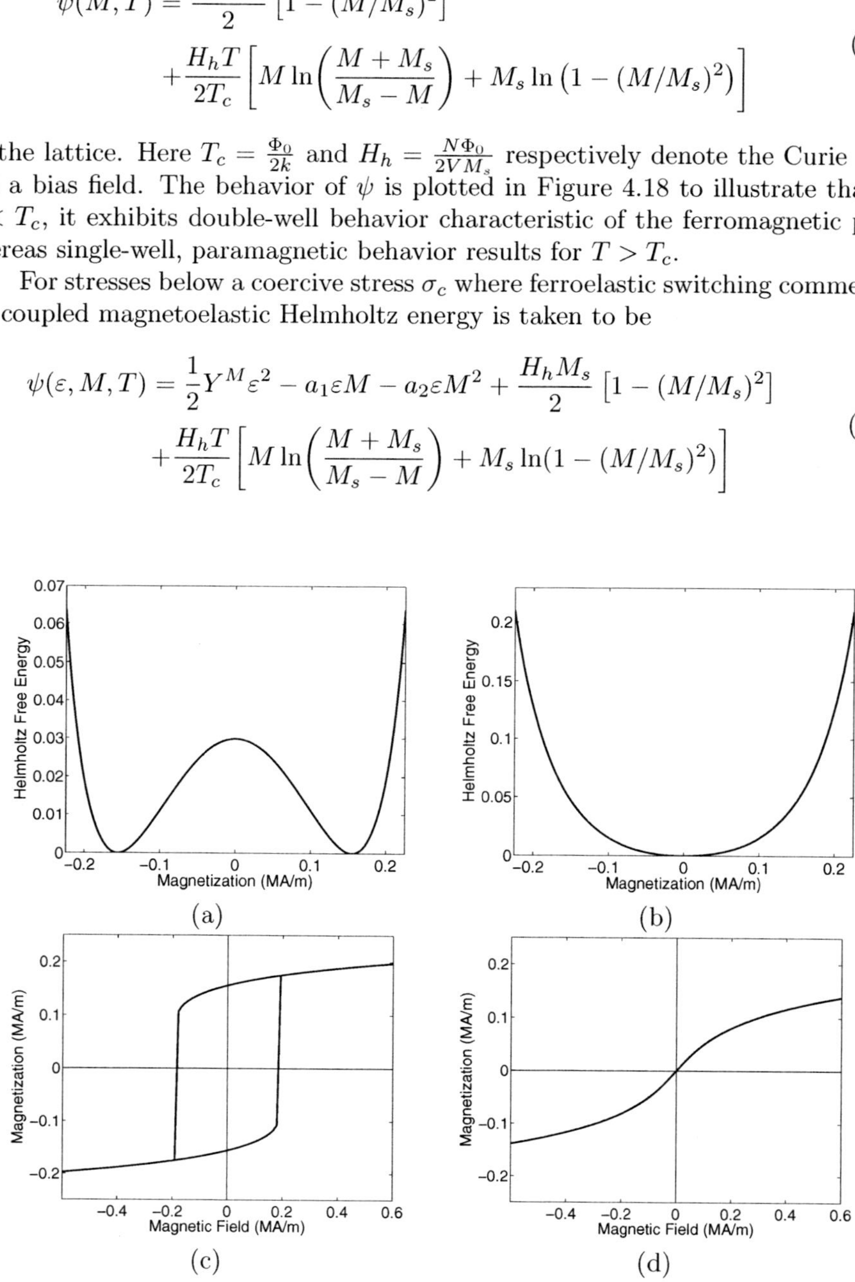

Figure 4.18. *Helmholtz energy specified by (4.27) for (a) $T < T_c$ and (b) $T > T_c$. Corresponding H-M relation given by (4.29) with (c) $T < T_c$ and (d) $T > T_c$.*

and the Gibbs energy is given by (4.25). The necessary conditions (4.26) subsequently yield the constitutive relations

$$\begin{aligned}\sigma &= Y^M\varepsilon - a_1 M - a_2 M^2 \\ \mu_0 H &= -a_1\varepsilon - 2a_2\varepsilon M - \frac{H_h}{M_s}M + \frac{H_h T}{T_c}\operatorname{arctanh}(M/M_s).\end{aligned} \tag{4.29}$$

It is observed that (4.29) includes the linear relations (4.21) but also incorporates the quadratic dependency (4.15) as well as nonlinear saturation behavior associated with the arctangent function. We note that these relations are similar to the nonlinear constitutive relations employed by Duenas, Hsu and Carman [139].

4.4.2 Truncated Polynomial Energy Relations

The Helmholtz relation (4.27) encapsulates the ferromagnetic-paramagnetic behavior associated with second-order phase transitions but also inherits restrictions associated with the underlying mean field assumptions. One technique for generalizing these relations is consideration of truncated series expansions with general coefficients.

Series expansion of the logarithmic components of (4.28) gives

$$\begin{aligned}\psi(\varepsilon, M, T) &= \frac{1}{2}Y^M\varepsilon^2 - a_1\varepsilon M - a_2\varepsilon M^2 + \frac{E_h M_s}{2}\left[1 - (M/M_s)^2\right] \\ &+ \frac{H_h T}{2T_c}\left[\frac{M_s^2}{2}\left(\frac{M}{M_s}\right)^2 + \frac{M_s^4}{4\cdot 3}\left(\frac{M}{M_s}\right)^4 + \frac{M_s^6}{6\cdot 5}\left(\frac{M}{M_s}\right)^6 + \cdots\right].\end{aligned}$$

Truncation after quartic components and reformulation in terms of general, positive, coefficients yields

$$\psi(\varepsilon, M, T) = \frac{1}{2}Y^M\varepsilon^2 - a_1\varepsilon M - a_2\varepsilon M^2 + \psi_0(T) + \alpha_1(T - T_c)M^2 + \alpha_2 M^4 \tag{4.30}$$

which is the magnetization equivalent of the polarization relation (2.36). Enforcement of the necessary conditions (4.26) in this case yields the nonlinear constitutive relations

$$\begin{aligned}\sigma &= Y^M\varepsilon - a_1 M - a_2 M^2 \\ \mu_0 H &= -a_1\varepsilon - a_2\varepsilon M + 2\alpha_1(T - T_c)M + 4\alpha_2 M^3.\end{aligned} \tag{4.31}$$

The behavior of ψ for $\varepsilon = 0$, $\psi_0(T) = 0$, $\alpha_1 > 0$, $\alpha_2 > 0$ and T above and below T_c is plotted in Figure 4.19(a) and (b) with the corresponding H-M behavior, obtained by inverting the converse relation (4.31), plotted in Figure 4.19(c) and (d). It is observed that this choice of energy functional provides a kernel which exhibits instantaneous transitions in the ferromagnetic regime and single-valued behavior for the paramagnetic phase.

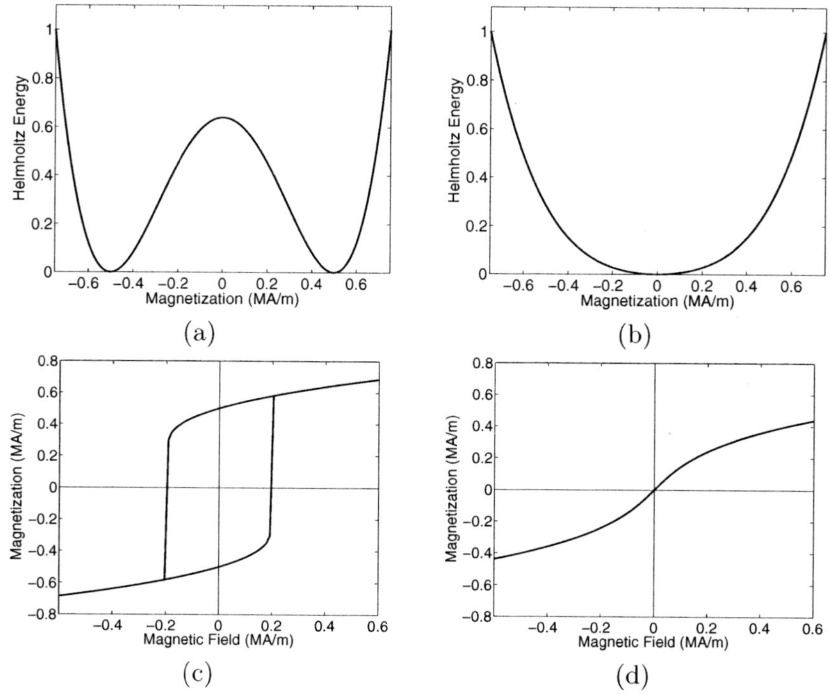

Figure 4.19. *Helmholtz energy (4.30) with $\varepsilon = 0$ for (a) $T < T_c$ and (b) $T > T_c$. Corresponding H-M relation given by (4.31) with (c) $T < T_c$ and (d) $T > T_c$.*

4.4.3 Piecewise Polynomial Relations

The Helmholtz relations (4.27) and (4.30) encapsulate the ferromagnetic double well potential and associated hysteresis through high-order polynomials derived either through statistical mechanics principles or phenomenological arguments. While efficient to formulate, their efficiency and accuracy can be encumbered by their high-order and limited definition. An alternative is to employ a piecewise quadratic definition for ψ.

In a manner analogous to that detailed in Section 2.3.3, Taylor expansions of ψ, given by (4.27), about the three equilibria for fixed $T < T_c$ motivate the Helmholtz relation

$$\psi(M) = \begin{cases} \frac{1}{2}\eta(M + M_R)^2 & , \ M \leq -M_I \\ \frac{1}{2}\eta(M - M_R)^2 & , \ M \geq M_I \\ \frac{1}{2}\eta(M_I - M_R)\left(\frac{M^2}{M_I} - M_R\right) & , \ |M| < M_I. \end{cases} \tag{4.32}$$

In the absence of applied stresses, the Gibbs energy is given by (4.24). The energy relation and hysteron resulting from the necessary condition $\frac{\partial G}{\partial M} = 0$ are plotted in Figure 4.20. It is observed that M_I and M_R denote the inflection point and positive magnetization which minimizes ψ. Because M_R minimizes G in the absence of an applied field H, it can also be interpreted as a local remanence magnetization.

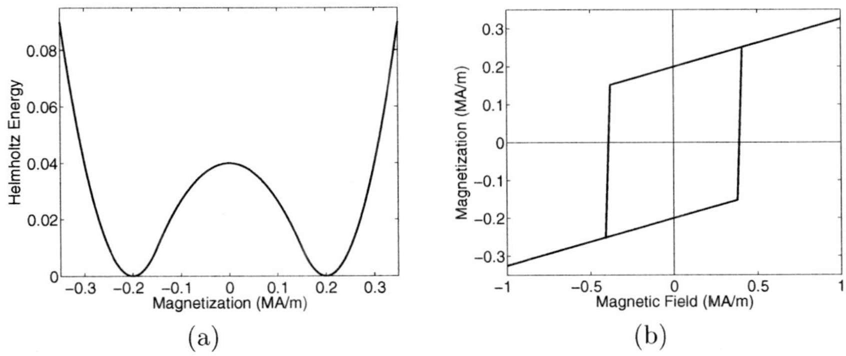

Figure 4.20. *(a) Helmholtz energy ψ given by (4.32) and (b) H-M relation resulting from the equilibrium condition* $\frac{\partial G}{\partial M} = 0 \Leftrightarrow H = \frac{\partial \psi}{\partial M}$.

In contrast to the kernels provided by the energy relations (4.28) and (4.30), the kernel resulting from the piecewise quadratic Helmholtz relation (4.32) does not saturate. Due to its simplicity, (4.32) is often employed in the homogenized energy framework of Section 4.7. When operating near saturation, however, (4.28) may provide greater accuracy due to the saturating nature of the kernel.

4.5 Preisach Models

The Preisach model for ferromagnetism has its genesis in the 1935 paper by Ferenc (Franz) Preisach [382] in which he proposed a model based on the concept that hysteresis in magnetic materials could be represented as a superposition of elementary kernels termed *hysterons*. In a series of papers beginning in 1952, Everett and colleagues independently initiated and investigated an alternative form of the framework [151–154] — they first cite the paper by Preisach in [152] where they point out that it was only recently being referenced in English texts. Whereas initially developed in the context of absorption processes, the authors point out that the method is applicable for magnetic, ferroelectric, chemical, biological and economic applications thus providing an initial indication regarding the breadth of the approach. During the 1970's, Krasnosel'skiĭ and his collaborators realized that the Preisach–Everett model encapsulated a rich mathematical framework which they investigated from an abstract perspective [268]. This established operator properties of the framework as well as addressed issues regarding model well-posedness and completeness (or lack thereof) of underlying function spaces. It is a testimonial to the breadth of the methodology that investigations focused on its physical interpretation, mathematical framework, and extension to a wide variety of physical applications still constitute a highly active research area.

A topic of contention regarding Preisach models centers around the question of whether or not they are based on physical principles. The original theory was based on physically-motivated mechanisms but did not provide an energy basis for predicting magnetic behavior. The mathematical foundations provided by Krasnosel'skiĭ et al. and subsequent theoreticians clarified ambiguities and demonstrated the broad applicability of the framework, but also emphasized it as a mathematical rather than physical entity. However, parallel investigations have also established that a broad range of physical insights are provided by the framework and are in fact necessary to extend it to accommodate the vast array of mechanisms which contribute to hysteretic phenomena. Hence the power of Preisach models is due to the synergy between physical and mathematical principles rather than an isolated dependence on either discipline.

The literature on Preisach techniques for modeling hysteresis in magnetic materials is vast and we refer the reader to [40,126,334] for a comprehensive treatment of this field. Rather than attempt to survey the topic, we focus on specific facets that are pertinent for model-based control design and relate to alternative models discussed in subsequent sections. In particular, we will provide a number of comparisons between Preisach models and the homogenized energy framework described in Section 4.7 which provides an energy basis for certain extended Preisach formulations [447]. The section is configured as follows.

Section 4.5.1: Classical Preisach Model. Categorized by the conditions of *congruency* and *deletion*, this formulation provides a basis for analysis that encompasses a number of materials and operating regimes. It has the following limitations, some of which are discussed in subsequent sections:

- Inaccurate representation of reversible material behavior;
- Does not characterize noncongruencies measured in various regimes;
- Model does not incorporate frequency, rate, stress or temperature dependent mechanisms;
- Classical model does not incorporate nonclosure properties including accommodation and after-effects;

Section 4.5.2: Specific Density Choices. The choice of normal and lognormal densities yields models having only 4–5 parameters, some of which have physical interpretations.

Section 4.5.3: Reversibility. Incorporated through a modification of the kernel.

Section 4.5.4: Noncongruency. The inclusion of field or magnetization components in the density yields noncongruent minor loops in accordance with measured data.

Section 4.5.5: Minor Loop Nonclosure. We summarize the manner through which accommodation and after-effects can be incorporated in the model.

Section 4.5.6: Generalized Kernels. The discussion summarizes energy-based kernels employed in the framework of Section 4.7, Krasnosel'skiĭ–Pokrovskiĭ kernels amenable to approximation and parameter estimation, and piecewise linear kernels which can be directly inverted for model-based control design.

Section 4.5.7: Model Inverses. Techniques to exactly or approximately invert Preisach models to provide inverse filters for linear control design are summarized.

4.5.1 Classical Preisach Model

We define the *classical Preisach model* as that determined by the conditions of *congruency* and *deletion* — which will be defined later in the section — and refer to models modified to incorporate additional physical criteria as *extended Preisach formulations*. We follow the discussion, but not the notation, from Mayergoyz [332–335] and refer the reader there for additional details. The resulting magnetization model can be compared with its ferroelectric counterpart in Section 2.4.2.

The fundamental element in the formulation is the relay or kernel $[k_s(v,\xi)](t)$ depicted in Figure 4.21. Here v and ξ respectively denote the input and initial state and $s = (s_1, s_2)$ is a point in the Preisach plane

$$S = \{(s_1, s_2) | s_1 \le s_2\} \tag{4.33}$$

depicted in Figure 4.22. The values s_1 and s_2 can be considered as thresholds at which the kernel switches. Hence $k_s(v,\xi)$ switches from -1 to $+1$ as v is increased past s_2 with an opposite switch occurring as v is decreased through s_1. It is observed that knowledge about the initial condition is lost once a switch occurs and the subsequent state is determined completely by the last threshold or switch. Finally, t can be interpreted as time or, more generally, as any parameterization specifying the behavior of v.

In a loose sense, the hysteron or kernel depicted in Figure 4.21 can be interpreted as representing the state of a single moment in a lattice. A negative moment corresponds to the -1 designation whereas $+1$ denotes a positive moment. As a result, some authors employ the moment strengths $\pm m$ rather than ± 1 in the definition of the relay.

Macroscopic hysteretic behavior is represented as a superposition of these kernels or hysterons,

$$f(t) = [\mathcal{F}(v,\xi)](t) = \iint\limits_S [k_s(v,\xi)](t)\nu(s)\,ds \tag{4.34}$$

where ν denotes a weight or density and $\mathcal{F}$ denotes the Preisach operator.[15] We

[15]The notation $[k_s(v,\xi)](t)$ is convenient when establishing theoretic properties of Preisach operators whereas $k_s(v(t),\xi)$ indicates the dependence of v on t. We employ the former to remain consistent with certain theoretic treatments of the framework and provide a common notation with the energy models discussed in Section 4.7.

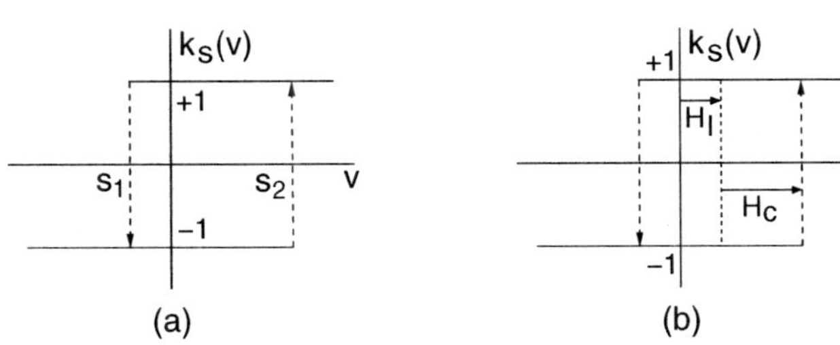

Figure 4.21. *(a) Classical Preisach relay with threshold values at s_1 and s_2 and (b) relay defined in terms of an interaction field H_I and local coercive field H_c.*

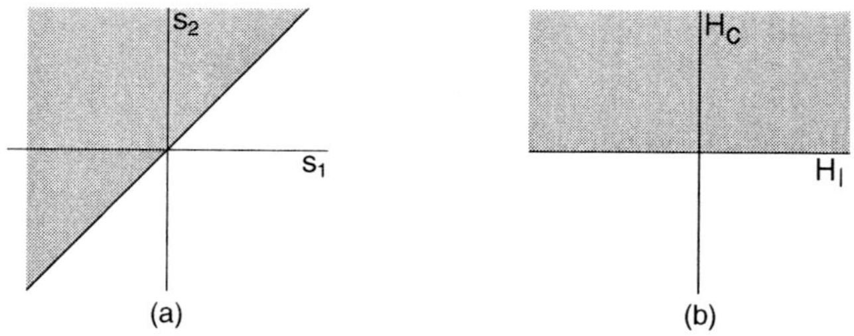

Figure 4.22. *(a) Preisach plane S given by (4.33) and (b) domain of integration for the magnetization relation (4.35).*

note that whereas the notation differs, the formulation (4.34) is identical to the definition

$$f(t) = \iint\limits_{\alpha \geq \beta} \widehat{\gamma}_{\alpha\beta} u(t) \mu(\alpha, \beta) d\alpha d\beta$$

employed by Mayergoyz if one makes the identifications: $\alpha = s_2$, $\beta = s_1$, $\widehat{\gamma}_{\alpha\beta} = k_s$, $\mu = \nu$ and $u = v$. However, the formulation (4.34) avoids ambiguity associated with previous definitions for the variables α, β, γ, u and v.

For magnetic systems, $u(t) = H(t)$ and $f(t) = M(t)$ whereas $u(t) = E(t)$ and $f(t) = P(t)$ for the ferroelectric models of Section 2.4.2. This provides an initial indication regarding the generality of the framework.

Coercive and Interaction Fields

It is observed that the majority of relays will be biased so that $s_1 \neq -s_2$. This can be interpreted as being due to the presence of an interaction field H_I resulting from adjacent particles. If we let H_c denote the local coercive field required to initiate a jump when $H_I = 0$, then we can change coordinates via the relations

$$H_I = \frac{s_1 + s_2}{2} \quad , \quad H_c = \frac{s_2 - s_1}{2}$$

or, equivalently,

$$s_2 = H_I + H_c \quad , \quad s_1 = H_I - H_c$$

as depicted in Figure 4.21(b). In this case, the magnetization can be represented as

$$M(t) = \int_0^\infty \int_{-\infty}^\infty [k_{H_I,H_c}(v,\xi)](t)\nu(H_I,H_c)\, dH_I dH_c \tag{4.35}$$

where the the Jacobian $J = 2$ has been incorporated in the density. The domain of integration is illustrated in Figure 4.22(b) and results from the property that $H_I \in \mathbb{R}$ whereas $H_c \in \mathbb{R}^+$ due to the condition that $s_1 \leq s_2$.

We introduce the formulation (4.35) for two reasons. (i) It indicates physical relations between the abstract definition of the relay, and (ii) it provides a formulation which permits direct comparison with models derived in Section 4.7 through energy principles.

Definition of the Preisach Kernel

Figure 4.21 illustrates the behavior of the multivalued relay but does not provide a definition. To uniquely define the state at a parameter value t, we define the initial condition by

$$[k_s(v,\xi)](0) = \begin{cases} -1 & \text{if} \quad v(0) \le s_1 \\ \xi & \text{if} \quad s_1 < v(0) < s_2 \\ +1 & \text{if} \quad v(0) \ge s_2 \end{cases} \tag{4.36}$$

where $\xi \in \{-1, 1\}$. The kernel is then defined by

$$[k_s(v,\xi)](t) = \begin{cases} [k_s(v,\xi)](0) & \text{if} \quad \tau(t) = \emptyset \\ -1 & \text{if} \quad \tau(t) \ne \emptyset \text{ and } v(\max \tau(t)) = s_1 \\ +1 & \text{if} \quad \tau(t) \ne \emptyset \text{ and } v(\max \tau(t)) = s_2 \end{cases} \tag{4.37}$$

where

$$\tau(t) = \{t_s \in (0,t] \,|\, v(t_s) = s_1 \text{ or } v(t_s) = s_2\} \tag{4.38}$$

defines times (parameter values) at which thresholds are reached. We noted that the definitions (4.36)–(4.38) are identical to (2.56)–(2.58) for ferroelectric materials.

Correspondence Between Kernels and Preisach Plane

The general relation (4.34) mathematically defines the Preisach operator. Much of its interpretation, however, is provided by the fact that the kernels k_s are in a one to one correspondence with points on the Preisach plane S. To motivate the manner through which geometric properties of the Preisach plane correspond to physical H-M behavior of ferromagnetic materials, we employ the physically-motivated assumption that ν has compact support and consider the triangular region

$$S_\triangle = \{(s_1, s_2) \,|\, \underline{s}_1 \le s_1 \le s_2 \le \overline{s}_2\} \tag{4.39}$$

depicted in Figure 4.23. This encompasses all regimes in which symmetric major loops are closed and hence does not significantly restrict the discussion. We also assume that all moments are negatively oriented at first so the material has initial magnetization $-M_s$. In the context of the Preisach plane, this implies that all kernels k_s are initialized to -1.

Symmetric Major Loops:

The process used to characterize a symmetric major loops is illustrated in Figure 4.23(a)–(d). Starting from negative saturation, the behavior associated with an increasing field is characterized by the threshold value s_2 required to switch from a negative to positive state — see Figure 4.21(a). Hence for the input field H_1 shown in Figure 4.23(b), all moments having a threshold below this value will switch, thus partitioning the Preisach plane into sets $S^+(t)$ and $S^-(t)$ of positively and negatively oriented moments separated by an interface line $L(t)$. Further field increase yields the saturation magnetization M_s. For $\frac{dv}{dt} = \frac{dH}{dt} < 0$, the state of the kernel k_s is determined by the left threshold s_1 so reduction of the field to $v = H_2$ yields a vertical interface line $L(t)$ as depicted in Figure 4.23(d).

Biased Minor Loops:

The same process characterizes biased minor loops as shown in Figure 4.23(e)–(g). Increase of the field to $v = H_3$ and subsequent decrease to $v = H_4$ respectively yields horizontal and vertical interface lines due to the dependence on s_2 for $\frac{dH}{dt} > 0$ and s_1 for $\frac{dH}{dt} < 0$. An important feature is noted when the field is further decreased

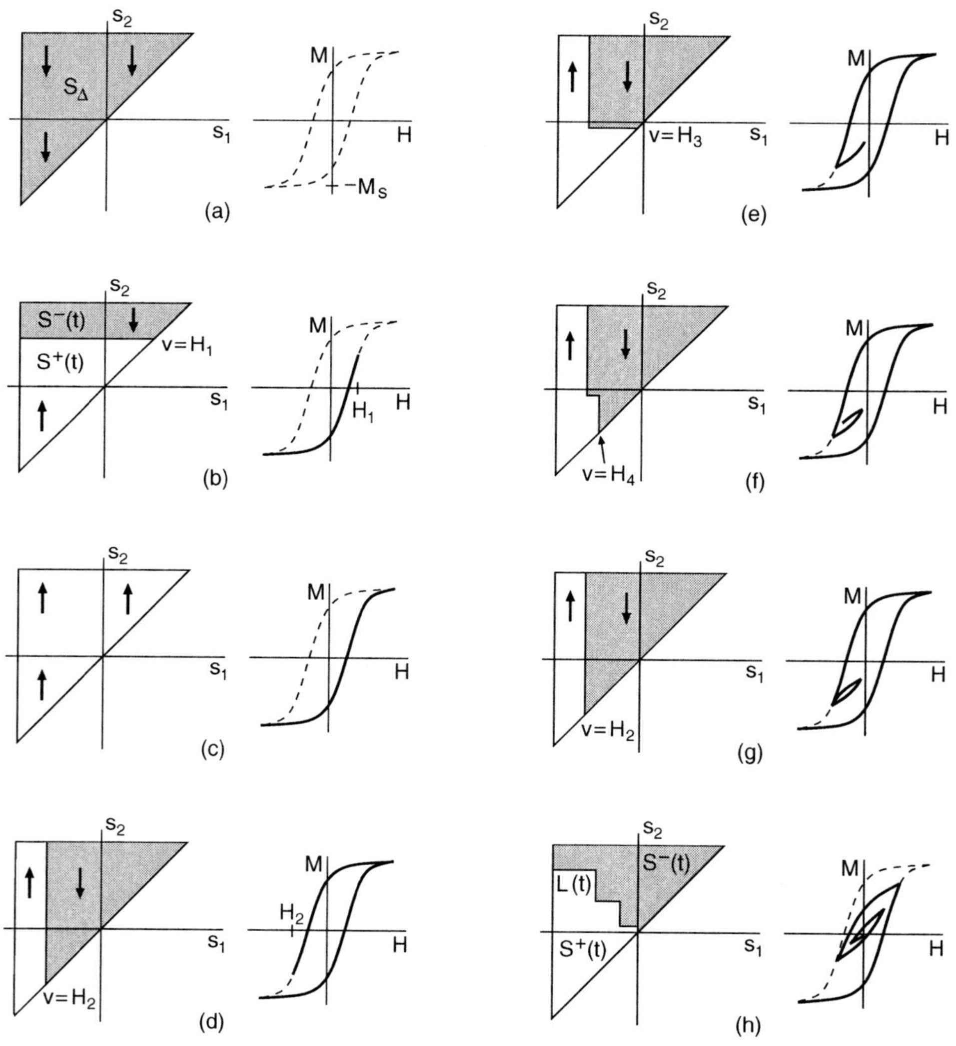

Figure 4.23. *Equivalence between the partitioning of the Preisach plane and H-M behavior in major loops, biased minor loops, and the demagnetization process.*

to $v = H_2$. The partitioning of the Preisach plane is returned to that of the symmetric major loop. This illustrates the *deletion* or *wiping out* property which has the following consequences. (i) Physically, it ensures the closure of biased minor loops which is experimentally observed in most quasistatic operating regimes. (ii) In the Preisach plane, all record of interior excursions are removed and the state is returned to that present when the loop was initiated. This is true for multiply nested loops and constitutes one of the necessary and sufficient conditions for classical Preisach representations and a property which must be relaxed for certain dynamic regimes.

Demagnetization Process:

The demagnetization process is illustrated in Figure 4.23(h). Physically, demagnetization is achieved by applying a decaying periodic field with mean $H = 0$. We note that application of the procedure with a set of DC field values yields the global anhysteretic curves discussed in Section 4.1.6.

Equivalent Model Definition

Figure 4.23 illustrates that for any parameterized input value $v(t)$, the Preisach plane can be partitioned into sets $S^+(t)$ and $S^-(t)$ separated by the interface line $L(t)$. We formally define these regions and then pose the model in terms of S^+ and S^- to provide a framework which facilitates implementation.

Definition 4.5.1. *Consider densities ν defined on the compact region $S_\triangle$ defined in (4.39). For a piecewise monotone input $v(t)$, $S^+(t), S^-(t)$ and $L(t)$ are defined by*

$$
\begin{aligned}
S^+(t) &= \{s = (s_1, s_2) \in S_\triangle \mid [k_s(v,\xi)](t) = 1\} \\
S^-(t) &= \{s = (s_1, s_2) \in S_\triangle \mid [k_s(v,\xi)](t) = -1\} \\
L(t) &= \overline{S}^+(t) \cap \overline{S}^-(t)
\end{aligned}
$$

where $\overline{S}^+$ and $\overline{S}^-$ denotes the closure of the sets.

It follows immediately that the Preisach operator can be formulated as

$$
f(t) = [\mathcal{F}(v,\xi)](t) = \iint\limits_{S^+(t)} \nu(s)ds - \iint\limits_{S^-(t)} \nu(s)ds. \tag{4.40}
$$

As detailed in Section 1.4 of Mayergoyz [334], the representation (4.40) can be exploited for numerical approximation of the model through explicit evaluation of the integrals.

Deletion, Congruency and a Representation Theorem

One of the strengths of the classical Preisach formulation is that it can be completely characterized in terms of two easily verified conditions: congruency and deletion. Consider first deletion (often termed *wiping out*) which is illustrated in Figure 4.23(d) and (g).

Definition 4.5.2. ***Deletion:*** *For any parameter t at which $v(t)$ achieves a local minimum, all vertices on $L(t)$ with s_2-coordinates greater than this minimum are deleted. Alternatively, each local maximum of $v(\cdot)$ deletes vertices on $L(t)$ having s_1-coordinates below the maximum.*

Intuitively, the deletion property states that once the state of the Preisach plane is returned to the state when the minor loop was initialized, all history of interior partitions is deleted. From a physical perspective, this implies that biased minor loops must close and, once closed, all record of interior excursions is eliminated.

For a number of materials operating in quasistatic or low frequency regimes, minor loop closure is observed and hence the property is realized. As illustrated in Section 4.1.7, however, thermal and rate-dependent effects such as accommodation and after-effects negate the deletion property thus motivating extended Preisach models.

Definition 4.5.3. ***Congruency:*** *All minor loops generated by inputs $v(t)$ oscillating between values v_m and v_M are congruent, regardless of the output level.*

Congruent minor loop behavior was illustrated in Figure 2.4.3 for ferroelectric materials and is depicted in Figure 4.24 in the context of the Preisach plane. Simply put, congruency can be attributed to the fact that congruent triangles S_M are added to the sets $S^+(t)$ and $S^-(t)$ employed in the formulation (4.40) thus producing

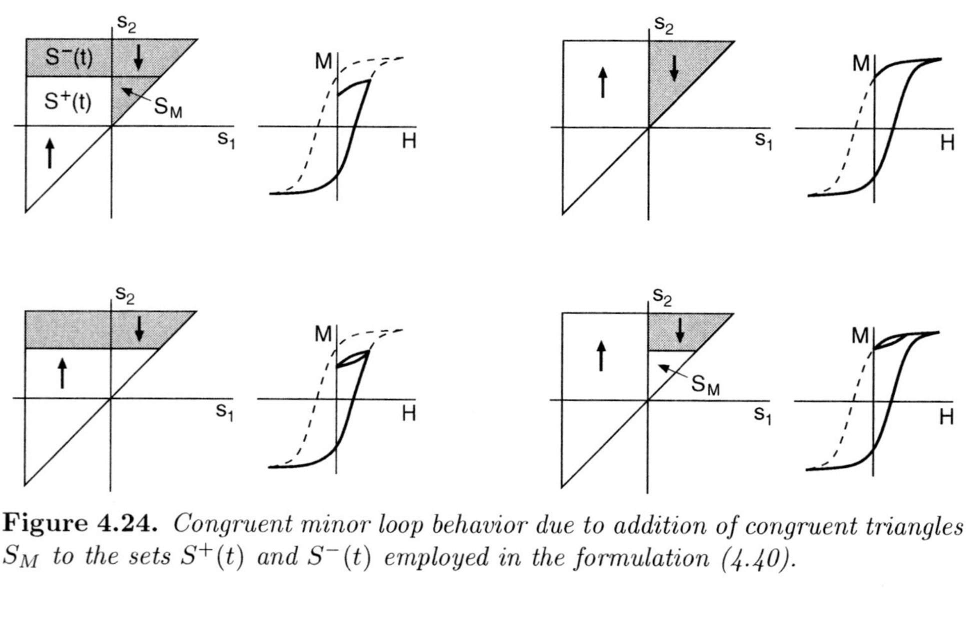

Figure 4.24. *Congruent minor loop behavior due to addition of congruent triangles S_M to the sets $S^+(t)$ and $S^-(t)$ employed in the formulation (4.40).*

equivalent increments in the output f. In the context of magnetization, this states that application of oscillatory fields bounded by H_m and H_M produces congruent minor loops regardless of the magnetization level.

Whereas reasonable for certain operating regimes, it is illustrated in Section 4.5.4 that congruency is often not achieved at high magnetization levels. This motivates consideration of extended Preisach formulations such as moving Preisach models.

The following theorem from Mayergoyz [334] establishes the equivalence between the deletion and congruency properties and classical Preisach representations.

Theorem 4.5.4. *Consider piecewise monotone inputs $v(\cdot)$. The deletion and congruency properties constitute necessary and sufficient conditions for a hysteresis nonlinearity to be represented by the classical Preisach formulation (4.34).*

One of the strengths of this result is the fact that it can be readily applied to experimental data to determine the adequacy of the model. For a given material and operating conditions, one simply needs to check these two conditions to ascertain whether the formulation will provide sufficient accuracy or if the extensions outlined in later sections must be considered.

Determination of $\nu(s)$

To construct the model (4.34) or (4.40), it is necessary to determine the density $\nu(s)$. One option is to employ least squares parameter estimation techniques analogous to those detailed in Section 2.6.6 for the homogenized ferroelectric energy model. Alternatively, one can employ the following approach to directly formulate ν in terms of the second derivative of f.

As detailed in Section 3 of [334], one can consider a negatively saturated material followed by the application of an increasing input to s_2' followed by a decreasing input to s_1'. This forms a *first-order transition* (*reversal*) curve as depicted

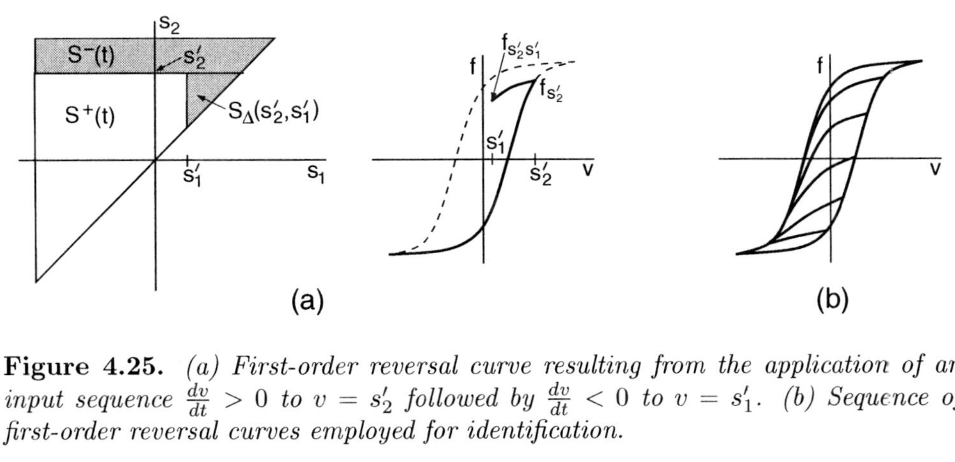

Figure 4.25. *(a) First-order reversal curve resulting from the application of an input sequence $\frac{dv}{dt} > 0$ to $v = s_2'$ followed by $\frac{dv}{dt} < 0$ to $v = s_1'$. (b) Sequence of first-order reversal curves employed for identification.*

in Figure 4.25. Now define the function

$$F(s_2', s_1') = \frac{1}{2}\left(f_{s_2'} - f_{s_1'}\right) \tag{4.41}$$

where $f_{s_2'}, f_{s_1'}$ denote the output values at the input turning points. By noting that the triangle $S_\triangle(s_2', s_1')$ is subtracted from S^+ and added to S^-, it follows that

$$F(s_2', s_1') = \iint\limits_{S_\triangle(s_2', s_1')} \nu(s)ds = \int_{s_1'}^{s_2'} \left[\int_{s_1}^{s_2'} \nu(s)ds_2\right] ds_1. \tag{4.42}$$

This implies that

$$\nu(s_2', s_1') = -\frac{\partial^2 F(s_2', s_1')}{\partial s_1' \partial s_2'} = \frac{1}{2}\frac{\partial^2 f_{s_2', s_1'}}{\partial s_1' \partial s_2'}. \tag{4.43}$$

In theory, one can then employ a sequence of first-order reversal curves, of the type depicted in Figure 4.25(b), to identify ν.

A significant practical difficulty lies in the fact that this necessitates taking two derivatives of data which is well known to be ill-posed in the sense that noise is significantly augmented. One possibility is to fit a smooth surface to the data to facilitate stable differentiation [190,227]. Alternatively, it may be necessary to employ least squares formulations analogous to those detailed in Section 2.6.6 to estimate ν. In both cases, the accuracy of ν will be determined by the 'richness' of data employed in the identification process. This is illustrated in Figure 4.25(b) where it is observed that a coarse set of reversal curves will produce a coarse estimated of ν.

4.5.2 Specific Density Choices

Formulation of the model (4.34) in terms of a general density ν provides significant flexibility and accuracy if a sufficiently comprehensive data set is employed for identification. However, a disadvantage associated with this generality is the fact that the model requires a large number of parameters, none of which are directly correlated with physical properties of the measured data — other than the second derivative. This model construction also necessitates a complete re-identification of parameters if material behavior changes due to operating conditions — e.g., hysteresis properties are dependent on temperature.

An alternative is to consider parameterized or functional relations for ν which incorporate *a priori* material properties and require a significantly reduced set of parameters. To illustrate, consider the formulation (4.35) posed in terms of local coercive fields H_c and interaction fields H_I. For certain materials, it is established in Della Torre [126] that use of the central limit theorem yields a normal or Gaussian distribution for H_I. The positivity of H_c is enforced by considering it to be either lognormally distributed or having a normal distribution with positive arguments. The latter choice yields the joint density

$$\nu(H_I, H_c) = \frac{M_s}{2\pi\sigma_c\sigma_I} e^{-(H_c - \overline{H}_c)^2/2\sigma_c^2} e^{-H_I^2/2\sigma_I^2} \tag{4.44}$$

where $\overline{H}_c$ denotes the mean local coercive field and σ_I^2, σ_c^2 are variances. In this case, the saturation magnetization M_s can be directly ascertained from the data and $\overline{H}_c$ can be approximated by the measured bulk coercivity. Moreover, σ_I and σ_c are qualitatively related to the degree of pre-remanent switching and slope at coercivity in a manner analogous to that described in Section 2.6.11. Hence the *a priori* density choice significantly facilitates model construction and updating.

The trade-off is a potential loss in accuracy for certain operating regimes. This is illustrated in Section 2.6.10 in the context of ferroelectric characterization using the related homogenized energy model.

4.5.3 Reversibility

In the classical Preisach model formulated in terms of the piecewise constant hysteron k_s depicted in Figure 4.21, reversible behavior occurs along the horizontal arm of the hysteron once switching occurs. Hence the classical Preisach model predicts that H-M curves will have zero slope once ascending and descending branches merge. As illustrated in Figure 4.26, this is in contrast with the physical behavior exhibited by many magnetic materials which is characterized by reversible H-M behavior with $\frac{dM}{dH} \downarrow 0$ as $H \to \infty$ (the notation $\downarrow$ indicates a limit from above).

This has been addressed by various authors by considering the magnetization to be the sum of irreversible and reversible components,

$$M = M_{irr} + M_{rev} = \iint_S [k_s(v,\xi)](t)\nu(s)ds + M_{rev}. \tag{4.45}$$

As discussed by Della Torre [126], one option for defining M_{ref} is to construct general hysterons comprised of irreversible and reversible components in the manner depicted in Figure 4.27. This yields a magnetization-dependent reversible component which exhibits properties of the Ising or Langevin kernels developed in Section 4.5.6 and employed in the homogenized energy model detailed in Section 4.7.

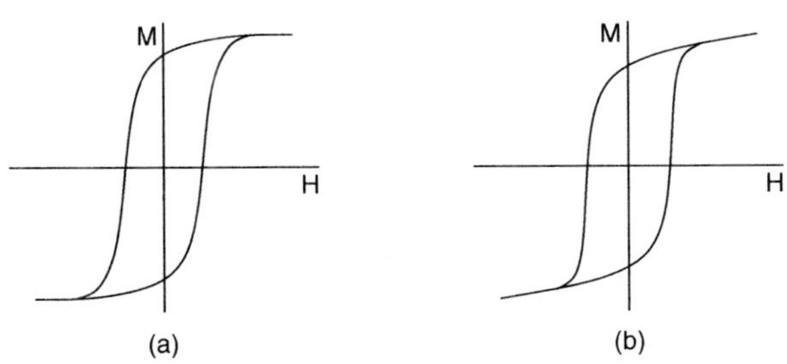

Figure 4.26. *(a) Behavior provided by the classical Preisach model (4.34) predicting reversible behavior with* $\frac{dM}{dH} = 0$ *and (b) physical behavior exhibited by magnetic materials where* $\frac{dM}{dH} \downarrow 0$ *as* $H \to \infty$.

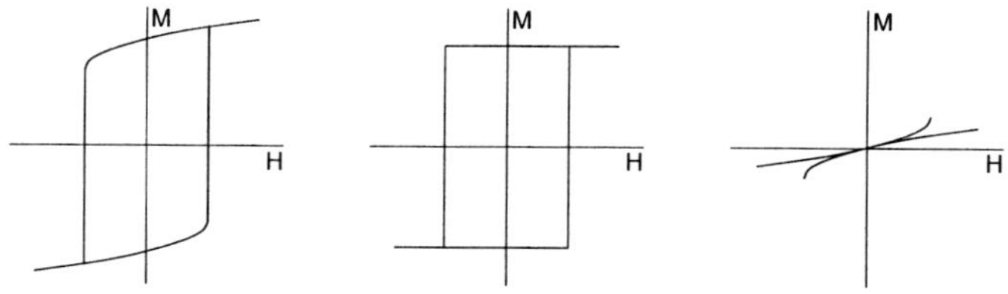

Figure 4.27. *General hysteron comprised of irreversible and reversible components.*

4.5.4 Noncongruency

It was emphasized in Section 4.5.1 that congruency is necessary for a classical Preisach formulation. Unfortunately, this condition is somewhat restrictive and a number of materials exhibit minor loops which are magnetization-dependent and hence noncongruent.

To provide a simple illustration of physical noncongruency, consider a single Stoner–Wohlfarth particle which exhibits the H-M behavior depicted in Figure 4.28. The behavior on each branch is reversible so the two minor loops between H_m and H_M are simply the upper and lower branches between these two values. Due to the saturation behavior exhibited by the particle, the slopes at all points between H_m and H_M will differ thus determining that the minor loops are noncongruent.

Whereas a variety of techniques have been proposed to circumvent this problem, most typically do so by replacing the density $\nu(s)$ by more general relations. Two which we summarize are nonlinear input-dependent densities $\nu(s, v)$ and output-dependent relations $\nu(s, f)$.

We consider first the former while recalling that in the context of ferromagnetic materials, v and f respectively designate the magnetic field H and magnetization M. Following the development in Chapter 2 of Mayergoyz [334], we consider the representation

$$f(t) = [\mathcal{F}(v,\xi)](t) = \iint_{S^+(t)} \nu(s, v(t))ds - \iint_{S^-(t)} \nu(s, v(t))ds + f_{rev} \tag{4.46}$$

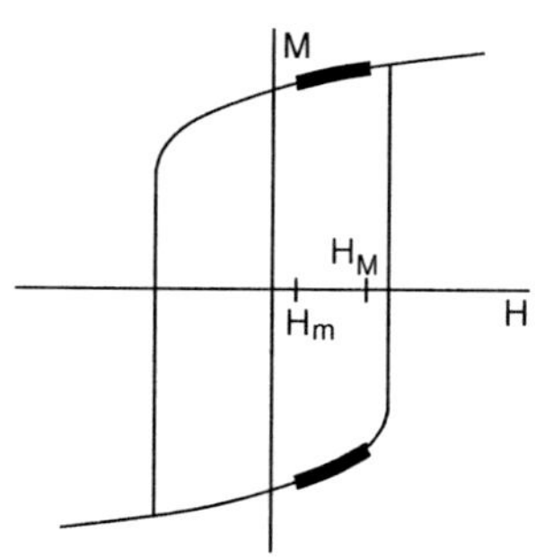

Figure 4.28. *The H-M behavior exhibited by a single Stoner–Wohlfarth particle and noncongruent minor loop behavior in the interval $[H_m, H_M]$ on the reversible upper and lower hysteresis branches.*

which is a nonlinear generalization of (4.40) that additionally incorporates the reversible effects discussed in Section 4.5.3. To construct the model, it is necessary to identify the input(field)-dependent density $\nu(s, v)$.

As detailed in Mayergoyz, this can be accomplished by considering second-order reversal curves generated when a monotone increasing input is applied following the monotone decreasing input used to generate the first-order curve — see Figure 4.29. The output value corresponding to an input v on the second-order reversal curve is denoted by $f_{s_2',s_1',v}$. By defining a function analogous to F in (4.41) and integrating over an appropriate region, one can show that

$$\nu(s_2', s_1', v) = \frac{1}{2}\frac{\partial^2 f_{s_2',s_1',v}}{\partial s_2' \partial s_1'} \tag{4.47}$$

which is analogous to (4.43).

Whereas (4.47) provides an explicit relation for ν, its implementation is hampered by the fact that a highly comprehensive set of measurements are required to accurately resolve ν. Furthermore, one must fit a smooth surface to the data to avoid the instabilities inherent to differentiating noisy data. Despite these difficulties, the nonlinear or input-dependent Preisach model (4.46) provides one technique for relaxing the congruency property for physical regimes for which noncongruency is observed.

Alternatively, one can employ mean field theory to express $\nu(s, f)$ in terms of the output $f = M$. This approach is motivated by applying central limit theorem arguments to an ensemble of randomly oriented particles to determine that the mean of the interaction field is

$$\overline{H}_i = \alpha M, \tag{4.48}$$

e.g., see [126].[16]

[16]The induced field $H_I = \alpha M$ follows from the mean theory in Section 4.4.1 and plays a fundamental role in the anhysteretic relations employed in the Jiles–Atherton theory of Section 4.6 and energy-based hysterons developed in Section 4.7. Hence it is central to a number of hysteresis models for ferromagnetic materials.

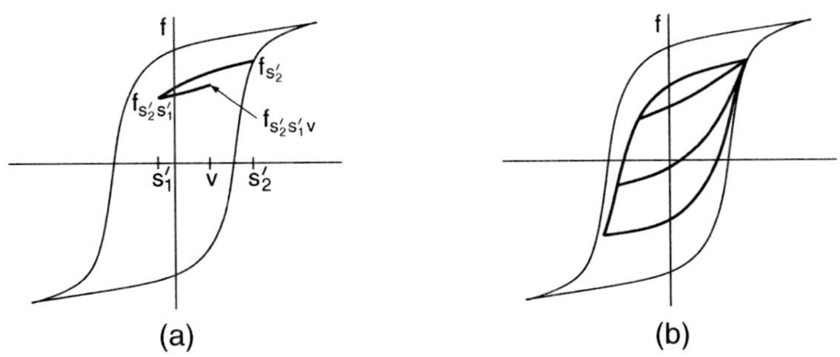

Figure 4.29. *(a) First and second-order reversal curves along the ascending hysteresis branch. (b) Family of second-order reversal curves employed in the identification process.*

Following the approach of Della Torre [126], the density ν in the general formulation (4.34) or (4.40) is taken to be

$$\nu(H_I, H_c, M) = \frac{M_s}{2\pi\sigma_c\sigma_I} e^{-(H_c-\overline{H}_c)^2/2\sigma_c^2} e^{-(H_I+\alpha M)^2/2\sigma_I^2} \tag{4.49}$$

which is a magnetization-dependent generalization of (4.44). Because the Gaussian distribution quantifying interaction field effects evolves with the magnetization, the model is termed a *moving Preisach model.*

Details regarding the latter model formulation and techniques for identifying ν are provided in [126,129]. Like the field-dependent model, this formulation provides a nonlinear technique for incorporating the noncongruency exhibited by certain materials in varied drive regimes.

4.5.5 Minor Loop Nonclosure

The classical Preisach model and the previously described extended Preisach formulations are derived under the assumptions that thermal relaxation mechanisms are negligible and hysteresis properties are rate-independent. For many operating regimes, these assumptions are reasonable and the previous models provide reasonable accuracy. In other cases, the after-effect and accommodation phenomena discussed in Section 4.1.7 cause nonclosure of minor loops thus negating the deletion property. We summarize briefly the nature of extensions developed to address these phenomena.

Magnetic After-Effects

There exist a number of techniques for incorporating after-effects in Preisach models and the reader is referred to [126, 334] for details regarding various approaches. To illustrate the issues, we summarize the extended framework proposed in [10, 126–129].

The crux of the formulation focuses on extending the definition of the hysteron $[k_s(v,\xi)](t)$ employed in (4.34) or (4.35) to incorporate thermal relaxation mechanisms modeled by the Arrhenius relation (4.6). To accomplish this, the energy levels required to overcome barriers are computed to obtain the differential equation

$$\mathcal{T}\frac{d[k_s(v,0)](t)}{dt} + [k_s(v,\xi)](t) = [k_s(v,\xi)](\infty) \tag{4.50}$$

for fixed input v, relaxation time $\mathcal{T}$ and steady state kernel $[k_s(v,\xi)](\infty)$. The solution of (4.50) is

$$[k_s(v,\xi)](t) = ([k_s(v,\xi)](0) - [k_s(v,\xi)](\infty))\, e^{-t/\mathcal{T}} + [k_s(v,\xi)](\infty) \tag{4.51}$$

where $[k_s(v,\xi)](0)$ denotes the initial state of the hysteron. The hysteretic output is subsequently modeled by the relation

$$f(t) = \iint_S [k(v,\xi)](t)\nu(s)\, ds$$

which is exactly the original formulation (4.34) — recall that $f(t) = M(t)$ and $v(t) = H(t)$ in the ferromagnetic model. The density $\nu(s)$ can either be treated as general parameter to be identified through the techniques described in Section 4.5.1 cr an *a priori* function such as (4.44) having a reduced number of physical parameters. It is observed that through the term $e^{-t/\mathcal{T}}$ in the kernel formulation (4.51), this extended model incorporates the relaxation and after-effect behavior depicted in Figure 4.10. Details regarding this model formulation can be found in [126].

Accommodation (Reptation)

The extension of Preisach models to incorporate the accommodation behavior depicted in Figure 4.11 requires that the deletion property be relaxed. This has been addressed through a variety of techniques but most involve the modification of the hysteron k_s to permit constant outputs other than ± 1. Details regarding possible extensions and associated identification procedures can be found in [126, 334].

4.5.6 Generalized Kernels

In Sections 4.5.3 and 4.5.5, we summarized certain modifications of the Preisach kernel made to accommodate reversibility and nonclosure properties exhibited by magnetic materials. These are by no means the only extensions that have been considered and we summarize here three other hysteron definitions which have been considered to address theoretical and practical considerations: (i) energy-based formulations, (ii) Krasnosel'skiĭ–Pokrovskiĭ (K–P) kernels, and (iii) general piecewise linear hysterons.

Energy-Based Kernels

The discussion in Section 4.4 focused on the construction of various formulations for the Helmholtz energy ψ and Gibbs energy G at the mesoscopic or lattice level. Enforcement of the necessary condition $\frac{\partial G}{\partial M} = 0$ or balancing G and the relative thermal energy kT/V through Boltzmann principles yields hysterons or kernels which form the basis for the homogenized energy model discussed in Section 4.7. It is noted there that the energy framework provides a basis for certain Preisach formulations thus motivating the consideration of energy-based kernels for extended Preisach representations.

To illustrate, consider the Helmholtz energy relation (4.27) derived through statistical mechanics principles and the piecewise quadratic approximation (4.32). In the absence of applied stresses, the corresponding Gibbs energy is given by (4.24).

Enforcement of the equilibrium condition $\frac{\partial G}{\partial M} = 0$ subsequently yields the hysteron relations

$$\overline{M}(H) = M_s \tanh\left(\frac{H + \alpha M}{a(T)}\right) \tag{4.52}$$

and

$$\overline{M}(H) = \frac{H}{\eta} + M_R \delta. \tag{4.53}$$

In the Ising relation, $\alpha = \frac{H_h}{\mu_0 M_s}$ and $a(T) = \frac{H_h T}{\mu_0 T_c}$ whereas $\delta = \pm 1$ in (4.53) with $\delta = 1$ for positive moments and $\delta = -1$ for negative moments. The two hysterons are depicted in Figure 4.30 along with the hysteron developed in Section 4.7.1 through Boltzmann principles. It is observed that the Ising kernel (4.52) incorporates the reversible behavior depicted in Figure 4.27 and noncongruency of the type exhibited by the Stoner–Wohlfarth H-M relation illustrated in Figure 4.28. It is illustrated in Section 4.7 that models employing the Boltzmann kernel depicted in Figure 4.30(c) can be used to characterize magnetic after-effects and accommodation since it incorporates thermal relaxation processes. Additional details regarding the energy formulation for these kernels is provided in Section 4.7.

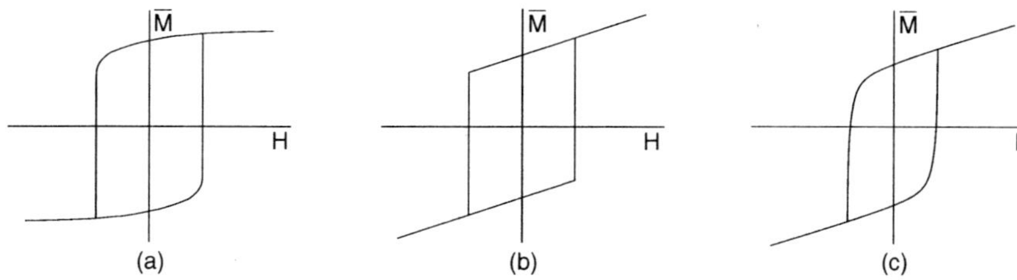

Figure 4.30. *Hysterons developed from energy principles. (a) $\overline{M}(H)$ given by (4.52), (b) $\overline{M}(H)$ specified by (4.53), and (c) $\overline{M}(H)$ given by (4.78) in Section 4.7.1.*

Krasnosel'skiĭ–Pokrovskiĭ (K–P) Kernels

A second choice of hysteron which provides improved continuity and approximation properties is based on the *play operators* proposed by Krasnosel'skiĭ and Pokrovskiĭ [268]. To define the hysteron, we consider translates $r(v-s_1)$ and $r(v-s_2)$ of a Lipschitz continuous ridge function $r(v)$. On each time interval $[t_{k-1}, t_k]$ where v is monotone, the K–P operator is defined recursively by

$$[k_s(v,\xi)](t) = [\mathcal{R}(v, \mathcal{R}_{k-1})](t) \quad , \quad t \in [t_{k-1}, t_k]$$

where

$$[\mathcal{R}(v,\mathcal{R}_{k-1})](t) = \begin{cases} \max\{\mathcal{R}_{k-1}, r(v(t) - s_2)\} & \text{if } v \text{ is non-decreasing} \\ \min\{\mathcal{R}_{k-1}, r(v(t) - s_1)\} & \text{if } v \text{ is non-increasing} \end{cases}$$

and

$$\mathcal{R}_k = \begin{cases} \mathcal{R}(v, \mathcal{R}_{k-1})(t_k) \ , & k = 2, \ldots, j \\ \mathcal{R}_0 = \xi \ , & k = 1, \xi \in \{-1, 1\} \end{cases}$$

defines the values of $\mathcal{R}$ at times t_k. Two choices for r are

$$r(v) = \begin{cases} -1 & , \quad v < -1 \\ v & , \quad -1 \le v \le 1 \\ 1 & , \quad v > 1 \end{cases} \tag{4.54}$$

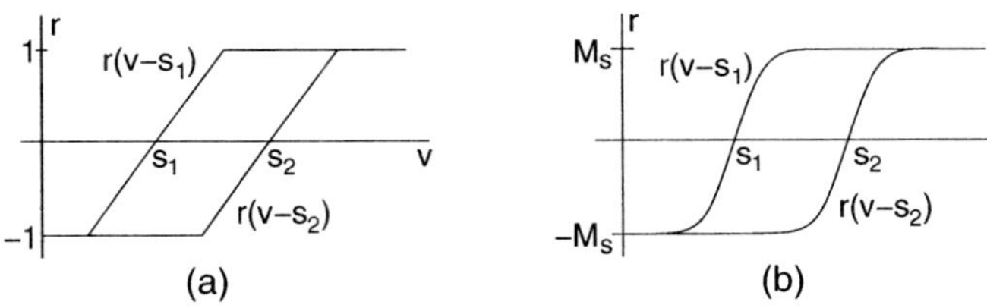

Figure 4.31. *Hysteresis envelops provided by ridge functions r given by (a) relation (4.54) and (b) relation (4.55).*

and

$$r(v) = M_s \tanh(av) \tag{4.55}$$

which yield the hysteresis envelops depicted in Figure 4.31. Further properties of r and k_s are depicted in Figures 2.22 and 2.23 of Section 2.4.

As detailed in [28,29,57,172,268,312], the K–P kernel provides additional continuity which facilitates aspects of approximation and identification. Additionally, the necessity and advantages of considering measure-based formulations (2.55) are provided in those references.

Piecewise Linear Kernel

A third choice for the kernel k_s addresses the issue of reversibility and provides a formulation which can be analytically inverted and employed in the manner depicted in Figure 4.32 as a filter for linear control design.

As depicted in Figure 4.33, this construction is piecewise linear and the threshold values s_1, s_2 are replaced by the points v_1, v_2, v_3 and v_4 which designate the switching between half-lines and line segments. When constructing the model, we employ the notation u and v from Figure 4.32 to indicate inputs and outputs as a prelude to constructing the inverse model in the next section.

Following the development in [477,478], a hysteresis loop is defined by the two half-lines

$$u(t) = \begin{cases} m_t v(t) + c_t \ , & v(t) > v_1 = \dfrac{c_t + m_\ell c_\ell}{m_\ell - m_t} \\ m_b v(t) + c_b \ , & v(t) < v_2 = \dfrac{c_b + m_r c_r}{m_r - m_b} \end{cases}$$

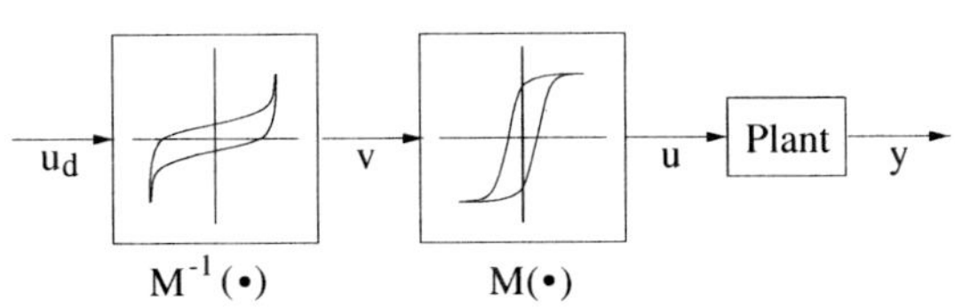

Figure 4.32. *Plant with actuator hysteresis $M(\cdot)$ and inverse filter $M^{-1}(\cdot)$ for the hysteresis.*

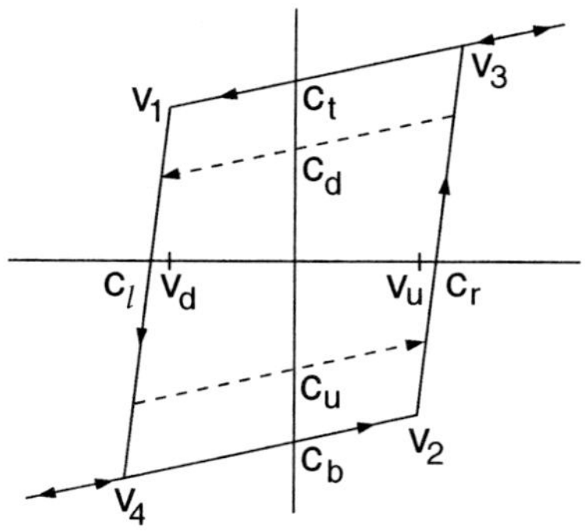

Figure 4.33. *Piecewise linear Preisach hysteresis kernel.*

and two line segments

$$u(t) = \begin{cases} m_r(v(t) - c_r) \ , & v_2 \le v(t) < v_3 = \dfrac{c_t + m_r c_r}{m_r - m_t} \\ m_\ell(v(t) - c_\ell) \ , & \dfrac{c_b + m_\ell c_\ell}{m_\ell - m_b} = v_4 < v(t) \le v_1 \end{cases}$$

where m_i denote slopes and c_i denote intercepts. The subscripts ℓ, r, b and t respectively indicate left, right, bottom and top.

The time derivatives of $u(t)$ and $v(t)$ are of constant sign along the line segments with $\frac{dv}{dt} > 0$ and $\frac{du}{dt} > 0$ for the segment $u(t) = m_r(v(t) - c_r)$. On the half line $u(t) = m_t v(t) + c_t$, $\frac{dv}{dt} < 0$ and $\frac{du}{dt} < 0$ if $v(t) < v_3$ but $\frac{dv}{dt}$ and $\frac{du}{dt}$ need only be the same sign if $v(t) > v_3$. Similar conditions apply for the remaining segments of the hysteresis model.

The line segments for inner hysteresis loops are defined by

$$u(t) = \begin{cases} m_t v(t) + c_d(t) \ , & \dfrac{dv}{dt} < 0 \\ m_b v(t) + c_u(t) \ , & \dfrac{dv}{dt} > 0 \end{cases}$$

where c_d and c_u are the intercepts for the downward and upward inner loop segments, respectively. As developed and detailed in [478], the combination of half-line and segment relations yields a dynamic model quantifying $\frac{du}{dt}(t)$ in terms of $\frac{dv}{dt}(t)$.

To obtain a discrete-time model suitable for numerical or experimental implementation, time derivatives are replaced by comparing $v(t)$ with v_d and v_u where $v_4 \le v_d \le v_u \le v_3$ and

$$v_d = \frac{m_\ell c_\ell + c_d}{m_\ell - m_t} \ , \qquad v_u = \frac{m_r c_r + c_u}{m_r - m_b} \ .$$

Due to symmetry, $m_t = m_b$ and the constants c_d and c_u are given by

$$c_d = c_u = u(t_{k-1}) - m_t v(t_{k-1}) \ .$$

As detailed in [477,478], the discrete-time hysteresis model is then given by

$$
[k_s(v)](t_k) = u(t_k) = \begin{cases} u(t_{k-1}) & \text{if } v(t_k) = v(t_{k-1}) \\ m_t v(t_k) + c_t & \text{if } v(t_k) \geq v_3, \text{ or} \\ & \text{if } m_t < m_b, \\ & u(t_{k-1}) = m_t v(t_{k-1}) + c_t \\ & \text{and } v(t_{k-1}) < v(t_k) < v_3 \\ m_b v(t_k) + c_b & \text{if } v(t_k) \leq v_4, \text{ or} \\ & \text{if } m_t > m_b, \\ & u(t_{k-1}) = m_b v(t_{k-1}) + c_b \\ & \text{and } v_4 < v(t_k) < v(t_{k-1}) \\ m_t v(t_k) + c_d & \text{if } v_d < v(t_k) < v(t_{k-1}) \\ m_b v(t_k) + c_u & \text{if } v_d(t_{k-1}) < v(t_k) < v_u \\ m_\ell (v(t_k) - c_\ell) & \text{if } v_d \geq v(t_k) \geq v_4 \\ m_r (v(t_k) - c_r) & \text{if } v_u \leq v(t_k) \leq v_3 \,. \end{cases} \tag{4.56}
$$

4.5.7 Model Inverses

As detailed in Section 2.6.7 in the context of homogenized energy models for ferroelectric materials, one strategy for linearizing hysteretic actuator behavior to facilitate linear control design is through the use of inverse constitutive models. This reduces the degree to which control designs must expend energy compensating for hysteretic and nonlinear behavior which improves control authority.

An advantage provided by Preisach models is the property that they can be either exactly or approximately inverted to provide inverse filters for linear control design. We refer the reader to [174] for details regarding the inversion of the Krasnosel'skiĭ–Pokrovskiĭ operator and [239–242, 427, 471, 472] for analysis, control design, and experimental validation regarding inverses of the classical Preisach operator. We summarize the inversion of the piecewise linear kernel k_s given by (4.56) to illustrate the process for that choice of kernel.

To provide a framework suitable for subsequent adaptive control design using the model inverse, we let $\widehat{m}_t(t)$ denote the time-varying estimate of m_t with similar notation for the estimates of the remaining parameters. We summarize here only those details necessary to indicate the philosophy employed when inverting this model and refer the reader to [477, 478] for a complete description regarding the inverse construction.

The inverse model with fixed parameter estimates can be defined by the two half-lines

$$
v(t) = \begin{cases} \dfrac{1}{\widehat{m}_t}(u_d(t) - \widehat{c}_t) &, \quad u_d(t) > u_1 \\ \dfrac{1}{\widehat{m}_b}(u_d(t) - \widehat{c}_b) &, \quad u_d(t) < u_2 \end{cases}
$$

and two line segments

$$v(t) = \begin{cases} \dfrac{1}{\widehat{m}_r} u_d(t) + \widehat{c}_r &, \quad u_2 \le u_d(t) < u_3 \\ \dfrac{1}{\widehat{m}_\ell} u_d(t) + \widehat{c}_\ell &, \quad u_4 < u_d(t) \le u_1 \end{cases}$$

where the corners of the enclosed region are

$$u_1 = \frac{\widehat{m}_\ell(\widehat{m}_t \widehat{c}_t + \widehat{c}_t)}{\widehat{m}_\ell - \widehat{m}_t} \quad , \quad u_2 = \frac{\widehat{m}_r(\widehat{m}_b \widehat{c}_r + \widehat{c}_b)}{\widehat{m}_r - \widehat{m}_b}$$

$$u_3 = \frac{\widehat{m}_r(\widehat{m}_t \widehat{c}_r + \widehat{c}_t)}{\widehat{m}_r - \widehat{m}_t} \quad , \quad u_4 = \frac{\widehat{m}_\ell(\widehat{m}_b \widehat{c}_\ell + \widehat{c}_b)}{\widehat{m}_\ell - \widehat{m}_b} .$$

The conditions on the temporal derivatives of u_d and v are analogous to those for the hysteresis model.

To characterize the discrete-time inverse, we define the u_d-intercepts

$$\widehat{c}_d(t_k) = u_d(t_{k-1}) - \widehat{m}_t v(t_{k-1})$$
$$\widehat{c}_u(t_k) = u_d(t_{k-1}) - \widehat{m}_b v(t_{k-1})$$

along with the points

$$u_{dr}(t_k) = \begin{cases} \dfrac{\widehat{m}_r(\widehat{m}_t \widehat{c}_r + \widehat{c}_d)}{\widehat{m}_r - \widehat{m}_t} &, \quad u_d(t_{k-1}) > u_d(t_k) \\ \dfrac{\widehat{m}_r(\widehat{m}_b \widehat{c}_r + \widehat{c}_u)}{\widehat{m}_r - \widehat{m}_b} &, \quad u_d(t_{k-1}) < u_d(t_k) \end{cases}$$

$$u_{d\ell}(t_k) = \begin{cases} \dfrac{\widehat{m}_\ell(\widehat{m}_t \widehat{c}_\ell + \widehat{c}_d)}{\widehat{m}_\ell - \widehat{m}_t} &, \quad u_d(t_{k-1}) > u_d(t_k) \\ \dfrac{\widehat{m}_\ell(\widehat{m}_b \widehat{c}_\ell + \widehat{c}_u)}{\widehat{m}_\ell - \widehat{m}_b} &, \quad u_d(t_{k-1}) < u_d(t_k) \end{cases}$$

as the coordinate points where an inner loop begins. For a given input u_d, the inverse output v is then defined by the relation

$$v(t_k) = \begin{cases} v(t_{k-1}) & \text{if } u_d(t_k) = u_d(t_{k-1}) \\ \frac{1}{\widehat{m}_t}(u_d(t_k) - \widehat{c}_t) & \text{if } u_d(t_k) \ge u_3 \\ \frac{1}{\widehat{m}_b}(u_d(t_k) - \widehat{c}_b) & \text{if } u_d(t_k) \le u_4 \\ \frac{1}{\widehat{m}_r} u_d(t_k) + \widehat{c}_r & \text{if } u_3 \ge u_d(t_k) \ge u_{dr} \\ \frac{1}{\widehat{m}_\ell} u_d(t_k) + \widehat{c}_\ell & \text{if } u_4 \le u_d(t_k) \le u_{d\ell} \\ \frac{1}{\widehat{m}_t}(u_d(t_k) - \widehat{c}_d) & \text{if } u_{d\ell} < u_d(t_k) < u_{dr} \text{ and } u_d(t_{k-1}) > u_d(t_k) \\ \frac{1}{\widehat{m}_b}(u_d(t_k) - \widehat{c}_u) & \text{if } u_{d\ell} < u_d(t_k) < u_{dr} \text{ and } u_d(t_{k-1}) < u_d(t_k). \end{cases}$$

Details regarding the continuous time inverse and adaptive control designs based on these constructs can be found in [477, 478].

4.6 Jiles–Atherton Model

The Jiles–Atherton model was originally proposed in two papers appearing in 1984 [250] and 1986 [251] with numerous extensions and generalizations appearing in subsequent literature. The crux of the theory is based on the characterization of reversible and irreversible domain wall losses relative to the equilibrium anhysteretic magnetization. This magnetic theory preceded and motivated the domain wall theory discussed in Chapter 2 for ferroelectric materials and Chapter 5 for ferroelastic compounds thus leading to one of the unified frameworks provided in Chapter 6 for ferroic compounds.

Because significant detail was provided in Chapter 2, we provide only a summary of the development for ferromagnetic materials and refer the reader to the original citations [250, 251] or the text [247] for additional discussion detailing the magnetic domain wall model. Further analysis of the model can be found in [240, 496].

4.6.1 Magnetization Model

The hysteretic and nonlinear H-M relation for magnetic materials is developed in three steps: (i) characterization of the anhysteretic magnetization M_{an}, (ii) quantification of the irreversible magnetization M_{irr}, and (iii) characterization of the reversible magnetization M_{rev}. The total magnetization is then given by $M = M_{irr} + M_{rev}$.

Anhysteretic Magnetization

The first step of the model construction focuses on characterizing the anhysteretic magnetization M_{an} discussed in Section 4.1.6. To motivate a model for M_{an}, consider the constitutive relation (4.29) which was derived by minimizing $G = \psi - \mu_0 HM$ for the Helmholtz energy ψ derived from statistical mechanics principles. For $\varepsilon = 0$, inversion of the direct relation yields

$$M_{an}(H) = M_s \tanh\left(\frac{H + \alpha M}{a(T)}\right) \tag{4.57}$$

where $\alpha = \frac{H_h}{\mu_0 M_s}$ and $a(T) = \frac{H_h T}{\mu_0 T_c}$. Here

$$H_e = H + \alpha M \tag{4.58}$$

represents the effective field acting on moments and hence $H_I = \alpha M$ is an interaction field equivalent to (4.48) employed in moving Preisach models. For parameters α and a chosen so that M_{an} is single-valued — see Figure 4.18 — the Ising relation (4.57) provides a model for the *global anhysteretic* curve.

Alternatively, one can employ Boltzmann theory, as detailed in Section 2.5.1 for ferroelectric materials or [88, 247] for ferromagnetic compounds, to derive the Langevin relation

$$M(H) = M_s \left[\coth(H/a) - (a/H)\right].$$

Formulation in terms of the effective field H_e yields the anhysteretic model

$$M_{an}(H) = M_s \left[\coth\left(\frac{H+\alpha M}{a}\right) - \left(\frac{a}{H+\alpha M}\right) \right] \tag{4.59}$$

employed by Jiles and Atherton.

Whereas simple to formulate, several subtleties pervade the anhysteretic relation. The first is the fact that it is derived in the absence of moment interactions and hence strictly holds for paramagnetic regimes but is employed when characterizing hysteresis due to moment interaction — i.e., ferromagnetic regimes. Hence the formulation in terms of H_e and use for ferromagnetic regimes should be considered as phenomenological. Of extreme importance for implementation is the manner through which H_e is evaluated. The implicit formulation of M_{an} in terms of $H_e = H + \alpha M_{an}$ yields the single-valued *global anhysteretic* depicted in Figure 4.9 whereas the relation (4.58) employing the *prevailing magnetization* M is employed in the Jiles–Atherton model to characterize *local anhysteretic* effects.

Irreversible Magnetization

As detailed in [251], the irreversible magnetization M_{irr} designates that component of the magnetization associated with irreversible domain wall motion — e.g., translation of domain walls pinned at inclusions. In a manner which motivated the detailed analysis leading to (2.73), Jiles and Atherton developed the expression

$$d\mathcal{E}_{irr} = k_p dM_{irr}$$

quantifying the energy required for domain wall movement.

To characterize the work performed in the magnetization process, it is necessary to extend the relation

$$W = \oint \mathbf{H} \cdot d\mathbf{M} = -\oint \mathbf{M} \cdot d\mathbf{H}_e,$$

which is valid for full cycles, to partial cycles. To do this, we employ the approach of Iyer and Krishnaprasad [240,496] and postulate that the work over a partial cycle can be formulated as

$$W = -\int \mathbf{M} \cdot d\mathbf{H}_e.$$

The energy supplied to the material is then comprised of a conservative component yielding the anhysteretic response and hysteresis losses thus yielding the relation

$$\int_0^{H_e} M_{irr}(H_e) dH_e = \int_0^{H_e} M_{an}(H_e) dH_e - k_p \int_0^{H_e} \frac{dM_{irr}}{dH_e} dH_e.$$

Differentiation, and enforcement of the property that domain walls exhibit reversible motion following field reversal until the anhysteretic is achieved, yields the relation

$$\frac{dM_{irr}}{dH} = \widetilde{\delta} \frac{M_{an} - M_{irr}}{k_p \delta - \alpha (M_{an} - M_{irr})} \tag{4.60}$$

where

$$\tilde{\delta} = \begin{cases} 1 \, , & \{dH < 0 \text{ and } M > M_{an}\} \text{ or } \{dH > 0 \text{ and } M < M_{an}\} \\ 0 \, , & \text{otherwise} \, . \end{cases}$$

Reversible Magnetization

The reversible magnetization M_{rev} quantifies moment reorientation in response to reversible domain wall bending. As detailed in [251], this component of the magnetization is modeled by

$$M_{rev} = c(M_{an} - M_{irr}) \tag{4.61}$$

where the parameter c quantifies the degree of reversibility.

Total Magnetization

The total magnetization is given by

$$M = M_{irr} + M_{rev} \tag{4.62}$$

where M_{rev} and M_{irr} are given by (4.61) and (4.60). The anhysteretic magnetization M_{an} can be specified by either (4.57) or (4.59) with the effective field H_e given by (4.58). The model can be implemented using Algorithm 2.5.1 developed for the analogous ferroelectric model. Details regarding the identification of parameters can be found in [253] or Section 2.5.2.

Alternatively, one can combine the expressions for M_{an}, M_{irr} and M_{rev} to formulate a single differential equation specifying M in terms of H. For the Ising model (4.57), this yields

$$\begin{aligned} \frac{dM}{dH} &= \mathcal{G}(H, M) \\ M(H_0) &= M_0 \end{aligned} \tag{4.63}$$

where

$$\begin{aligned} \mathcal{G}(H, M) = \frac{1}{1 - \frac{cM_s\alpha}{a}\text{sech}^2\left(\frac{H+\alpha M}{a}\right)} \cdot \Bigg\{ & \frac{\tilde{\delta}\left[M_s \tanh\left(\frac{H+\alpha M}{a}\right) - M\right]}{k_p\delta - \tilde{\alpha}\left[M_s \tanh\left(\frac{H+\alpha M}{a}\right) - M\right]} \\ & + \frac{cM_s}{a}\text{sech}^2\left(\frac{H+\alpha M}{a}\right) \Bigg\}. \end{aligned} \tag{4.64}$$

One obtains an analogous function $\mathcal{G}$ if the Langevin anhysteretic relation (4.59) is employed in lieu of the Ising relation.

4.6.2 Inverse Magnetization Model

The primary advantage afforded by the formulation (4.63) is the fact that it can be used to construct an inverse magnetization model via the complementary differential equation

$$\frac{dM^{-1}}{dH} = \frac{1}{\mathcal{G}(M^{-1},H)}$$
$$M^{-1}(H_0) = \widehat{M_0}. \tag{4.65}$$

The performance when used as an inverse filter to linearize nonlinear and hysteretic actuator behavior is illustrated in Section 4.6.4. It is noted in Section 2.5.3 that the accuracy of both the forward model (4.62) or (4.63) and inverse relation (4.65) is highly dependent upon the fecundity of initial conditions. Hence robust or adaptive control techniques are recommended if employing this framework for model-based control designs.

4.6.3 Magnetomechanical Constitutive Relations

It was noted in Section 4.1.8 that the magnetomechanical effect is comprised of two fundamental mechanisms: (i) magnetic-to-elastic coupling so that changes in magnetization produce strains, and (ii) elastic-to-magnetic interactions which provide magnetostrictive materials with sensor capabilities. Whereas the two mechanisms are intrinsically coupled, we focus primarily on the former which is predominant in actuator designs.

To incorporate the quadratic M-ε dependence quantified by (4.15) — which characterizes strains due primarily to rotation as will be the case in a prestressed Terfenol-D transducer — we employ the Gibbs relations (4.25) with the Helmholtz relation

$$\psi(\varepsilon, M) = \frac{1}{2}Y^M\varepsilon^2 - a_2\varepsilon M^2$$

which results from (4.28) if we retain the elastic and quadratic electromechanical energy relations. The equilibrium condition $\frac{\partial G}{\partial \varepsilon} = 0$ yields the constitutive relation

$$\sigma = Y^M\varepsilon - a_2M^2. \tag{4.66}$$

For actuator characterization, one would employ (4.62) or (4.63) to quantify the nonlinear H-M relation. The resulting magnetization then acts as an input to (4.66). This yields a direct relation between H and σ and provides a framework for constructing differential equation models for distributed structures as detailed in Chapter 7.

The converse coupling between stresses and magnetization is highly complex and, in general, lies outside the Jiles–Atherton framework. Certain stress effects on the anhysteretic magnetization can be incorporated by employing the relation $\varepsilon = s^M\sigma$, where s^M denotes the compliance, in (4.29) to obtain

$$M_{an}(\sigma, H) = M_s\tanh(H_e(\sigma, H)/a(T)).$$

The effective field in this case is

$$H_e(\sigma, H) = H + (\alpha + s^M a_2 \sigma)M.$$

It is demonstrated in [75] that use of an analogous effective field expression in the Langevin anhysteretic model provides a hysteresis model capable of quantifying both field and prestress mechanisms in Terfenol-D.

However, the model does not incorporate the stress-induced decay to the anhysteretic depicted in Figure 4.15. Hence one must consider other theories — e.g., the magnetomechanical model of Jiles [249] or homogenized energy model which incorporate stress — to characterize the magnetomechanical behavior depicted in Figures 4.16 and 4.17.

4.6.4 Model Properties and Experimental Validation

Attributes of the Jiles–Atherton framework were discussed in Section 2.5.6 in the context of the ferroelectric model. A primary advantage of the approach is its efficiency for characterizing symmetric major loop behavior. A significant disadvantage in the context of control design is the fact that it does not guarantee closure of biased minor loops in quasistatic operating regimes. The reason for this is the property that M_{irr} is driven toward M_{an} thus yielding behavior of the type depicted in Figure 2.29. Jiles addressed this in [248] by using *a priori* knowledge of turning points to enforce closure thus providing the flexibility required for comprehensive material characterization. However, this technique cannot be applied in feedback control design since turning points are determined by the measured system response which is typically unknown in advance. Hence for applications in which nonclosure proves problematic, one may need to employ the Preisach models described in Section 4.5 or homogenized energy models of Section 4.7 for model-based control design.

Experimental Validation

To illustrate attributes of the forward model (4.62) or (4.63) and inverse model (4.65), we consider characterization and inverse filtering for a Terfenol-D transducer similar to that depicted in Figure 1.11. As detailed in [451], the Terfenol-D ($Tb_{0.3}Dy_{0.7}Fe_{1.92}$) rod had a length of 0.254 m, a diameter of 0.020 m and was composed of four laminae to minimize eddy current losses. A prestress of approximately $\sigma_0 = 1.0$ ksi, applied via Belleville washers, maintained the rod in a state of compression.

During operation, the current $I(t)$ to the solenoid and voltage $V(t)$ from a sensing pickup coil were measured. The field was subsequently computed using the relation $H(t) = nI(t)$ where n denotes the number of coils per unit length. From the induced voltage in the pickup coil, the Faraday–Lenz law was used to compute $\frac{dB}{dt}$. Integration then yielded the induction and magnetization $M = \frac{1}{\mu_0}B - H$.

A least squares fit to data collected at 1 Hz yielded the parameters $M_s = 7.65 \times 10^5$ A/m, $a = 1.5 \times 10^5$ A/m, $\alpha = 0.032$, $k_p = 5000$ A/m and $c = 0.18$. As noted in Section 2.5 and detailed in [253], initial estimates for M_s, k_p and c can be obtained from the maximum measured magnetization, the coercive field, and the ratio of the

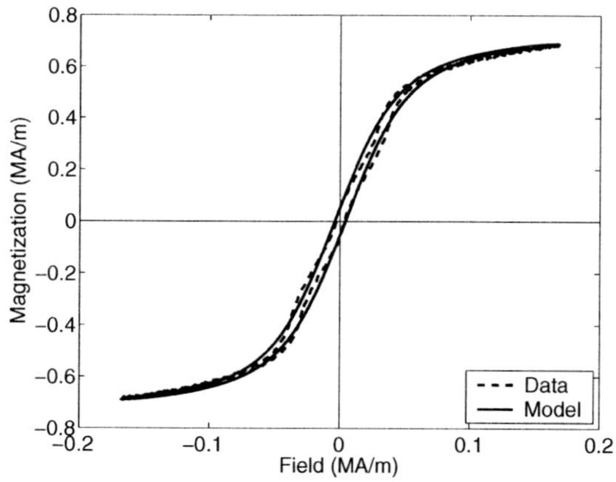

Figure 4.34. *Experimental H-M data from the Terfenol-D transducer and model fit provided by (4.63).*

slopes $\frac{dM}{dH}$ at field reversal. The data and model fit are plotted in Figure 4.34. The model does not incorporate the low field concavity changes, which are attributed to crystalline anisotropies, but it accurately quantifies both the measured hysteresis and saturation nonlinearities. Additional examples illustrating major and minor loop properties of the model for characterizing Terfenol-D transducers operating in more hysteretic regimes can be found in [75, 119, 120].

Secondly, the inverse model (4.65) is used to construct an inverse filter to linearize the transducer in the manner depicted in Figure 4.35. A 1 Hz triangular input u_d was applied to the inverse filter to generate a signal v which was input to the transducer. The output magnetization u from the transducer was then measured to determine the accuracy of the compensation technique. The input signal u_d is compared with the physical transducer response in Figure 4.36(a) and the prescribed and measured fields are plotted in Figure 4.36(b). It is observed that whereas u exhibits slight deviations due to unmodeled domain rotation effects in the burst region, the overall fit is accurate. Hence the inverse filter accurately compensates for hysteresis and constitutive nonlinearities and thus provides a technique for linearizing the actuator response to permit linear control design.

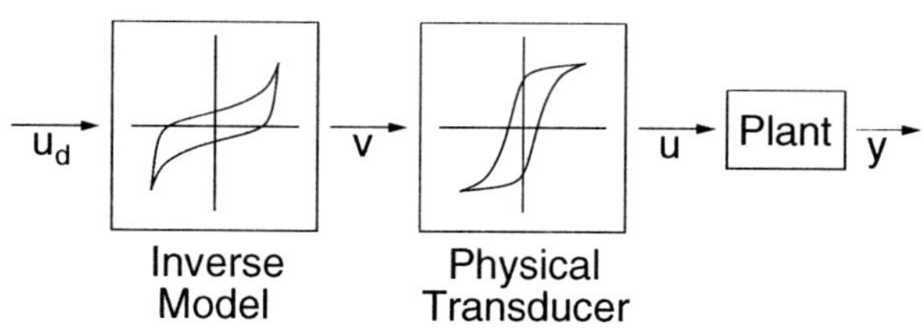

Figure 4.35. *Inverse filtering to linearize the nonlinear and hysteretic response of a magnetostrictive transducer.*

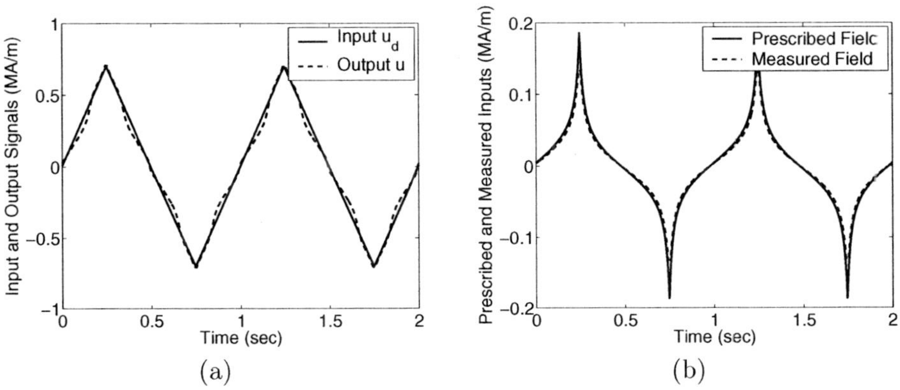

Figure 4.36. *(a) Input magnetization u_d to the inverse model (4.65) and measured output magnetization u from the Terfenol-D transducer. (b) Field v prescribed by the inverse model as input to the transducer and field measured in the transducer.*

4.7 Homogenized Energy Model

The homogenized energy model combines energy analysis at the lattice level, theory of thermally activated processes, and stochastic homogenization techniques to provide a highly comprehensive characterization framework. The models guarantee closure of biased minor loops for quasistatic drive regimes but will characterize the after-effects and accommodation phenomena discussed in Section 4.1.7. The framework incorporates various rate and temperature-dependences, characterizes the anhysteretic magnetization, quantifies reversible effects, and provides mechanisms for incorporating the magnetomechanical effects described in Section 4.1.8. Depending upon the choice of energy functional, it also provides both congruent and noncongruent minor loops thus providing significant flexibility for a range of materials and operating regimes.

The magnetic homogenized energy framework is analogous to that described in Section 2.6 for ferroelectric materials and hence we provide only a summary of the development for magnetic materials. Several aspects of the theory originated in the context of shape memory alloys and analogous theory for ferroelastic compounds is provided in Chapter 5. Hence the theory provides a unified framework for characterizing hysteresis and constitutive nonlinearities in a broad class of ferromagnetic, ferroelectric and ferroelastic — collectively designated as ferroic — compounds. The breadth of the framework evokes comparisons with the Preisach theory summarized in Section 4.5, and it is illustrated in Section 4.7.10 that it provides an energy basis for certain extended Preisach models.

4.7.1 Local Magnetization Models

We consider first the development of hysteretic and anhysteretic models at the mesoscopic or lattice level. As in Section 4.4.1, we initially consider a *homogeneous* lattice of volume V comprised of $N = N_+ + N_-$ cells, each of which contains one magnetic moment s_i. This provides a reasonable framework for constructing single crystal H-M relations and provides a basis for the nonhomogeneous material models developed in Section 4.7.2

Helmholtz and Gibbs Energies

Appropriate Helmholtz and Gibbs energy relations for this regime were described in Section 4.4.1. Mean field theory yields the Helmholtz energy expression

$$\begin{aligned}\psi(M,T) &= \frac{H_h M_s}{2}\left[1-(M/M_s)^2\right] \\ &+\frac{H_h T}{2T_c}\left[M\ln\left(\frac{M+M_s}{M_s-M}\right)+M_s\ln(1-(M/M_s)^2)\right]\end{aligned} \tag{4.67}$$

whereas approximation about the equilibria provides the isothermal relation

$$\psi(M) = \begin{cases} \frac{1}{2}\eta(M+M_R)^2 & , \; M \leq -M_I \\ \frac{1}{2}\eta(M-M_R)^2 & , \; M \geq M_I \\ \frac{1}{2}\eta(M_I-M_R)\left(\frac{M^2}{M_I}-M_R\right) & , \; |M| < M_I. \end{cases} \tag{4.68}$$

Here H_h, M_s, T_c, M_I, M_R and η respectively denote a bias field, the local saturation magnetization, the Curie point, an inflection point, the local remanence magnetization, and the slope $\frac{dH}{dM}$ after switching. We focus primarily on the piecewise quadratic relation but include the statistical mechanics relation since it motivates the analysis and it provides a hysteron which yields *noncongruent* minor loops.

The Gibbs relations

$$G(H,M,T) = \psi(M,T) - \mu_0 HM \tag{4.69}$$

and

$$G(\sigma,H,\varepsilon,M,T) = \psi(M,T) - \mu_0 HM - \sigma\varepsilon \tag{4.70}$$

respectively incorporate the work due to applied magnetic fields and stresses.[17] The minimization of G in the presence and absence of thermal kinetic effects provides local models for the average magnetization.

Local Magnetization — Negligible Thermal Activation

For operating regimes in which relaxation times are small compared with drive frequencies, the local average magnetization is determined through direct minimization of G. In the absence of applied stresses, enforcement of the necessary condition $\frac{\partial G}{\partial M} = 0$ for ψ given by (4.67) yields

$$\overline{M}(H) = M_s \tanh\left(\frac{H+\alpha M}{a(T)}\right) \tag{4.71}$$

where $\alpha = \frac{H_h}{\mu_0 M_s}$ and $a(T) = \frac{H_h T}{\mu_0 T_c}$. As detailed in Section 4.7.9, formulation in terms of the effective field $H_e = H + \alpha M$ provides the hysteron with the capacity for characterizing noncongruent minor loop behavior.

[17]See Footnote 14 on page 184.

Alternatively, use of the piecewise quadratic Helmholtz relation (4.68) yields the piecewise linear local magnetization relation

$$\overline{M}(H) = \frac{\mu_0}{\eta}H + M_R\delta. \tag{4.72}$$

Here $\delta = 1$ for positively oriented moments and $\delta = -1$ for those with negative orientation. The expression (4.72) illustrates why M_R can be interpreted as a local or lattice-level remanence value whereas η/μ_0 is the reciprocal slope after switching. To quantify δ in terms of initial moment configurations and previous switches, we employ Preisach notation — e.g., see (4.37) — and take

$$[\overline{M}(H; H_c, \xi)](t) = \begin{cases} [\overline{M}(H; H_c, \xi)](0) & , \tau(t) = \emptyset \\ \frac{\mu_0}{\eta}H - M_R & , \tau(t) \neq \emptyset \text{ and } H(\max \tau(t)) = -H_c \\ \frac{\mu_0}{\eta}H + M_R & , \tau(t) \neq \emptyset \text{ and } H(\max \tau(t)) = H_c. \end{cases} \tag{4.73}$$

Here

$$[\overline{M}(H; H_c, \xi)](0) = \begin{cases} \frac{\mu_0}{\eta}H - M_R & , H(0) \leq -H_c \\ \xi & , -H_c < H(0) < H_c \\ \frac{\mu_0}{\eta}H + M_R & , H(0) \geq H_c. \end{cases} \tag{4.74}$$

denotes the initial moment distribution and transition times are designated by

$$\tau(t) = \{t_s \in (0, t] \mid H(t_s) = -H_c \text{ or } H(t_s) = H_c\} \tag{4.75}$$

— see page 101 for further details regarding this notation. The dependence of $\overline{M}$ on the local coercive field

$$H_c = \frac{\eta}{\mu_0}(M_R - M_I) \tag{4.76}$$

is indicated as a prelude to the discussion in Section 4.7.2 where H_c is assumed distributed to accommodate material nonhomogeneities.

The two hysterons $\overline{M}(H)$ are depicted in Figure 4.37(a) and (b) to illustrate their behavior.

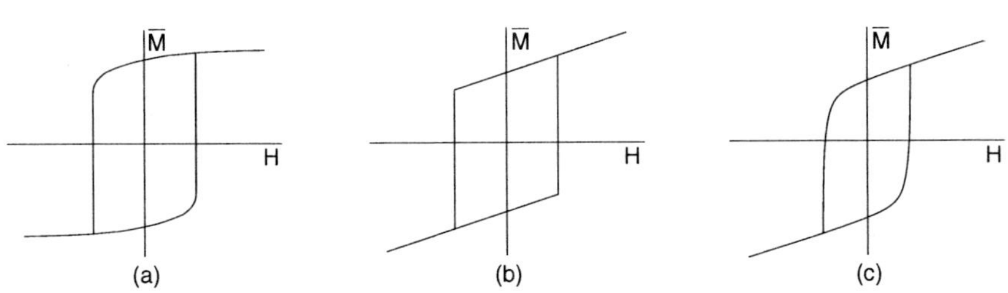

Figure 4.37. *Hysterons quantifying local H-M behavior in homogeneous materials. Negligible thermal activation: (a) $\overline{M}$ given by (4.71) and (b) $\overline{M}$ given by (4.72); (c) thermal activation: $\overline{M}$ specified by (4.78).*

Local Magnetization — Thermal Activation

The phenomena of accommodation (reptation) and magnetic after-effects (relaxation), discussed in Section 4.1.7, illustrate regimes in which thermal activation mechanisms must be incorporated and hence the hysterons (4.71) and (4.72) prove inadequate. To incorporate requisite thermal mechanisms, it is necessary to balance G with the relative thermal energy kT/V and entropic effects. As detailed in Section 2.6 for ferroelectric materials and [198] for general systems, minimization of energy relations which incorporate relative kinetic energy contributions $K = \frac{kTN}{2V}$, subject to state constraints, yields Boltzmann relations of the form

$$\mu(G) = Ce^{-GV/kT} \tag{4.77}$$

quantifying the probability of obtaining the energy level G. The integration constant C is specified to ensure integration to unity.

The local average magnetization is this regime is

$$\overline{M} = x_+ \langle M_+ \rangle + x_- \langle M_- \rangle \tag{4.78}$$

where x_+ and x_- respectively denote the fractions of moments having positive and negative orientations and $\langle M_+ \rangle, \langle M_- \rangle$ are the associated average magnetizations. As detailed in [433, 434], the latter are quantified by the general relations

$$\langle M_+ \rangle = \int_{M_I}^{\infty} M\mu(G)\, dM \quad , \quad \langle M_- \rangle = \int_{-\infty}^{-M_I} M\mu(G)\, dM.$$

Evaluation of the integration constant yields

$$\langle M_+ \rangle = \frac{\displaystyle\int_{M_I}^{\infty} Me^{-G(H,M)V/kT} dM}{\displaystyle\int_{M_I}^{\infty} e^{-G(H,M)V/kT} dM} \ , \ \langle M_- \rangle = \frac{\displaystyle\int_{-\infty}^{-M_I} Me^{-G(H,M)V/kT} dM}{\displaystyle\int_{-\infty}^{-M_I} e^{-G(H,M)V/kT} dM}. \tag{4.79}$$

The evolution of moment fractions are quantified by the differential equations

$$\begin{aligned} \dot{x}_+ &= -p_{+-}x_+ + p_{-+}x_- \\ \dot{x}_- &= -p_{-+}x_- + p_{+-}x_+ \end{aligned} \tag{4.80}$$

which can be simplified to

$$\dot{x}_+ = -p_{+-}x_+ + p_{-+}(1 - x_+)$$

through the identity $x_+ + x_- = 1$. Here

$$p_{+-} = \frac{1}{\mathcal{T}(T)} \frac{\displaystyle\int_{M_I-\epsilon}^{M_I} e^{-G(H,M)V/kT} dM}{\displaystyle\int_{M_I-\epsilon}^{\infty} e^{-G(H,M)V/kT} dM} \tag{4.81}$$

and

$$p_{-+} = \frac{1}{\mathcal{T}(T)} \frac{\displaystyle\int_{-M_I}^{-M_I+\epsilon} e^{-G(H,M)V/kT} dM}{\displaystyle\int_{-\infty}^{-M_I+\epsilon} e^{-G(H,M)V/kT} dM} \tag{4.82}$$

respectively denote the likelihoods that moments switch from positive to negative, and conversely. In these relations, ϵ is a small positive constant and $\mathcal{T}$ denotes the material-dependent relaxation time so that $\omega = \frac{1}{\mathcal{T}}$ quantifies the frequency at which moments attempt to switch.[18]

The behavior of the hysteron specified by (4.78) is compared in Figure 4.37 with the Ising and piecewise linear kernels resulting from direct minization of G. It is observed that if kT/V is large compared with G, a significant number of moments attain the energy required to reverse orientation before the coercive fields are achieved. This provides a gradual transition and decreased coercive values for the thermally activated material. Furthermore, it is detailed in [433] and Section 2.6 that the relation (4.78) converges to the piecewise linear relation (4.72) or (4.73) in the limit $kT/V \to 0$ of negligible relative thermal energy.

Local Anhysteretic Relation

It was noted in Section 4.1.6 that the anhysteretic magnetization M_{an} can be experimentally obtained at a DC field value H_0 by applying a decaying AC field centered at H_0. From a theoretical perspective, M_{an} represents the locus of magnetization values that would occur in a material devoid of inclusions. Since $\omega = \frac{1}{\mathcal{T}}$ quantifies the frequency at which moments attempt to switch, M_{an} can also be theoretically interpreted as the magnetization achieved when relaxation times $\mathcal{T}(T)$ are sufficiently small compared with the period of drive frequencies that moments achieve *global* equilibria.

We illustrate the latter theoretical interpretation in the context of the local magnetization model (4.78) derived under the assumption that G and kT/V are balanced through the relation (4.77). The condition of moment equilibrium yields $\dot{x}_+ = \dot{x}_- = 0$ in (4.80) which in turn implies that equilibrium solutions $\bar{x}_+$ and $\bar{x}_-$ satisfy the relation

$$\frac{\bar{x}_+}{\bar{x}_-} = \frac{p_{-+}}{p_{+-}}. \tag{4.83}$$

To demonstrate the ramification of (4.83), consider first the case when $H = 0$. From (4.81) and (4.82), it follows immediately that $p_{-+} = p_{+-}$ and hence $\bar{x}_+ = \bar{x}_-$. From the conservation relation $\bar{x}_+ + \bar{x}_- = 1$, it is deduced that $\bar{x}_+ = \bar{x}_- = \frac{1}{2}$, regardless of the initial conditions $x_+(0)$ and $x_-(0)$. The rate at which the relations converge to equilibrium values is determined by the relaxation time $\mathcal{T}(T)$ with smaller values of $\mathcal{T}$ producing more rapid equilibration.

To ascertain the resulting anhysteretic magnetization for $H = 0$, we note that the symmetry of (4.79) implies that $\langle \overline{M}_+ \rangle = -\langle \overline{M}_- \rangle$ at equilibrium. When

[18]Details regarding approximate formulations for p_{+-} and p_{-+} in terms of point evaluations in the numerator are provided in Remark 2.6.1 on page 105.

combined with the fact that $\bar{x}_+ = \bar{x}_- = \frac{1}{2}$, (4.78) yields

$$\overline{M}_{an}(H = 0) = 0.$$

The field-dependence of $p_{+-}, p_{-+}, \langle M_+ \rangle$ and $\langle M_- \rangle$ precludes a similar exploitation of symmetries for $H \neq 0$. However, $\overline{M}_{an}(H)$ can be easily computed by numerically approximating (4.78) with sufficiently small $\mathcal{T}$. The Gibbs energy G at the field value $H_0 = 2000$ A/m, unnormalized density $\mu(G) = e^{-GV/kT}$, and resulting local anhysteretic magnetization obtained with $\mathcal{T} = 1.0 \times 10^{-13}$ sec and relative thermal energies $kT/V = 5.0 \times 10^5$ and $kT/V = 7.14 \times 10^6$ are plotted in Figure 4.38.

It is observed that when kT/V is significant compared with G, thermal fluctuations produce switching between wells thus yielding a gradual anhysteretic transition between positive and negative saturation magnetizations. As kT/V becomes increasingly small, the local anhysteretic magnetization $\overline{M}_{an}$ provided by (4.78) converges to

$$\begin{aligned} \overline{M}(H) &= \frac{\mu_0}{\eta} H + M_R \delta \\ \delta &= \text{sign}(H). \end{aligned} \tag{4.84}$$

The limiting relation (4.84) can be interpreted as the locus of magnetization values that would occur in the absence of inclusions — which is manifested by $H_c = 0$ in the local model.

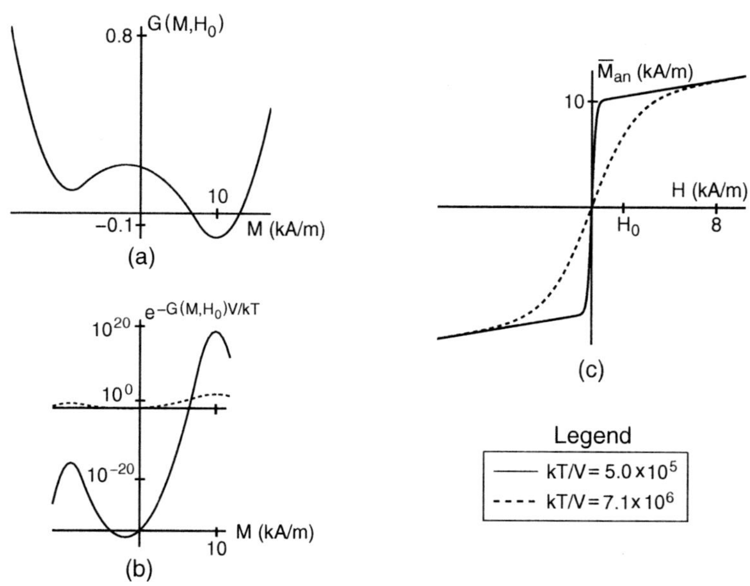

Figure 4.38. *(a) Gibbs energy G and (b) unnormalized Gaussian densities $\mu(G) = e^{-GV/kT}$ with $H_0 = 2000$ A/m. (c) Anhysteretic magnetization $\overline{M}_{an}$ given by (4.78) with $\mathcal{T} = 1.0 \times 10^{-13}$ sec.*

4.7.2 Homogenized Macroscopic Magnetization Model

For homogeneous, single crystal materials with constant effective fields, the *local* magnetization relations (4.71), (4.72) or (4.78) can be extended throughout the material to provide *global* H-M models. The models in this form prove adequate for characterizing the hysteretic behavior of certain materials such as the uniaxial nickel-iron film, magnetically annealed core of cobalt ferrous ferrite and manganese-magnesium ferrite illustrated on page 298 of Craik and Tebble [105]. However, the transitions provided by these local models are too steep to provide accurate characterization for general polycrystalline magnetic materials thus motivating the stochastic homogenization techniques discussed here.

An implicit assumption made in Section 4.4.1 when deriving Helmholtz energy relations for *homogeneous* materials is that the exchange energy Φ_0 and exchange integral $\mathcal{J}$ are constant throughout the lattice. This implies that the bias field $H_h = \frac{N\Phi_0}{VM_s}$, Curie point $T_c = \frac{\Phi_0}{k}$, mean field coefficient $\alpha = \frac{H_h}{\mu_0 M_s}$ and coefficient $a(T) = \frac{H_h T}{\mu_0 T_c}$ are single parameter values. This in turn implies that M_I, M_R and H_c in the approximating piecewise quadratic model are uniform throughout the lattice. However, for polycrystalline materials which exhibit material nonhomogeneities, variable grain orientations, nonuniform stress distributions, and variations due to texture, these assumptions are overly simplistic, and macroscopic models must incorporate the inherent variability in $\mathcal{J}$ and hence Φ_0.

At the quantum level, the variability of $\mathcal{J}_{ij}$ in (4.17) is incorporated by modeling the overlap of electron wave functions. Whereas this motivates *why* we should anticipate $\mathcal{J}$ and Φ_0 to be variable, it does not provide an efficient technique for constructing effective parameters which incorporate this variability. Instead, we consider model parameters related to Φ_0 to be manifestations of underlying statistical distributions which we identify for a given material or transducer design. Statistical homogenization in this manner yields low-order macroscopic models which incorporate fundamental physics through the mesoscopic energy relations but are sufficiently efficient for transducer design and model-based control implementation.

We first note that variability in Φ_0 implies variability in α which turn yields variable effective fields $H_e = H + \alpha M$. To facilitate implementation, we assume that $H_e = H + H_I$ is distributed about the applied field with an underlying density ν_2 characterizing the variability in the induced field H_I.

It is demonstrated in [433, 434] that consideration of the local coercive field H_c as a manifestation of an underlying density consistently incorporates inherent variability in η, M_I and M_R. We denote the density associated with H_c by ν_1.

As in (2.113), we assume that ν_1 and ν_2 satisfy the decay criteria

$$\begin{aligned} &\text{(i)} \quad \nu_1(x) \text{ defined for } x > 0, \\ &\text{(ii)} \quad \nu_2(-x) = \nu_2(x), \\ &\text{(iii)} \quad |\nu_1(x)| \le c_1 e^{-a_1 x} \quad , \quad |\nu_2(x)| \le c_2 e^{-a_2|x|} \end{aligned} \tag{4.85}$$

for positive c_1, a_1, c_2, a_2. These assumptions enforce the physical properties that local coercive fields are positive, low-field Rayleigh loops are symmetric [40], and local coercive and interaction fields decay as a function of distance.

The resulting macroscopic magnetization model is then given by

$$
\begin{aligned}
[M(H)](t) &= \int_0^\infty \int_{-\infty}^\infty \nu_1(H_c)\nu_2(H_I)[\overline{M}(H+H_I;H_c,\xi)](t)\,dH_I\,dH_c \\
&= \int_0^\infty \int_{-\infty}^\infty \nu(H_c,H_I)[\overline{M}(H+H_I;H_c,\xi)](t)\,dH_I\,dH_c
\end{aligned}
\tag{4.86}
$$

where ξ denotes the initial distribution of moments and $\overline{M}$ is given by (4.71), (4.72) or (4.78).[19] For implementation purposes, the decay criteria (4.85) are invoked to truncate the domain to a compact region Ω_2 and Gauss–Legendre or Newton–Cotes quadrature rules are employed to obtain the discretized model

$$
\begin{aligned}
[M(H)](t) &= \sum_{i=1}^{N_i}\sum_{j=1}^{N_j} \nu_1(H_{c_i})\nu_2(H_{I_j})[\overline{M}(H_{I_j}+H;H_{c_i},\xi_j)](t)v_i w_j \\
&= \sum_{i=1}^{N_i}\sum_{j=1}^{N_j} \nu(H_{c_i},H_{I_j})[\overline{M}(H_{I_j}+H;H_{c_i},\xi_j)](t)v_i w_j .
\end{aligned}
\tag{4.87}
$$

Here H_{I_j}, H_{c_i} denote the abscissas associated with respective quadrature formulae and v_i, w_j are the respective weights. As in the Preisach models of Section 4.5, t can be interpreted as time or, more generally, as simply a parameter governing the increase or decrease of H.

The implementation of the model (4.87) requires the identification of the general densities ν_1 and ν_2 or joint density $\nu = \nu_1 \cdot \nu_2$. As noted in Section 4.5 in the context of Preisach models, the identification procedure can be simplified significantly if functional or parameterized models are employed for ν_1 and ν_2. For example,

$$
\begin{aligned}
\nu_1(H_c) &= c_1 e^{-[\ln(H_c/\overline{H}_c)/2c]^2} \\
\nu_2(H_I) &= c_2 e^{-H_I^2/2b^2}
\end{aligned}
\tag{4.88}
$$

satisfy the physical criteria (4.85) and have been employed in both Preisach and homogenized energy models. As detailed [126] and noted in (2.118), the mean and variance of the lognormal distribution satisfy the properties

$$
\langle H_c\rangle \approx \overline{H}_c \quad , \quad \sigma \approx 2\overline{H}_c c \tag{4.89}
$$

if $\overline{H}_c$ is large compared with c. This implies that the measured coercive field can be employed as an initial estimate for $\overline{H}_c$. Finally, it is illustrated in Section 4.7.8 that an alternative choice for ν_1 is a normal density restricted to positive arguments [434].

Remark 4.7.1. *To simplify the model formulation and provide notation consistent with Preisach models, the densities ν_1 and ν_2 are unnormalized and incorporate the scaling necessary to achieve the correct saturation behavior. As noted in*

[19]The use of the operator notation $[\overline{M}(H+H_I;H_c,\xi)](t)$ and $[M(H)](t)$ facilitates analysis and is consistent with the Preisach notation in Section 4.5. This should be interpreted as $\overline{M}(H(t)+H_I;H_c,\xi)$ and $M(H(t))$ which indicates the explicit dependence of H on t.

Remark 4.7.2 of the next section, normalization proves necessary if correlating constants when one density is absent in the formulation but in general is unnecessary.

4.7.3 Anhysteretic Magnetization Model

It was illustrated in Section 4.7.1 that for the piecewise quadratic Gibbs energy G, the local anhysteretic magnetization $\overline{M}_{an}$ followed from (4.78) for thermally active regimes or (4.83) in the absence of magnetic after-effects. These local relations can be directly extended to provide global models which characterize the anhysteretic behavior discussed in Section 4.1.6.

Because local coercive fields play no role in the anhysteretic material behavior, the global anhysteretic model is

$$M_{an}(H) = \int_0^\infty \nu_2(H_I)\overline{M}_{an}(H+H_I)dH_I. \tag{4.90}$$

For thermally inactive regimes, the kernel is given by

$$\begin{aligned} \overline{M}_{an}(H+H_I) &= \frac{\mu_0}{\eta}(H+H_I) + M_r\delta(H;H_I) \\ \delta(H;H_I) &= \text{sign}(H+H_I) \end{aligned} \tag{4.91}$$

whereas one would employ the kernel (4.78) to incorporate magnetic after-effects. The parameterization of H with respect to t has been dropped since history plays no role in the anhysteretic response.

To further illustrate the relation between the hysteresis model (4.86) and anhysteretic model (4.90), we consider the specific density choices (4.88) where $c_1 = \widetilde{c}_1/I_1$ and $c_2 = \widetilde{c}_1/b\sqrt{2\pi}$ are expressed in terms of the normalization constants

$$I_1 = \int_0^\infty e^{-[\ln(H_c/\overline{H}_c)/2c]^2} dH_c \quad , \quad b\sqrt{2\pi} = \int_{-\infty}^\infty e^{-H_I^2/2b^2} dH_I.$$

The anhysteretic magnetization is then given by

$$\begin{aligned} M_{an}(H) &= \frac{\widetilde{c}_1}{I_1}\cdot\frac{\widetilde{c}_2}{b\sqrt{2\pi}}\int_0^\infty\int_{-\infty}^\infty \overline{M}(H+H_I)e^{-[\ln(H_c/\overline{H}_c)/2c]^2}e^{-H_I^2/2b^2}\,dH_I dH_c \\ &= \widetilde{C}\int_{-\infty}^\infty \overline{M}(H+H_I)e^{-H_I^2/2b^2}\,dH_I \end{aligned} \tag{4.92}$$

where $\widetilde{C} = \frac{\widetilde{c}_1\widetilde{c}_2}{b\sqrt{2\pi}}$.

Remark 4.7.2. *If solely quantifying anhysteretic material behavior, one can treat the constant $\widetilde{C}$ as a material parameter to be identified whereas if correlating modeled hysteretic and anhysteretic properties, one should identify the constants $\widetilde{c}_1$ and $\widetilde{c}_2$. Hence the normalization constants associated with the densities ν_1 and ν_2 must be included if correlating scaling constants in the anhysteretic and hysteretic models.*

Numerical Example

To illustrate the manner through which the anhysteretic magnetization is quantified by this framework, we numerically simulate the experimental procedure used to obtain M_{an} using the full hysteresis model (4.86), and compare with the value predicted by the anhysteretic model (4.90). Specifically, we applied the periodic and subsequently decaying AC field depicted in Figure 4.39(a) to the model (4.86) to simulate the experimental procedure used to obtain M_{an} or B_{an} at the DC field $H_0 = 2000$ A/m. The parameter values from Table 4.2 of Section 4.7.8 were employed so the result, $B_{an} = 0.5544$ tesla, which is plotted as $*$ in Figure 4.39(b), is representative of steel. A comparison with the corresponding value of $B_{an} = 0.5704$ tesla predicted by the anhysteretic model (4.90), which is denoted by $\circ$, illustrates that the two approaches yield identical results to within computational precision. The locus of points computed using (4.90) completes the comparison between the predicted anhysteretic and hysteretic responses for the material. Hence the modeling framework developed to quantify hysteresis in ferromagnetic materials also quantifies the anhysteretic behavior in a natural manner.

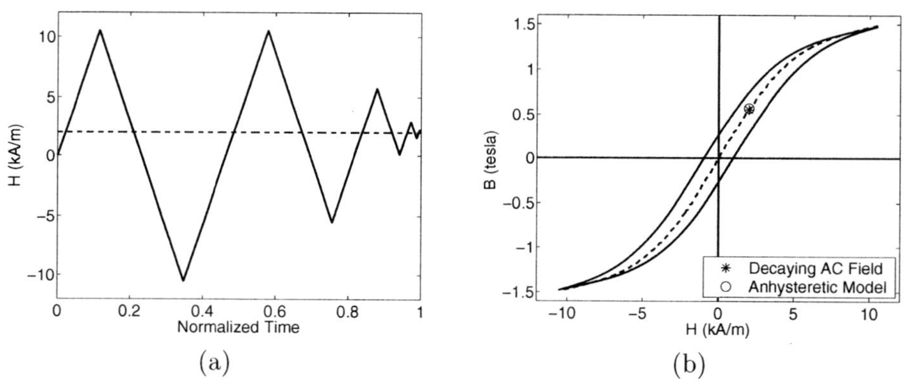

Figure 4.39. *(a) Field input to the hysteresis model (4.86) to generate the hysteretic response and anhysteretic value at $H_0 = 2000$ A/m. (b) Value of B_{an} generated by the decaying AC field ($*$) and anhysteretic model ($\circ$), and full anhysteretic curve (– – –) given by (4.90).*

4.7.4 Parameter Identification and Model Implementation

Significant detail is provided in Sections 2.6.5 and 2.6.6 regarding identification and implementation techniques for the analogous ferroelectric model; hence we simply summarize the implementation algorithm for model construction using the piecewise linear kernel (4.72). We also provide algorithms for implementing the anhysteretic model (4.90).

Hysteresis Model

To avoid the use of computationally inefficient `if-then` statements, we employ a matrix version

$$M(H) = \frac{\mu_0}{\eta} H + M_R \Delta(H; H_c, H_I)$$

of the local relation (4.72). As in (2.119), we define the matrices

$$\Delta_{init} = \begin{bmatrix} -1 & \cdots & -1 & 1 & \cdots & 1 \\ \vdots & & \vdots & \vdots & & \vdots \\ -1 & \cdots & -1 & 1 & \cdots & 1 \end{bmatrix}_{N_i \times N_j}, \quad \mathcal{H}_c = \begin{bmatrix} H_{c_1} & \cdots & H_{c_1} \\ \vdots & & \vdots \\ H_{c_{N_i}} & \cdots & H_{c_{N_i}} \end{bmatrix}_{N_i \times N_j}$$

$$\mathcal{H}_k = \begin{bmatrix} H_k + H_{I_1} & \cdots & H_k + H_{I_{N_j}} \\ \vdots & & \vdots \\ H_k + H_{I_1} & \cdots & H_k + H_{I_{N_j}} \end{bmatrix}_{N_i \times N_j}$$

and weight vectors

$$V^T = \left[v_1 \nu_1(H_{c_1}), \; \cdots \; , v_{N_i} \nu_1(H_{c_{N_i}}) \right]_{1 \times N_i}$$

$$W^T = \left[w_1 \nu_2(H_{I_1}), \; \cdots \; , w_{N_j} \nu_2(H_{I_{N_j}}) \right]_{1 \times N_j}$$

The magnetization $M_k = M(H_k)$ at a field value H_k is then computed using the following algorithm.

Algorithm 4.7.3.

$\Delta_{prev} = \Delta_{init}$

for $k = 1 : N_k$

$\quad \Delta = \text{sgn}(\mathcal{H}_k + \mathcal{H}_c . * \Delta_{prev})$

$\quad \overline{M} = \frac{\mu_0}{\eta} \mathcal{H}_k + M_R \Delta$

$\quad M_k = V^T \overline{M} W$

$\quad \Delta_{prev} = \Delta$

end

Anhysteretic Model

The implementation of the discretized anhysteretic model is significantly easier since it does not require updating Δ to retain a history of moment switches due to local coercive fields. One can employ either the matrices and vectors defined for the hysteresis model, or a reduced set of vectors which reflects the fact that the coercive density integrates to the constant $I_1 = \int_0^\infty \nu_1(H_c) dH_c$. The two *equivalent* approaches are illustrated in the following algorithms where

$$h_k = \left[H_k + H_{I_1}, \ldots , H_k + H_{I_{N_j}} \right]_{1 \times N_j}$$

in Algorithm 4.7.5. Algorithm 4.7.4 retains the direct correlation with the hysteresis model whereas Algorithm 4.7.5 is more efficient to implement since it requires vector rather than matrix multiplication.

Algorithm 4.7.4.

for $k = 1 : N_k$

$$\Delta = \text{sgn}(\mathcal{H}_k)$$
$$\overline{M}_{an} = \tfrac{\mu_0}{\eta}\mathcal{H}_k + M_R\Delta$$
$$M_k = V^T\overline{M}_{an}W$$

end

Algorithm 4.7.5.

for $k = 1 : N_k$

$$\Delta = \text{sgn}(h_k)$$
$$\overline{M}_{an} = \tfrac{\mu_0}{\eta}h_k + M_R\Delta$$
$$M_{an_k} = I_1\overline{M}_{an}W$$

end

4.7.5 Inverse Model

An important property of the previously described Preisach and Jiles–Atherton models is the fact that they can be approximately inverted to provide inverse filters for linear control design. As depicted in Figure 4.32 and 4.35, this linearizes the hysteretic transducer behavior so that control effort can be focused on tracking or stabilization criteria rather than solely on linearizing transducer dynamics.

It is illustrated in Section 2.6.7 and Appendix D that the homogenized energy model also provides highly effective inverse filters for subsequent use in linear control design. In contrast with the Jiles–Atherton inverse which requires the approximation of a differential equation, the homogenized energy model inverse, specified in Algorithm 2.6.8 or Algorithm D.1.1, is algebraic in nature. Hence it is insensitive to initial conditions. Furthermore, formulation in terms of the joint density ν yields a *linearly parameterized* model which permits consideration of *linear adaptive techniques* for both the model and model inverse.

4.7.6 Thermal Evolution

Temperature changes in magnetostrictive transducer materials are due to several mechanisms including ohmic heating in the solenoid, which is transmitted to the Terfenol-D rod through conduction, Joule heating due to eddy currents, and potential internal heating due to the reorientation of moments. To accommodate future designs, we also incorporate possible convection mechanisms analogous to those considered for piezoceramic and shape memory alloy transducers.

To model the temperature transition in the magnetic material, we employ the notation of Section 2.6.8 and let $\bar{c}, h_c, \mathcal{M}, \Omega, \lambda, \ell$ and $T_E(t)$ respectively denote the specific heat, a convection coefficient, the mass of the Terfenol-D rod, the surface area of the material, the thermal conductivity of the solenoid, the path length of conduction, and the time varying temperature of the adjacent environment An energy balance then yields the differential equation

$$\bar{c}\mathcal{M}\dot{T}(t) = -\Omega\left[h_c + \lambda/\ell\right]\left[T - T_E(T)\right] + J(t) - \left[h_+\dot{x}_+ + h_-\dot{x}_-\right] \tag{4.93}$$

governing the temperature evolution. A comparison of (4.93) with its ferroelectric counterpart (2.132) reveals that the general temperature relations are identical.

To quantify the conductive and convective contributions, it is necessary to specify $T_E(t)$ either experimentally or through additional models. For example, a fully coupled thermal model for a Terfenol-D transducer would require the measurement or characterization of ohmic heating in the solenoid and heat transmission through the device to specify $T_E(t)$. Similarly, the Joule component $J(t)$ can be determined either empirically or though eddy current relations. Finally, the last term in (4.93) quantifies potential changes in temperature due to the reorientation of moments. The relevance of this term for specific operating conditions should be established through validation experiments.

4.7.7 Magnetomechanical Constitutive Relations

To construct constitutive relations suitable for actuator and sensor characterization, it is necessary to incorporate facets of the magnetomechanical effects discussed in Section 4.1.8. A complete incorporation of the coupled mechanical coupling is highly complex and constitutes a current research topic. Hence we consider here only the incorporation of magnetostriction, or free strains, due to moment rotation as will be the case for prestressed Terfenol-D transducers. As noted in (4.15), the magnetostriction in this case exhibits a quadratic dependence on magnetization as indicated by the Terfenol-D data in Figure 4.15.

To incorporate this quadratic magnetomechanical coupling, we employ the magnetoelastic Helmholtz relation

$$\psi(\varepsilon, M) = \psi(M) + \frac{1}{2}Y^M\varepsilon^2 - a_2\varepsilon M^2 \tag{4.94}$$

and

$$G(\sigma, H, \varepsilon, M) = \psi(M, \varepsilon) - \mu_0 HM - \sigma\varepsilon \tag{4.95}$$

where $\psi(M)$ is given by (4.68), Y^M denotes Young's modulus at constant magnetization, and a_2 is a positive magnetoelastic coupling coefficient.

For regimes in which magnetic after-effects are negligible, enforcement of the necessary condition $\frac{\partial G}{\partial M} = 0$ yields hysterons $\overline{M}$ analogous to those derived in Section 2.6.9 for ferroelectric materials. For small strains, one can obtain reasonable accuracy using the relations (4.72) or (4.73) which neglect the effects of strains on M. To incorporate magnetic after-effects, the relation (4.78) is employed using the Gibbs relation (4.95).

Enforcement of the minimization criterion $\frac{\partial G}{\partial \varepsilon} = 0$ and combination with the magnetization relation (4.86) yields the constitutive relations

$$\begin{aligned} \sigma &= Y^M \varepsilon - a_2 M^2 \\ [M(H)](t) &= \int_0^\infty \int_{-\infty}^\infty \nu_1(H_c)\nu_2(H_I)[\overline{M}(H+H_I;H_c,\xi)](t)\, dH_I\, dH_c. \end{aligned} \tag{4.96}$$

Kelvin–Voigt damping can be incorporated by employing the converse relation

$$\sigma = Y^M \varepsilon + c\dot{\varepsilon} - a_2 M^2 \tag{4.97}$$

which is based on the assumption that stress is proportional to a linear combination of strain and strain rate when $M = 0$.

Remark 4.7.6. *The quadratic M-ε behavior encapsulated in (4.96) provides adequate characterization for various materials and drive regimes but does not provide a comprehensive description of the converse magnetomechanical effect. For example, it does not directly characterize the hysteretic M-ε Terfenol-D behavior shown in Figure 4.15(b). Aspects of this phenomenon have been demonstrated in [120] to be quantified by distributed models characterizing the dynamics displacements of a Terfenol-D rod. It has also been hypothesized that formulation of the strain model in terms of the phase fractions x_+ and x_-, and inclusion of 90° switching will be necessary to fully quantify the converse H-ε behavior. Finally, the formulation of (4.96) is based on the assumption that stresses lie below the coercive stress σ_c which precludes switching. The extension of the theory to include the magnetomechanical switching phenomena discussed in Section 4.1.8 is under present investigation.*

4.7.8 Material and Device Characterization

To illustrate attributes of the framework, we consider the characterization of anhysteretic and hysteretic steel data, Terfenol-D data exhibiting biased H-M behavior, and H-ε data from a Terfenol-D transducer. In the first two cases, the lognormal and normal densities (4.88) were employed in the magnetization model (4.86) whereas a normal induced field density was employed in the third example. Quasistatic operating regimes and negligible after-effect and accommodation phenomena permitted the use of the piecewise linear kernel (4.72) and implementation Algorithm 4.7.3.

Steel H-B Characterization

We first consider the characterization of data from a steel specimen having a length of 6 cm and cross-sectional area of 1 cm as reported by Jiles and Atherton [250]. The anhysteretic H-B_{an} response and hysteretic H-B data collected at four input levels are plotted in Figure 4.40.

The parameters $M_R, \eta, b, \widetilde{C} = c_1 \cdot c_2 \cdot I_1$ and $\overline{H}_c, c$ were respectively estimated through least squares fits to the anhysteretic and major loop data yielding the values summarized in Table 4.2 and fits shown in Figure 4.40. Measured periodic fields having lower amplitudes were subsequently input to the model — using the *same*

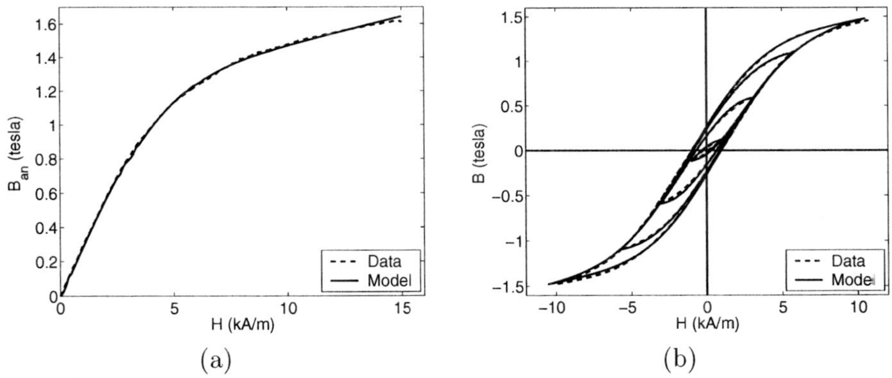

Figure 4.40. *Steel data from [250] collected under $\sigma = 0$ prestress conditions. (a) Anhysteretic data and model fit. (b) Hysteresis data, major loop model fit, and minor loop predictions.*

parameter values — to obtain the symmetric minor loop predictions. For both the anhysteretic and hysteretic models, the expression (4.5) is employed to relate B, H and M. Additional details regarding this example can be found in [433].

Parameters	η/μ_0	M_R (A/m)	b (A/m)	$\widetilde{C}$	$\overline{H}_c$ (A/m)	c (A/m)
Values	6.5	5.4×10^3	3521.4	0.0190	250	0.75

Table 4.2. *Parameter values identified for the anhysteretic model (4.92) and hysteresis model (4.86).*

Terfenol-D Transducer: H-M Characterization

To illustrate the biased minor loop capabilities of the framework in the absence of after-effects, we summarize the characterization of data collected at 0.2 Hz from the Terfenol-D transducer depicted in Figure 1.11.

To construct the model, the measured coercive field value $H_c = 6158$ A/m and reciprocal slope $\frac{dH}{dM} = 6.5$ after saturation were used to obtain initial estimates for $\overline{H}_c$ and η/μ_0. The final parameter values $M_R = 8.7 \times 10^4$ A/m, $\eta/\mu_0 = 7$, $H_c = 2000$ A/m, $c = 0.65$ A^2/m^2, $b = 1.58 \times 10^4$ A/m, $C = c_1 \cdot c_2 = 1.98 \times 10^{-8}$ were then obtained through a least squares fit to the symmetric major loop data yielding the model fit illustrated in Figure 4.41. Biased, periodic fields were subsequently input to the model, using the same parameter values, to obtain the minor loop predictions. We note that when predicting the minor loop responses, the starting magnetization is determined from the symmetric loop fit; hence the accuracy of the minor loops is highly dependent on the accuracy of the symmetric major loop.

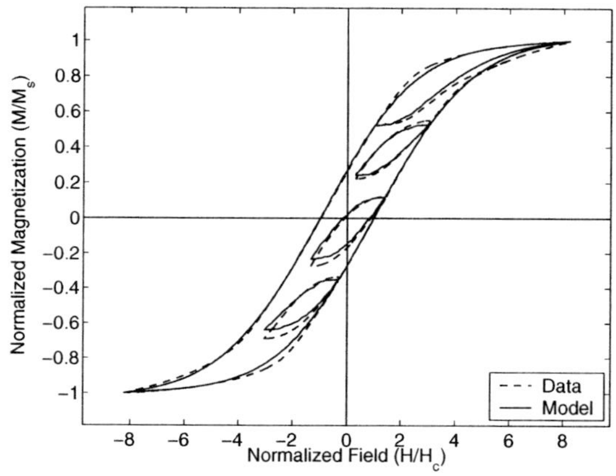

Figure 4.41. *Terfenol-D data from [449] and model response provided by (4.86).*

Terfenol-D Transducer: H-ε Characterization

Whereas the characterization of the H-M behavior of a Terfenol-D transducer is of fundamental importance, it does not quantify the strains which are ultimately employed in the structural and structural-acoustic applications summarized in Section 1.3. To accomplish this, we employ the constitutive relations (4.96) to characterize the H-ε behavior of a Terfenol-D transducer driven at 1 Hz with a prestress of 1 ksi (6.9 MPa). Two input levels yields the moderate and high drive level magnetization and strain data plotted in Figure 4.42 — details regarding the transducer design and data collection techniques are provided in [119].

As detailed in [434], parameters were estimated through a least squares fit to the high drive level data producing the model response plotted in Figure 4.42. The model with the same parameter values was then used to predict the moderate drive level behavior.

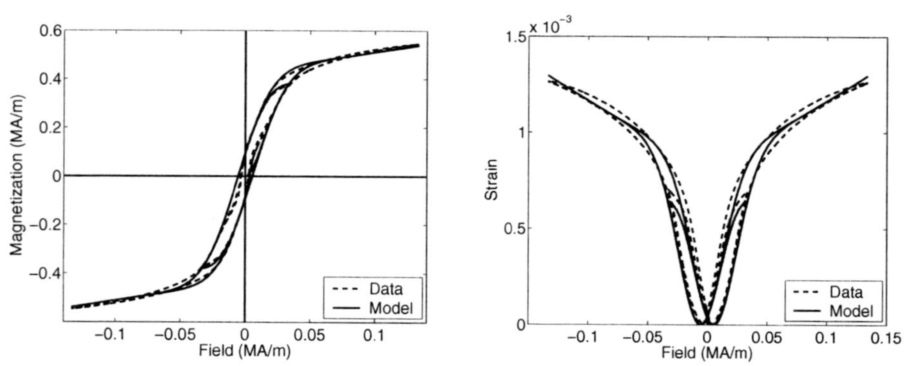

(a) (b)

Figure 4.42. *Experimental data (– – –) from [119] and model response (——): (a) field-magnetization relation and (b) field-strain relation.*

The discrepancies between the model fit and data are attributed primarily to two mechanisms. The concavity change in the low-field magnetization data is typically associated with crystalline anisotropies inherent to Terfenol-D but not presently incorporated in the underlying Gibbs relations. The discrepancy in the tips of the strain loop indicate limitations in the quadratic constitutive relation as summarized in Remark 4.7.6. Despite these issues, the model provides sufficient accuracy for material and device characterization for a wide range of operating conditions as well as model-based control design.

4.7.9 Model Attributes

Model Construction

The homogenized energy framework characterizing the H-M and H-ε behavior in ferromagnetic compounds is analogous to that constructed in Chapter 2 to quantify ferroelectric E-P and E-ε behavior. We refer the reader to Section 2.6.11 for a complete summary of the ferroelectric model and note that the ferromagnetic model is constructed in a similar manner.

Model Attributes

Physical Nature of Parameters

The magnetization model (4.86), formulated in terms of the lognormal and normal densities (4.88), contains the parameters $M_R, \eta/\mu_0, \overline{H}_c, b, c$ and $C = c_1 \cdot c_2$ which must be determined for a specific compound. As detailed in [433, 434], M_R and C can be combined into a single parameter which simply scales the magnetization resulting from a given field input. For the piecewise linear kernel (4.72), bulk measurements of the reciprocal slope $\frac{dH}{dM}$ at field reversals provide initial estimates for η/μ_0. In the lognormal density ν_1, $\overline{H}_c$ is asymptotically related to the coercive field for the bulk material whereas the standard deviation $\sigma \approx 2\overline{H}_c c$ quantifies the variability in coercive fields. Hence materials with steep transitions at coercivity yield small values of c whereas large values are required when characterizing materials with gradual transition behavior. Finally, the parameter b in the normal interaction field density ν_2 quantifies the degree of switching which occurs before remanence is reached. Materials with nearly linear H-M relations at remanence yield small values of b whereas large values are required to accommodate significant pre-remanence switching. Exploitation of these physical properties can greatly facilitate parameter estimation and model updating to accommodate changing operating conditions.

The discretized model (4.87) employing general densities requires the estimation of $N_i + N_j$ parameters if estimating the component densities ν_1 and ν_2 and $N_i \cdot N_j$ parameters if the polarization is formulated in terms of the joint density ν. The price paid for exploiting this generality is the necessity of identifying a large number of nonphysical parameters thus necessitating the constrained minimization algorithms described in Section 2.6.6. For many applications, however, the higher accuracy provided by the general densities justifies the additional effort required to identify parameters. It should be noted that once the parameters are identified for a given material or transducer design, the two formulations provide *equivalent efficiency for model-based control design* since the required number of operations in the forward and inverse algorithms are the same in both cases.

Reversibility

The monotone increasing behavior of the energy-based hysterons, as illustrated in Figure 4.37, implies that the model incorporates both irreversible and reversible behavior throughout the drive range. As a result, the model accurately characterizes the reversible post-switching behavior exhibited by the Terfenol-D data in Figure 4.42. The model response is replotted in Figure 4.43 to illustrate this property which constitutes a required extension in Preisach models.

Noncongruency

It was noted in Section 4.5.4 that various magnetic systems, including single particles having the H-M behavior depicted in Figure 4.28, exhibit noncongruency in the sense that minor loops having the same bounding field values but different magnetizations will have different shapes. The congruency or noncongruency provided by the energy is dependent upon the choice of hysteron or kernel $\overline{M}$.

Consider first the model constructed using the piecewise linear hysteron $\overline{M}$ specified by (4.73) with general densities ν_1 and ν_2. Both densities are independent of magnetization which facilitates implementation and yields highly efficient forward and inverse algorithms. It also produces congruent minor loops at differing magnetization levels as illustrated in Figure 4.44(a). For a large number of applications, this provides sufficient accuracy for both material characterization and model-based control design. However, for regimes in which noncongruent behavior is significant, it may be necessary to consider the kernel (4.71) or include magnetization-dependency in the interaction field density in a manner analogous to that employed for moving Preisach models [126, 334].

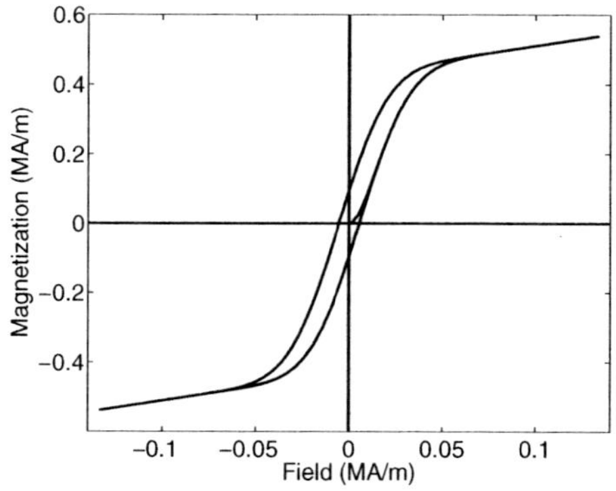

Figure 4.43. *Reversible post-switching behavior exhibited by the model (4.86) when characterizing the Terfenol-D behavior.*

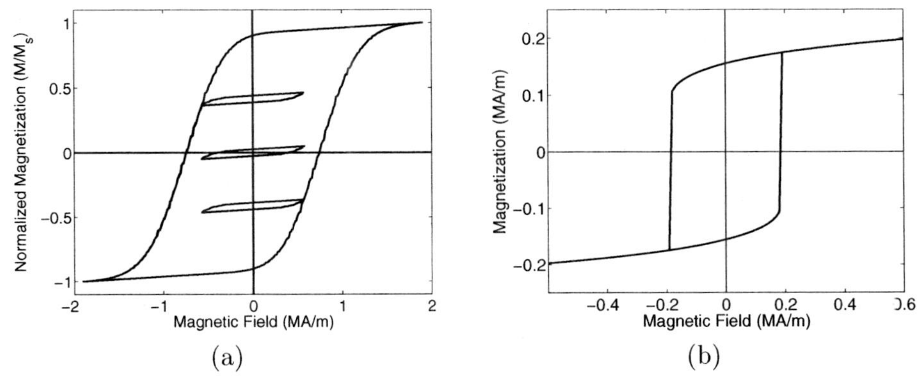

Figure 4.44. *(a) Congruent minor loop behavior exhibited by the model employing the kernel (4.73) and (b) noncongruent behavior of the kernel (4.71).*

The kernel (4.71),

$$\overline{M}(H) = M_s \tanh\left(\frac{H + \alpha M}{a}\right),$$

automatically incorporates magnetization-dependence in the effective field $H_e = H + \alpha M$ due to the inclusion of moment interactions in the mean field assumptions. As illustrated in Figure 4.44(b), the noncongruency exhibited at magnetization levels just below the coercive field are analogous to those depicted in Figure 4.28 for single Stoner–Wohlfarth particles and indicate the capability of the resulting macroscopic model to accommodate noncongruency. As with moving Preisach models which address noncongruency, macroscopic models employing the kernel (4.71) are nonlinear which increases the cost of implementation.

Minor Loop Closure, After-Effects, and Accommodation

The closure or nonclosure of biased minor loops is dictated by whether or not thermal activation processes are included in the kernel. Direct minimization of the Gibbs energy G to obtain the hysterons (4.71) or (4.73) in the absence of thermal activation yields closed minor loops, as illustrated in the examples of Section 4.7.8, whereas the kernel (4.78) — which incorporates both the Gibbs and relative thermal energy kT/V — provides a model that permits nonclosure of minor loops.

As detailed in Section 4.1.7, magnetic after-effects due to atomic diffusion or thermally-induced switching provide one mechanism leading to minor loop nonclosure. It is noted in Section 4.5.5 that characterization of after-effects using Preisach models requires extension of the theory to incorporate the relaxation mechanisms — e.g., via Arrhenius relations. The homogenized energy model employing the kernel (4.78) automatically quantifies magnetic after-effects due to the fact that it incorporates both the Gibbs and relative thermal energies. This is illustrated in Figure 4.45 where a discontinuous step input field is applied to the model to simulate the after-effect phenomenon depicted in Figure 4.10. This highlights an advantage of the energy-based framework.

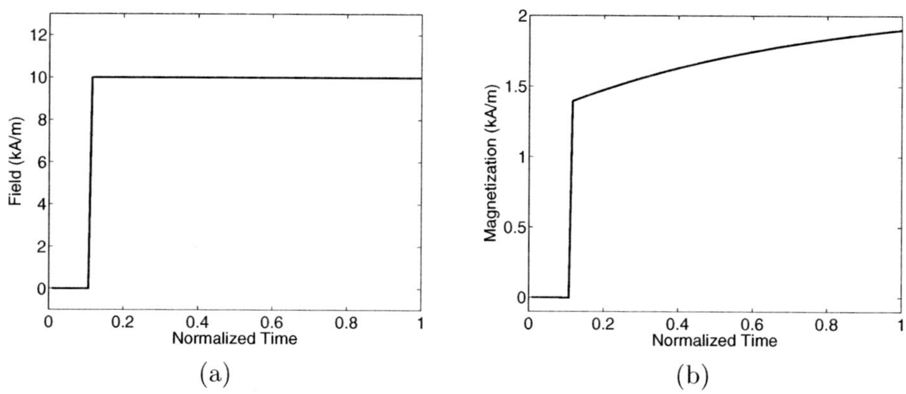

Figure 4.45. *(a) Input field H and (b) response of the model employing the kernel (4.78) which incorporates thermal activation.*

A second phenomenon that produces minor loop nonclosure is the accommodation or reptation process depicted in Figure 4.11. This is due to equilibration of moments during initial field cycles and produces a training or breaking in period analogous to the initial plastic deformations observed with SMA — see pages 249–251. For certain operating regimes, this *magnetic relaxation* process produces macroscopic behavior similar to that manifested by *thermal relaxation*. In such cases, the homogenized energy model employing the kernel (4.78) can be used to *phenomenologically* quantify the process. This is illustrated in Figure 4.46 where a periodic field is input to the model to simulate the manner in which the H-M relation accommodates to an equilibrium trajectory. We emphasize that use of the present model to characterize accommodation (reptation) exploits phenomenological similarities between thermal and magnetic relaxation processes and extension of the theory to incorporate the energy mechanisms associated with magnetic accommodation remains an open research topic.

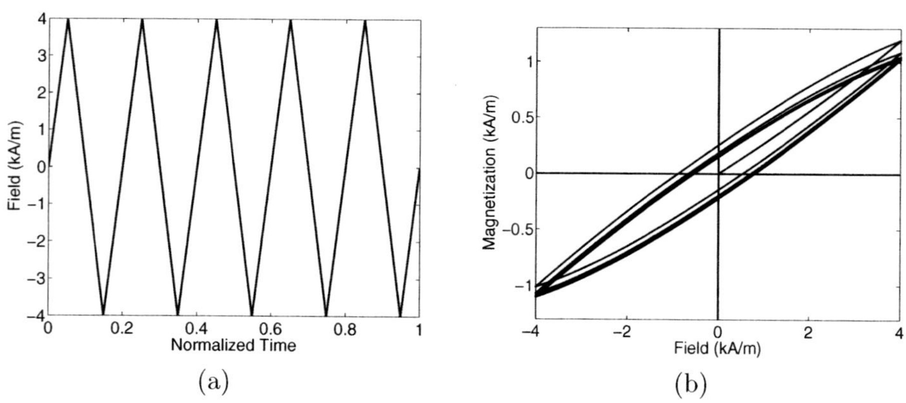

Figure 4.46. *(a) Input field H and (b) model response with $\overline{M}$ given by (4.78) as it accommodates to an equilibrium trajectory.*

Temperature and Stress-Dependencies

The use of the Helmholtz energy relation $\psi(M, T)$ derived from statistical mechanics principles yields an Ising kernel or hysteron $\overline{M}$ which incorporates certain temperature dependencies. When combined with the temperature evolution relation (4.93), this provides an initial qualitative mechanism for incorporating temperature-dependencies including the second-order ferromagnetic to paramagnetic phase transition depicted in Figure 4.18. It should be noted, however, that the Ising relation can prove overly restrictive when quantifying fine-scale temperature effects or changes over a broad temperature range. For such applications, phenomenological temperature mechanisms analogous to those discussed in Chapter 5 for shape memory alloys may need to be incorporated. Further details illustrating the incorporation of temperature-dependencies are provided in Chapter 3 in the context of relaxor ferroelectric compounds.

The constitutive relations (4.96) quantify the converse effect governing the manner through which magnetization changes influence stresses or strains in the *absence* of ferroelastic switching. They also quantify one of the mechanisms through which strains influence the magnetization kernel $\overline{M}$. For stresses larger than the coercive stress σ_c, ferroelastic switching produces phenomena of the type illustrated in Figures 4.16–4.14 which are not characterized by the present formulation.

Software

MATLAB m-files for implementing the homogenized energy model for ferromagnetic materials can be found at the website `http://www.siam.org/books/fr32`.

4.7.10 Comparison with Other Models

Jiles–Atherton Model

The Ising relation (4.71) obtained by minimizing the Gibbs energy constructed through statistical mechanics tenets provides one choice for the anhysteretic magnetization in the Jiles–Atherton model. It is also analogous, and agrees through first-order terms, with the Langevin relation (4.59) employed in the original formulation. The difference in theories lies in the manner through which energy relations are constructed and losses are incorporated.

Stoner–Wohlfarth Theory

The original Stoner–Wohlfarth theory quantified the rotation of noninteracting, single-domain particles having uniaxial anisotropy [464]. While the model has received widespread use in the magnetic recording industry, its use for general material characterization was originally limited by the fact that it did not incorporate moment interactions. A number of recent extensions to both the theory and underlying philosophy have substantially improved its utility. The model was extended to include cubic anisotropies by Lee and Bishop [289] whereas Armstrong [13], Clark, Savage and Spano [94], and Jiles and Thoelke [252] extended the theory to quantify magnetoelastic effects in Terfenol-D. Certain mean field effects are incorporated in [17, 378] while pinning losses are incorporated in [299] where it is illustrated that this latter mechanism is necessary to achieve physical minor loop behavior. Finally, relations between the Stoner–Wohlfarth theory and micromagnetic models are detailed in [4].

The hysteresis kernels in the homogenized energy theory are analogous to those predicted by the Stoner–Wohlfarth model when the applied magnetic field is aligned with the easy axis of particles. This is consistent with the assumption in the proposed theory that moment alignment occurs in two diametrically opposite directions. The assumptions differ from the original Stoner–Wohlfarth model in that moment interactions are incorporated whereas anisotropy energies are neglected.

Preisach Models

Given the generality of the homogenized energy theory, it is not surprising that it bears certain similarities to the Preisach models discussed in Section 4.5. Initial analysis detailing the manner through which this framework provides an energy basis for Preisach theory was presented in [447] and this topic is under present investigation. We summarize here the relation between the homogenized energy framework and the classical and extended Preisach models discussed in Sections 4.5.1–4.5.5.

Classical Preisach Model

The classical Preisach model is actually a special case of the thermally inactive magnetization model in the limit $\eta \to \infty$. To illustrate, consider the Helmholtz energy (4.68) with $M_I = M_R - \frac{s_2 \mu_0}{\eta}$ and $M_R = 1$. In the absence of thermal activation, minimization of $G(H, M) = \psi(M) - \mu_0 HM$ yields

$$[\overline{M}(H; H_c, \xi)](t) = \begin{cases} [\overline{M}(H; H_c, \xi)](0) & , \tau(t) = \emptyset \\ \frac{\mu_0}{\eta} H - 1 & , \tau(t) \neq \emptyset \text{ and } H(\max \tau(t)) = -H_c \\ \frac{\mu_0}{\eta} H + 1 & , \tau(t) \neq \emptyset \text{ and } H(\max \tau(t)) = H_c \end{cases}$$

where the initial moment distribution and transition times are defined in (4.74) and (4.75). From (4.76), the coercive field is $H_c = s_2$. The limit $\eta \to \infty$ yields

$$[\overline{M}(H; H_c, \xi)](t) = \begin{cases} [\overline{M}(H; H_c, \xi)](0) & , \tau(t) = \emptyset \\ -1 & , \tau(t) \neq \emptyset \text{ and } H(\max \tau(t)) = -s_2 \\ +1 & , \tau(t) \neq \emptyset \text{ and } H(\max \tau(t)) = s_2 \end{cases}$$

which is precisely the definition (4.36) for the piecewise constant kernel $[k_s(H, \xi)](t)$ employed in the classical Preisach model in the absence of interaction fields — see Figure 4.47.

We now consider the incorporation of distributed coercive and interaction fields. Formulation in terms of a joint density ν yields the model

$$[M(H)](t) = \int_0^\infty \int_{-\infty}^\infty \nu(H_c, H_I)[\overline{M}(H + H_I; H_c, \xi)](t)\, dH_I\, dH_c$$

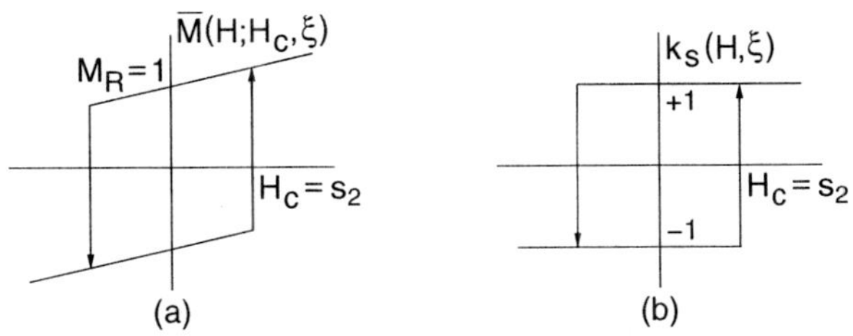

Figure 4.47. *(a) Hysteron* $\overline{M} = \frac{\mu_0}{\eta}H + \delta$ *derived through minimization of* $G = \psi - \mu_0 HM$, ψ *given by (4.68), with* $M_I = M_R - \frac{s_2\mu_0}{\eta}$ *and* $M_R = 1$. *(b) Hysteron* $k_s(H,\xi) = \overline{M}(H;H_c,\xi)$ *in the limit* $\eta \to \infty$.

which is equivalent to the Preisach formulation (4.35) in the limit $\eta \to \infty$.

Whereas this framework provides an energy basis for classical Preisach models, the limiting process required to achieve hysterons having zero slope yields a highly distorted energy landscape. The strength of this approach lies in the basis that it provides for extended Preisach models.

Specific Density Choices

The energy and Preisach frameworks both employ *a priori* parameterized functions for densities to facilitate model construction and updating. Choices include normal representations ν_2 for the interaction field and either normal or lognormal densities ν_1 for the local coercive field. Hence joint densities of the form

$$\nu(H_c, H_I) = \frac{M_s}{2\pi\sigma_c\sigma_I} e^{-(H_c-\overline{H}_c)^2/2\sigma_c^2} e^{-H_I^2/2\sigma_I^2}$$

or

$$\nu(H_c, H_I) = Ce^{-[\ln(H_c/\overline{H}_c)/2\sigma_c]} e^{-H_I^2/2\sigma_I^2}$$

have been employed in both frameworks. Discussion relating coefficients in the resulting homogenized energy model to properties of measured data is provided in Section 4.7.9 and analogous discussion for Preisach models can be found in [126].

Reversibility

As noted in Section 4.5.3, the incorporation of reversible H-M behavior constitutes one of the extensions to classical Preisach theory which has been developed to accommodate post-switching behavior. One technique to accomplish this in Preisach models is to modify hysterons in the manner depicted in Figure 4.27 and summarized in Figure 4.48. As illustrated in Figures 4.43 and Figure 4.48(b), reversibility is an inherent property of the energy-based hysterons which permits the characterization capabilities illustrated for Terfenol-D in Figure 4.42.

Noncongruency

The incorporation of noncongruency in Preisach models has spawned a number of extensions including several which employ input or output-dependent densities.

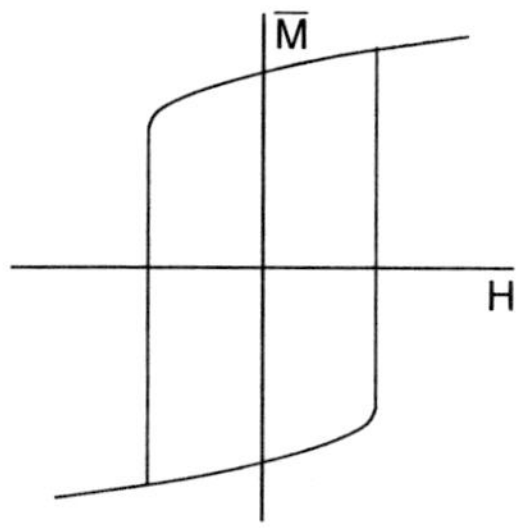

(a)

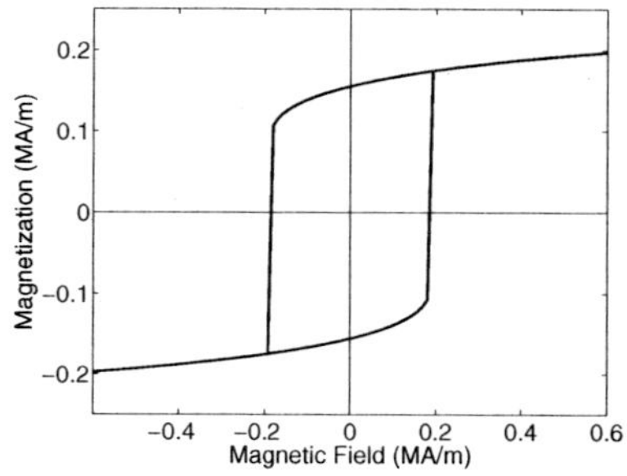

(b)

Figure 4.48. *(a) Preisach hysteron modified to incorporate reversibility and (b) hysteron obtained through minimization of $G = \psi - \mu_0 HM$ with ψ given by (4.68).*

This includes consideration of the density (4.49) which employs the central limit theorem to formulate the mean of the interaction field as $\overline{H}_i = \alpha M$.

The incorporation of noncongruency in the homogenized energy framework differs in the sense that interaction fields $H_I = \alpha M$, $\alpha = \frac{H_h}{\mu_0 M_s}$, are incorporated in the kernel

$$\overline{M}(H, M) = M_s \tanh\left(\frac{H + \alpha M}{a}\right),$$

rather than the density ν, thus yielding the macroscopic model

$$[M(H, M)](t) = \int_{-\infty}^{\infty}\int_{-\infty}^{\infty} \nu(H_h, H_c)\overline{M}(H + H_I, M; H_c)\, dH_h dH_c$$

where the bias field $H_h = \frac{N\Phi_0}{VM_s}$ is distributed to incorporate variation in the exchange energy Φ_0.

The inclusion of magnetization-dependency in the kernel rather than the density results from the energy basis for the theory and represents one of the fundamental differences between the energy and Preisach frameworks. Similar observations pertain to the inclusion of temperature, stress and rate-dependence which occur in the kernel of the homogenized energy model and typically occur in the parameters or densities of Preisach formulations.

Minor Loop Closure, After-Effects, and Accommodation

The deletion property (also termed *wiping out*) constitutes one of the two conditions established by Mayergoyz as necessary and sufficient for classical Preisach models. As noted in Section 4.5.5, the deletion property guarantees closure of biased minor loops and hence it must be relaxed for materials or operating regimes where after-effects or accommodation are significant. One technique for incorporating relaxation processes in Preisach models is through the solution of differential equations which incorporate the Arrhenius relation

$$\mathcal{T}(T) = \mathcal{T}_0 e^{W/kT}$$

where W denotes the energy required to overcome barriers.

The homogenized energy framework incorporates these phenomena in a natural manner by balancing the Gibbs energy G and relative thermal energy kT/V through either the Boltzmann probability relation

$$\mu(G) = Ce^{-GV/kT}$$

or the constrained minimization framework detailed in Section 2.6.2. Hence the model (4.86) employing the kernel (4.78) can be directly employed to characterize after-effects as illustrated in Figure 4.45. Moreover, it is proven in [433] and Section 2.6.3 that (4.78) converges to the piecewise linear kernel (4.73) in the limit $kT/V \to 0$ of negligible thermal activation. In this latter regime, the model provides deletion and guarantees biased minor loop closure as illustrated in the validation examples of Section 4.7.8. Hence the framework incorporates both the Preisach deletion property and extensions necessary to characterize thermally induced magnetic after-effects. Finally, it was noted on page 234 of Section 4.7.9 that for regimes in which atomic diffusion or magnetic relaxation processes produce macroscopic behavior analogous to that resulting from thermal relaxation, the homogenized energy framework can be used to phenomenologically quantify diffusion after-effects and accommodation (reptation).

Temperature and Stress-Dependencies

Because of the energy basis, the homogenized free-energy framework provides mechanisms for incorporating certain temperature and stress-dependent effects in the kernel or basis. Further details regarding the development of temperature-dependent models are provided in Chapter 3 in the context of relaxor ferroelectric materials. The incorporation of magnetoelastic energy contributions to provide models for the coupled magnetomechanical effects discussed in Section 4.1.8 is in its nascency but initial results indicate strong promise for the technique.

In contrast, extensions to Preisach theory to incorporate these effects have focused primarily on the use of lookup tables or vector-valued weights or densities — e.g., densities of the form $\nu(r, s, T)$ are employed in [460] when modeling a Stoner–Wohlfarth particle system and [269] to incorporate temperature-dependence in a thermoplastic Preisach model. In some regimes, this can hinder efficiency when employing Preisach models for system design or model-based control design.

Complementary Frameworks

We have illustrated that the classical Preisach model, characterized by the properties of congruency and deletion, is a special case of the homogenized energy framework in the limit $\eta \to \infty$ and absence of thermal activation mechanisms. Furthermore, construction of the framework by minimizing a combination of the Gibbs energy G and relative thermal energy kT/V provides a kernel which incorporates reversibility and noncongruency, and yields a model which characterizes after-effects, accommodation, certain temperature and stress-dependencies, and provides deletion in the limit of negligible thermal activation. Due to its energy genesis, input and output dependencies tend to occur in the kernel of the energy framework whereas they are typically incorporated in the weights or densities of Preisach formulations. The former can facilitate implementation since it relies solely on input and output measurements rather than requiring vector-valued weights or lookup tables.

In summary, the energy framework provides a broad energy basis for characterizing hysteresis and constitutive nonlinearities whereas Preisach theory contributes a rich and mature mathematical framework combined with physical and empirical principles. It is expected that theory exploiting the advantages of the combined frameworks will significantly impact the development of comprehensive characterization techniques for ferromagnetic, ferroelectric and ferroelastic compounds.

Chapter 5

Model Development for Shape Memory Alloys

Shape memory alloys (SMA) are characterized by solid state displacive transformations between austenite and martensite phases in response to mechanical, thermal, electromagnetic and ultrasonic inputs. This provides the materials with the capability for sustaining and recovering from strains up to 10% which imbues them with unique actuator and potential sensor capabilities in smart material systems. The inherent thermal-mechanical-(magnetic) coupling and hysteresis associated with the phase transformations also pose significant modeling challenges which must be addressed to exploit the full transducer capabilities of the materials.

Regions in which austenite and martensite variants segregate are termed *ferroelastic domains* in analogy with the domain structures consisting of aligned moments or dipoles in ferromagnetic or ferroelectric materials. To complete this analogy, *ferroelastic materials* are categorized as those exhibiting ferroelastic domains, thus completing the trilogy of materials collectively termed *ferroic compounds*. While focusing on SMA, we will consider aspects of the model development for the broader class of ferroelastic materials to provide common characterization frameworks for ferroic materials as summarized in Chapter 6.

We note that the use of the term ferroelastic in this context is related to but more specific than the description in Section 2.1 of ferroelastic switching in materials such as $PbTiO_3$ and PZT. There, the term referred to 90° dipole rotation resulting from applied forces whereas we additionally associate the presence of high symmetry austenite and lower symmetry martensite phases and related domain structures in the present discussion. The two concepts are compatible if one associates 90° switches with austenite to martensite transformations, thus providing one of the unifying aspects of the ferroic theory in Chapter 6.

It was noted in Section 1.4.1 that a number of materials, including alloys comprised of nickel-titanium (NiTi), copper-aluminum-nickel (CuAlNi), copper-zinc-aluminum (CuZnAl) and iron-manganese-silicon (FeMnSi) exhibit shape memory effects. We will focus on Nitinol (NiTi)[20] due to its superior structural and mem-

[20] More generally, one can consider Ni_xTi_{1-x} where shape memory effects are present for x in the approximate interval 0.47 to 0.51.

ory capabilities but note that much of the description and analysis applies to the other compounds.

Modeling Frameworks

Due to both the fundamental and complex nature of shape memory effects and their importance in applications employing SMA, model development for these compounds has been addressed from a number of perspectives. Rather than provide a complete summary of modeling frameworks, we provide a brief categorization which relates to some of the frameworks discussed in this chapter. The reader is referred to [38,413] for a comprehensive summary of models for bulk SMA compounds and [323] for references pertaining to SMA films.

Micromechanical Models

Micromechanical models quantify the effects of phase transformations at the microscopic scale through the construction of elastic, thermal and chemical free energy relations which are subsequently used to quantify local strains and phase transformation criteria. These local relations are then averaged or homogenized to derive macroscopic models for transducer design and control applications.

A number of micromechanical theories have been derived based on various assumptions regarding the specific energy functionals and degree to which grain variations, lattice structure, and polycrystallinity are incorporated. These include the models of Gao, Huang and Brinson [178,225], Goo and Lexcellent [188], Lu and Weng [304], and Patoor and Berveiller [377]. While micromechanical theory provides important insights regarding the multiscale nature of materials and manner through which microscopic energy behavior influences macroscopic effective parameters, the complexity of models presently precludes real-time implementation for device optimization or model-based control design. Hence, like the analogous micromagnetic theory, we employ micromechanical energy relations as starting points for constructing low-order mesoscopic and macroscopic models that are sufficiently efficient for implementation in smart material applications.

Mesoscopic or Lattice-Level Models

Mesoscopic or lattice models occupy the next level in the multiscale hierarchy. These frameworks are typically comprised of energy relations constructed for a representative lattice cell or control volume in combination with homogenization of some form to drive macroscopic constitutive relations. Models in this genre include the domain wall theory summarized in Section 5.4 and the statistical mechanics models of Müller and Wilmanski [353] and Achenbach, Müller and Seelecke [1,2,409, 410,413] which motivated both the homogenized energy SMA models in Section 5.5 and the thermally activated ferroelectric models in Section 2.6 and ferromagnetic models in Section 4.7. We note that the theory of Govindjee and Hall [193] employs similar evolution laws to define phase fractions under the assumption of isothermal operating conditions.

Macroscopic Models

Macroscopic models are based on the assumption that points in the macroscopic material are representative of the underlying phase mixture rather than segregated into distinct and identifiable phases. This framework includes a broad spectrum of models ranging from theory based on irreversible thermodynamic principles to phenomenological Preisach models.

Early models in this vein were proposed by Falk starting in 1980 [157–159]. This theory, which is summarized in Section 5.2.1, is based on Landau–Devonshire theory for ferroelectric and ferromagnetic compounds and yields energy relations analogous to those summarized in Sections 2.3.1 and 4.4.2.

A number of later models employed internal variables (order parameters) e to characterize the underlying structure and phase composition subsequently employed in macroscopic state space representations. The models differ in their choice of internal variables, kinetic equations governing the evolution of internal variables, underlying energy relations and resulting state equations, and degree to which energy conservation (1st law of thermodynamics), entropy balance and the Clausius–Duham inequality (2nd law of thermodynamics) are invoked to ensure thermodynamic consistency. Early models of this type include those of Tanaka et al. [473, 474] who employed the phase fraction of martensite, $e = x_M$, as the internal variable and strain ε and temperature T as input or control variables — the collection of internal and control variables is denoted by $s = \{\varepsilon, T, x_M\}$. This theory was extended in 1990 by Liang and Rogers [297], and in 1993 by Brinson [55] who employed the twinned and detwinned martensite fractions to characterize quasiplastic behavior. A model for pseudoelasticity was proposed by Ivshin and Pence in 1994 based on thermodynamic analysis and employing the input set $s = \{\sigma, T, x_A\}$, where $x_A = 1 - x_M$ denotes the austenite fraction [238]. In the same year, Boyd and Lagoudas presented a 3-D model based on irreversible thermodynamic tenets [50] which was extended in a later series of papers by Bo and Lagoudas to incorporate back and drag stress as internal variables to characterize plastic strains and the two-way shape memory effect [44–46, 279].

As detailed in Sections 2.4 and 4.5, Preisach models have the advantage of generality and a mature mathematical framework, and hence it is not surprising that a number of researchers have employed them for SMA applications — a partial list includes Gorbet, Wang and Morris [190], Hughes and Wen [227], Huo [229], Ktena et al. [272], Lagoudas and Bhattacharyya [278], Ortín [369], and Webb, Kurdila and Lagoudas [504]. We include a description of the Preisach models for SMA in Section 5.3 since it provides one of the unified frameworks for ferroic compounds. However, physical interpretation of the Preisach model for SMA lags significantly behind its ferromagnetic counterpart so we focus more heavily on the physically-based homogenized energy framework discussed in Section 5.5. It is anticipated that this latter theory will provide an energy basis for certain extended Preisach theories for SMA — in a manner analogous to that discussed in Section 4.7.10 — which will unify and strengthen both approaches.

Chapter Organization

- Section 5.1 – A summary of SMA properties is provided to illustrate issues which must be addressed in models.
- Section 5.2 – We summarize high-order energy relations to provide a basis for subsequent models.
- Section 5.3 – Preisach models: Preisach relations quantifying σ-ε behavior are analogous to the models in Sections 2.4 and 4.5.
- Section 5.4 – Domain wall theory: This theory is based on the quantification of equilibrium, irreversible and reversible strains and yields a model analogous to the ferroelectric and ferromagnetic models discussed in Sections 2.5 and 4.6.
- Section 5.5 – Homogenized energy theory: This framework is based on the Müller-Achenback-Seelecke theory for thermally activated processes in SMA which preceded and motivated the corresponding ferroelectric and ferromagnetic models in Sections 2.6 and 4.7. This is the most comprehensive of the three approaches considered in this chapter and it is constructed with the goal of providing an energy-based framework which is sufficiently efficient to permit real-time implementation in smart material design and control applications.

5.1 Physical Properties of Shape Memory Alloys

The behavior of shape memory alloys is dictated by transformations from a high symmetry austenite phase to lower symmetry martensite phases. These phase transformations can be induced by heat, stresses, electromagnetic fields, and ultrasonic forces but we will focus primarily on temperature and stress-induced transformations while noting that the development and characterization of ferromagnetic shape memory alloys (FSMA) constitutes an active and growing research area.

To illustrate the nature of phase transformations, we consider the austenitic NiTi lattice cell depicted in Figure 5.1. In response to an applied stress, each of the 6 face-diagonal planes can shear in 2 directions and shift in 2 directions as indicated by the arrows. This yields 24 possible martensite variants in 3-D which are characterized by changes from 90° to approximately 96° in the lattice. The situation in 1-D is significantly simpler since there are only two martensite variants, M^- and M^+ as shown in Figure 5.2.[21] Throughout this discussion, we will focus on the uniaxial case due to its tractability and prevalence in applications employing SMA wires, films and rods. However, designs employing 2-D and 3-D SMA plates and shells are becoming increasingly prevalent — e.g., see the SMA chevron in Figure 1.23 — thus motivating the development of higher-dimensional models feasible for real-time implementation.

[21] There is also an intermediary R-phase (rhombohedral-phase) but we focus on the austenite and martensite phases since they are primarily responsible for shape memory mechanisms.

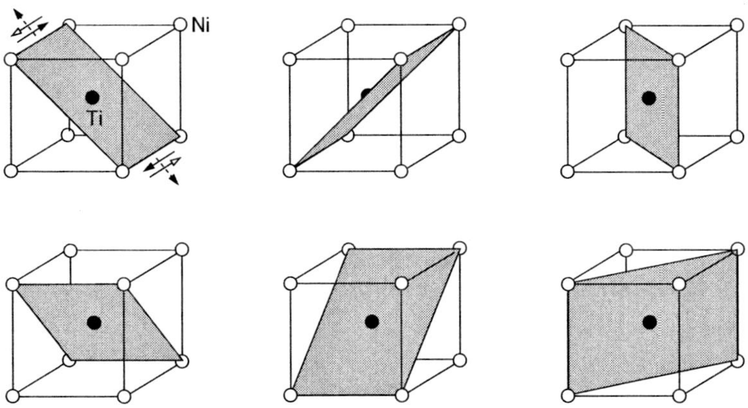

Figure 5.1. *Lattice cell of NiTi illustrating the six face-diagonal planes. In the manner illustrated for the first plane, each can shift in the direction of the dashed arrows or shear in the direction of the open and closed arrows to produce a total of 24 martensite variants.*

Temperature-Induced Phase Transformations

In the absence of an applied stress, austenite is stable at high temperatures whereas the martensite variants are equilibrium states at low temperatures. To illustrate, consider the temperature-induced phase transformations depicted in Figure 5.3(a) and (b). When a material in the austenite phase is cooled, the martensite transformation commences at the temperature M_s and is completed at M_f. Twinned or self-accommodated martensite is energetically favorable in the absence of a load so that negligible macroscopic strains are produced. Subsequent heating produces an austenite transformation through the temperature interval A_s–A_f thus returning the material to its original, symmetric crystallographic state.

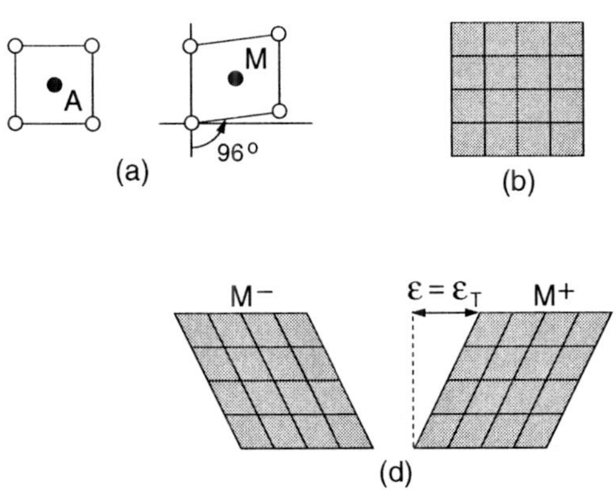

Figure 5.2. *(a) Austenite to martensite transformation in 1-D. (b) Uniaxial lattice element in the austenite phase, (c) twinned martensite, and (d) detwinned martensite configuration.*

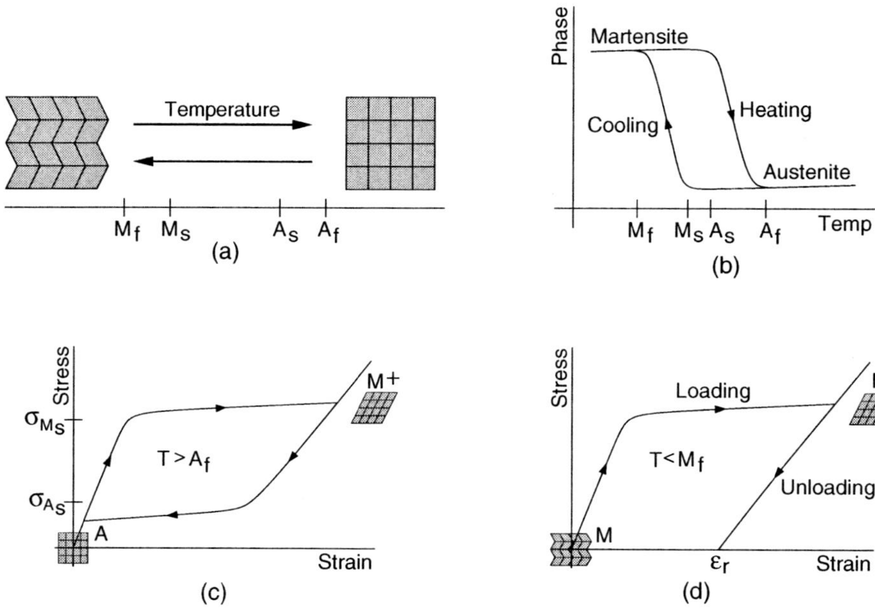

Figure 5.3. *(a) Temperature-induced phase transformation between austenite and twinned martensite in the absence of an applied load. (b) Hysteretic relation between austenite and martensite as a function of temperature. (c) Stress-induced phase transformation and pseudoelastic behavior for $T > A_f$. (d) Quasiplastic behavior and residual strain ε_r generated when $T < M_f$.*

Stress-Induced Phase Transformations

Transformations between the austenite and martensite phases can also be induced by applied stresses. Consider first the application of a tensile stress ($\sigma > 0$) for fixed $T > A_f$. For stresses less than a transformation stress σ_{M_s}, the stress-strain relation is approximately linear, as shown in Figure 5.3(c),[22] and can be modeled by Hooke's law — hence $\sigma_A = Y_A \varepsilon$ where Y_A denotes the Young's modulus of austenite. If the stress is increased above σ_{M_s}, the material transforms to detwinned martensite where it exhibits elastic behavior, $\sigma = Y_M(\varepsilon - \varepsilon_r)$, until stresses are reduced to σ_{A_s} where the reverse transformation commences.[23] In this high temperature regime, full shape recovery is observed upon unloading and the phenomenon is termed *pseudoelastic* or, more specifically, *superelastic*.

[22]In the physics and mathematics literature, it is conventional to respectively plot stress and strain on the abscissa and ordnate of figures to indicate their input-output relation. In many smart materials and engineering applications, however, transducers operate in a mode of strain or displacement control so the convention is reversed. We employ primarily the latter convention but warn the reader that both are common in the literature.

[23]Note that Y_A and Y_M are typically different. Representative values for NiTi are $Y_A = 70 \times 10^3$ MPa and $Y_M = 30 \times 10^3$ MPa. Similarly, other material properties such as the density, electric resistivity and specific heat differ for the two phases thus necessitating consideration of phase-dependent material coefficients.

We now consider the same loading and unloading process for fixed $T < M_f$. The application of stresses $\sigma > \sigma_{M_s}$ in this regime produces detwinning since the variant aligned with the stress is energetically favored. Unloading in this case produces a remanent strain ε_r since the material remains in the detwinned martensite state. This low temperature material behavior is sometimes termed *quasiplastic* to differentiate it from plastic strains characterized by permanent deformations.

Figure 5.3(c) and (d) illustrates the stress-strain behavior of SMA for tensile stresses and temperatures below the transition temperature T_2, discussed in Section 2.3.1, above which only elastic behavior is observed. To complete the repertoire, we illustrate in Figure 5.4 the material behavior for both compressive and tensile stresses and high-temperature elastic regimes. Details regarding the transition behavior between regimes will be provided in Section 5.2.3 when discussing the construction of appropriate Helmholtz energy relations.

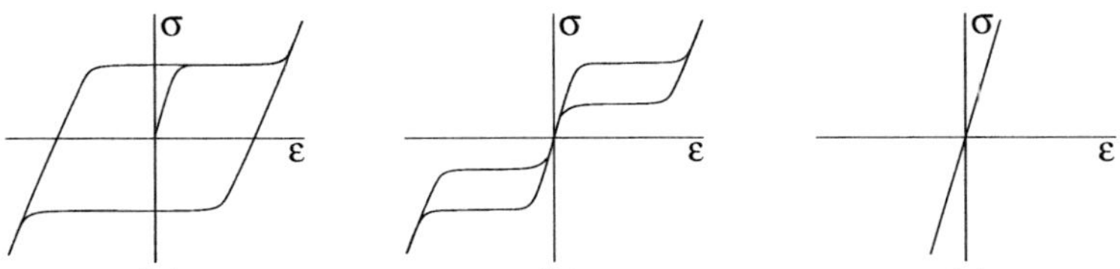

Figure 5.4. *(a) Ferroelastic material behavior for $T < M_f$, (b) pseudoelastic behavior for $A_f < T < T_2$, and (c) elastic response for $T > T_2$.*

Shape Memory Effect

The combination of temperature and stress-induced phase transformations provide shape memory alloys with their memory capabilities. When the detwinned material is unloaded and heated above A_f, the residual strain ε_r is recovered and the material returns to its original shape. The recovery of stress-induced strains through heating constitutes the *shape memory effect (SME)*. The complete shape memory process is depicted in Figure 5.5, and the pseudoelastic behavior and shape

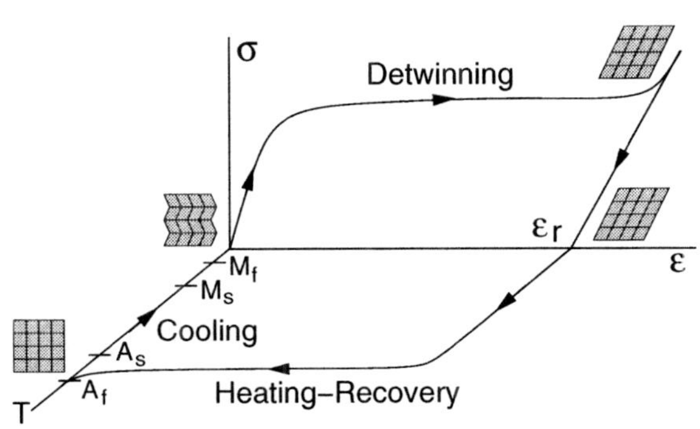

Figure 5.5. *Shape memory effect in a uniaxial SMA.*

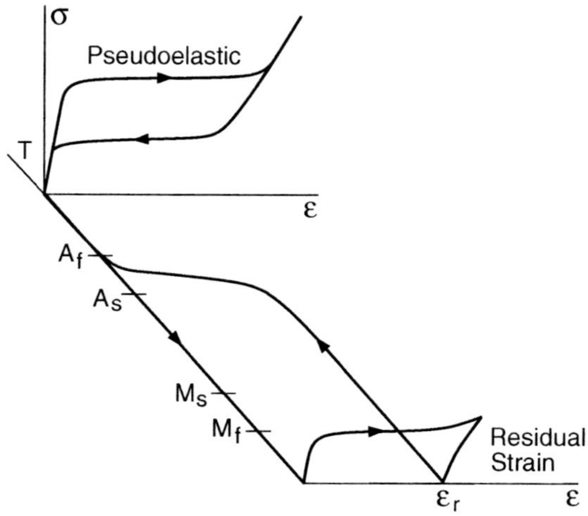

Figure 5.6. *Pseudoelastic behavior and shape memory effect in uniaxial SMA due to stress and temperature-induced phase transformations.*

memory effects due to stress and temperature-induced phase transformations are summarized in Figure 5.6.

The shape memory effect can be exploited for device design by inducing the desired geometry or performance capabilities in the twinned martensite configuration. This can be accomplished by placing the heated, austenitic, material in a mold and letting it cool to the twinned martensite phase — which has the same geometry — or by constructing the product directly for the twinned material. The device or apparatus is subsequently packaged or stressed which produces detwinning and generates residual strains. The original configuration is recovered by heating it to austenite which has the geometry originally induced in the twinned martensite material.

This process is illustrated in Figure 5.7 which depicts the molding of wire in the shape WY as it is cooled from austenite to twinned martensite. The wire is subsequently crushed into a ball and released ($\sigma = 0$), thus rendering it unrecognizable. Subsequent heating with a lighter or electric current returns the emblem to its original shape.

While the example provides a prototype to illustrate the process, it is doubtful that groups other than the Wyoming Chamber of Commerce will support its development! To illustrate the process for aeronautic and aerospace applications, we summarize two additional examples.

The first demonstrates the manner through which Nitinol pipe or hose couplers are constructed. During the manufacturing process, a TiNiFe alloy tube is constructed at low temperature ($\approx -150^o$ C) to have a diameter approximately 4% smaller than the pipe or hose to be joined. While maintained at this low temperature, and hence in the martensite phase, a tapered plug is inserted in the tube and used to increase the diameter to a dimension larger than its mate. In the phase diagram of Figure 5.5, this represents the application and release of stress which

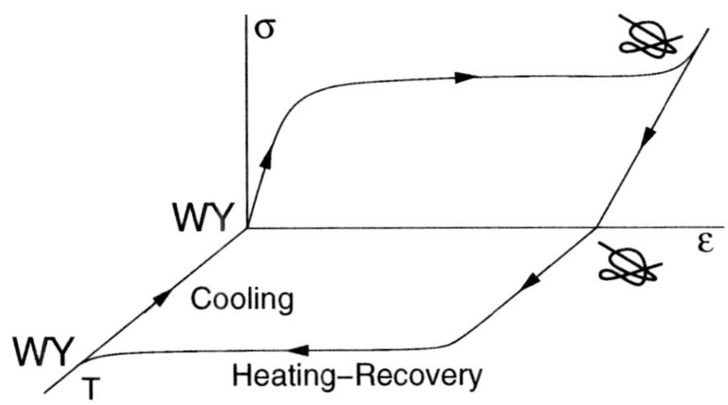

Figure 5.7. *Fabrication of the emblem* WY *in the twinned martensite phase, stressing it into a ball, and recovery of the original shape by heating.*

detwins the material. The two pipes are subsequently joined and allowed to warm to ambient conditions where the austenite phase is stable. The coupling tube returns to its original diameter thus providing a highly effective seal. As noted in Section 1.4.2, applications exploiting this technique include the joining of hydraulic lines on F-14 fighter jets.

An aerospace application entails the construction of large aerospace structures — e.g., modular antennas — in the twinned martensite phase followed by packaging to fit in a rocket faring. Once launched in space, either solar or applied heat is used to recover strains induced during packaging to achieve the original shape. This process can be visualized in Figure 5.7 by replacing WY by a modular antenna.

Material Training and the Two-Way Shape Memory Effect

The previously described shape memory effects are *one-way* in the sense that once residual strains are recovered through heating, they cannot be reintroduced when material is cooled without additional application of stress. Hence the emblem WY in Figure 5.7 remains upon cooling rather than returning to a crumbled ball. This is in contrast with *two-way shape memory effect* which can be introduced through training or plastic deformations that produce material or stress anisotropies favoring certain martensite variants.

Plastic Strains and Material Training

As detailed in [45], the generation and evolution of plastic strains in shape memory alloys differs fundamentally from the process observed in conventional metals. In the latter, plastic strains are typically introduced by loads in excess of the elastic limit whereas in SMA, plasticity is introduced by phase transformations — hence it is unavoidable in operating regimes which involve austenite–martensite transformations. The magnitude of the effect is illustrated in Figure 5.8 where the first two cycles of NiTi data, reported in [45], are depicted. In this figure, the material starts in the austenite phase at point a and is subsequently driven through the points b–e by two cooling and heating cycles. The difference between a and c

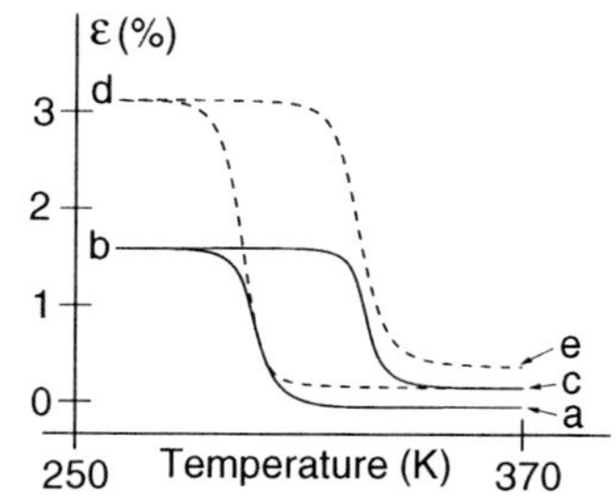

Figure 5.8. *Depiction of first two cycles of NiTi data from [45]: Cycle 1 (——), Cycle 2: (- - -).*

represents the plastic strain incurred during the first cycle whereas c to e is that in the second cycle. The results in [45] demonstrate that this accumulation of plastic strains can continue in excess of 2000 cycles and hence it must be accommodated in the design process for SMA transducers.

One technique to accomplish this is to run the training period until the plastic strain behavior stabilizes and then employ the material in a transducer. If employed before stabilization occurs, material models and model-based control designs must incorporate accumulating plastic strains since they significantly influence the material behavior. Additionally, material properties such as the austenite/martensite start and finish temperatures evolve during the training process thus necessitating inclusion of these mechanisms in models if employing the materials in training regimes.

Two-Way Shape Memory Effect

An additional consequence of the plastic strains which accumulate during training is the generation of internal stresses and material asymmetries which produce preferential formation of specific martensite variants. As a result, the materials return to a biased martensite state when temperatures are cooled from A_f to M_f rather than twinned martensite configuration depicted in Figure 5.5. This produces what is termed a *two way shape memory effect* (TWSM) or (TWSME) in which the material exhibits a stable high temperature shape in austenite and a *different* low temperature shape when cooled to martensite — in the absence of applied stresses. As illustrated in Figure 5.9, the recovered two-way strain ε_{tw} is a function of the number and nature of training cycles which limits to zero in the absence of training. Hence while the shape memory effect and pseudoelastic mechanisms are intrinsic properties of SMA, the two-way shape memory effect is an acquired characteristic. Details regarding the underlying physical mechanisms and experimental quantification of the effect of training under different conditions on the TWSME can be found in [45, 212, 459, 481, 493, 494].

Determination of the design ramifications of the two-way shape memory effect constitute a present research topic since the effect provides the capability for thermal cycling between two configurations without secondary springs or restoring mechanisms. To illustrate, consider the curved beam depicted in Figure 5.10 which

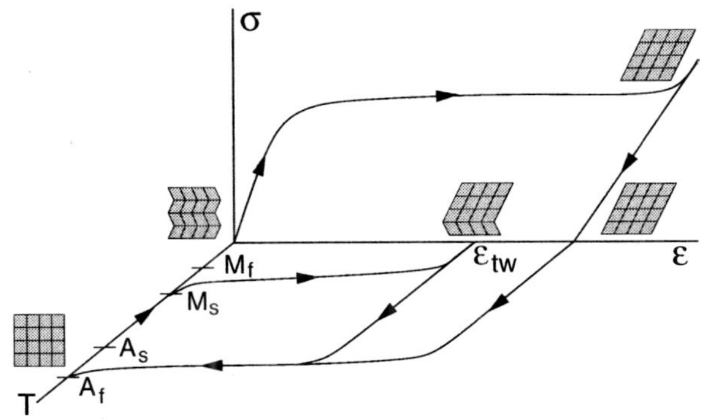

Figure 5.9. *Two-way shape memory effect manifested as a recovered strain ε_{tw} when the material is heated to $T > A_f$ and cooled to $T < M_f$.*

might be considered as a prototype for flow control, valve design, or morphing the shape of an airfoil. The one-way SME provides the capability for returning to the initial shape when heated but secondary restoring mechanisms are required to cycle between shapes or configurations. However, the device with two-way SME provides this capability thus simplifying the design and improving efficiency. Details regarding the use of TWSME for potential airfoil design can be found in [467].

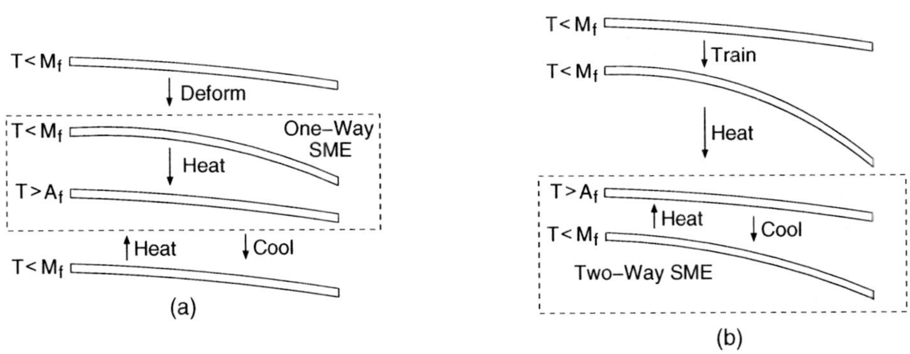

Figure 5.10. *(a) One-way SME produces a shape change only when initially heated and thus requires restoring mechanism to achieve bi-directional shape changes. (b) The two-way SME provides the capability to achieve bi-directional shape changes without secondary restoring mechanisms.*

Rate-Dependence and Minor Loops

The rate at which stresses or displacements are realized can significantly affect the hysteretic behavior of SMA. As illustrated in Figure 5.11(a) through a depiction of data from Shaw and Kyriakides [425], a slow displacement rate of $\dot{\delta}/L = 4 \times 10^{-3}\ s^{-1}$ yields nearly instantaneous A-M transformations whereas the hysteresis

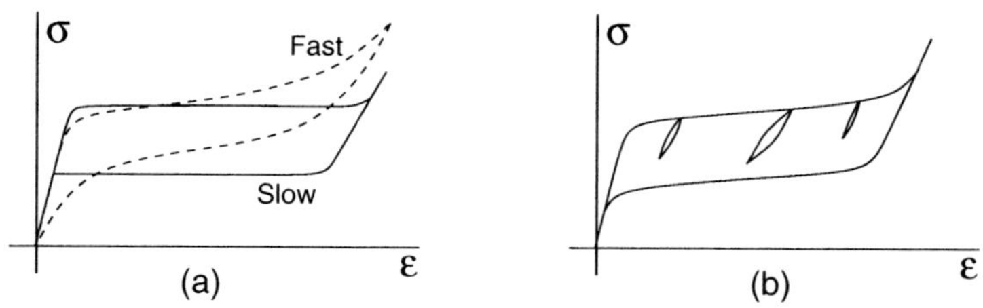

Figure 5.11. *(a) Effect of strain rate on the pseudoelastic material behavior (after [425]): $\dot{\delta}/L = 4 \times 10^{-3}$ s^{-1} (——), $\dot{\delta}/L = 4 \times 10^{-2}$ s^{-1} (- - -). (b) Minor loop behavior observed in the pseudoelastic regime.*

is skewed at the higher rate of $\dot{\delta}/L = 4 \times 10^{-2}$ s^{-1}. Furthermore, it is shown in [425] that nearly identical results are obtained in air and water at the slower rate whereas the responses differ significantly at $\dot{\delta}/L = 4 \times 10^{-2}$ s^{-1}. These rate-dependent phenomena are attributed to the manner through which temperature changes are induced at the different rates and their effect on nucleation processes during the loading and unloading processes. Accommodation of these effects in model design is important in applications involving transient or rate-dependent behavior.

An additional feature that must be incorporated in models is biased subloops due to incomplete phase transformations [475]. During quasistatic operation, biased loops typically close in the manner depicted in Figure 5.11(b). As frequencies increase, however, nonclosure analogous to the accommodation (reptation) behavior discussed in Section 4.1.8 for magnetic materials is observed and must be addressed by models.

The effects of rate-dependence can be introduced in minor loops by performing experiments or employing devices in a manner which dictates that minor loops are traversed at the same *frequency* as major loops, thus guaranteeing that major loops are traversed at a higher rate than minor loops. Hence care should be taken to ensure either uniform *rates* when traversing minor loops or inclusion of rate effects in models.

Ferroelastic Domains

Ferroelectric domains are defined as regions in ferroelastic crystals distinguished by strain states having the same crystallographic orientation, and the transition regions between domains are termed *domain walls* [402]. As shown by Roytburd [397], ferroelastic domains form to minimize internal energy in much the same manner indicated in Chapters 2 and 4 for ferroelectric and ferromagnetic domains. In a manner similar to their ferroelectric and ferromagnetic buddies, it is energetically favorable for ferroelastic domain walls to form at pinning sites comprised of material or stress nonhomogeneities, and domain wall motion due to local phase transformations is considered one source of nonlinearities and hysteresis in the macroscopic material response [293, 352, 402].

In SMA, ferroelastic domains consist of regions comprised of austenite or like martensite variants as illustrated for the uniaxial case in Figure 5.12. Here the

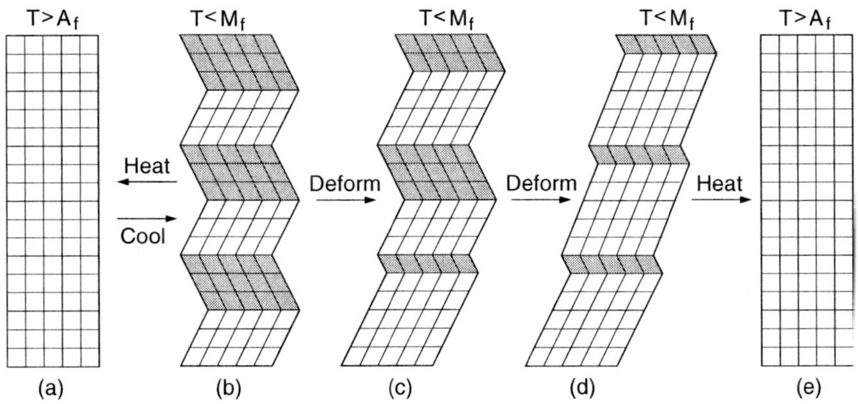

Figure 5.12. *(a) Single A-domain at* $T > A_f$ *and (b)* M^+, M^-*-domains for* $T < M_f$. *(c), (d) Growth of* M^+*-domains through domain wall movement in response to an applied stress* σ. *(e) Return to A-domain upon heating to* $T > A_f$.

application of stress causes M^+-domains to grow at the expense of M^--domains. The losses associated with this process will be exploited in the model of Section 5.4.

The analogies between ferroelastic, ferroelectric and ferromagnetic domain wall processes are obvious and are exploited in Chapter 6 when constructing one of the unified characterization frameworks for ferroic materials.

Hysteresis Losses

In Sections 2.8 and 4.1.5, we showed that losses incurred in ferroelectric and ferromagnetic materials are proportional to the area of the hysteresis loop. The same is true for ferroelastic materials. In this case, integration of the elastic energy

$$U = \boldsymbol{\sigma} \cdot \boldsymbol{\varepsilon}$$

over a complete hysteresis cycle yields the work relation

$$W = \oint \boldsymbol{\sigma} \cdot d\boldsymbol{\varepsilon}.$$

Hence the amount of elastic energy converted to potential energy or heat is proportional to the area of the hysteresis loop.

This has important ramifications for the civil and aerospace applications discussed in Section 1.4 where SMA tendons or wires are considered for vibration suppression in buildings, bridges and aerospace structures. To provide optimal damping without significantly augmenting stiffness, one would choose dimensions and operating conditions in a manner which *maximizes hysteresis* in the pseudoelastic regime. This is in direct contrast with many applications which strive to suppress the hysteretic nature of transducers.

5.2 Energy Relations

We summarize high-order energy relations to provide both initial models and a basis for macroscopic models described in later sections. The discussion will focus primarily on the construction of various Helmholtz energy functionals $\psi(\varepsilon, T)$ where ε denotes the uniaxial shear strain which, as depicted in Figure 5.2(d), has an equilibrium value $\varepsilon = 0$ for austenite and $\varepsilon = \varepsilon_T$ for martensite in the stress-free state. The Gibbs energy is subsequently defined by

$$G(\sigma, \varepsilon, T) = \psi(\varepsilon, T) - \sigma\varepsilon; \tag{5.1}$$

see Section 2.2.3 for details motivating the formulation of G in terms of both the order parameter ε and the conjugate stress σ. Analogous development for ferroelectric and ferromagnetic materials can be found in Sections 2.3 and 4.4.

5.2.1 First-Order Phase Transitions and Falk's Model

In Section 2.3.1, it was noted that like ferroelectric materials such as PZT, shape memory alloys exhibit first-order phase transitions. This motivated Falk to consider even polynomial expansions of the form

$$\psi(\varepsilon, T) = \psi_0(T) + \alpha_1(T - T_0)\varepsilon^2 - \alpha_2\varepsilon^4 + \alpha_3\varepsilon^6 \tag{5.2}$$

to characterize temperature and strain components of the Helmholtz energy [157–159]. Here $\alpha_1, \alpha_2, \alpha_3$ are positive constants and T_0 denotes the Curie temperature. A typical representation for $\psi_0(T)$ is

$$\psi_0(T) = -TS$$

where

$$S = c\rho \ln(T/T_R) + \eta$$

denotes the specific entropy, η is an entropy constant, ρ is the density, T_R is a reference temperature, and c denotes the specific heat capacity.

We first note that (5.2) is analogous to the Devonshire polarization relation (2.38) if one equates the order parameters ε and P. Secondly, the necessary condition

$$\frac{\partial G}{\partial \varepsilon} = 0 \Leftrightarrow \sigma = \frac{\partial \psi}{\partial \varepsilon} \tag{5.3}$$

implies that shear strains ε adjust according to the constitutive relation

$$\sigma = 2\alpha_1(T - T_0)\varepsilon - 4\alpha_2\varepsilon^3 + 6\alpha_3\varepsilon^5$$

in response to an applied shear stress σ.

To incorporate nonlocal effects such as interfacial energies or domain wall effects, Falk also incorporated a gradient component ε_x^2 to yield the augmented Helmholtz energy relation

$$\widetilde{\psi}(\varepsilon, \varepsilon_x, T) = \psi(\varepsilon, T) + \frac{\gamma}{2}\varepsilon_x^2.$$

We point out that inclusion of the gradient component follows from Ginzburg–Landau theory similar to that summarized in Section 2.7 which, in addition to incorporating physical nonlocal effects, serves to regularize the solution since it acts as a Tikhonov functional.

We now illustrate the development of a PDE model quantifying the longitudinal displacement u of an SMA rod of length ℓ and cross-sectional area $\mathcal{A}$ which is subjected to a body force $f(t,x)$. Noting that u and ε are related by

$$\varepsilon(t,x) = \frac{\partial u}{\partial x}(t,x),$$

the total free energy at time t is given by

$$U(t) = \mathcal{A} \int_0^\ell \left[\psi(\varepsilon(t,x),T) + \frac{\gamma}{2}\varepsilon_x^2(t,x) - \rho u(t,x) f(t,x) \right] dx$$

and the kinetic energy is

$$K(t) = \mathcal{A} \int_0^\ell \frac{\rho}{2} u_t^2(t,x) dx.$$

Application of Hamilton's principle with the Lagrangian $\mathcal{L} = K - U$ subsequently yields the PDE

$$\rho u_{tt} - \frac{\partial}{\partial x}\left(\frac{\partial \psi}{\partial \varepsilon}(\varepsilon, T) \right) + \gamma u_{xxxx} = \rho f(t,x).$$

The boundary conditions will depend on the specific configuration of the rod.

The second component of the model consists of the evolution equation used to characterize heat transduction. Rather than give a cursory summary here, we refer the reader to Section 5.5.2 where this is discussed in the context of the homogenized energy model. Additional details regarding the Falk models as well as well-posedness analysis and approximation techniques can be found in [57].

5.2.2 Higher-Order Energy Relations

The sixth-order expansion (5.2) represents the lowest-order truncation of the even infinite series which incorporates first-order transition behavior but higher-order expansions can be considered. The tradeoff for potentially improved accuracy is an increase in the number of coefficients to be identified and decreased stability due to the oscillatory nature of high-order polynomials.

As detailed in [322], Helmholtz or Landau relations of the form

$$\psi(\varepsilon, T) = \psi_0(T) + \alpha_1(T - T_0)\varepsilon^2 - \alpha_2\varepsilon^4 + \sum_{j=3}^{m} \alpha_j \varepsilon^{2j} \tag{5.4}$$

with $m = 4$ and 5 prove advantageous when constructing the domain wall models described in Section 5.4. Regardless of the order m, the Gibbs energy is specified by (5.1) and stress-strain relations are provided by the necessary conditions (5.3).

5.2.3 Piecewise Polynomial Relations

The third choice for the Helmholtz energy is the C^1 piecewise polynomial

$$\psi(\varepsilon, T) = \begin{cases} \frac{Y_M}{2}(\varepsilon + \varepsilon_T)^2 & , \quad \varepsilon \leq -\varepsilon_M(T) \\ -\frac{E_0(T)}{2}(\varepsilon + \varepsilon_0(T))^2 + \psi_0(T) & , \quad -\varepsilon_M(T) < \varepsilon < -\varepsilon_A(T) \\ \frac{Y_A}{2}\varepsilon^2 + \Delta\beta(T) & , \quad |\varepsilon| \leq \varepsilon_A(T) \\ -\frac{E_0(T)}{2}(\varepsilon - \varepsilon_0(T))^2 + \psi_0(T) & , \quad \varepsilon_A(T) < \varepsilon < \varepsilon_M(T) \\ \frac{Y_M}{2}(\varepsilon - \varepsilon_T)^2 & , \quad \varepsilon \geq \varepsilon_M(T) \end{cases} \tag{5.5}$$

where Y_M and Y_A denote the Young's moduli for austenite and martensite. As shown in Figure 5.13, the temperature-dependent inflection points $\pm\varepsilon_M(T), \pm\varepsilon_A(T)$ delineate the transition from convex regions — which represent stable phases — to concave regions representing unstable states. The maxima of the concave parabolae occur at the temperature-dependent points $(\pm\varepsilon_0(T), \psi_0(T))$, and the parameter $E_0(T)$ is chosen to ensure C^1 continuity. The austenite minimum has the height

$$\Delta\beta(T) = \beta_A(T) - \beta_M(T) \tag{5.6}$$

where

$$\beta_\alpha(T) = c_\alpha(T - T_R) + u_R - TS_\alpha \tag{5.7}$$

represent chemical (nonelastic) free energies [324, 413, 424] and $\alpha = A, M^+, M^-$. Here $c_\alpha, \rho_\alpha, u_\alpha$ and T_R respectively denote specific heat capacities, densities, phase-dependent internal energy constants, and the temperature of the reference state from which energies are computed. Furthermore,

$$S_\alpha = c_\alpha \rho_\alpha \ln(T/T_R) + \eta_\alpha \tag{5.8}$$

are specific entropies and η_α are phase-dependent entropy constants. We point out that the phase-dependent component TS_α of (5.7) is precisely the averaged term $\psi_0(T) = -TS$ employed by Falk in the polynomial expansion (5.2).

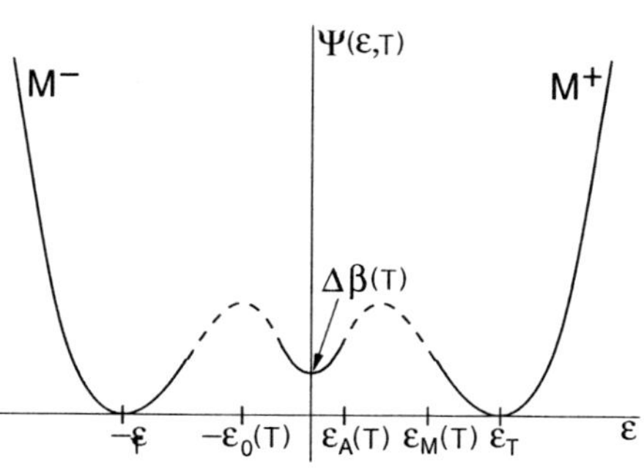

Figure 5.13. *Piecewise quadratic Helmholtz energy (5.5) for fixed T.*

First-Order Behavior of Energy Relation

We illustrate here the manner in which the piecewise quadratic Helmholtz energy relation (5.5) quantifies the first-order phase transition phenomena described in Section 2.3.1. SMA material behavior in temperature intervals delineated by the transition temperatures $T_0 < T_c < T_1 < T_2$ can be summarized as follows.

$T < T_0$: Material has only martensite phases and exhibits ferroelastic behavior as well as shape memory effects.

$T_0 < T < T_1$: The austenite phase nucleates at $T = T_0$, is metastable[24] for $T_0 < T < T_c$, and is absolutely stable for $T_c < T < T_1$. Hence the Curie point T_c delineates the transition from absolute martensite stability to absolute austenite stability.

$T_1 < T < T_2$: Only austenite is observed for $\sigma = 0$ but martensite variants can be induced by sufficiently large stresses. The material behavior is pseudoelastic.

$T > T_2$: The material exhibits elastic austenite behavior.

For various temperatures within these intervals, the Helmholtz energy and σ-ε behavior, dictated by the condition $\sigma = \frac{\partial \psi}{\partial \varepsilon}$ resulting from the minimization of $G = \psi - \sigma\varepsilon$, are depicted in Figure 5.14. It is observed that ψ and G incorporate

[24]Recall that metastable states are those associated with relative rather than absolute minima.

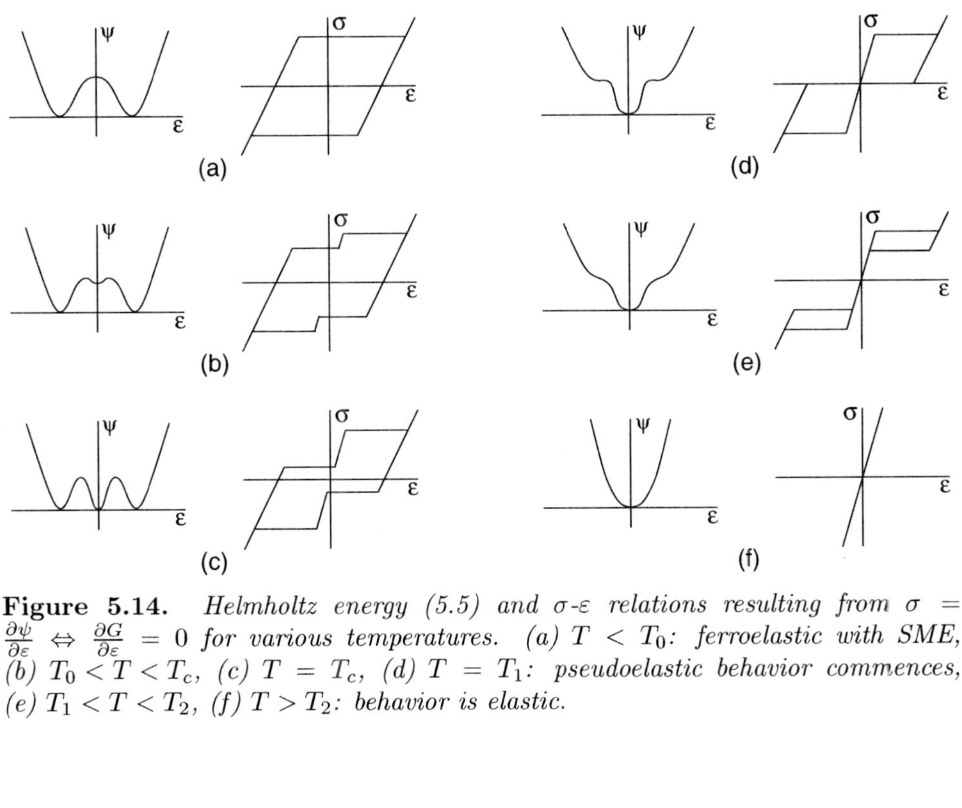

Figure 5.14. *Helmholtz energy (5.5) and σ-ε relations resulting from $\sigma = \frac{\partial \psi}{\partial \varepsilon} \Leftrightarrow \frac{\partial G}{\partial \varepsilon} = 0$ for various temperatures. (a) $T < T_0$: ferroelastic with SME, (b) $T_0 < T < T_c$, (c) $T = T_c$, (d) $T = T_1$: pseudoelastic behavior commences, (e) $T_1 < T < T_2$, (f) $T > T_2$: behavior is elastic.*

the ferroelastic, pseudoelastic and shape memory effects shown in Figures 5.3 and 5.4 as well as the first-order phase transition phenomena described in Section 2.3.1.

Because this regime neglects both thermal effects and material nonhomogeneities, phase transitions are instantaneous so that $A_s = A_f$ and $M_s = M_f$. For $\sigma = 0$, it is observed that $T_0 = M_s = M_f$ and $T_1 = A_s = A_f$.

Additional details regarding the behavior of ψ and G will be provided in Section 5.5 where these energy relations are employed to construct the homogenized free energy framework for SMA.

5.3 Preisach Models

Due to their generality, it is not surprising that a number of researchers have employed Preisach models to characterize both the pseudoelastic behavior and shape memory effects inherent to SMA. The concept, if not the implementation, of employing Preisach representations for hysteretic stress-strain characterization can be traced back to Everett and Whitton in 1952 [154] but the primary development has occurred within the last 15 years. In 1989, Huo developed a four parameter model for SMA [229] while Ortín employed a classical Preisach approach in 1992 to model pseudoelastic Cu-Zn-Al behavior [369]. Hughes and Wen [227], Gorbet, Wang and Morris [190], and Webb, Kurdila and Lagoudas [504] developed models from the perspective of control implementation and addressed issues such as data smoothing to permit accurate identification of densities and approximate model inversion to facilitate linear control design. Finally, Banks, Kurdila and Webb [28,29] addressed a number of theoretical issues regarding the well-posedness of both the forward and inverse operators. This provides only a partial summary of Preisach model development for SMA and the reader is referred to the references in the chapter introduction for additional citations.

Significant detail regarding magnetic Preisach models was provided in Section 4.5 so we provide here only a brief overview of the framework employed to characterize pseudoelastic stress-strain behavior in SMA. The reader should not assume from the brevity of this discussion that Preisach models for SMA are more limited than their ferroelectric and ferromagnetic buddies; rather, the pertinent framework has been established in previous chapters in the context of polarization and magnetization models.

In Section 4.5.1, the general Preisach operator

$$f(t) = \iint_S [k_s(v,\xi)](t)\nu(s)ds$$

was defined on the Preisach plane

$$S = \{(s_1, s_2) | s_1 \leq s_2\}.$$

Here ν denotes a general density to be identified, k_s is the kernel having values of ± 1 as depicted in Figure 4.21 and defined in (4.37), and $v(t)$ denotes the input.

For applications in which the stress σ and strain ε constitute the input and output, an appropriate Preisach representation is

$$\varepsilon(t) = \iint_S [k_s(\sigma, \xi)](t)\nu(s)ds.$$

By considering first-order reversal curves and defining

$$F(s_2', s_1') = \frac{1}{2}\left(f_{s_2'} - f_{s_1'}\right) = \frac{1}{2}\left(\varepsilon_{s_2'} - \varepsilon_{s_1'}\right),$$

the theory in Section 4.5.1 can be used to show that

$$\nu(s_2', s_1') = -\frac{\partial^2 F(s_2', s_1')}{\partial s_1' \partial s_2'} = \frac{1}{2}\frac{\partial^2 \varepsilon_{s_2', s_1'}}{\partial s_1' \partial s_2'}$$

thus providing a technique for identifying ν. Gorbet, Wang and Morris point out that this procedure is ill-posed for data, since measurement errors are augmented by differentiation so they fit a surface $\widetilde{\varepsilon}$ to the data before differentiating to estimate ν. Details illustrating the performance of the model for pseudoelastic material characterization, including biased minor loops of the type illustrated in Figure 5.11(b) can be found in Ortín [369].

The development and physical interpretation of Preisach theory for SMA is less mature than magnetic Preisach theory but several correlations between the energy-based micro and mesoscale material behavior and the mathematical Preisach framework have been established. Lagoudas and Bhattacharyya have investigated the use of micromechanics principles to evaluate the Preisach distributions for polycrystalline materials [278] and, as detailed in Section 4.7.10, the homogenized energy framework has been shown to provide an energy basis for certain extended Preisach representations. It is anticipated that the continued investigation of Preisach and energy frameworks will improve both our fundamental understanding of material mechanisms and provide increasingly efficient models for real-time applications.

5.4 Domain Wall Models

The domain wall theory comprises the second unified framework that we present for ferroic compounds. This theory was proposed by Jiles and Atherton in 1984 and 1986 for magnetic materials [250, 251] and was extended to ferroelectric materials in 1999 [438]. In 1988, Tuszyński et al. employed this approach to characterize ferroelastic hysteresis in $LiCsSO_4$ [484] and Massad and Smith later extended the theory to general ferroelastic compounds in 2003 [322].

As with the Preisach discussion, detailed analysis of the framework in previous chapters — Section 2.5 for ferroelectric materials and Section 4.6 for ferromagnetic compounds — permits an abbreviated discussion for ferroelastic materials. We simply summarize the model and refer the reader to [322] for details regarding its motivation, development and implementation.

5.4.1 Model Development

The domain wall theory is based on the quantification of energy dissipated when domain walls move in the manner depicted in Figure 5.12. In ferromagnetic and ferroelectric materials, domain wall movement is due to moment and dipole reorientation as respectively quantified by the magnetostatic and electrostatic energies. In SMA, domains are comprised of segregated austenite and martensite variants, and reorientation is quantified by the elastic energy.

The model is comprised of three components: (i) the equilibrium strain ε_{an}, (ii) irreversible strains ε_{irr} due to domain wall translation, and (iii) reversible strains ε_{rev} due to domain wall bending.

Equilibrium Strains

The equilibrium strains ε_{an} minimize the Gibbs energy

$$G(\sigma, \varepsilon, T) = \psi(\varepsilon, T) - \sigma_e \varepsilon$$

for the Helmholtz energy (5.4) and effective stress σ_e. We employ the notation ε_{an} to provide unity with the magnetization and polarization models but note that in general, ε_{an} will be hysteretic.

As detailed in [168, 187], effective stresses have the general form

$$\sigma_e = \sigma + \sigma_I$$

where σ_I denotes interaction stresses. To incorporate general interactions, we consider effective fields of the form

$$\sigma_e = \sigma + \alpha\varepsilon \tag{5.9}$$

where α denotes a mean field constant.

Evaluating the necessary condition $\frac{\partial G}{\partial \varepsilon} = 0$ subsequently yields the relation

$$\sigma_e = 2\alpha_1(T - T_0)\varepsilon_{an} - 4\alpha_2\varepsilon_{an}^3 + \sum_{j=3}^{m} 2j\alpha_j\varepsilon_{an}^{2j-1} \tag{5.10}$$

implicitly quantifying the behavior of ε_{an}. This characterizes the instantaneous switching which occurs in homogeneous materials devoid of inclusions.

Irreversible Strains

The second component is the irreversible strain ε_{irr} which accompanies domain wall translation. By employing the elastic energy relation

$$U = \int \sigma_e(\varepsilon)d\varepsilon$$

to quantify the work required to reorient austenite and martensite variants, it is shown in [322] that ε_{irr} can be expressed as

$$\varepsilon_{irr} = \varepsilon_{an} - \delta k_p \frac{d\varepsilon_{irr}}{d\sigma_e}$$

where k_p is a material constant and $\delta = \text{sign}(d\sigma)$ ensures that the energy required to break pinning sites opposes changes in the stress. Reformulation and enforcement of reversible behavior within the equilibrium region bounded by ε_{an} yields the differential equation

$$\frac{d\varepsilon_{irr}}{d\sigma} = \widetilde{\delta}\frac{\varepsilon_{an}(\sigma_e) - \varepsilon_{irr}(\sigma)}{\delta k_p + \alpha[\varepsilon_{an}(\sigma_e) - \varepsilon_{irr}(\sigma)]}$$

which is analogous to (2.76) and (4.60). Here $\widetilde{\delta} = 1$ for all values (σ, ε) that lie outside the region bounded by the equilibrium curves $(\sigma, \varepsilon_{an})$ and $\widetilde{\delta} = 0$ otherwise.

Reversible Strains

The reversible strains ε_{rev} incorporate reversible domain changes and are given by

$$\varepsilon_{rev} = c(\varepsilon_{an} - \varepsilon_{irr}).$$

The constant c quantifies the degree of reversibility and necessarily has values in the interval $[0, 1]$.

Total Strains

The total macroscopic strains are then given by

$$\begin{aligned} \varepsilon &= \varepsilon_{irr} + \varepsilon_{rev} \\ &= (1 - c)\varepsilon_{irr} + c\varepsilon_{an}. \end{aligned} \tag{5.11}$$

It is observed that when $c = 1$, the total strains are precisely the equilibrium values specified by (5.10) whereas $c = 0$ yields solely irreversible strains. Additional details regarding the implementation and identification of parameters in (5.11) can be found in [322].

5.4.2 Model Properties and Validation

Model Properties

The relation between the equilibrium strains specified by (5.10) and the full strain model (5.11) is illustrated in Figure 5.15 for four temperatures. The parameters for these simulations are $T_0 = 273$ K, $k_p = 10$ MJ/m^3, $\alpha = 7 \times 10^3$ MPa, $\alpha_1 = 0.638 \times 10^3$ MPa/K, $\alpha_2 = 4.908 \times 10^6$ MPa and $\alpha_3 = 6.108 \times 10^8$ MPa, and $c = 0.50$. It is observed the equilibrium relations derived through minimization of the Gibbs energy yield instantaneous austenite-martensite switching whereas the full model, which incorporates pinning losses, yields more gradual transitions as measured in both single crystal and polycrystalline materials. By constructing the polynomial kernel (5.10) in a manner which incorporates first-order transition behavior, both the equilibrium strains and full model encapsulate the transition from ferroelastic to pseudoelastic behavior inherent to SMA as temperatures are increased through the Curie temperature T_0. Additional details can be found in [322].

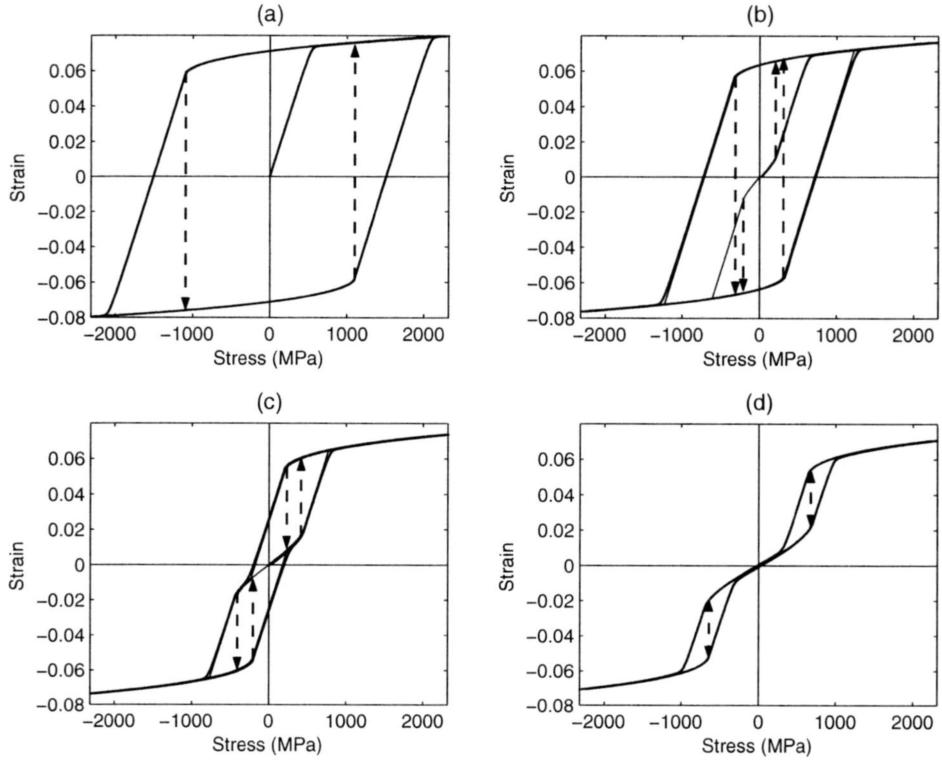

Figure 5.15. *Simulated behavior of the full model (——) and equilibrium strains (- - -) at (a) $T = 272$ K, (b) $T = 283$ K, (c) $T = 291$ K, and (d) $T = 298$ K.*

Experimental Validation

The performance of the model when characterizing the major and minor loop behavior of polycrystalline NiTi is illustrated in Figure 5.16. The data from Bundara et al. [65] corresponds to tensile experiments on a polycrystalline NiTi wire having a 55.0 at.% Ti composition. To minimize the effects of aging or training discussed in Section 5.1, the stabilized hysteresis loop was measured after 22 cycles at 295 K. Following the major loop measurement, minor loops were generated by partial loading, complete unloading cycles.

The Curie temperature for the material is $T_0 = 288$ K and a least squares fit to the data yielded the parameter values $k_p = 20.39$ MJ/m^3, $c = 0.90$, $\alpha = 698.8$ MPa, $\alpha_1 = 8.430 \times 10^3$ MPa/K, $\alpha_2 = 1.059 \times 10^8$ MPa, $\alpha_3 = 8.775 \times 10^{10}$ MPa, $\alpha_4 = -3.224 \times 10^{13}$ MPa and $\alpha_5 = 4.465 \times 10^{15}$ MPa and model fit shown in Figure 5.16. As detailed in [322], model construction with $m = 3$, so that ψ is the Falk expression (5.2), yields overly rapid transition behavior thus motivating the consideration of the higher-order Helmholtz relation (5.4). Further details regarding both this example and the characterization of single-crystal NiTi data can be found in [322].

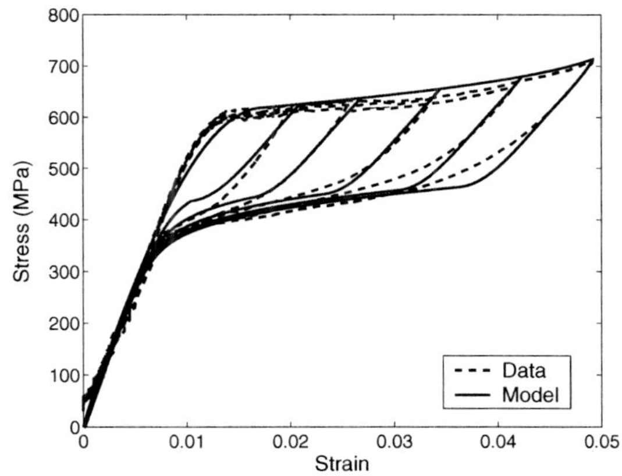

Figure 5.16. *Characterization of polycrystalline data from Bundara et al. [65].*

For this operating regime, the model accurately characterizes both major and minor loop behavior. However, it was noted in Sections 2.5.6 and 4.6.4 that one of the limitations of the framework is its inability to provide biased minor loop closure — e.g., of the type depicted in Figure 5.11(b) — without *a priori* knowledge of turning points. For general characterization which includes these regimes, the homogenized energy model described in Section 5.5 is preferable.

5.5 Homogenized Energy Framework

The homogenized energy framework has its genesis in the theory of thermally activated processes developed for SMA by Müller, Achenbach and Seelecke [1,2,353, 374,409,413]. The original theory focused on balancing Gibbs and thermal energies through Boltzmann principles in single crystal compounds. In [374], initial effects of polycrystallinity were incorporated by integrating over a distribution of lattice and stress orientations thus providing a catalyst for much of the distributional analysis which followed. Models incorporating distributions in the relative and effective stresses were developed in [323,324] for SMA thin films with analogous models for bulk SMA developed in [411]. This has produced a modeling framework for SMA which characterizes ferroelastic and pseudoelastic behavior in addition to inherent shape memory effects. Additionally, it accommodates biased minor loops due to partial phase transformations and incorporates dynamic effects, temperature evolution, and after-effects or thermal relaxation.

To achieve the efficiency necessary for transducer design and model-based control implementation, a hierarchical or multiscale approach is employed. Energy relations are developed at the mesoscale, or lattice level, to incorporate fundamental physics. The effects of lattice variations, polycrystallinity, and variable stresses are subsequently incorporated by assuming that parameters such as relative and effective stresses are manifestations of underlying distributions rather than constants. Homogenization in this manner yields low-order macroscopic models that are highly efficient to implement.

It was noted in Sections 2.6 and 4.7 that the Müller-Achenbach-Seelecke theory preceeded and motivated the development of analogous frameworks for ferroelectric and ferromagnetic materials. As summarized in Chapter 6, the homogenized free energy theory thus provides a unified and highly comprehensive framework for characterizing the hysteretic and nonlinear constitutive behavior of a broad class of ferroic compounds.

5.5.1 Mesoscopic Model

We summarize here the theory developed in [323, 324, 374, 410–413] for uniaxial SMA behavior. The material is considered to be comprised of austenite A and the martensite variants M^+, M^-, and the resulting macroscopic models are appropriate for wires, rods and certain thin film configurations.

We treat a lattice volume V of mass ν as the fundamental element in the model and let $x_A(t), x_+(t)$ and $x_-(t)$ respectively denote the volume fraction of A, M^+ and M^- layers in the SMA. The phase fractions constitute internal variables which necessarily satisfy the conservation relation

$$x_A + x_+ + x_- = 1 \tag{5.12}$$

over all time. To consolidate notation, we let α generically refer to the A, M^+ and M^- variants so that (5.12) can be reformulated as

$$\sum_\alpha x_\alpha(t) = 1.$$

Helmholtz and Gibbs Energies

The piecewise quadratic Helmholtz and Gibbs energy relations summarized in Section 5.2 are used to quantify the equilibrium strains in the lattice element. As shown in Figure 5.17, the fundamental order parameter is the shear strain ε which has the value $\varepsilon = 0$ for austenite and the equilibrium value $\varepsilon = \varepsilon_T$ for martensite in a stress-free state. In the absence of applied stresses, the internal energy of the lattice is quantified by the C^1 Helmholtz relation

$$\psi(\varepsilon, T) = \begin{cases} \frac{Y_M}{2}(\varepsilon + \varepsilon_T)^2 & , \quad \varepsilon \leq -\varepsilon_M(T) \\ -\frac{E_0(T)}{2}(\varepsilon + \varepsilon_0(T))^2 + \psi_0(T) & , \quad -\varepsilon_M(T) < \varepsilon < -\varepsilon_A(T) \\ \frac{Y_A}{2}\varepsilon^2 + \Delta\beta(T) & , \quad |\varepsilon| \leq \varepsilon_A(T) \\ -\frac{E_0(T)}{2}(\varepsilon - \varepsilon_0(T))^2 + \psi_0(T) & , \quad \varepsilon_A(T) < \varepsilon < \varepsilon_M(T) \\ \frac{Y_M}{2}(\varepsilon - \varepsilon_T)^2 & , \quad \varepsilon \geq \varepsilon_M(T) \end{cases}$$

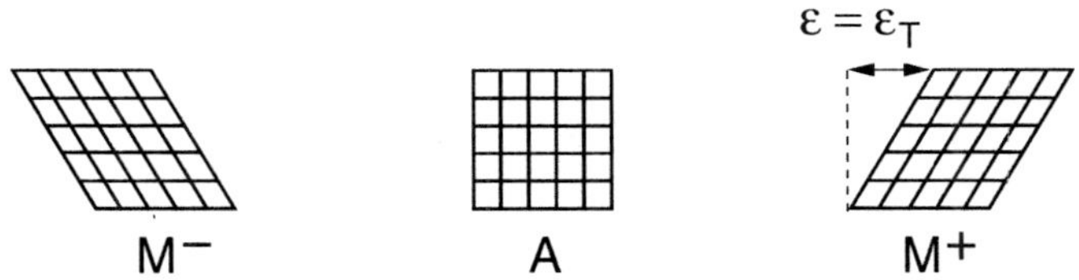

Figure 5.17. *Lattice element exhibiting the martensite M^-, M^+ and austenite A equilibrium configurations.*

where Y_M, Y_A denote the linear Young's moduli for the martensite and austenite phases, $\varepsilon_M(T), \varepsilon_A(T)$ and their negatives denote the inflection points, and the austenite minimum height $\Delta B(T)$ is defined in (5.6). The construction of ψ is illustrated in Figure 5.13 and its behavior for temperatures in the ferroelastic, pseudoelastic, and elastic regimes is shown in Figure 5.14.

To incorporate the work due to applied stresses, we employ the Gibbs energy

$$G(\sigma, \varepsilon, T) = \psi(\varepsilon, T) - \sigma\varepsilon \tag{5.13}$$

which is illustrated in Figure 5.18 at a fixed temperature $T \geq A_f$ chosen so that the material exhibits the austenite phase for $\sigma = 0$. As σ is increased, the landscape distorts until the critical stress σ_M when the stable austenite equilibrium ceases to be a local minimum and M^+ becomes the stable phase. It will remain such until the stress is decreased to a second critical value σ_A. Here the local martensite minimum disappears and the material returns to the austenite phase.

Local Stress-Strain Relations — Negligible Thermal Activation

For operating regimes in which relaxation times are small compared with drive frequencies, local stress-strain relations follow directly from the necessary conditions

$$\frac{\partial G}{\partial \varepsilon} = 0 \quad , \quad \frac{\partial^2 G}{\partial \varepsilon^2} > 0 \tag{5.14}$$

associated with the minization of G. Due to the quadratic definition of G, the local stress-strain behavior is linear in the absence of thermal activation, as depicted in Figures 5.14 and 5.18(b). To incorporate the switching history, we employ the Preisach notation

$$[\bar{\varepsilon}(\sigma, T; \sigma_R, \xi)](t) = \begin{cases} [\bar{\varepsilon}(\sigma, T; \sigma_R, \xi)](0) & , \ \tau(t) = \emptyset \\ \frac{\sigma}{Y_M} + \varepsilon_T & , \ \tau(t) \neq \emptyset \ , \ \sigma(\max \tau(t)) = \sigma_A(T) \\ \frac{\sigma}{Y_A} & , \ \tau(t) \neq \emptyset \ , \ \sigma(\max \tau(t)) = \sigma_M(T) \end{cases} \tag{5.15}$$

to quantify the local average strains due to positive applied loads — similar expressions hold for compressive stresses. Here

$$[\bar{\varepsilon}(\sigma, T; \sigma_R, \xi)](0) = \begin{cases} \frac{\sigma}{Y_A} & , \ \sigma(0) \leq \sigma_M(T) \\ \xi & , \ \sigma_M(T) < \sigma(0) < \sigma_A(T) \\ \frac{\sigma}{Y_M} + \varepsilon_T & , \ \sigma(0) \geq \sigma_A(T) \end{cases} \tag{5.16}$$

where $\xi = \frac{\sigma}{Y_A}$ or $\xi = \frac{\sigma}{Y_M} + \varepsilon_T$ depending on the initial strain configuration and

$$\tau(t) = \{t_s \in (0,t] \,|\, \sigma(t_s) = \sigma_M \text{ or } \sigma(t_s) = \sigma_A\},$$

denotes the set of transition times.

It follows that critical stress values are defined by

$$\begin{aligned} \sigma_A(T) &= Y_A \varepsilon_A(T) \\ \sigma_M(T) &= Y_M[\varepsilon_M(T) - \varepsilon_T]. \end{aligned} \tag{5.17}$$

When constructing macroscopic models in Section 5.5.3, it proves advantageous to distribute the relative stress

$$\sigma_R(T) = \sigma_M(T) - \sigma_A(T) \tag{5.18}$$

since it can be correlated with properties of the measured data. For this reason, we include it as a parameter in the strain relations (5.15) and (5.16).

The stress-strain behavior dictated by the equilibrium conditions (5.14) for six different temperatures is depicted in Figure 5.14 to illustrate that the piecewise quadratic Helmholtz and Gibbs energy relations incorporate the necessary ferroelectric, pseudoelectric and shape memory effects in addition to the temporal behavior associated with materials that exhibit first-order phase transitions.

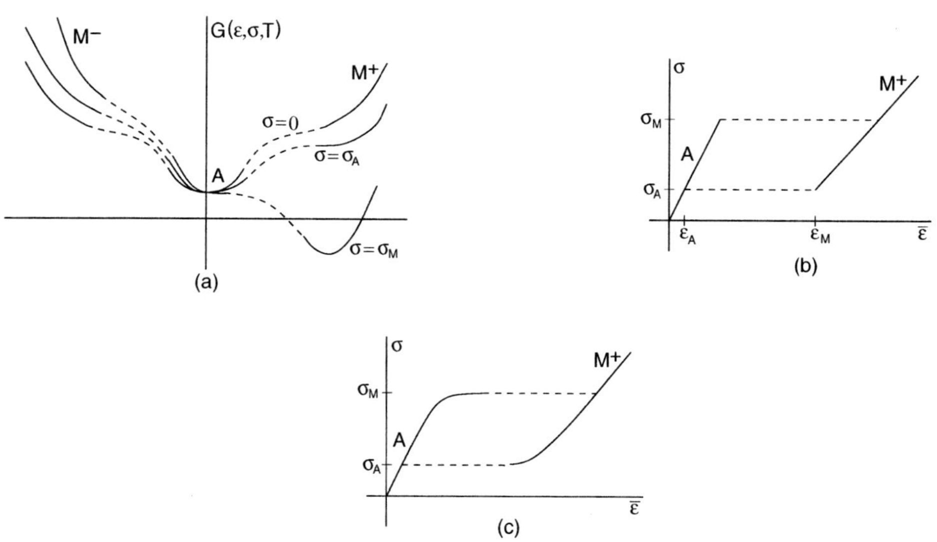

Figure 5.18. *(a) Gibbs energy (5.13) for a fixed temperature T in the austenite range with $\sigma = 0$, $\sigma = \sigma_A$ and $\sigma = \sigma_M$. (b) Local stress-strain relation produced by the equilibrium condition (5.14) for fixed temperature T and negligible relative thermal energy kT/V. (c) Local stress-strain relation (5.20) which incorporates thermally activated phase transitions.*

Local Stress-Strain Relations — Thermal Activation

To quantify the local average strains $\overline{\varepsilon}$ for operating regimes in which thermal activation is significant, it is necessary to balance the Gibbs energy G with the relative thermal energy kT/V through the Boltzmann relation

$$\mu(G) = Ce^{-GV/kT}. \tag{5.19}$$

Physically, large values of kT/V mean that with higher probability, layers will achieve the energy required to exit a local minimum before the stress values σ_A or σ_M are reached. This produces the gradual transitions and decreased transition stress values depicted in Figure 5.18(c).

As detailed in [324,410,413], the likelihood $p_{A\pm}$ that austenite will transform to $M^{\pm}$ and the likelihoods $p_{\pm}$ that $M^{\pm}$ will transform either to austenite or the other martensite variant are given by

$$p_{A\pm}(\sigma,T) = \frac{1}{\mathcal{T}(T)} \frac{\int_{\pm\varepsilon_A^\epsilon} e^{-G(\sigma,\varepsilon,T)V/kT} d\varepsilon}{\int_{\chi_A} e^{-G(\sigma,\varepsilon,T)V/kT} d\varepsilon}$$

$$p_{\pm}(\sigma,T) = \frac{1}{\mathcal{T}(T)} \frac{\int_{\pm\varepsilon_M^\epsilon} e^{-G(\sigma,\varepsilon,T)V/kT} d\varepsilon}{\int_{\chi_{M\pm}} e^{-G(\sigma,\varepsilon,T)V/kT} d\varepsilon}.$$

Here $\mathcal{T}(T) = \mathcal{T}_1\sqrt{V/kT}$ denotes the relaxation time, $\chi_A(T) = (-\varepsilon_A(T), \varepsilon_A(T))$, $\chi_{M^+}(T) = (\varepsilon_M(T), \infty)$ and $\chi_{M^-}(T) = (-\infty, -\varepsilon_M(T))$ respectively denote regions over which austenite, M^+ and M^- are stable, and $\pm\varepsilon_A^\epsilon, \pm\varepsilon_M^\epsilon$ denote ϵ intervals about the transition strains $\pm\varepsilon_A, \pm\varepsilon_M$. Note that in practice, one often employs the relations

$$p_{A\pm}(\sigma,T) = \frac{1}{\mathcal{T}(T)} \frac{e^{-G(\sigma,\pm\varepsilon_A,T)V/kT}}{\int_{\chi_A} e^{-G(\sigma,\varepsilon,T)V/kT} d\varepsilon}$$

$$p_{\pm}(\sigma,T) = \frac{1}{\mathcal{T}(T)} \frac{e^{-G(\sigma,\pm\varepsilon_M,T)V/kT}}{\int_{\chi_{M\pm}} e^{-G(\sigma,\varepsilon,T)V/kT} d\varepsilon}.$$

which can be obtained using a midpoint approximate to the integrals in the numerator and incorporating the factor of 2ϵ in $\mathcal{T}_1$.

The expected strains due to austenite and $M\pm$ variants are given by

$$\langle\varepsilon_-\rangle = \int_{\chi_{M-}} \varepsilon\mu(G(\sigma,\varepsilon,T))\, d\varepsilon$$

$$\langle\varepsilon_+\rangle = \int_{\chi_{M+}} \varepsilon\mu(G(\sigma,\varepsilon,T))\, d\varepsilon$$

$$\langle\varepsilon_A\rangle = \int_{\chi_A} \varepsilon\mu(G(\sigma,\varepsilon,T))\, d\varepsilon$$

where μ is defined in (5.19) and the Gibbs energy is specified in (5.13).

The evolution of phase fractions is governed by the rate laws

$$\dot{x}_-(t) = p_{A-}x_A(t) - p_-x_-(t)$$
$$\dot{x}_+(t) = p_{A+}x_A(t) - p_+x_+(t)$$
$$\dot{x}_A(t) = p_-x_-(t) - p_{A-}x_A(t) + p_+x_+(t) - p_{A+}x_A(t)$$

which can be reduced to

$$\dot{x}_-(t) = -(p_- + p_{A-})x_-(t) - p_{A-}x_+(t) + p_{A-}$$
$$\dot{x}_+(t) = -(p_+ + p_{A+})x_+(t) - p_{A+}x_-(t) + p_{A+}$$

through the conservation relation (5.12).

To compute the local average strain for the reference volume, we make the assumption that thermal strains are small compared to mechanical strains and retain only the latter component. As discussed in [323,324], this assumption is valid for bulk materials but may need to be modified for some SMA thin films. Hence for regimes in which thermal activation or relaxation mechanisms are significant, the local average strains are given by

$$\bar{\varepsilon} = \langle \varepsilon_- \rangle x_- + \langle \varepsilon_+ \rangle x_+ + \langle \varepsilon_A \rangle x_A. \tag{5.20}$$

The stress-strain behavior encapsulated in (5.20) is compared in Figure 5.18 with that resulting solely from the minization of G. In the latter case, some layers achieve sufficient thermal energy to switch in advance of the coercive stresses σ_M, σ_A thus producing mollified transitions.

5.5.2 Thermal Evolution

The quantification of thermal processes is analogous to that considered in (2.132) for ferroelectric compounds and (4.93) for ferromagnetic materials, and includes convective and conductive mechanisms, Joule heating and heat transduction due to phase transitions. As in Sections 2.6.8 and 4.7.6, we let $h_c, T_E, \rho_\alpha^e, c_\alpha, \Omega, \mathcal{A}, \mathcal{M}, \lambda$ and ℓ respectively denote the convection coefficient, temperature of the surrounding environment, phase-dependent electric resistivity, phase-dependent specific heat, SMA surface area, SMA cross-sectional area, the mass of the SMA, the thermal conductivity of the surrounding medium, and the length of any conduction paths. An energy balance then yields the differential equation

$$\mathcal{M}\bar{c}(t)\dot{T}(t) = -\Omega[h_c + \lambda/\ell][T - T_e(t)] + J(t) - \sum_\alpha h_\alpha \dot{x}_\alpha \tag{5.21}$$

which quantifies the temperature evolution (see [324, 410, 413, 425]). The first term on the right hand side characterizes heat exchanged with the surrounding environment through convection and conduction in a manner analogous to (2.132) and (4.93) for ferroelectric and ferromagnetic materials. For an input current $I(t)$, the relation

$$J(t) = \bar{\rho}^e(t)\frac{I^2(t)}{\mathcal{A}}, \tag{5.22}$$

where $\bar{\rho}^e(t) = \sum_\alpha \rho_\alpha^e x_\alpha(t)$ is the average resistivity per unit length, quantifies heat generated through Joule heating. The final term in (5.21) accounts for heat generated or lost during phase transformations. As discussed in [324], the specific enthalpies h_α have the form

$$h_\alpha = G_\alpha + TS_\alpha$$

where G_α are local minima of the Gibbs relation (5.13) and S_α denotes the specific entropies defined in (5.8). Finally, the average specific heat is given by

$$\bar{c}(t) = \sum_\alpha c_\alpha x_\alpha(t).$$

We note that $\bar{c}$ and $\bar{\rho}^e$ are assumed to be constant in the ferroelectric and ferromagnetic relations whereas their time-dependence in (5.21) and (5.22) reflects their dependence on evolving phase distributions in shape memory alloys.

5.5.3 Macroscopic Model

The mesoscopic strain relations (5.15) and (5.20) were developed under the assumption of uniform material properties in the lattice element. For homogeneous single crystal compounds, they can be extrapolated to provide macroscopic relations quantifying the stress-strain behavior of SMA. Their accuracy will be determined in part by the transition behavior of the material. For materials and operating regimes in which the austenite-martensite transitions occurring at σ_{M_s} and σ_{A_s} are close to instantaneous, the relations will be fairly accurate. Otherwise, homogenized models employing (5.15) or (5.20) as kernels are required to provide the mollified transitions indicative of material or stress nonhomogeneities.

To incorporate the effects of polycrystallinity, material nonhomogeneities, and lattice variations across grain boundaries, we consider either the transformation stress σ_A or relative stress $\sigma_R = \sigma_M - \sigma_A$, given by (5.18), to be manifestations of underlying distributions. Because $\sigma_M \geq \sigma_A$, the density ν_1 for σ_R is defined only for positive arguments. One can treat this as either an unparameterized density to be identified using the techniques of Section 2.6.6 or prescribe a functional or parameterized form to facilitate implementation. The latter strategy includes the lognormal representation

$$\nu_1(\sigma_R) = c_1 e^{-[\ln(\sigma_R/\bar{\sigma}_R)/2c]^2} \tag{5.23}$$

which is analogous to the coercive and magnetic field densities (2.117) and (4.88).

Secondly, the applied stress is augmented by local interaction stresses σ_I so that the effective stress at the lattice level has the form

$$\sigma_e = \sigma + \sigma_I.$$

Rather than apply the mean field relation $\sigma_I = \alpha\varepsilon$ in (5.9), which would yield a nonlinear model, we consider σ_I and hence σ_e to be statistically distributed with an underlying density ν_2. Like ν_1, ν_2 can be treated as an unparameterized density to provide high fidelity characterization or a parameterized functional form to facilitate model identification and updating. In the latter case, two representations which have been employed are the normal density

$$\nu_2(\sigma_I) = c_2 e^{-\sigma_I^2/2b^2} \tag{5.24}$$

and Laplace relation

$$\nu_2(\sigma_I) = c_2 e^{-|\sigma_I|/b}. \tag{5.25}$$

The macroscopic relation

$$[\varepsilon(\sigma, T)](t) = \int_0^\infty \int_{-\infty}^\infty \nu_1(\sigma_R)\nu_2(\sigma_I)[\overline{\varepsilon}(\sigma + \sigma_I, T; \sigma_R, \xi)](t)\, d\sigma_I d\sigma_R \tag{5.26}$$

quantifies bulk strains due to input stresses and evolving temperatures. The kernel $\overline{\varepsilon}$ is given by (5.15) or (5.20) and ν_1, ν_2 are either general densities satisfying the positivity, symmetry and integrability conditions

$$\begin{aligned} &\text{(i)} \quad \nu_1(x) \text{ defined for } x > 0,\\ &\text{(ii)} \quad \nu_2(-x) = \nu_2(x),\\ &\text{(iii)} \quad |\nu_1(x)| \le c_1 e^{-a_1 x} \quad , \quad |\nu_2(x)| \le c_2 e^{-a_2|x|}, \end{aligned} \tag{5.27}$$

for positive c_1, a_1, c_2, a_2, or *a priori* functions such as (5.23), (5.24) or (5.25). Comparison of (5.26) with the polarization model (2.114) and magnetization model (4.86) illustrates that all have the same form — hence the homogenized energy theory provides a unified characterization framework for a broad class of ferroic compounds.

Due to the uniformity of the models, we direct the reader to Sections 2.6.4 and 4.7.2 for details regarding the discretization, identification of parameters, and implementation of the model. Further details concerning the correlation of model parameters with measured data properties and numerical techniques for implementing the model are provided in [323, 324].

5.5.4 Model Attributes and Material Characterization

Simulated Model Behavior

The capability of the model to characterize superelastic behavior, the shape memory effect, and biased minor loops is illustrated in Figures 5.19–5.22 from [321, 323]. Details regarding the examples can be found in these references.

Figure 5.19 illustrates the stress-strain behavior produced by (5.26) with the temperature evolution governed by (5.21). In contrast to the instantaneous transition produced by the local kernel (5.20), which is depicted in Figure 5.18(c), the gradual transition in Figure 5.19 reflects the effects of material nonhomogeneities as quantified by the densities ν_1 and ν_2. The simulation also reflects the role played by the enthalpy component $\sum_\alpha h_\alpha \dot{x}_\alpha$ in (5.21) which quantifies the heat generated and absorbed during phase transformations. Part of this heat is transferred to the

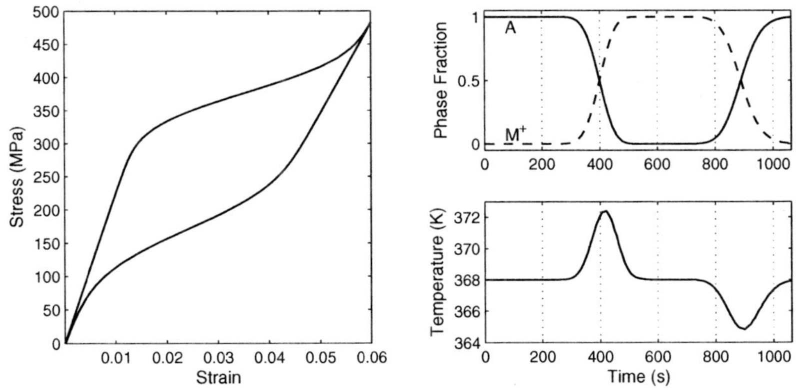

Figure 5.19. *Superelastic model behavior at $T = 258$ K including internal temperature changes due to heat generation and absorption during phase transformations.*

environment via convection while the remainder produces a change in internal temperature which in turn changes the temperature-dependent transformation stresses $\sigma_A(T)$ and $\sigma_M(T)$ in (5.17). This serves to further mollify the transition.

The shape memory effect is illustrated in Figures 5.20 and 5.21 for fixed and variable stress regimes. In both cases, the external temperature was controlled by convection at a rate of 0.1 K/s. While transition enthalpies still affect the internal temperature, their magnitude is small compared with the changes due to convection. It is observed that the transition temperatures A_s, A_f, M_s, M_f are distinct in contrast with the instantaneous transitions which result solely from the minimization of the Gibbs energy (5.13).

The simulated effects of Joule heating and convective cooling for a 10 μm thin film SMA are shown in Figure 5.22. Starting at an ambient temperature of 303 K,

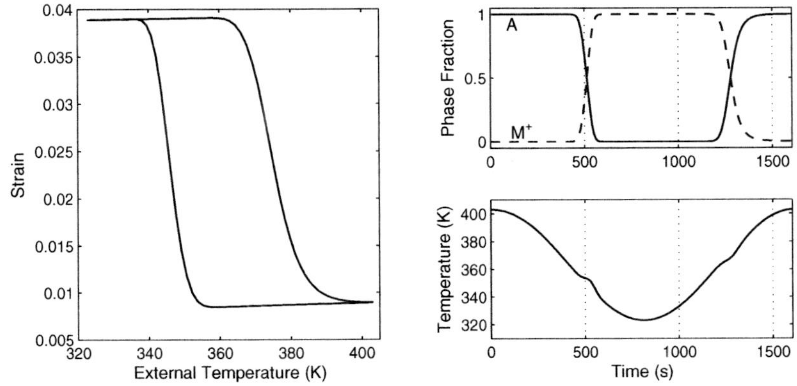

Figure 5.20. *Shape memory effect at the fixed stress $\sigma = 200$ MPa. The temperature plot reflects both heating and cooling due to convection and heat generated and absorbed during phase transformations.*

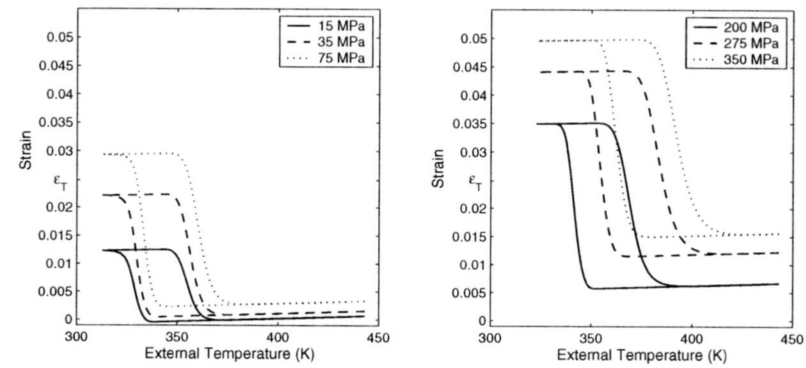

Figure 5.21. *Stress-dependence of the thermal hysteresis loop.*

Joule heating via the component $J(t)$ in (5.21) was applied for 10 ms to raise the temperature 50 K and induce a partial phase transformation from M_+ to austenite. Cooling via convection until $t = 0.05$ seconds yielded a decrease in x_A and increase in x_+ thus producing the initial leg of the minor loop. Subsequent heating past A_f completed the transformation to austenite. The heating rate was sufficiently slow to allow approximate closure of the biased minor loop — faster heating rates can produce accommodation (reptation) behavior analogous to that simulated in Figure 4.10 for magnetic materials. The second cooling cycle ended before M_s was reached and it was not until the third cooling cycle that $M_s \approx 345$ K was achieved thus initiating the phase transition back to martensite.

The slope of the temperature curve during the third cooling cycle illustrates an important property of thin film SMA. Before M_s was achieved at $t \approx 0.14$ seconds, $\dot{x}_A = \dot{x}_M \approx 0$ and $\dot{T}$ in (5.21) is dominated by the convection component $-\Omega h_c[T - T_E(t)]$. When the phase transformation commences, temperature evolu-

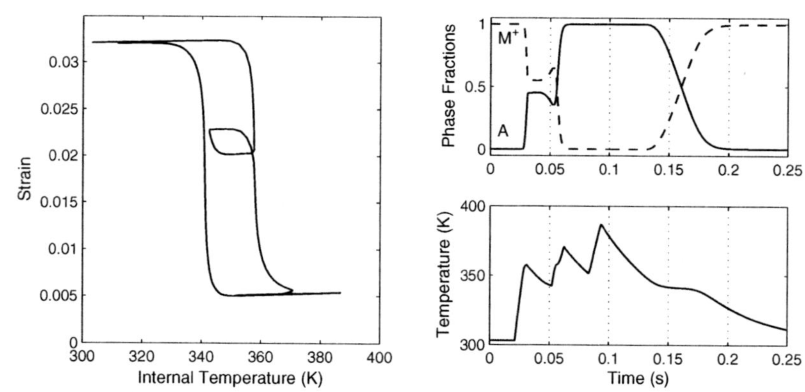

Figure 5.22. *Simulated Joule heating and cooling via forced convection in a thin film SMA to produce a minor loop due to a partial transformation.*

tion is governed by the relation

$$\mathcal{M}\bar{c}(t)\dot{T}(t) = -\Omega h_c[T - T_e(t)] - \sum_{\alpha} h_\alpha \dot{x}_\alpha$$

and the cooling rate is reduced by the exchange of transformation enthalpies. This yields the diminished slope $\dot{T}$ in the time interval corresponding to $[M_s, M_f]$ and negligible cooling when $x_A = x_M$. The dependence of $\dot{T}$ on $\dot{x}_\alpha$ during the phase transformation corroborates the observation in [310] that the response of thin film SMA actuators is limited by the rate of phase transitions rather than the cooling time. This has significant implications for design since it indicates potential rate limits for thin film SMA transducers.

Material Characterization

We summarize two examples from [321, 323] demonstrating the performance of the model when characterizing SMA foil and thin film. Additional examples demonstrating the characterization of bulk SMA can be found in [448, 449].

NiTiFe Foil

We illustrate first the characterization of 500μm thick NiTiFe foil manufactured by Furukawa Techno Material as reported in [328]. The data plotted in Figure 5.23 was collected at a strain rate of 0.01 /s (equivalent to 12.67 MPa/s) which is sufficiently fast to produce self-heating and highly sloped pseudoelastic transitions as depicted in Figure 5.11(a). To minize the accumulation of plastic strains, the multi-loop data shown in 5.23 was collected after 100 training cycles.

Published values yielded the parameter specifications A_s = 234 K, A_f = 281 K, M_s = 280 K, M_f = 244 K, ρ = 6450 kg/m^3, and the parameters Y_A =

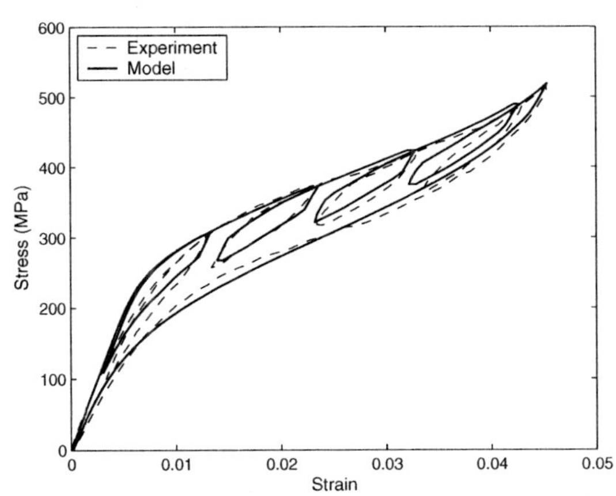

Figure 5.23. *Experimental NiTiFe data from [328] and model fit and predictions. Parameters were estimated through a least squares fit to the bounding loop data and the resulting model was used to predict the biased minor loop behavior.*

37.2 GPa, $Y_M = 25.5$ GPa, $\varepsilon_T = 0.0253$ were computed from the bounding loop data. The remaining parameter values $\overline{\sigma}_R = 70$ MPa, $V = 6903$ nm^3, $\mathcal{T} = 11.5$ ms, $c_A = c_M = 2.9025$ MJ/m^{-3}K^{-1}, $b = 55$ MPa, $c = 52$ MPa and $h_c = 10$ Wm^{-2}K^{-1} were estimated through a least squares fit to the bounding pseudoelastic curve.

The resulting model fit is shown in Figure 5.23. The model with the same parameters was then used to *predict* the minor loop behavior. The capability of the model to accurately characterize biased minor loops due to partial phase transformations is necessary for control designs required to accommodate transient dynamics.

NiTi Thin Film

Secondly, we illustrate the characterization of data from a 8 μm thick NiTi film constructed at the UCLA Active Materials Laboratory [517]. Data collected at 353 K and 298 K is plotted in Figure 5.24. It is observed that the high temperature data exhibits pseudoelastic behavior with a slight residual strain due to plastic deformations whereas the low temperature data exhibits a transformation from twinned (self-accommodating) martensite to detwinned martensite and has a residual strain of approximately 3.4%.

The parameters summarized in [321, 323] were obtained from manufacturer specifications in combination with a least squares fit to the 353 K data. The resulting model fit at 353 K and prediction at 298 K illustrate that the framework naturally incorporates the transition between pseudoelastic and quasiplastic behavior which occurs as temperatures are decreased or increased through the transition temperature T_1.

Software

MATLAB m-files for implementing the homogenized energy SMA model can be found at the website `http://www.siam.org/books/fr32`.

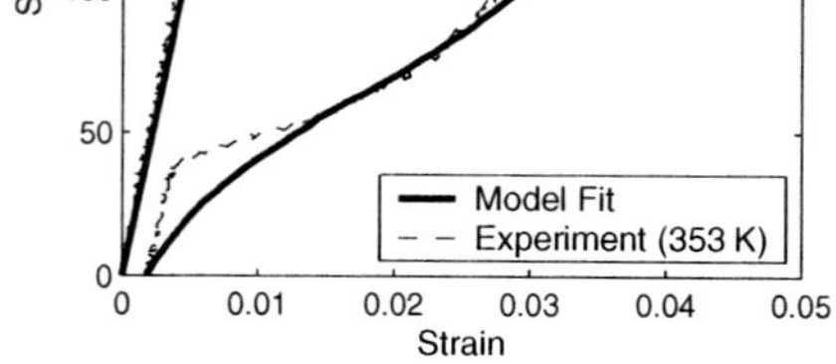

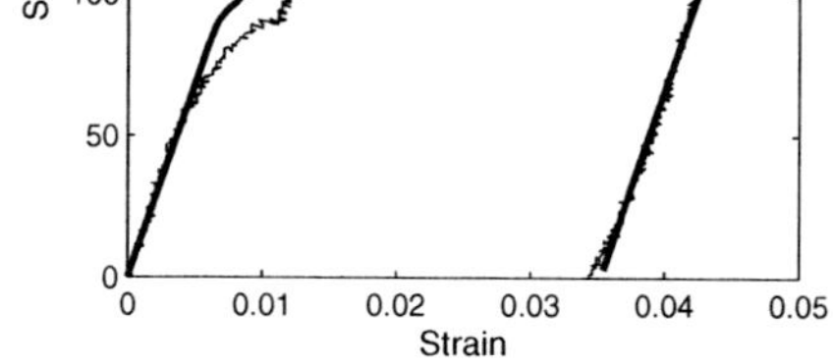

Figure 5.24. *Data from [517]. (a) Model fit at* $T = 353$ *K, and (b) model prediction at* $T = 298$ *K.*

Chapter 6

Unified Modeling Frameworks for Ferroic Compounds

Throughout the last five chapters, we have highlighted material properties common to ferroelectric, ferromagnetic and ferroelastic materials and indicated ways that shared physical mechanisms can be exploited to develop unified modeling frameworks. In this chapter, we summarize three unified frameworks for characterizing hysteresis and constitutive nonlinearities in this combined class of materials based on their shared physical attributes. This is important for a number of reasons. From a fundamental perspective, this quantifies similarities and differences in the underlying physical mechanisms in a manner which promotes our understanding of the materials. From a design perspective, it provides criteria for designing new materials having augmented performance capabilities and provides a framework for analyzing hybrid designs employing multiple materials. These unified characterization frameworks also provide a basis for constructing model-based control algorithms which are viable for a wide range of nonlinear and hysteretic actuator and sensor designs.

We refer to the collective class of ferromagnetic, ferroelectric and ferroelastic materials as ferroic compounds — as with ferroelectric and ferroelastic, the prefix "ferro," meaning iron, was borrowed from ferromagnetism and used to indicate materials having properties analogous to ferromagnetic traits originally observed in iron-based compounds. However, all four designations have long ago outgrown their iron-origin and now refer to compounds exhibiting behavior associated with domain structures. Hence ferroic compounds encompass a broad range of metals, ceramics, composites, films and biomedical materials. For additional details regarding the exploitation of unified properties of ferroic materials for smart material applications, we refer the reader to Newnham [363] and Wadhawan [502].

The degree of similarity shared by ferroelectric, ferromagnetic and ferroelastic compounds depends on your visual acuity. At the microscopic level, the material share analogous energy mechanisms but have highly varied physical properties, thus prohibiting the development of micro-level unified characterization frameworks. The similarities appear at the domain and macroscopic scales and it is there that unified models can be constructed.

Historical Perspective

Recognition of the duality between electric and magnetic phenomena predates Maxwell and the unified theory of electromagnetism provides one of the foundations of classical physics. However, it was not until the early 20th century that analogies between ferromagnetic and ferroelectric materials received sustained investigation. The term "ferroelectric" can be traced to Erwin Schrodinger in 1912 and similarities between ferromagnetic and ferroelectric phenomena were increasingly recognized in the 1920's–1940's as materials such as Rochelle salt were analyzed and potassium dihydrogen phosphate, KH_2PO_4 (KDP), and barium titanate, $BaTiO_3$, were discovered to be ferroelectric. The concept of an order parameter, proposed by Bragg and Williams in 1934 and 1935 [52, 53], was generalized by David Landau in 1937 when he proposed a theory for phase transitions in a range of materials based on both thermodynamic and symmetry principles [281]. This theory was subsequently expanded by Ginzburg [184, 283] and Devonshire [134] to encompass a variety of ferromagnetic and ferroelectric compounds. As detailed in Sections 2.3.1 and 4.4.2, the Landau–Ginzburg–Devonshire theory provides truncated series representations in terms of the order parameter — polarization P for ferroelectric materials and magnetization M for ferromagnetic compounds — and its gradient, which characterize certain temperature-dependent mechanisms. Because the series quantify the material free energy, minimization with respect to the order parameter yields constitutive relations governing aspects of the local E-P, H-M behavior.

During the 1950's, several researchers began to investigate the analogies between hysteretic magnetic and electric phenomena and stress–strain properties of compounds such as AuCd which had recently been shown to exhibit shape memory effects. For example, Everett and Whitten alluded to the possibility of characterizing ferroelastic hysteresis with a Preisach-like representation [154] and Nye discussed certain physical mechanisms in the context of inherent crystal properties [365].

Then, as now, the term ferroelastic encompassed related but not totally identical phenomena. In the context of ferroelectric or ferromagnetic materials, the 90° dipole or moment switching due to applied stresses is typically referred to as ferroelastic switching whereas in the context of shape memory materials, it indicates the presence of ferroelastic domains comprised of austenite or martensite variants. While seemingly disparate, the two concepts are compatible if one associates 90° switches with austenite to martensite phase transformations.

In a series of papers starting in 1980, Falk developed models for SMA which employed the strain as the order parameter and included the strain gradient to incorporate nonlocal effects [157–159]. These models are analogous to the Landau–Ginzburg–Devonshire theory for ferroelectric and ferromagnetic compounds, and the combined theory provides a unified framework for ferroic compounds. Subsequent unified characterization frameworks include the rheological theory of Soukhojak and Chiang [454], Preisach representations, domain wall models, and homogenized energy models. We summarize the latter three in Sections 6.2–6.4.

In addition to these general modeling frameworks, a number of investigations have focused on analysis and classification issues associated with ferroic compounds. In 1970, Aizu presented a unified treatment of symmetry properties of ferroelectric, ferromagnetic and ferroelastic compounds which he collectively referred to as ferroic compounds [8]. Unification through symmetry analysis and resulting group formulations have grown extensively since then and provide an important classification framework for ferroic compounds. The reader is referred to the text by Wadhawan [502] for a comprehensive discussion of this topic along with further details regarding the historical development of ferroic theory.

6.1 Physical Properties

The constituent material behavior of ferroelectric, ferromagnetic, and ferroelastic compounds has been discussed in Sections 2.1, 4.1, and 5.1 and we summarize here only those properties shared by all three classes of compounds. As noted in Table 6.1, analogies in behavior occur at subdomain, domain and macroscopic levels, but the latter two are of primary importance when constructing unified characterization frameworks.

Fundamental Properties

The physical mechanisms which produce hysteresis and constitutive nonlinearities in ferroic materials differ significantly at subdomain scales. In ferroelectric and ferromagnetic materials, dipole and moment reorientation constitute two microscopic mechanisms which produce macroscopic hysteresis. While these processes can be conceptually correlated, the underlying physical mechanisms differ significantly — ferroelectricity is due to the ionic structure of materials whereas ferromagnetism results from interactions between magnetic moments associated with electron spins. Furthermore, hysteresis in shape memory alloys is due to solid state displacive phase transformations between austenite and martensite variants. Hence the construction of unified energy-based models at subdomain scales is not considered feasible.

Domain Properties

Domains in ferroelectric, ferromagnetic and ferroelastic materials consist of regions congregated by dipoles, moments, and austenite or martensite variants having the same orientations. The transition regions between domains are termed

Ferroelectric	Ferromagnetic	Ferroelastic
Polarization	Magnetization	Strain
Electric field	Magnetic field	Stress
Paraelectric phase	Paramagnetic phase	Austenite phase
Ferroelectric phase	Ferromagnetic phase	Martensite phase
Domain walls	Bloch or Neel walls	Variant boundaries
Devonshire theory	Mean field theory	Landau theory
Micromechanical theory	Micromagnetic theory	Ginzburg–Landau theory

Table 6.1. *Analogies between physical properties of ferroic materials.*

domain walls. Whereas domain nucleation in all three classes of compounds can be attributed to energy minimization, the specific nature of underlying energy mechanisms produce different domain morphologies as depicted in Figure 6.1 and 6.2. In ferroelectric materials, 180° and 90° domains form to minimize electrostatic and ferroelastic energy whereas the minimization of magnetostatic and ferroelastic energy is accompanied by flux closure in the formation of ferromagnetic domains. Finally, minimization of elastic energy produces twinned ferroelastic domains.

For all three classes of materials, energy dissipation associated with domain wall movement constitutes one source of hysteresis and constitutive nonlinearities, and quantification of these loss mechanisms provides the basis for the unified domain wall theory summarized in Section 6.3.

Whereas the homogenized energy framework of Section 6.4 does not explicitly exploit the domain properties of ferroic materials, it incorporates domain properties in two fundamental ways. (i) The assumption that moments or dipoles are aligned with the field or diametrically opposite to it — which underlies the construction of the ferroelectric and ferromagnetic Helmholtz energy relations — is based on 180° switching, and theory to construct similar energy relations which quantify 90° switching has been initiated in [24]. When combined with the Helmholtz relations for SMA, this provides a framework for characterizing the local domain behavior within a grain. (ii) Variations due to polycrystallinity, which yields grains with differing orientations, and other general stress and material nonhomogeneities are incorporated by the homogenization process which involves integration over a distribution of material properties. This provides an efficient technique for incorporating

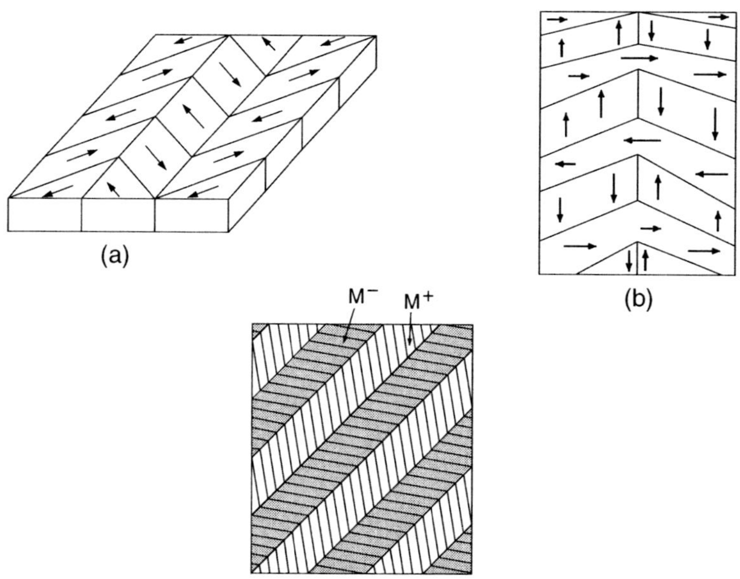

Figure 6.1. *Twinned 90° and 180° domain structure for (a) $BaTiO_3$ and (b) iron. (c) Twinned martensite structure for the SMA NiTi.*

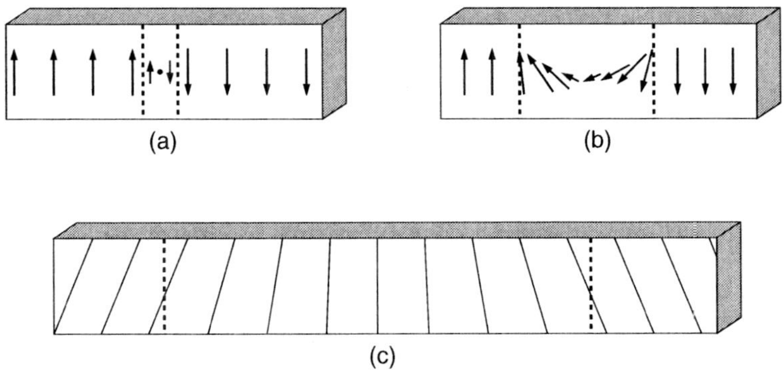

Figure 6.2. *Domain walls in (a) ferroelectric, (b) ferromagnetic, and (c) ferroelastic materials.*

effective domain properties in a manner which yields macroscopic models having the potential for real-time implementation.

When developing series relations for the free energy, the quantification of transitions within domain walls necessitates inclusion of M_x, P_x or ε_x gradient terms as proposed by Landau and Lifschitz for ferromagnetic materials [282] and later extended by Ginzburg. As noted in Section 2.7, inclusion of these gradient terms also serves to regularize models in the manner of a Tikhonov functional.

While shared properties at the domain level provide the basis for unified characterization frameworks, care must be taken not to overextend these analogies. For example, it was noted in Section 4.1.2 that at least three fundamental physical criteria differentiate ferroelectric and ferromagnetic domains processes. (i) The magnitude of elastic stresses on 90° formation in the two classes are typically quite different. (ii) The presence of quantum exchange interactions in ferromagnetic materials produces thicker domain walls since the exchange energy in this case is significant compared with the anisotropy energy. (iii) Ferroelectric materials have electrons which carry charge whereas the analogous construct in ferromagnetic materials is absent. Similar differences exist when comparing ferroelastic domains with their ferroelectric and ferromagnetic counterparts. For example, it is noted in [199] that ferroelastic domain walls are typically an order of magnitude thicker than ferromagnetic domain walls as depicted in Figure 6.2.

In summary, the analogies between domain properties of ferroic materials provide an important starting point for constructing unified characterization frameworks but the differences must be understood if quantifying fundamental or fine-scale properties of the materials.

Macroscopic Properties

A manifestation of the microscopic and domain properties of ferroic materials is the presence of hysteresis and constitutive nonlinearities in the relation between input fields and stresses and output polarization, magnetization and strains. This is illustrated in Figure 6.3 for $BaTiO_3$, PZT, Terfenol-D and SMA with numerous other examples provided in previous chapters.

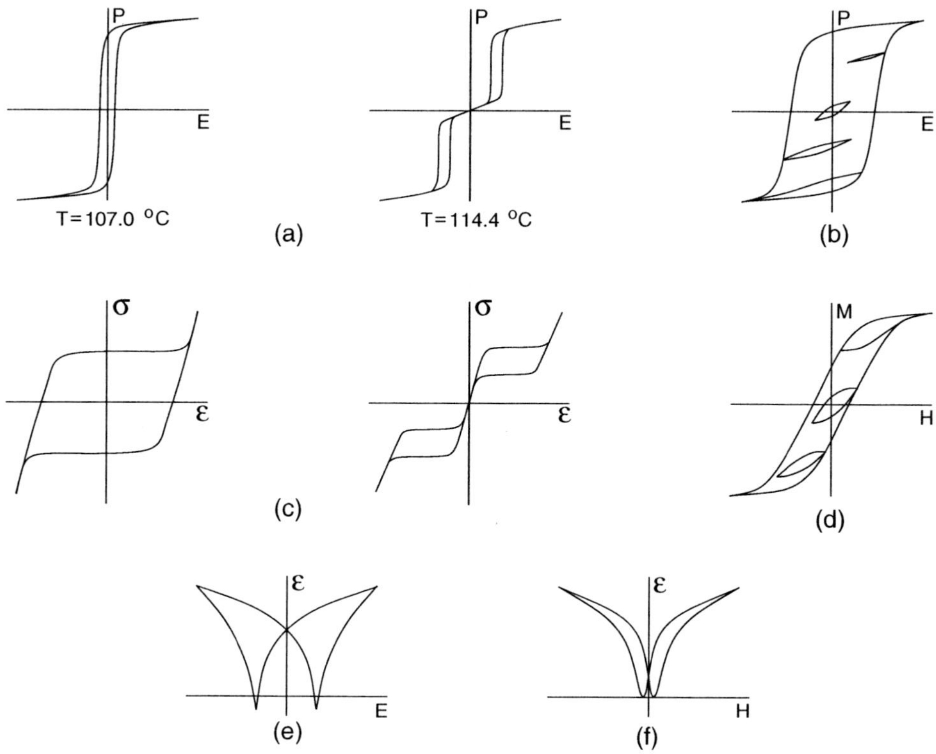

Figure 6.3. *(a) $BaTiO_3$ behavior in the neighborhood of the Curie point $T_c = 107^o$ C (after [338]). (b) Major and minor loop behavior of PZT5H, (c) SMA behavior in the ferroelastic and pseudoelastic regimes. (d) Major and minor loops exhibited by Terfenol-D. (e) Field-strain butterfly behavior for PZT, and (f) Terfenol-D.*

It was noted in Section 2.3.1 that shape memory alloys and certain ferroelectric materials, including $BaTiO_3$ and PZT, exhibit first-order phase transition behavior whereas ferromagnetic materials and other ferroelectric compounds (e.g., Rochelle salts and KH_2PO_4) exhibit second-order transition behavior. Transducers employing ferroelectric and ferromagnetic components are typically employed well below the Curie point to enhance performance so their transition behavior is often irrelevant for smart system design. However, SMA are typically employed in pseudoelastic regimes which necessitates the accommodation of certain first-order transition behavior.

As noted in Section 2.1 and illustrated in Figures 2.12 and 6.4, the shape memory effect, which is traditionally considered in the realm of shape memory alloys, is also exhibited by soft PZT compounds. Furthermore, the investigation of ferromagnetic shape memory alloys (FSMA) makes this a fundamental mechanism for certain magnetic compounds, thus providing an additional unifying phenomenon for ferroic materials.

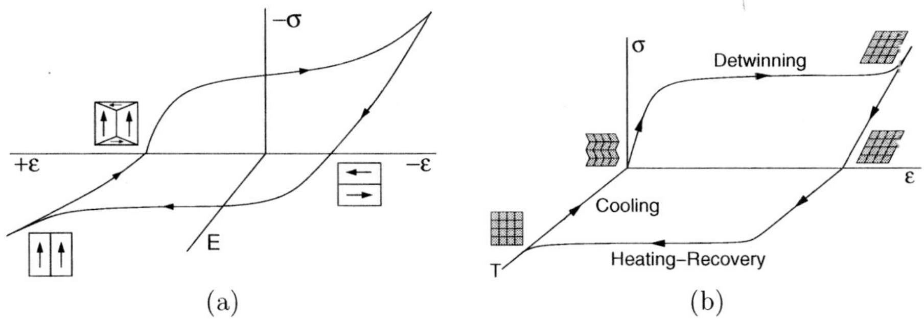

Figure 6.4. *(a) Shape memory effect (SME) in soft PLZT compounds due to stress-induced* 90° *switching and field-induced repolarization. (b) SME in SMA due to a stress-induced transformation from twinned (self-accomodated) martensite to detwinned martensite followed by a temperature-induced martensite-austenite transformation.*

Fundamental Energy Relations and Hysteresis Losses

The fundamental energy relations for quantifying hysteresis losses and constructing Gibbs energy relations are the electrostatic energy

$$U = \mathbf{E} \cdot \mathbf{P}, \tag{6.1}$$

magnetostatic energy

$$U = \mu_0 \mathbf{H} \cdot \mathbf{M}, \tag{6.2}$$

and elastic energy

$$U = \boldsymbol{\sigma} \cdot \boldsymbol{\varepsilon}. \tag{6.3}$$

As discussed in Jiles [247], one can eliminate the magnetic permeability μ_0 in (6.2) and completely unify the notation with $U = \mathbf{H} \cdot \mathbf{M}$ in the Kennelly and Gaussian (CGS) systems of magnetic units. However, this leads to contradictions in the Sommerfeld notation introduced in Chapter 4 which we employ due to its acceptance by the International Union for Pure and Applied Physics (IUPAP) and increasing prevalence within the magnetics community. Hence we live with the additional factor of μ_0 in the magnetic relation.

The work required for traversal of a hysteresis loop in the three cases is

$$W = \oint \mathbf{E} \cdot d\mathbf{P}$$

$$W = \mu_0 \oint \mathbf{H} \cdot d\mathbf{M}$$

$$W = \oint \boldsymbol{\sigma} \cdot d\boldsymbol{\varepsilon}.$$

Hence the amount of energy which is converted to potential energy or heat is proportional to the area of the hysteresis loop. This has significant ramifications for applications involving vibration suppression — e.g., vibration attenuation in civil structures using SMA tendons — since it requires maximal hysteresis rather than the typical goal of minimizing hysteretic effects.

Order Parameters, Conjugate Fields, and Polynomial Helmholtz Relations

The concept of an order parameter was employed by Bragg and Williams [52, 53] when quantifying the concentration of constituent alloys in binary alloys such as CuAu and was generalized by Landau in his theory for phase transitions in solids [281]. For our discussion, we define an order parameter e as a scalar or vector-valued variable of state which characterizes the difference between two phases in the sense that the thermal average is nonzero below the Curie point or transition temperature T_c and zero above it [41, 482].[25] Furthermore, we let φ denote external fields that are thermodynamically conjugate to e.

For ferroelectric, ferromagnetic and ferroelastic materials, natural choices for e are respectively the polarization P, magnetization M, and strain ε as summarized in Table 6.2. The corresponding external fields φ are taken to be the electric field E, magnetic field H, and stress σ. We point out that the transition of e from nonzero to zero behavior should be considered in the absence of an applied field. Finally, we note that order parameters limit continuously to zero (with discontinuous derivative at zero) for materials exhibiting second-order phase transitions and discontinuously for first-order materials as depicted in Figures 2.14, 2.17 and 6.5. Details regarding the ramifications of the order of phase transitions on the statistical mechanics and thermodynamic properties of materials can be found in [187, 482, 524].

For temperatures T, we denote the Helmholtz energy by $\psi(e, T)$ and the Gibbs energy by $G(e, \varphi, T)$.[26] In the absence of applied fields, the Helmholtz relations provide a natural measure of energy and, under the assumption of differentiability, thermodynamic equilibria are provided by the necessary condition

$$\frac{\partial \psi}{\partial e} = 0.$$

In the presence of an applied field, the Gibbs relation

$$G(e, \varphi, T) = \psi(e, T) - \varphi e \tag{6.4}$$

is used to quantify the total energy, and equilibrium states are specified by the necessary conditions

$$\frac{\partial G}{\partial e} = 0 \quad , \quad \frac{\partial^2 G}{\partial e^2} > 0. \tag{6.5}$$

[25] The terminology is motivated by early analysis of ferromagnetic materials where the order parameter $e = M$ was used to delineate the paramagnetic phase, with *disordered* spins, from the *ordered* ferromagnetic phase. As indicated in the introduction of Chapter 5, there exist close ties between the order parameter and internal variables employed in a number of models and we refer the reader to Section 4.7 of [331] for details correlating these concepts. Finally, details regarding the connection between the composition of materials exhibiting first and second-order phase transitions and long- and short-range order parameters can be found in [263, 466].

[26] See Section 2.2.3 for details motivating the formulation of G in terms of both the order parameter e and conjugate field φ.

Class of Compounds	Order Parameter e	Conjugate Field φ
Ferroelectric	P	E
Ferromagnetic	M	H
Ferroelastic	ε	σ

Table 6.2. *Order parameters and conjugate fields for ferroic materials.*

Note that in formulating (6.4), we have incorporated μ_0 into the magnetic Helmholtz energy to unify notation. The relation (6.5) can be interpreted as providing conditions dictating how the order parameter adjusts to balance the internal energy with work due to the applied field. It thus provides local constitutive relations quantifying the dependence of P, M, ε on E, H, σ in the absence of thermal activation.

To illustrate the construction of ψ, we summarize the polynomial energy relations developed by Landau, Devonshire and Falk as detailed in Sections 2.3.1, 4.4.2 and 5.2.2. For materials exhibiting first-order transitions — e.g., SMA and PZT — the Helmholtz energy is taken to be

$$\psi(e,T) = \psi_0(T) + \alpha_1(T - T_0)e^2 - \alpha_2 e^4 + \alpha_3 e^6 \tag{6.6}$$

where $\alpha_1, \alpha_2, \alpha_3$ are positive constants and T_0 denotes the Curie temperature. Solution of

$$\varphi = 2\alpha_1(T - T_0)e - 4\alpha_2 e^3 + 6\alpha_3 e^5, \tag{6.7}$$

resulting from (6.5), yields a local constitutive relation specifying the manner in which $e = P, \varepsilon$ depends on $\varphi = E, \sigma$. For materials such as iron or Terfenol-D which exhibit second-order transitions, ψ is given by

$$\psi(e,T) = \psi_0(T) + \alpha_1(T - T_c)e^2 + \alpha_2 e^4 \tag{6.8}$$

where T_c is the Curie point or transition temperature.[27] Enforcement of the necessary conditions (6.5) yields a constitutive relation analogous to (6.7).

Whereas the formulations (6.6) and (6.8) are highly compact and unified, they impose overly restrictive limitations due to their high-order polynomial nature. Alternative Helmholtz energy choices are summarized in Section 6.4 where they provide a basis for the unified homogenized energy framework.

[27] As noted in Section 2.3.1, the Curie point and Curie temperature are often nearly identical in second-order materials but can differ by more than 10 °C in materials exhibiting first-order phase transitions.

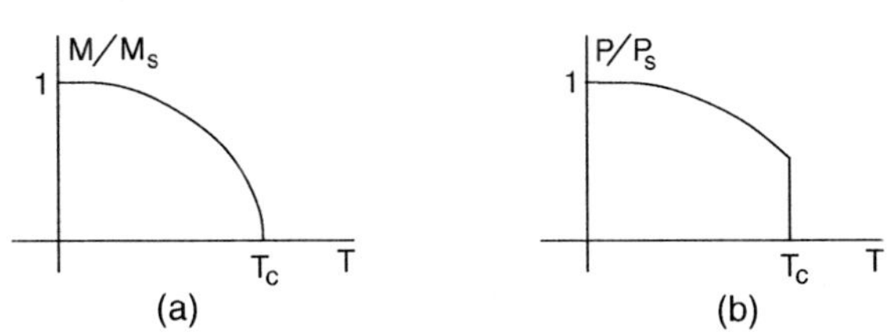

Figure 6.5. *Temperature-dependence of the order parameter for (a) a second-order phase transition and (b) first-order phase transition.*

6.2 Preisach Representations

The general mathematical nature of Preisach representations makes them a natural candidate for constructing a unified characterization framework for ferroic compounds. This appears to have been first recognized by Everett and Whitten in 1952 although they did not implement the model in this context [154]. Originally developed by Preisach for ferromagnetic hysteresis in 1935 [382], this approach was applied by Huo to SMA in 1989 [229] and ferroelectric materials by Sreenan, Salvady and Naganathan in 1993 [399]. In 1997, Hughes and Wen discussed the unified framework provided by the Preisach theory for PZT, magnetic compounds, and SMA as well as provided inverse filters appropriate for subsequent control design [227]. Other than linear theory, which has very limited applicability for SMA, Preisach theory constitutes the earliest unified framework developed for characterizing hysteresis and constitutive nonlinearities in ferroic compounds.

Details regarding the formulation of Preisach models for ferroelectric, ferromagnetic and ferroelastic compounds are provided in Sections 2.4, 4.5 and 5.3 and we provide here only a brief summary of a unified representation for ferroic compounds.

Consider the Preisach plane

$$S = \{(s_1, s_2) | s_1 \leq s_2\}$$

and kernels k_s having values of ± 1 and thresholds at s_1 and s_2 as shown in Figure 6.6. The kernels are defined by

$$[k_s(\varphi, \xi)](t) = \begin{cases} [k_s(\varphi, \xi)](0) & \text{if} \quad \tau(t) = \emptyset \\ -1 & \text{if} \quad \tau(t) \neq \emptyset \text{ and } \varphi(\max \tau(t)) = s_1 \\ +1 & \text{if} \quad \tau(t) \neq \emptyset \text{ and } \varphi(\max \tau(t)) = s_2 \end{cases}$$

where

$$[k_s(\varphi, \xi)](0) = \begin{cases} -1 & \text{if} \quad \varphi(0) \leq s_1 \\ \xi & \text{if} \quad s_1 < \varphi(0) < s_2 \\ +1 & \text{if} \quad \varphi(0) \geq s_2 \end{cases}$$

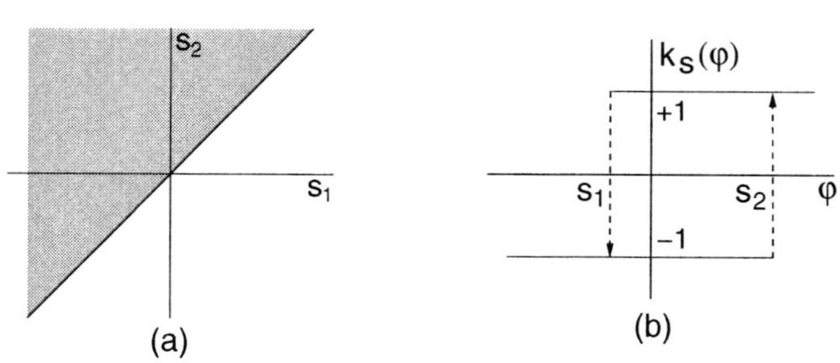

Figure 6.6. *(a) Preisach plane S and (b) classical Preisach relay with threshold values at s_1 and s_2.*

defines the initial state of the kernel in terms of $\xi \in \{-1, 1\}$, and

$$\tau(t) = \{t_s \in (0, t] \mid \varphi(t_s) = s_1 \text{ or } \varphi(t_s) = s_2\}$$

defines times (parameter values) at which thresholds are reached.

Recalling that $e = P, M, \varepsilon$ denotes order parameters and $\varphi = E, H, \sigma$ are associated conjugate fields, the polarization, magnetization or strain due to an applied field or stress is quantified by the representation

$$e(t) = \iint_S [k_s(\varphi, \xi)](t)\nu(s)\, ds. \tag{6.9}$$

Here ν denotes a material-dependent density to be identified through the techniques discussed in Section 4.5.

To provide a framework which includes related models such as that of Prantl and is amenable to convergence analysis, it is advantageous to consider the more general formulation

$$e(t) = \iint_S [k_s(\varphi, \xi)](t)\, d\mu(s) \tag{6.10}$$

where $\mu \in \mathbb{M}$ is a signed Borel measure in the set $\mathbb{M}$ of all finite, signed Borel measures on S. The representation (6.9) is commonly employed in applications but (6.10) provides a broader analytic foundation for the framework.

The advantages and disadvantages of Preisach theory, as well as extensions required to accommodate several physical phenomena, are detailed in Section 4.5 in the context of magnetic materials where the theory is most complete. It is illustrated in Section 4.7.10 that the homogenized energy framework provides an energy basis for certain extended Preisach models, and it is anticipated that the synergy between the two frameworks will be increasingly exploited to unify the theory of ferroic compounds.

6.3 Domain Wall Theory

The domain wall framework originated with the magnetic models of Jiles and Atherton in 1984 and 1986 [250,251]. Tuszyński et al. employed the approach for $LiCsSO_4$ in 1998 [484] and a comprehensive theory for ferroelastic compounds was published in 2003 [322]. The theory was extended to ferroelectric materials in 1999 [438, 442] and first proposed as a unified characterization framework for ferroic compounds in 1999 [441]. Details regarding the development of the theory for ferroelectric, ferromagnetic and ferroelastic materials can be found in Sections 2.5, 3.3, 4.6 and 5.4.

In the domain wall framework, polarization, magnetization and strain relations are constructed in three steps: (i) quantification of equilibrium values in the absence of pinning sites or inclusions, (ii) quantification of irreversible energy dissipation during domain wall translation, and (iii) quantification of reversible changes due to domain wall bending. In the original magnetic model development, the equilibrium states yielded anhysteretic magnetization relations. We retain the notation but generalize the terminology since the ferroelectric and ferroelastic relations are often hysteretic. The domain and domain wall structures which motivate this framework are depicted in Figures 6.1 and 6.2.

Equilibrium Relations in the Absence of Inclusions

The equilibrium relations are obtained by minimizing the Gibbs energy (6.4) for various choices of the Helmholtz energy ψ. Hence they represent equilibrium values of the order parameter e in response to an applied conjugate field φ when mean field interactions are incorporated.

As detailed in Sections 2.3.2 and 4.4.1, one technique for constructing ψ is to apply statistical mechanics principles to a homogeneous lattice volume V comprised of $N = N_- + N_+$ cells. The use of mean field approximations to simplify exchange interactions and Boltzmann theory to quantify entropic effects yields the general Helmholtz relation

$$\psi(e,T) = \frac{\varphi_h e_s}{2}\left[1-(e/e_s)^2\right] + \frac{\varphi_h T}{2T_c}\left[e\ln\left(\frac{e+e_s}{e_s-e}\right) + e_s\ln(1-(e/e_s)^2)\right]. \quad (6.11)$$

The Curie point $T_c = \frac{\Phi_0}{2k}$ and bias field $\varphi_h = \frac{N\Phi_0}{2Ve_s}$ are defined in terms of the Boltzmann's constant k, average reorientation energy Φ_0, and saturation value e_s. For magnetic materials, we will typically incorporate μ_0 into the relations for T_c and φ_h to unify notation.

Expansion of (6.11) yields an even-powered series which can be truncated after m terms to yield a second Helmholtz expression

$$\psi(e,T) = \psi_0(T) + \alpha_1(T-T_c)e^2 - \alpha_2 e^4 + \sum_{j=3}^{m}\alpha_j e^{2j} \quad (6.12)$$

which is a higher-order form of the Landau–Devonshire–Falk expansions (6.6) and (6.8).

Minimization of G with ψ given by (6.11) yields the Ising relation

$$e_{an}(\varphi) = e_s\tanh\left(\frac{\varphi+\alpha e}{a(T)}\right) \quad (6.13)$$

where $\alpha = \frac{\varphi_h}{e_s}$ and $a(T) = \frac{\varphi_h T}{T_c}$. We note that

$$\varphi_e = \varphi + \alpha e$$

provides an effective field in which the applied field φ is augmented by the interaction field $\varphi_I = \alpha e$.

Alternatively, minimization of $G = \psi - \varphi_e e$ with ψ given by (6.12) yields

$$\varphi_e = 2\alpha_1(T-T_c)e_{an} - 4\alpha_2 e_{an}^3 + \sum_{j=3}^{m} 2j\alpha_j e_{an}^{2j-1} \quad (6.14)$$

which implicitly defines e_{an} in terms of φ_e.

A third choice for the equilibrium relation is the Langevin expression

$$e_{an} = e_s \left[\coth\left(\frac{\varphi + \alpha e}{a}\right) - \left(\frac{a}{\varphi + \alpha e}\right) \right]. \tag{6.15}$$

As detailed in Section 2.5.1, this expression is derived by integrating over a continuum of moment or dipole orientations whereas the Ising relation (6.13) is based on the assumption that dipoles or moments are limited to orientations in the direction of or diametrically opposite to the applied field.

The magnetic and polarization models typically employ e_{an} given by (6.13) or (6.15) whereas e_{an} specified by (6.14) is employed in the SMA model.

Domain Wall Model

The irreversible component e_{irr} quantifies the polarization, magnetization or strains generated when the equilibrium values e_{an} are augmented by energy dissipated during domain wall translation. Use of the electrostatic, magnetostatic, or elastic energy relations (6.1)–(6.3) to account for reorientation inherent to domain wall movement yields the differential equation

$$\frac{de_{irr}}{d\varphi} = \widetilde{\delta} \frac{e_{an} - e_{irr}}{\delta k_p + \alpha[e_{an} - \varepsilon_{irr}]}.$$

Here k_p is a material-dependent loss coefficient, $\delta = \text{sign}(d\varphi)$ ensures that energy is always dissipated, and $\widetilde{\delta}$ enforces reversible behavior until the equilibrium curve is reached — e.g., see (2.77).

The final component of the model is the reversible component

$$e_{rev} = c(e_{an} - e_{irr})$$

which incorporates the reversible effects of domain wall bending. Here c is a coefficient with values between 0 and 1.

The complete model

$$\begin{aligned} e &= e_{irr} + e_{rev} \\ &= (1-c)e_{irr} + ce_{an} \end{aligned}$$

quantifies the polarization, magnetization or strains due to applied fields or stresses. It is noted that in the limiting case $c = 1$, the solution is the equilibrium value whereas $c = 0$ indicates no reversible behavior.

The advantage of the domain wall model is its efficiency and qualitative accuracy when characterizing major loop behavior. Its primary deficit is the fact that it requires *a priori* knowledge of turning points to guarantee closure of biased minor loops as typically observed in quasistatic operating regimes with negligible accommodation or reptation. While not a crucial issue for material characterization, this does limit its use for feedback control design where turning points are determined by state measurements or estimates. For such applications or general characterization including relaxation or thermal processes, the homogenized energy framework is recommended.

6.4 Homogenized Energy Framework

The homogenized energy framework originated with the theory of thermally activated processes developed by Müller, Achenbach and Seelecke for SMA [1,2,353,374, 409,413]. The original theory focused on homogeneous, single crystal compounds with initial effects due to polycrystallinity considered in [374]. The framework was subsequently extended to ferromagnetic compounds [434], ferroelectric materials [450], SMA thin films [323,324], and polycrystalline bulk SMA compounds [411] in 2002–2004.

In addition to greatly expanding the scope of the framework, the latter extensions had three important ramifications. (i) They established the necessity of treating certain parameters as manifestations of underlying distributions rather than constants to accommodate material nonhomogeneities and variable effective fields. (ii) The extensions led to the formulation of a unified framework for characterizing hysteresis and constitutive nonlinearities in a broad class of ferroic compounds [448,449]. (iii) As detailed in Section 4.7.10, the extended framework unifies attributes shared by a range of macroscopic theories — e.g., Jiles–Atherton domain wall theory, Stoner-Wohlfarth magnetic theory, Puglisi and Truskinovski plasticity theory [384–386] — and provides an energy basis for certain extended Preisach formulations [447]. The fundamental difference between this framework and Preisach theory lies in the fact that through the energy basis, mechanisms necessary to incorporate thermal activation and relaxation, reversibility, and various temperature, rate and stress effects are directly incorporated in the kernel or hysteron as compared with Preisach models which typically incorporate these effects through vector-valued densities. This facilitates model construction and implementation when characterizing hysteresis in the presence of multiple inputs or transient environmental conditions.

Significant discussion regarding the framework has been provided in Sections 2.6, 3.4, 4.7 and 5.5, and we provide here only a summary of the unified formulation for ferroic compounds. As compiled in Table 6.2, the conjugate field $\varphi = E, H, \sigma$ and order parameter $e = P, M, \varepsilon$ constitute the inputs and outputs in the constituent material classes.

6.4.1 Mesoscopic Model

We consider first the formulation of Helmholtz and Gibbs energy relations and construction of local average polarization, magnetization, and strain relations at the lattice level under the assumption of homogeneous material properties. By employing the order parameter and conjugate fields, the polarization and magnetization models can be completely unified in the absence of phase transitions. The strain model for ferroelastic compounds falls within the same modeling framework but some individual definitions differ slightly due to the inherent phase transformations. While notation can be established to formulate all three models in terms of unified definitions and expressions, the required generality obscures rather than clarifies the framework so we omit it and consider ferroelastic materials separately.

Helmholtz and Gibbs Energy Relations

Ferroelectric and Ferromagnetic Materials

One choice for the Helmholtz energy is the relation (6.11) employed in the construction of the domain wall framework. Although it has the advantage of being based on statistical mechanics principles, the relation is overly restrictive for some regimes. An alternative is to employ Taylor expansions about the equilibria, as detailed in Section 2.3.3, to construct the piecewise quadratic Helmholtz relation

$$\psi(e) = \begin{cases} \frac{1}{2}\eta(e+e_R)^2 & , \; e \leq -e_I \\ \frac{1}{2}\eta(e-e_R)^2 & , \; e \geq e_I \\ \frac{1}{2}\eta(e_I - e_R)\left(\frac{e^2}{e_I} - e_R\right) & , \; |e| < e_I \end{cases} \tag{6.16}$$

where e_I, e_R and η respectively denote the inflection point, local remanence, and reciprocal slope after switching. As shown in Figure 2.30, the kernel or hysteron resulting from (6.16) is piecewise linear with slope $\frac{1}{\eta}$ after switching whereas that derived from (6.11) saturates to the value e_s as dictated by the Ising relation (6.13).

The Gibbs energy relation

$$G(\varphi, e, T) = \psi(e, T) - \varphi e \tag{6.17}$$

subsequently incorporates the work due to applied conjugate fields.

Ferroelastic Materials

When constructing Helmholtz energy relations for ferroelastic materials, it is necessary to incorporate the phase transition between high temperature austenite and low temperature martensite behavior. This is in contrast to ferroelectric and ferromagnetic compounds which are typically employed well below their Curie point to optimize performance.

A continuously differentiable choice for the Helmholtz energy is

$$\psi(e,T) = \begin{cases} \frac{Y_M}{2}(e+e_T)^2 & , \; e \leq -e_M(T) \\ -\frac{E_0(T)}{2}(e+e_0(T))^2 + \psi_0(T) & , \; -e_M(T) < e < -e_A(T) \\ \frac{Y_A}{2}e^2 + \Delta\beta(T) & , \; |e| \leq e_A(T) \\ -\frac{E_0(T)}{2}(e-e_0(T))^2 + \psi_0(T) & , \; e_A(T) < e < e_M(T) \\ \frac{Y_M}{2}(e-e_T)^2 & , \; e \geq e_M(T). \end{cases} \tag{6.18}$$

Here Y_M, Y_A denote the Young's moduli for the martensite and austenite phases, e_M, e_A are inflection points, $(\pm e_0(T), \psi_0(T))$ denote maxima of the concave parabolae, $E_0(T)$ is chosen to ensure C_1 continuity and $\Delta B(T)$ denotes the austenite minimum height.

The Gibbs energy is again given by (6.17). The behavior of the Helmholtz and Gibbs energy relations is illustrated in Figures 5.13, 5.14 and 5.18.

Mesoscopic Model — Negligible Thermal Activation

For operating regimes in which thermal activation is negligible or relaxation times are small compared with drive frequencies, local constitutive relations follow directly from the minimization of the Gibbs energy.

Ferroelectric and Ferromagnetic Materials

Enforcement of (6.5) with ψ given by (6.16) yields the general relation

$$\overline{e} = \frac{\varphi}{\eta} + \delta e_R$$

where $\delta = 1$ on the upper branch of the hysteron and $\delta = -1$ on the lower branch. Initial dipole/moment orientations and subsequent switching are formalized by the notation

$$[\overline{e}(\varphi; \varphi_c, \xi)](t) = \begin{cases} [\overline{e}(\varphi; \varphi_c, \xi)](0) & , \; \tau(t) = \emptyset \\ \frac{\varphi}{\eta} - e_R & , \; \tau(t) \neq \emptyset \text{ and } \varphi(\max \tau(t)) = -\varphi_c \\ \frac{\varphi}{\eta} + e_R & , \; \tau(t) \neq \emptyset \text{ and } \varphi(\max \tau(t)) = \varphi_c \, . \end{cases} \tag{6.19}$$

Here

$$\varphi_c = \eta(e_R - e_I) \tag{6.20}$$

denotes local coercive points at which local minima of G are eliminated,

$$[\overline{e}(\varphi; \varphi_c, \xi)](0) = \begin{cases} \frac{\varphi}{\eta} - e_R & , \; \varphi(0) \leq -\varphi_c \\ \xi & , \; -\varphi_c < \varphi(0) < \varphi_c \\ \frac{\varphi}{\eta} + e_R & , \; \varphi(0) \geq \varphi_c \end{cases}$$

quantifies initial dipole or moment orientations in terms of the parameter $\xi = \frac{\varphi}{\eta} \pm e_R$, and

$$\tau(t) = \{t_s \in (0, t] \,|\, \varphi(t_s) = -\varphi_c \text{ or } \varphi(t_s) = \varphi_c\}$$

specifies the transition times. The behavior of $\overline{e}$ designated in this manner is illustrated in Figures 2.31 and 4.37.

Ferroelastic Materials

For tensile stresses, the minimization of G with ψ given by (6.18) yields

$$[\overline{e}(\varphi; \varphi_c, \xi)](t) = \begin{cases} [\overline{e}(\varphi; \varphi_c, \xi)](0) & , \; \tau(t) = \emptyset \\ \frac{\varphi}{Y_M} + e_T & , \; \tau(t) \neq \emptyset \text{ and } \varphi(\max \tau(t)) = \varphi_M \\ \frac{\varphi}{Y_A} & , \; \tau(t) \neq \emptyset \text{ and } \varphi(\max \tau(t)) = \varphi_A \end{cases} \tag{6.21}$$

where

$$[\bar{e}(\varphi;\varphi_c,\xi)](0) = \begin{cases} \frac{\varphi}{Y_A} & , \varphi(0) \le \varphi_M \\ \xi & , \varphi_M < \varphi(0) < \varphi_A \\ \frac{\varphi}{Y_M} + e_T & , \varphi(0) \ge \varphi_A \end{cases}$$

and

$$\tau(t) = \{t_s \in (0,t] \,|\, \varphi(t_s) = \varphi_M \text{ or } \varphi(t_s) = \varphi_A\}.$$

Similar expressions hold for compressive stresses. Additional discussion regarding this notation is provided on page 101.

As illustrated in Figure 5.14, this energy functional and resulting kernel quantify the ferroelastic, pseudoelastic, shape memory effect and first-order transition behavior associated with SMA.

Mesoscopic Model — Thermal Activation

To incorporate thermal activation processes which produce relaxation and reptation behavior, it is necessary to balance the Gibbs energy G and relative thermal energy kT/V through the Boltzmann relation

$$\mu(G) = Ce^{-GV/kT}$$

which quantifies the probability of achieving an energy level G. The constant C is specified to be unity when μ is integrated over all admissible energy levels.

Ferroelectric and Ferromagnetic Materials

As detailed in Sections 2.6.2 and 4.7.1, the incorporation of thermal activation processes yields the mesoscopic relation

$$\bar{e} = x_+ \langle e_+ \rangle + x_- \langle e_- \rangle \tag{6.22}$$

where the dipole/moment fractions are specified by

$$\begin{aligned} \dot{x}_+ &= -p_{+-}x_+ + p_{-+}(1 - x_+) \\ x_- &= 1 - x_+ \,. \end{aligned}$$

The average polarization/magnetization values are given by

$$\langle e_\pm \rangle = \int_{\chi_\pm} e\mu(G(e,\varphi,T))\, de$$

and the transition likelihoods are

$$p_\pm = \frac{1}{\mathcal{T}(T)} \mu(G(\pm e_I, \varphi, T)).$$

Here $\mathcal{T}$ denotes the relaxation times and integration intervals are $\chi_+ = (e_I, \infty)$ and $\chi_- = (-\infty, -e_I)$.[28] As illustrated in Figures 2.31 and 4.37, the inclusion of

[28]See Remark 2.6.1 on page 105 for details justifying the point evaluation of the continuous density μ in the definition of $p_\pm$.

thermal activation mechanisms mollifies the behavior of the kernel and reduces the local coercive behavior since some dipoles/moments achieve the energy necessary to switch in advance of local coercive fields $\pm\varphi_c$ given by (6.20).

Ferroelastic Materials

The same process yields the local model

$$\bar{e} = x_+ \langle e_+ \rangle + x_- \langle e_- \rangle + x_A \langle e_A \rangle \tag{6.23}$$

where the phase fractions are governed by

$$\begin{aligned}
\dot{x}_-(t) &= -(p_- + p_{A-})x_-(t) - p_{A-}x_+(t) + p_{A-} \\
\dot{x}_+(t) &= -(p_+ + p_{A+})x_+(t) - p_{A+}x_-(t) + p_{A+} \\
x_A &= 1 - x_+ - x_- \,.
\end{aligned}$$

The expected strain values are

$$\langle e_\pm \rangle = \int_{\chi_{M\pm}} e\mu(G(e,\varphi,T))\,de \quad , \quad \langle e_A \rangle = \int_{\chi_A} e\mu(G(e,\varphi,T))\,de$$

and the transition likelihoods are

$$p_\pm = \frac{1}{\mathcal{T}(T)}\mu(G(\pm e_M,\varphi,T)) \quad , \quad p_{A\pm} = \frac{1}{\mathcal{T}(T)}\mu(G(\pm e_A,\varphi,T)).$$

The integration intervals in this case are $\chi_A = (-\varepsilon_A, \varepsilon_A), \chi_{M^+} = (\varepsilon_M, \infty)$ and $\chi_{M^-} = (-\infty, -\varepsilon_M)$. Figure 5.18 illustrates the mollified behavior of $\bar{e}$ given by (6.23) as compared with (6.21) obtained solely through the minimization of G.

6.4.2 Thermal Evolution

Temperature changes due to convection, conduction, dipole/moment switching or phase changes, and Joule heating can be quantified by the general relation

$$\bar{c}\mathcal{M}\dot{T}(t) = -\Omega[h_c + \lambda/\ell][T(t) - T_E(t)] + J(t) - \sum_\alpha h_\alpha \dot{x}_\alpha. \tag{6.24}$$

Here $\bar{c}, \mathcal{M}, h_c, \Omega, \lambda, \ell$ and $T_E(t)$ respectively denote the specific heat for the actuator material, the mass of the actuator, a convection coefficient, the surface area of the material, the thermal conductivity of the surrounding medium, the interval over which conduction occurs, and the time varying temperature of the adjacent environment. The components $-h_c\Omega[T(t) - T_E(t)]$ and $-\lambda\Omega[T(t) - T_E(t)]/\ell$ respectively quantify the effects of convection and conduction, and $J(t)$ characterizes Joule heating mechanisms. The final component quantifies heat transduction due to dipole/moment switching or phase transitions so that h_α represents specific enthalpies. Details regarding the motivation and construction of constituent components in the relation can be found in Sections 2.6.8, 4.7.6 and 5.5.2.

6.4.3 Unified Macroscopic Framework

The local relations $\overline{e}$ quantify the polarization, magnetization, or strains at the lattice level based on the assumption of homogeneous material properties. For certain single crystal compounds, these local relations can be extrapolated to provide reasonably accurate macroscopic constitutive relations. For general material characterization, however, it is necessary to incorporate the effects of polycrystallinity and variable grain orientations, variations due to texture, nonuniform stress distributions, and variable interaction and effective fields. We incorporate these effects by employing the local relations $\overline{e}$ as kernels in representations based on the assumption that certain local properties are manifestations of underlying distributions rather than constants. Stochastic homogenization in this manner provides a unified framework for material characterization, hybrid transducer design, and unified control design for systems employing ferroelectric, ferromagnetic or ferroelastic actuators and sensors.

Let $\varphi_c = E_c, H_c$ or σ_R denote general coercive or relative fields and let $\varphi_I = E_I, H_I, \sigma_I$ denote interaction fields. The behavior of φ_c and φ_I is quantified by the densities ν_1 and ν_2. To satisfy positivity, symmetry, and integrability conditions, we assume that $\nu_1 \geq 0$ and $\nu_2 \geq 0$ satisfy the relations

$$
\begin{aligned}
&\text{(i)} \quad \nu_1(x) \text{ defined for } x > 0,\\
&\text{(ii)} \quad \nu_2(-x) = \nu_2(x),\\
&\text{(iii)} \quad |\nu_1(x)| \leq c_1 e^{-a_1 x} \quad , \quad |\nu_2(x)| \leq c_2 e^{-a_2 |x|},
\end{aligned}
$$

for positive c_1, a_1, c_2, a_2. Furthermore, we let $\overline{e} = \overline{P}, \overline{M}, \overline{\varepsilon}$ denote the local average polarization, magnetization, or strains defined in (6.19), (6.21), (6.22) or (6.23). Initial dipole, moment or strain configurations are designated by ξ.

The hysteresis and constitutive nonlinearities inherent to ferroic compounds can then be quantified by the general relation

$$
\begin{aligned}
[e(\varphi)](t) &= \int_0^\infty \int_{-\infty}^\infty \nu_1(\varphi_c)\nu_2(\varphi_I)[\overline{e}(\varphi + \varphi_I; \varphi_c, \xi)](t)\, d\varphi_I\, d\varphi_c \\
&= \int_0^\infty \int_{-\infty}^\infty \nu(\varphi_c, \varphi_I)[\overline{e}(\varphi + \varphi_I; \varphi_c, \xi)](t)\, d\varphi_I\, d\varphi_c.
\end{aligned}
\tag{6.25}
$$

Comparison with (6.9) gives an initial illustration of how the framework provides an energy basis for certain extended Preisach models as discussed in Section 4.7.10.

As detailed in Sections 2.6.4, 4.7.2, and 5.5.3, ν_1, ν_2 or the joint density $\nu = \nu_1 \cdot \nu_2$, can be treated as unparameterized densities to provide high accuracy characterization or as parameterized functional forms to facilitate parameter identification and model updating. For the latter option, representations which have been employed for a number of materials are

$$
\begin{aligned}
\nu_1(\varphi_c) &= c_1 e^{-[\ln(\varphi_c/\overline{\varphi}_c)/2c]^2}\\
\nu_2(\varphi_I) &= c_2 e^{-\varphi_I^2/2b^2}
\end{aligned}
$$

where the lognormal mean and variance satisfy the properties

$$\langle \varphi_c \rangle \approx \overline{\varphi}_c \quad , \quad \sigma \approx 2\overline{\varphi}_c c \tag{6.26}$$

if c is small compared with $\overline{\varphi}_c$. When combined with properties of the normal density, (6.26) provides a means of correlating model parameters with quantitative and qualitative properties of the data.

6.4.4 Model Attributes

The combination of (6.25) with the thermal evolution relation (6.24) provides a highly comprehensive characterization framework for a broad range of ferroic materials. Implementation techniques and numerous properties of the framework have been elucidated in Sections 2.6, 4.7 and 5.5 and we refer the reader to those sections for additional details and experimental validation examples illustrating attributes of the framework. Likewise, the reader is referred to Section 4.7.10 for details regarding the manner through which the homogenized energy framework provides an energy basis for various extended Preisach representations — the ramifications of this bear repeating, however, since the correlation of the two frameworks serves to unify the physical and mathematical understanding of ferroic material behavior. Finally, we note that this theory is still young and present investigations are focused on a number of facets which will continue to expand its breadth for smart material characterization, design and control applications.

Chapter 7

Rod, Beam, Plate and Shell Models

Chapters 2–6 focus on the development of models which characterize both the approximately linear low drive behavior and the nonlinear and hysteretic high drive properties of ferroelectric, relaxor ferroelectric, ferromagnetic and shape memory alloy compounds. In this chapter, we employ the linear and nonlinear constitutive relations to construct distributed models for wire, rod, beam, plate and shell-like structures arising in smart material applications. To motivate issues associated with model development, we summarize several applications detailed in Chapter 1 in terms of these five structural classes.

Shells

Shells comprise the most general structural class that we consider and actually subsume the other material classes. A fundamental attribute of shell-like structures is the property that in-plane and out-of-plane motion are coupled due to curvature. This adds a degree of complexity and yields systems of coupled equations in resulting models.

Several applications from Chapter 1 which exhibit shell behavior are summarized in Figure 7.1. The cylindrical actuator employed as an AFM stage is wholly comprised of PZT whereas the cylindrical shell employed as a prototype for noise control in a fuselage is constructed from aluminum with surface-mounted PZT patches utilized as actuators and possible sensors. Whereas both involve cylindrical geometries, the latter requires that models incorporate the piecewise inputs and changes in material properties associated with the patches. The THUNDER transducer and SMA-driven chevron involve more general shells having noncylindrical reference surfaces. THUNDER transducers constructed with wide PZT patches have a doubly-curved final geometry due to the mismatch in thermal properties of the PZT and steel or aluminum backing material. Within the region covered by the patch, the device exhibits an approximately constant radius of curvature in the coordinate directions whereas the uncovered tabs remain flat. The geometry of the chevron is even more complex and is ultimately governed by the design of the underlying jet engine.

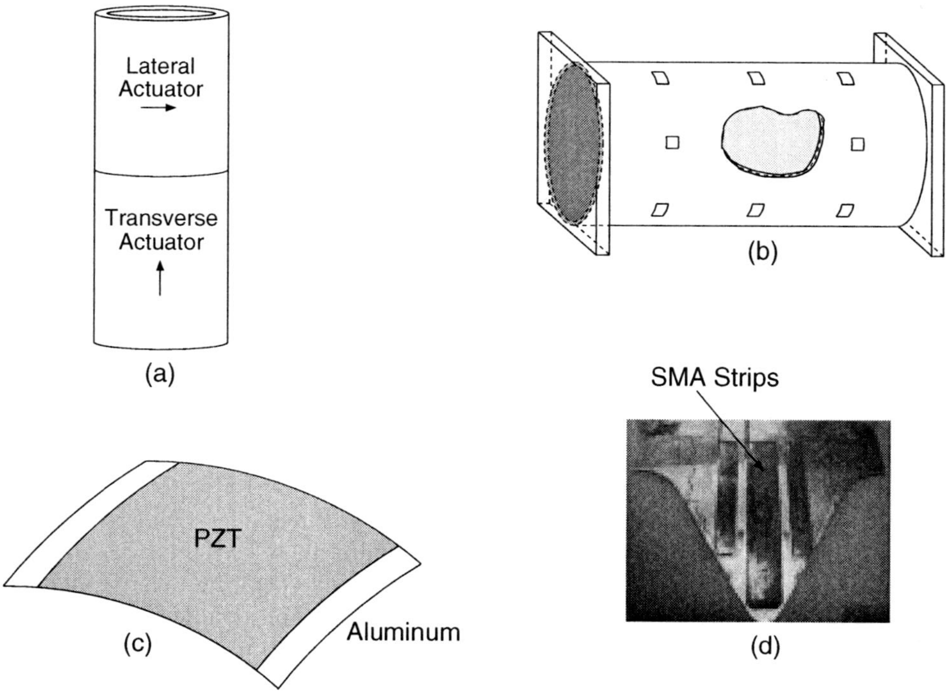

Figure 7.1. *(a) Cylindrical PZT actuator employed for nanopositioning in an atomic force microscope (AFM). (b) Structural acoustic cavity used as a prototype for noise control in a fuselage. (c) THUNDER transducer considered for flow control, synthetic jets and high speed valve design. (d) SMA-driven chevron employed to reduce jet noise and decrease drag.*

For the drive levels employed in the structural acoustic application, linear approximations to the E-ε behavior prove sufficiently accurate and models are constructed using the linear constitutive relations developed in Section 2.2. Present AFM designs with cylindrical stages also use linear constitutive relations with robust feedback laws employed to mitigate hysteresis and creep. This proves successful at low drive frequencies but the push to very high drive frequencies for applications involving real-time product diagnostics or biological monitoring has spawned research focused on model-based control design in a manner which accommodates the inherent hysteresis. Finally, the nonlinear and hysteretic behavior illustrated in Figures 1.6 and 1.23 demonstrate that nonlinear models are required to achieve the high drive capabilities of THUNDER transducers and SMA-drive chevrons.

Plates

Plates can be interpreted as shells having infinite radius of curvature — equivalently, zero curvature — and hence they comprise a special class of shell structures. Thus plate models can be employed as an approximation for shells when the curvature is negligible or for characterizing inherently flat structures whose width is

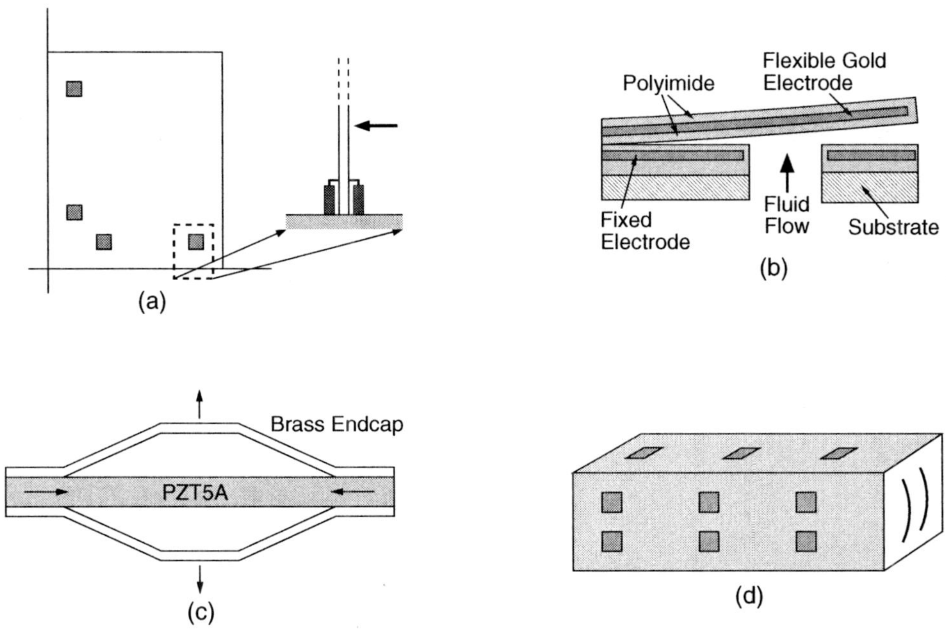

Figure 7.2. *(a) Control of a plate using Terfenol-D transducers as a prototype for general vibration control. (b) Cross-section of the MEMs actuator depicted in Figure 1.27 for microfluidic control and (c) cross-section of the PZT cymbal actuator depicted in Figure 1.7. (d) PZT patches employed for attenuating structure-borne noise in a duct.*

significant compared with the length. For flat plate structures that are symmetric through the thickness, in-plane and out-of-plane motion are inherently decoupled which simplifies both the formulation and approximation of resulting models.

Several smart material applications involving plate-like structures are depicted in Figure 7.2. Because plates incorporate 2-D behavior while avoiding curvature-induced coupling between in-plane and out-of-plane motion, they provide an intermediate level prototype for formulating and testing vibration reduction or control strategies as depicted for magnetostrictive transducers in Figure 7.2(a). The MEMs and cymbal actuators in (b) and (c) typically have widths that are significant when compared with the length and hence exhibit plate-like dynamics. The structural acoustic system depicted in Figure 7.2(d) is analogous to its cylindrical counterpart in Figure 7.1(a) and is employed as a prototype for flat ducts.

As with shells, these applications involve PZT, Terfenol-D, and potentially PMN and SMA, operating in both linear and highly nonlinear and hysteretic regimes. It will be shown in subsequent sections that the same kinematic equations can be employed in both cases, with the linear or nonlinear constitutive behavior incorporated through the models developed in Chapters 2–6.

Membranes

Membranes are a special case of shell or plate constructs in which stiffness effects are approximated in various senses or are considered negligible. Hence the resulting models are generalized 2-D analogues of familiar 1-D string models.

Due to their thinness, several of the semicrystalline, amorphous, and ionic polymers discussed in Section 1.5 yield structures that exhibit membrane behavior. To illustrate, consider the use of ionic polymers for biological or chemical detection or PVDF for membrane mirror design as depicted in Figure 7.3. A third example is provided by the SMA films and membranes discussed in Section 1.4 for use in MEMs and biomedical applications. In all three cases, membrane models which incorporate constitutive nonlinearities and hysteresis are necessary for device characterization. It is expected that as the focus on polymers and SMA thin films continues to grow, an increasing number of smart material systems will be characterized by linear and nonlinear membrane models.

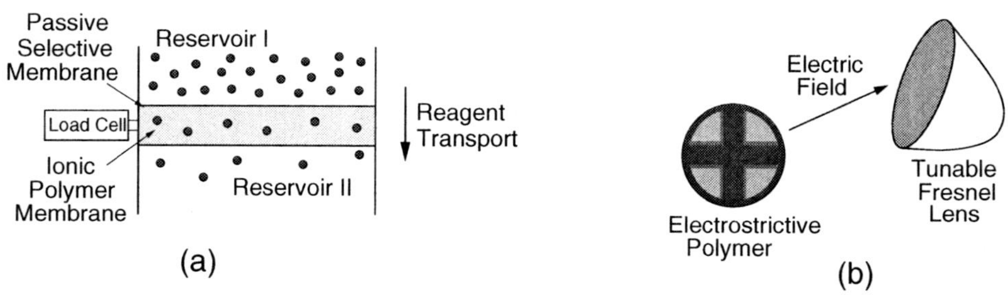

Figure 7.3. *(a) Chemical detection using chemical-specific permeable ionic polymer membranes. (b) Membrane mirror constructed from PVDF.*

Beams

Beams comprise a subset of shells and plates whose widths are small compared with lengths. This permits motion in the width direction to be neglected which reduces the dimensionality of models.

Some smart material applications involving flat and curved beam dynamics are depicted in Figure 7.4. The thin beam depicted in Figure 7.4(a) provides a theoretical, numerical and experimental prototype for model development and control design as well as a technological prototype for evolving unimorph designs. The polymer unimorph depicted in Figure 7.4(b) is presently being considered for applications ranging from pressure sensing to flow control and it represents a geometry where the reference surface differs from the middle surface [122]. The THUNDER transducer in Figure 7.4(c) exhibits negligible curvature or motion in the width direction and hence is modeled by curved beam relations in the region covered by PZT coupled with a flat beam model for the tabs. As noted in Section 1.5, the electrostrictive MEMs device depicted in Figure 7.4(d) is being investigated for use in electrical relays and switches, optical and infrared shutters, and microfluidic valves.

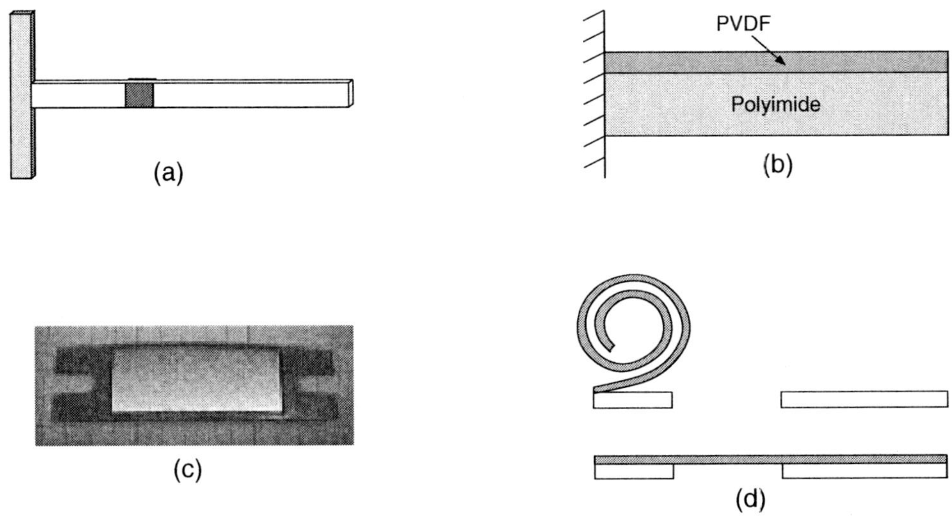

Figure 7.4. *(a) Thin beam with surface-mounted PZT patches employed as a prototype for vibration control. (c) Polymer unimorph comprised of PVDF and polyimide presently considered for pressure sensing and flow control. (c) Curved THUNDER transducer whose width is small compared with the length. (d) Electrostrictive MEMs actuator employed as a high speed shutter.*

As with shells and plates, both linear and nonlinear input behavior must be accommodated in the constitutive relations. Furthermore, both the THUNDER and MEMs actuators can exhibit very large displacements in certain drive regimes. This necessitates consideration of nonlinear kinematic models which incorporate both high-order strain-displacement terms and consider force and moment balancing in the context of the deformed reference line.

Rods

In both beams and rods, motion is considered with respect to the reference or neutral line and hence models are 1-D. The difference is that beams exhibit out-of-plane motion whereas rod dynamics are solely in-plane. From the perspective of model development, beam models are constructed using both moment and force balancing whereas in-plane force balancing is required when constructing rod models. Due to the geometric coupling associated with curved beams, resulting models have a rod component quantifying in-plane dynamics. We summarize here several smart material applications which solely exhibit rod dynamics without the bending (transverse or out-of-plane) motion associated with beams.

PZT, SMA, and magnetostrictive transducers employed in rod configurations are depicted in Figure 7.5. The stacked PZT actuators employed as x- and y-stages in atomic force microscopes (AFM) provide the highly repeatable set point placement required for positioning the sample to within nanometer accuracy. In this configuration, d_{33} or in-plane motion is utilized thus motivating the development of rod models having boundary conditions commensurate with the devise design. As

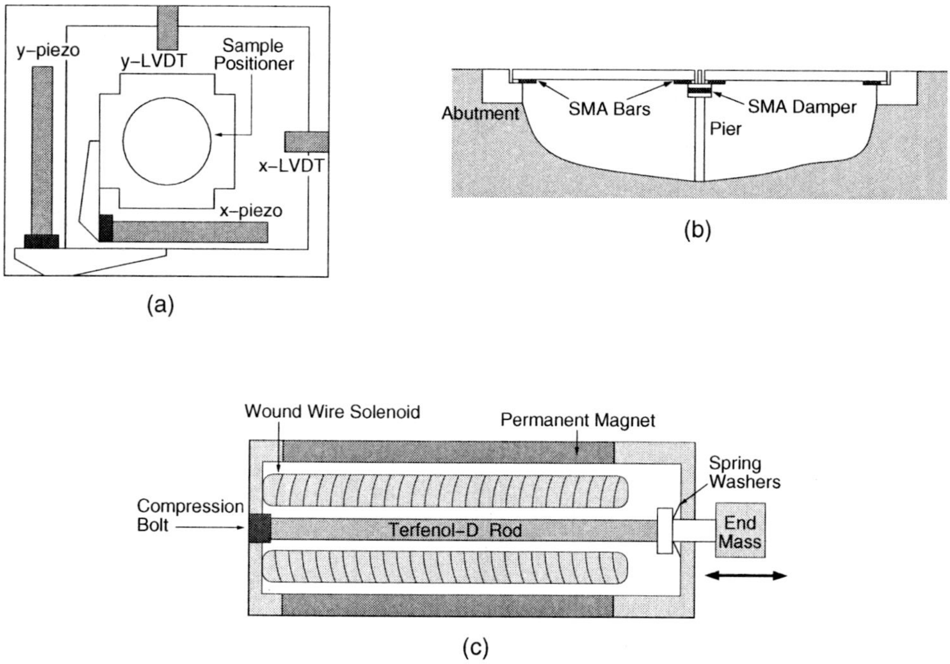

Figure 7.5. *(a) Stacked PZT actuator employed as x- and y-stages in an AFM. (b) SMA bars to reduce lateral displacements in a bridge and (c) cross-section of a magnetostrictive transducer employing a Terfenol-D rod.*

illustrated in Figure 1.10, the field-displacement relation exhibits hysteresis which is incorporated via the constitutive relations developed in Chapter 2.

The SMA rod employed to reduce displacements and vibrations in bridge abutments relies on energy dissipated in the pseudoelastic phase and hence is designed for maximal hysteresis. In this case, the constitutive relations from Chapter 5 are used to quantify the σ-ε behavior when constructing rod models.

Finally, present magnetostrictive transducer designs employ field inputs to a solenoid to rotate moments and produce in-plane motion in a Terfenol-D rod. This produces significant force capabilities but necessitates the use of the constitutive relations developed in Chapter 4 to incorporate the hysteresis and constitutive nonlinear shown in Figure 1.13.

Wires and Tendons

The final structural family that we mention are wires or tendons. Like rods, the wire motion under consideration is due to in-plane forces or stresses. The difference lies in the property that unlike rods, wires maintain their geometry only when subjected to tensile stresses — compressive stresses cause them to crumple in the manner depicted in Figure 5.7.

In present smart material systems, wires or tendons occur primarily in SMA constructs, but there they are very common. Two prototypical examples illustrat-

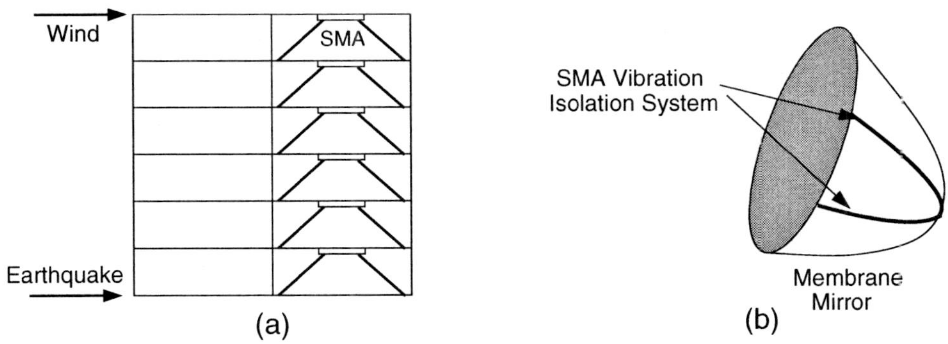

Figure 7.6. *(a) SMA tendons to attenuate earthquake or wind-induced vibrations in a building and (b) SMA tendons for vibration suppression in a membrane mirror.*

ing their use for vibration attenuation in civil or aerospace structures are illustrated in Figure 7.6. In both cases, maximal energy dissipation occurs when the design ensures maximal pseudoelastic hysteresis loops thus necessitating the use of nonlinear constitutive relations when constructing distributed models. As detailed in Section 1.4, SMA wires and tendons exploiting the shape memory effect (SME) are presently employed in numerous biomedical applications including orthodontics and catheters, and are under consideration for a wide range of future biomedical, aeronautic, aerospace and industrial applications. A crucial component necessary for the continued developed of SMA devices is the formulation and efficient numerical approximation of distributed models which accommodate the inherent hysteresis and constitutive nonlinearities.

Model Hierarchies

The cornerstones of distributed wire, rod, beam, plate and shell models are the linear and nonlinear constitutive relations developed in Chapters 2–6 and we summarize these in Section 7.1 as a prelude to subsequent model development. In Section 7.2, we summarize the four assumptions established by Love which provide the basis for constructing linear moment and force relations and strain-displacement relations.

When constructing distributed models for the various structural classes, there are several strategies. The first is to develop the models in a hierarchical manner starting with the simplest case of rods and finishing with shells. Alternatively, one can employ the fact that shell models subsume the other classes and consider first this very general regime — rod, beam and plate models then follow as special cases. The latter strategy emphasizes the unified nature of the development but obscures the details. For clarity, we thus employ a third strategy. We consider the development of rod models in Section 7.3 from both Newtonian and Hamiltonian perspectives. This illustrates the use of the linear and nonlinear constitutive relations from Chapters 2–6 when constructing distributed models from force balance or energy principles. In Sections 7.4 and 7.5, we summarize the development of flat beam and plate models to illustrate the manner through which moment balancing yields fourth-order models. The coupling between in-plane and out-of-plane motion, inherent to curved structures are addressed in Section 7.6 in the context of general shell models. Special cases, which include cylindrical shells and curved beams are addressed in Section 7.7. Additionally, we summarize the manner in which the general shell framework encompasses rod, beam, and plate models. In Section 7.8, we relax the Love criteria to obtain linear Timoshenko and Mindlin-Reissner models and nonlinear von Kármán relations. The chapter concludes with the formulation of an abstract analysis framework in Section 7.9. Numerical approximation techniques for various structural models are presented in Chapter 8.

7.1 Linear and Nonlinear Constitutive Relations

The linear and nonlinear constitutive relations developed in previous chapters provide the basis for incorporating the coupled and typically nonlinear hysteretic behavior inherent to ferroelectric, ferromagnetic and shape memory alloy compounds. We summarize relevant constitutive relations as a prelude to distributed model development in later sections.

7.1.1 Ferroelectric and Relaxor Ferroelectric Materials

We summarize linear constitutive relations developed in Section 2.2 and nonlinear hysteretic relations resulting from the homogenized energy framework of Section 2.6. Additional nonlinear relations resulting from Preisach and domain wall theory can be found in Sections 2.4 and 2.5.

Linear Constitutive Relations

For low drive regimes, linear constitutive relations for 1-D and 2-D geometries were summarized in Section 2.2.5. We summarize the relations for voltage inputs derived through the approximation $V = EL$ where L denotes the distance through which the field is propagated. For d_{31} motion, $L = h$ is the thickness of the actuator whereas $L = \ell$ is the actuator length for d_{33} inputs. Note that linear constitutive relations for alternative input variables can be found in Tables 2.1 and 2.2.

1-D Relations: Beams

Damped linear constitutive relations appropriate for beam models are

$$\begin{aligned} \sigma &= Y\varepsilon + c\dot{\varepsilon} - Y\frac{d_{31}}{h}V \\ P &= Yd_{31}\varepsilon + \chi\frac{V}{h} \end{aligned} \tag{7.1}$$

where Y and c denote the Young's modulus and Kelvin–Voigt damping coefficients and χ is the dielectric susceptibility.

1-D Relations: Rods

Rods employ d_{33} inputs so one employs $\frac{d_{33}}{\ell}$ rather than $\frac{d_{31}}{h}$ in the converse relation.

2-D Relations: General Shells

For general shell models, we let $\varepsilon_\alpha, \sigma_\alpha$ and $\varepsilon_\beta, \sigma_\beta$ denote normal strains and stresses in the α and β directions and let $\varepsilon_{\alpha\beta}, \sigma_{\alpha\beta}$ denote shear strains and stresses. The Poisson ratio is denoted by ν. Linear constitutive relations for this regime are

$$\begin{aligned}
\sigma_\alpha &= \frac{Y}{1-\nu^2}(\varepsilon_\alpha + \nu\varepsilon_\beta) + \frac{c}{1-\nu^2}(\dot{\varepsilon}_\alpha + \nu\dot{\varepsilon}_\beta) - \frac{Yd_{31}}{1-\nu}\frac{V}{h} \\
\sigma_\beta &= \frac{Y}{1-\nu^2}(\varepsilon_\beta + \nu\varepsilon_\alpha) + \frac{c}{1-\nu^2}(\dot{\varepsilon}_\beta + \nu\dot{\varepsilon}_\alpha) - \frac{Yd_{31}}{1-\nu}\frac{V}{h} \\
\sigma_{\alpha\beta} &= \frac{Y}{2(1+\nu)}\varepsilon_{\alpha\beta} + \frac{c}{2(1+\nu)}\dot{\varepsilon}_{\alpha\beta} \\
P &= Yd_{31}\varepsilon + \chi\frac{V}{h}
\end{aligned} \tag{7.2}$$

— see [33] for details. For homogeneous, isotropic materials, electromechanical coupling does not produce significant twisting and hence piezoelectric effects are neglected in the shear relation. Note that d_{33} effects can be incorporated in the manner described for rods.

2-D Relations: Cylindrical Shell and Plates

The relations for cylindrical shells and plates are special cases of (7.2). For cylindrical shells in which x and θ delineate the longitudinal and circumferential coordinates, one employs $\alpha = x$ and $\beta = \theta$. For flat plates, we will use the coordinates $\alpha = x$ and $\beta = y$.

Nonlinear Constitutive Relations

As detailed in Section 2.1, constitutive nonlinearities and hysteresis are inherent to the E-P relation due to dipole rotation and energy dissipation during domain wall movement. Moreover, 90° dipole switching due to certain stress inputs can produce the ferroelastic hysteresis depicted in Figures 2.11 and 2.12. We restrict our discussion to stress levels below the coercive stress σ_c but note that ferroelastic switching must be accommodated in certain high stress regimes — e.g., THUNDER in various configurations exhibits ferroelastic switching. Initial extensions to the theory to incorporate 90° ferroelastic switching are provided in [24].

1-D Relations: Rods and Beams

For poled materials operating about the bias polarization $P_0 = P_R$, extension of (2.135) to include Kelvin–Voigt damping yields the 1-D constitutive relations

$$\begin{aligned}
\sigma &= Y\varepsilon + c\dot{\varepsilon} - a_1(P - P_R) - a_2(P-P_R)^2 \\
[P(E,\varepsilon)](t) &= \int_0^\infty \int_{-\infty}^\infty \nu_1(E_c)\nu_2(E_I)[\overline{P}(E+E_I,\varepsilon;E_c,\xi)](t)\,dE_I\,dE_c
\end{aligned} \tag{7.3}$$

where ν_1 and ν_2 are densities satisfying the conditions (2.113). For moderate strain levels, the kernel $\overline{P}$ is given by (2.89), (2.90) or (2.99) whereas the relations in Section 2.6.9 can be employed if strains are significant. The elastic constitutive relation incorporates both linear piezoelectric and quadratic electrostrictive effects and hence characterizes a broad range of ferroelectric and relaxor ferroelectric behavior. Furthermore, the coefficients a_1 and a_2 can be chosen to incorporate either the longitudinal or transverse inputs analogous to d_{33} or d_{31} inputs in linear regimes. Finally, we note that one can employ more general bias polarizations P_0, including $P_0 = 0$, if operating about points other than the remanence.

Remark 7.1.1. *The inclusion of strain behavior in the polarization model yields nonlinear stress-strain relations and hence will yield distributed models having a nonlinear state-dependence. For actuator applications, the strain-dependence in $\overline{P}$ and hence P is typically small compared with the field-dependence and is generally neglected — this yields constitutive relations and distributed models have a linear state-dependence but a nonlinear and hysteretic input-dependence. For sensor applications, this direct effect is retained to incorporate the effects of ε, σ on E, P or V.*

2-D Relations: Shells

The development of constitutive relations for shells combines the linear elastic relations (7.2) and nonlinear inputs from (7.3). For $\widetilde{P} = P - P_R$, this yields

$$
\begin{aligned}
\sigma_\alpha &= \frac{Y}{1-\nu^2}(\varepsilon_\alpha + \nu\varepsilon_\beta) + \frac{c}{1-\nu^2}(\dot{\varepsilon}_\alpha + \nu\dot{\varepsilon}_\beta) - \frac{1}{1-\nu}\left[a_1\widetilde{P} + a_2\widetilde{P}^2\right] \\
\sigma_\beta &= \frac{Y}{1-\nu^2}(\varepsilon_\beta + \nu\varepsilon_\alpha) + \frac{c}{1-\nu^2}(\dot{\varepsilon}_\beta + \nu\dot{\varepsilon}_\alpha) - \frac{1}{1-\nu}\left[a_1\widetilde{P} + a_2\widetilde{P}^2\right] \\
\sigma_{\alpha\beta} &= \frac{Y}{2(1+\nu)}\varepsilon_{\alpha\beta} + \frac{c}{2(1+\nu)}\dot{\varepsilon}_{\alpha\beta} \\
[P(E,\varepsilon)](t) &= \int_0^\infty \int_{-\infty}^\infty \nu_1(E_c)\nu_2(E_I)[\overline{P}(E+E_I,\varepsilon;E_c,\xi)](t)\,dE_I\,dE_c
\end{aligned}
\tag{7.4}
$$

where $\alpha = x, \beta = \theta$ for cylindrical shells and $\alpha = x, \beta = y$ for flat plates.

7.1.2 Ferromagnetic Materials

The development of constitutive relations for ferromagnetic materials is analogous to that for ferroelectric compounds and we summarize here only the 1-D relations employed for rod models.

Linear Constitutive Relations

Linear constitutive relations formulated in terms of the input variable pair (ε, H) can be obtained by posing the elastic relation in (4.23) as a function of ε

or by employing a magnetic Gibbs energy relation analogous to the electric Gibbs energy in Table 2.1 of Section 2.2. Inclusion of Kelvin–Voigt damping yields

$$\begin{aligned} \sigma &= Y\varepsilon + c\dot{\varepsilon} - aM \\ M &= Yd_{31}\varepsilon + \chi H \end{aligned} \tag{7.5}$$

where χ is the magnetic susceptibility. These piezomagnetic relations should be employed only in low to moderate drive regimes where hysteresis and quadratic magnetostrictive effects are negligible.

Nonlinear Constitutive Relations

For the homogenized energy model, incorporation of Kelvin–Voigt damping, operation about a bias magnetization M_0 — which can be the remanence value M_R — and inclusion of linear σ-M behavior in (4.96) yields the constitutive relations

$$\begin{aligned} &\sigma = Y\varepsilon + c\dot{\varepsilon} - a_1(M - M_0) - a_2(M - M_0)^2 \\ &[M(H)](t) = \int_0^\infty \int_{-\infty}^\infty \nu_1(H_c)\nu_2(H_I)[\overline{M}(H + H_I; \varepsilon, H_c, \xi)](t)\, dH_I\, dH_c. \end{aligned} \tag{7.6}$$

Here ξ denotes the initial moment distribution and the kernel $\overline{M}$ is given by (4.71), (4.72) or (4.78). As noted in Remark 7.1.1, the general kernel depends on ε, thus producing nonlinear constitutive relations and nonlinear rod models. For actuator models, this direct effect can be neglected since it is small compared with the field-dependence.

We note that if employing the Preisach or Jiles–Atherton models, one would replace the H-M model in (7.6) by (4.34) or (4.62).

7.1.3 Shape Memory Alloys

Like ferroelectric and ferromagnetic compounds, the constitutive behavior of shape memory alloys can be characterized through a number of techniques including high-order polynomials which quantify the inherent first-order transition behavior, Preisach models, domain wall theory, and homogenized free energy theory. The use of polynomial-based stress-strain relations to derive a 1-D distributed model for an SMA rod was illustrated in Section 5.2.1 with details given in [57]. We summarize here the macroscopic homogenized energy relations from Section 5.5 and we refer the reader to Chapter 5 for details regarding the other theories.

For densities ν_1 and ν_2 satisfying the decay criteria (5.27), the dependence of strains on stresses and temperature is quantified by (5.26),

$$[\varepsilon(\sigma, T)](t) = \int_0^\infty \int_{-\infty}^\infty \nu_1(\sigma_R)\nu_2(\sigma_I)[\overline{\varepsilon}(\sigma + \sigma_I, T; \sigma_R, \xi)](t)\, d\sigma_I d\sigma_R, \tag{7.7}$$

where $\sigma_R = \sigma_M - \sigma_A$ denotes the relative stress and the kernel $\overline{\varepsilon}$ is given by (5.15) or (5.20). The temperature evolution is governed by (5.21).

For a number of 1-D applications, (7.7) can be directly employed to characterize the pseudoelastic behavior and shape memory effects inherent to SMA wires and rods. For applications in which SMA is employed an actuator or is coupled to an adjacent structure, the relation quantifies the nonlinear and hysteretic constitutive behavior in a manner which can be coupled with structural constitutive relations to construct system models.

7.2 Linear Structural Assumptions

Whereas the input-dependence is often nonlinear and hysteretic as characterized by the constitutive relations, classical theory can often be employed when balancing forces and moments, and constructing the strain-displacement relations employed in distributed models. We summarize here four assumptions established by Love which form the foundation of classical shell theory [301] — and hence are fundamental for the subclasses of rods, beams and plates. Relaxation of these assumptions yields the coupled and nonlinear models summarized in Section 7.8.

1. *The shell thickness h is small compared with the length ℓ and radius of curvature R.* This permits the development of thin shell models and encompasses a broad range of civil, aerospace, aeronautic, industrial and biomedical structures and devices. As detailed in [145,364], this criterion is generally satisfied if $h/R < \frac{1}{20}$ to $\frac{1}{10}$.

2. *Small deformations.* For small deformations, higher powers in strain-displacement relations can be neglected and kinematic and equilibrium conditions are developed in relation to the unperturbed shell neutral surface. This condition may not hold for large displacements of the type depicted in Figure 1.29 and 7.4 for an electrostatic MEMs actuator. Relaxation of this condition yields the nonlinear von Kármán model summarized in Section 7.8.

3. *Transverse normal stresses σ_z are negligible compared to the normal stresses $\sigma_\alpha, \sigma_\beta$.* As detailed in [292], this assumption leads to certain contradictions regarding the retention of stresses but yields models which provide reasonable accuracy for a wide range of applications.

4. *Lines originally normal to the reference or neutral surface remain straight and normal during deformations as depicted in Figure 7.7(a).* This is referred to as the Kirchhoff hypothesis and is a generalization of the Euler hypothesis for thin beams which asserts that plane sections remain plane. For coupled in-plane and out-of-plane motion, this implies that strains ε at a point z in the thickness direction can be expressed as

$$\varepsilon = e + \kappa(z - z_n) \tag{7.8}$$

   where z_n denotes the position of the neutral surface and e, κ are the in-plane strain and curvature changes at the neutral surface as depicted in Figure 7.8. For moderate to thick structures, the relaxation of this hypothesis yields the Timoshenko beam model and Mindlin–Reissner plate model which include rotational effects and shear deformation.

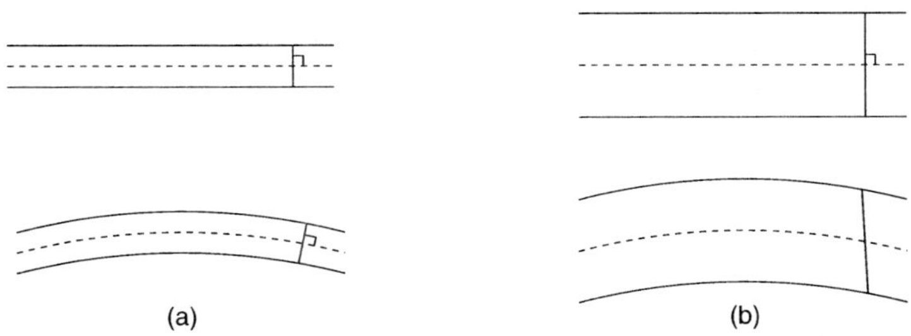

Figure 7.7. *Behavior of normal lines to the neutral surface during bending. (a) Lines remain normal in thin structures in accordance with Assumption 4 and (b) non-normal response in thick structures due to transverse shear strains.*

Remark 7.2.1. *Through Assumptions 3 and 4, the second-order 3-D elasticity problem is reduced to a 2-D problem formulated in terms of a reference or neutral surface. This yields fourth-order models for the transverse motion and leads to an imbalance with the in-plane relations which remain second-order. However, the efficiency gained by reducing dimensions typically dominates the added complexity associated with approximating the fourth-order relations in weak form.*

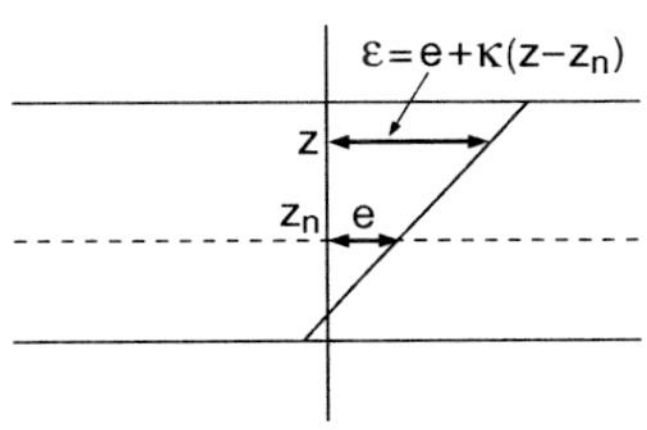

Figure 7.8. *Strain profile posited by Assumption 4 and comprised of an in-plane component* e *and bending component* κz.

7.3 Rod Models

To illustrate the construction of distributed models from both Newtonian and Hamiltonian principles, we consider first models which quantify the in-plane dynamics of the rod structures depicted in Figure 7.5. A prototypical geometry comprised of a homogeneous rod of length ℓ and cross-sectional area A is shown in Figure 7.9. The density, Young's modulus and Kelvin–Voigt damping coefficient are denoted by ρ, Y and c.[29] The longitudinal displacement in the x-direction and distributed force per unit length are denoted by u and f. Finally, the end at $x = 0$ is considered fixed whereas we consider a mass m_ℓ and boundary spring with stiffness k_ℓ and

[29] From Tables 1.1 and 4.1 on pages 28 and 165, it is noted that representative Young's moduli for PZT and Terfenol-D are 71 GPa and 110 GPa whereas representative densities are 7600 kg/m^3 and 9250 kg/m^3. However, parameter values for a specific device, including damping coefficients, are typically estimated through a least squares fit to data.

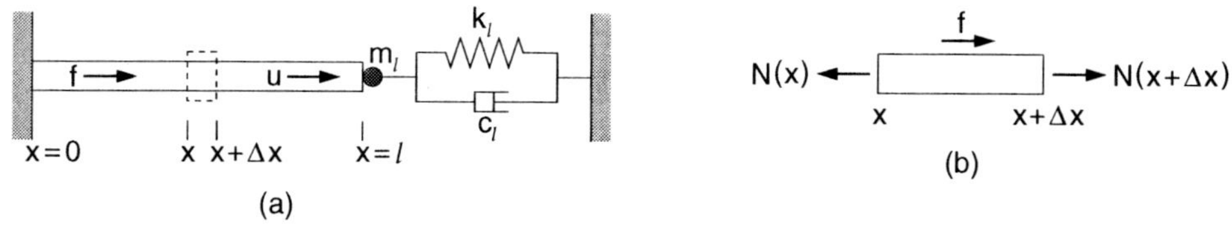

Figure 7.9. *(a) Rod of length ℓ and cross-sectional area A with a fixed end at $x = 0$ and energy dissipating boundary conditions at $x = \ell$. (b) Infinitesimal element considered when balancing forces.*

damping coefficient c_ℓ at $x = \ell$. The latter incorporates the energy dissipation and mass associated with prestress mechanisms and loads in a Terfenol-D transducer or elastic mechanisms connected to AFM stages.

7.3.1 Newtonian Formulation

To quantify the dynamics of the rod, we consider a representative infinitesimal element $[x, x + \Delta x]$ as depicted in Figure 7.9(b). In-plane force resultants are denoted by $N(t, x)$ and $N(t, x + \Delta x)$ where

$$N(t,x) = \int_A \sigma \, dA = \sigma(t,x)A \tag{7.9}$$

since the rod is assumed uniform and homogeneous.

The balance of forces for the element gives

$$\int_x^{x+\Delta x} \rho A \frac{\partial^2 u}{\partial t^2}(t,s)ds = N(t, x+\Delta x) - N(t,x) + \int_x^{x+\Delta x} f(t,s)ds$$

$$\Rightarrow \lim_{\Delta x \to 0} \frac{1}{\Delta x} \int_x^{x+\Delta x} \rho A \frac{\partial^2 u}{\partial t^2}(t,s)ds = \lim_{\Delta x \to 0} \frac{N(t,x+\Delta x) - N(t,x)}{\Delta x}$$

$$+ \lim_{\Delta x \to 0} \frac{1}{\Delta x} \int_x^{x+\Delta x} f(t,s)ds$$

which yields

$$\rho A \frac{\partial^2 u}{\partial t^2} = \frac{\partial N}{\partial x} + f \tag{7.10}$$

as a strong formulation of the model. The resultant is evaluated using (7.9) with σ specified by the various linear and nonlinear constitutive relations summarized in Section 7.1.

A necessary step when evaluating these relations is to relate in-plane strains ε and the longitudinal displacements u. For the geometry under consideration, the relation follows directly from the definition of the strain as the displacement relative to the initial length of an infinitesimal element; hence

$$\varepsilon = \lim_{\Delta x \to 0} \frac{\Delta u}{\Delta x} = \frac{\partial u}{\partial x}. \tag{7.11}$$

Boundary and Initial Conditions

It follows from the assumption of a fixed-end condition at $x = 0$ that

$$u(t, 0) = 0. \tag{7.12}$$

The balance of forces at $x = \ell$, in the manner detailed in [120], yields the second boundary condition

$$N(t, \ell) = -k_\ell u(t, \ell) - c_\ell \frac{\partial u}{\partial t}(t, \ell) - m_\ell \frac{\partial^2 u}{\partial t^2}(t, \ell). \tag{7.13}$$

Note that this energy-dissipating boundary condition reduces to the free-end condition

$$N(t, \ell) = 0$$

in the absence of an end mass and damped, elastic restoring force. Moreover, it is observed that if one divides by k_ℓ and takes $k_\ell \to \infty$ to model an infinite restoring force, the dissipative boundary condition (7.13) converges to the fixed-end condition (7.12). The boundary conditions can thus be summarized as

$$\begin{aligned} &u(t, 0) = 0 \\ &N(t, \ell) = -k_\ell u(t, \ell) - c_\ell \frac{\partial u}{\partial t}(t, \ell) - m_\ell \frac{\partial^2 u}{\partial t^2}(t, \ell). \end{aligned}$$

Finally, initial conditions are specified to be

$$\begin{aligned} &u(0, x) = u_0(x) \\ &\frac{\partial u}{\partial t}(0, x) = u_1(x). \end{aligned}$$

Strong Formulation of the Model

We summarize here rod models for stacked PZT actuators operating in linear and nonlinear input regimes with constitutive behavior quantified by (7.1) and (7.3). The magnetic models are completely analogous and follow directly form the constitutive relations (7.5) and (7.6).

PZT Rod Model — Linear Inputs

$$\begin{aligned} &\rho A \frac{\partial^2 u}{\partial t^2} - YA \frac{\partial^2 u}{\partial x^2} - cA \frac{\partial^3 u}{\partial x^2 \partial t} = f - YA \frac{d_{31}}{h} \frac{\partial V(t)}{\partial x} \\ &P = Y d_{31} \frac{\partial u}{\partial x} + \chi \frac{V}{h} \end{aligned} \tag{7.14}$$

PZT Rod Model — Hysteretic and Nonlinear Inputs

$$\rho A \frac{\partial^2 u}{\partial t^2} - YA\frac{\partial^2 u}{\partial x^2} - cA\frac{\partial^3 u}{\partial x^2 \partial t} = f - a_1 A \frac{\partial(P(t) - P_R)}{\partial x} - a_2 A \frac{\partial(P(t) - P_R)^2}{\partial x}$$

$$[P(E, \frac{\partial u}{\partial x})](t) = \int_0^\infty \int_{-\infty}^\infty \nu_1(E_c)\nu_2(E_I)[\overline{P}(E + E_I, \frac{\partial u}{\partial x}; E_c, \xi)](t)\, dE_I\, dE_c. \tag{7.15}$$

In the polarization relation, the densities ν_1 and ν_2 satisfy the conditions (2.113) with a possible choice given by (2.117). The kernel $\overline{P}$ is given by (2.89), (2.90) or (2.99). As detailed in Remark 7.1.1, the strain-dependence in the polarization is typically neglected in actuator models but may need to be retained for sensor characterization.

Weak Formulation of the Model

The strong formulation of the model, derived via force balancing or Newtonian principles, illustrates in a natural manner the forced dynamics of the rod. However, it has two significant disadvantages from the perspective of approximation. First, the second derivatives in x necessitate the use of cubic splines, cubic Hermite elements, or high-order difference methods to construct a semi-discrete system. Secondly, the neglect of direct electromechanical/magnetomechanical effects to create a linear model in u leads to spatial derivatives of *spatially invariant* voltage and polarization terms $V(t)$ and $P(t)$. This produces a Dirac distribution at $x = \ell$ which will curtail the convergence of modal methods applied to the strong formulation of the model.

Both problems can be alleviated by considering a weak or variational formulation of the model developed either via integration by parts or Hamiltonian energy principles as summarized in Section 7.3.2. We emphasize that the designation "weak form" refers to the fact that underlying assumptions regarding differentiability are weakened in the sense of distributions rather than indicating a form having diminished utility. Conversely, the energy basis provided by the Hamiltonian formulation, in combination with the fact that reduced differentiability requirements make the weak form a natural setting for numerical approximation, imbues the weak model formulation with broader applicability than the strong formulation in a number of applications.

To construct a weak formulation of the model via integration by parts, we consider states $\xi(t) = (u(t,\cdot), u(t,\ell))$ in the state space

$$X = L^2(0,\ell) \times \mathbb{R}$$

with the inner product

$$\langle \Phi_1, \Phi_2 \rangle_X = \int_0^\ell \rho A \phi_1 \phi_2 dx + m_\ell \varphi_1 \varphi_2 \tag{7.16}$$

where $\Phi_1 = (\phi_1, \varphi_1), \Phi_2 = (\phi_2, \varphi_2)$ with $\varphi_1 = \phi_1(\ell)$, $\varphi_2 = \phi_1(\ell)$. Test functions ϕ are required to satisfy the essential boundary condition (7.12) at $x = 0$ but not the natural condition (7.13) at $x = \ell$ so the space of test functions is taken to be

$$V = \left\{\Phi = (\phi, \varphi) \in X \,|\, \phi \in H^1(0,\ell), \phi(0) = 0, \phi(\ell) = \varphi\right\}$$

with the inner product

$$\langle \Phi_1, \Phi_2 \rangle_V = \int_0^\ell YA\phi_1'\phi_2' dx + k_\ell \varphi_1 \varphi_2. \tag{7.17}$$

Consider the general relation (7.10). Multiplication by $\phi \in H_0^1(0,\ell) = \{\phi \in H^1(0,\ell) \,|\, \phi(0) = 0\}$ and integration by parts in space yields the weak form

$$\int_0^\ell \rho A \frac{\partial^2 u}{\partial t^2} \phi dx + \int_0^\ell N \frac{d\phi}{dx} dx - N(t,\ell)\phi(\ell) = \int_0^\ell f\phi dx$$

where $N(t,\ell)$ is given by (7.13). For nonlinear and hysteretic inputs, the weak formulation of the model is thus

$$\begin{aligned}
&\int_0^\ell \rho A \frac{\partial^2 u}{\partial t^2} \phi dx + \int_0^\ell \left[YA \frac{\partial u}{\partial x} + cA \frac{\partial^2 u}{\partial x \partial t} \right] \frac{d\phi}{dx} dx \\
&\quad = \int_0^\ell f\phi dx + A\left[a_1(P - P_R) + a_2(P - P_R)^2\right] \int_0^\ell \frac{d\phi}{dx} dx \\
&\quad - \left[k_\ell u(t,\ell) + c_\ell \frac{\partial u}{\partial t}(t,\ell) + m_\ell \frac{\partial^2 u}{\partial t^2}(t,\ell) \right] \phi(\ell)
\end{aligned} \tag{7.18}$$

which must hold for all $\phi \in V$. The polarization is specified by (7.15) or (2.114).

Equivalent analysis is used to construct the weak formulation of the PZT model with linear inputs or equivalent models for rods in ferromagnetic transducers.

7.3.2 Hamiltonian Formulation

Alternatively, one can employ calculus of variations and fundamental energy relations to derive a weak formulation of the model. This is most easily motivated in the case of conservative forces so we consider initially a regime for which $c = m_\ell = k_\ell = c_\ell = 0$ as well as $F = P = 0$. Hence we consider an elastic rod that is fixed at $x = 0$ and free at $x = \ell$. The space of test functions is $V = H_0^1(0,\ell)$ with the inner product (7.17) employing $k_\ell = 0$.

As detailed in Appendix C, two fundamental energy relations are the Lagrangian

$$\mathcal{L} = K - U$$

and the total energy

$$\mathcal{H} = K + U \tag{7.19}$$

where K and U respectively denote the kinetic and potential energies. It is shown in Section C.3 that for conservative systems, the Hamiltonian — which is the Legendre transform of $\mathcal{L}$ — is exactly the total energy specified in (7.19) thus providing one of the correlations between Lagrangian and Hamiltonian theory.

Lagrangian mechanics, which we will employ here, is based on variational principles — extremals of functionals — whereas Hamiltonian mechanics relies directly on total energy principles. The former leads to natural computational frameworks whereas the the latter provides a basis for developing some of the deeper theoretical results associated with celestial, quantum and statistical mechanics. The combined field of Lagrangian and Hamiltonian mechanics provides one of the pillars of classical physics and we refer the reader to [15,319] for details regarding the fundamental physics and Weinstock [505] for application of Lagrangian theory to elastic systems analogous to that considered here.

The reader is cautioned that terminology can be confusing. For example, Hamilton's principle formulated in terms of the Lagrangian $\mathcal{L}$ is fundamental to Lagrange dynamics, the variational basis for which was discovered by Hamilton [204]!

For the rod, the kinetic and potential energies are

$$\begin{aligned} K &= \frac{1}{2}\rho A \int_0^\ell u_t^2(t,x)dx \\ U &= \frac{1}{2}A \int_0^\ell \sigma\varepsilon dx = \frac{1}{2}YA \int_0^\ell u_x^2(t,x)dx \end{aligned} \tag{7.20}$$

so that

$$\mathcal{L} = \frac{1}{2}A \int_0^\ell \left[\rho u_t^2 - Y u_x^2\right] dx.$$

The integral of $\mathcal{L}$ over an arbitrary time interval $[t_0, t_1]$,

$$\mathcal{A}[u] = \int_{t_0}^{t_1} \mathcal{L} dt,$$

is termed the *action* or *action integral* and provides the functional at the heart of Hamilton's principle.

Hamilton's Principle

Hamilton's principle can be broadly state in this context as follows: "for the arbitrary time interval $[t_0, t_1]$, the motion u of the rod renders the action integral stationary when compared with all admissible candidates $\widehat{u} = u + \epsilon\Theta$ for the motion." As detailed in Section C.2, this yields the requirement that

$$\left.\frac{d}{d\epsilon}\mathcal{A}[u+\epsilon\Theta]\right|_{\epsilon=0} = 0 \tag{7.21}$$

for all admissible Θ.

To quantify the class of admissible perturbations, consider variations of the form

$$\widehat{u}(t,x) = u(t,x) + \epsilon\eta(t)\phi(x)$$

where η and ϕ satisfy

$$\begin{aligned} &\text{(i)} \quad \eta(t_0) = \eta(t_1) = 0 \\ &\text{(ii)} \quad \phi \in V = H_0^1(0,\ell). \end{aligned} \tag{7.22}$$

The first criterion guarantees that

$$\widehat{u}(t_0, x) = u(t_0, x) \; , \; \widehat{u}(t_1, x) = u(t_1, x),$$

as depicted in Figure 7.10, whereas the second assumption guarantees that $\widehat{u}(t, \cdot) \in H_0^1(0, \ell)$ so that candidates satisfy the essential boundary condition and have sufficient smoothness to permit evaluation of the potential energy.

The condition (7.21) then yields

$$\begin{aligned} 0 &= \int_{t_0}^{t_1} \int_0^\ell [\rho A u_t \Theta_t - Y A u_x \Theta_x] \, dx dt \\ &= -\int_{t_0}^{t_1} \eta(t) \int_0^\ell [\rho A u_{tt} \phi + Y A u_x \phi_x] \, dx dt \end{aligned} \tag{7.23}$$

which implies that

$$\rho A \int_0^\ell u_{tt} \phi dx + Y A \int_0^\ell u_x \phi_x dx = 0 \tag{7.24}$$

for all $\phi \in V$. Integration by parts, in combination with condition (i) of (7.22), was employed in the second step of (7.23).

We first note that (7.24) is identical to (7.18) if one takes $c = P = f = 0$ and $m_\ell = c_\ell = k_\ell = 0$ in the latter formulation. Moreover, if u exhibits the additional smoothness $u(t, \cdot) \in H_0^1(0, \ell) \cap H^2(0, \ell)$, integration by parts yields the strong form (7.14) or (7.15) with the simplifying parameter choices. However, the weakened smoothness requirement $u(t, \cdot) \in H_0^1(0, \ell)$ is natural from an energy perspective and advantageous for approximation.

Secondly, inclusion of the elastic and inertial boundary components k_ℓ, m_ℓ, distributed force f and nonlinear polarization components $a_1(P - P_R)$ and $a_2(P - P_R)^2$ can be accomplished using an augmented action integral

$$\mathcal{A}[u] = \int_{t_0}^{t_1} [K - U + F_{nc}] \, dt \tag{7.25}$$

and *extended Hamilton's principle* as detailed in Section 6-7 of Weinstock [505]. Here F_{nc} directly incorporates the nonconservative distributed force f and linear or nonlinear polarization inputs when low-order strain effects are neglected in the polarization model.

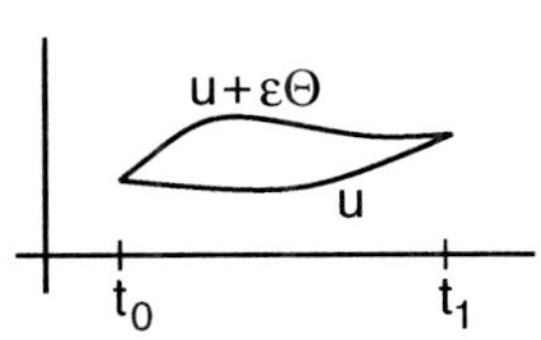

Figure 7.10. *Admissible variations of the motion considered in Hamilton's principle.*

The incorporation of Kelvin–Voigt and boundary damping is more difficult in the variational formulation since they involve derivatives of the displacement which constitutes a generalized coordinate. Hence the incorporation of nonconservative internal damping provides an example of when integration of the strong formulation obtained through force balancing proves an easier strategy for obtaining a weak formulation of the model than direct application of variational principles. Even in this case, however, the consideration of energy or variational principles provides the natural function spaces for constructing the weak formulation and developing approximation techniques as detailed in Section 8.2.

7.3.3 Device Characterization

We illustrate here the performance of the rod model (7.18) for characterizing the displacements shown in Figures 7.11 and 7.12 which were generated by the AFM stage depicted in Figure 7.5. The nonlinear field-polarization relation is characterized by the homogenized energy model (7.15) or (2.114) with general densities ν_1 and ν_2 identified via the parameter estimation techniques detailed in Section 2.6.6. The polarization P_k at each measured field value $E_k = E(t_k)$ was subsequently input to the rod model (7.18) approximated in the manner discussed in Section 8.1.

Figures 7.11 and 7.12 illustrate the data and model fits obtained at four drive levels and four input frequencies. The behavior in Figure 7.11 represents nested mi-

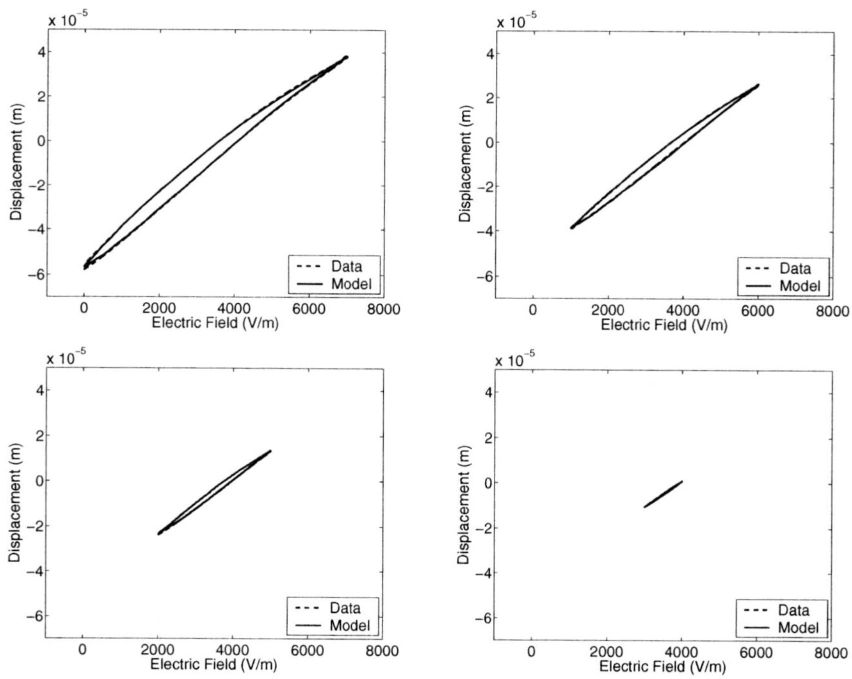

Figure 7.11. *Data and model fit for a stacked PZT actuator employed in the AFM stage depicted in Figure 7.5 at 0.1 Hz.*

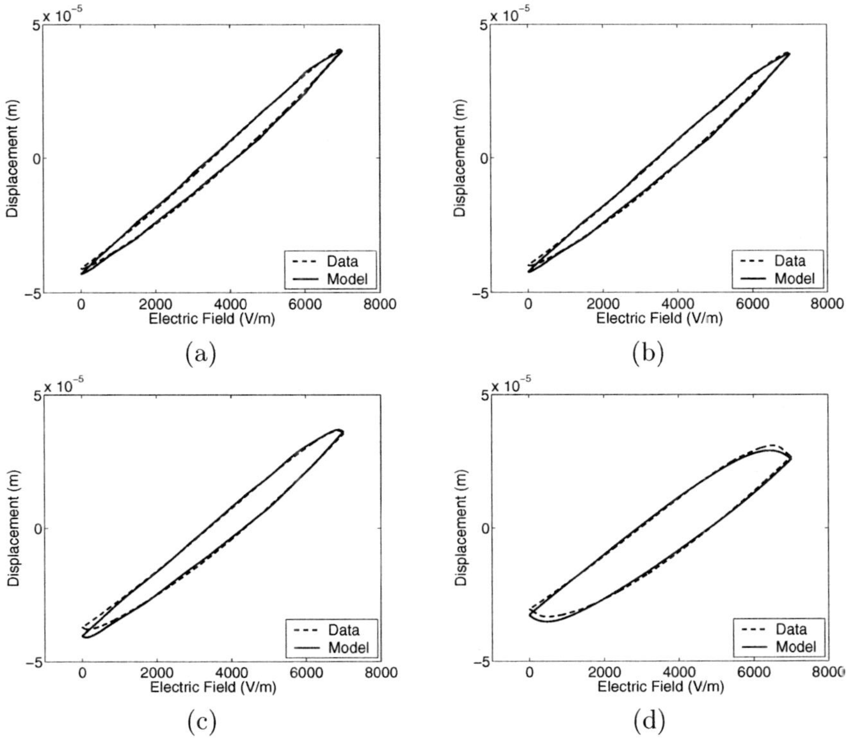

Figure 7.12. *Use of the polarization model (7.15) and rod model (7.18) to characterize the frequency-dependent behavior of a stacked PZT actuator employed in the AFM stage: (a) 0.28 Hz, (b) 1.12 Hz, (c) 5.58 Hz, and (d) 27.9 Hz.*

nor loop behavior which is plotted separately to demonstrate the model's accuracy. Figure 7.12 illustrates that the hysteretic PZT behavior exhibits frequency and rate-dependence even within the 0.1–0.5 Hz range. This necessitates the incorporation of dynamic input behavior — which is one of the hallmarks of the homogenized energy framework — when characterizing and developing model-based control designs for broadband applications. Details regarding the characterization and robust control design for this AFM application can be found in [210].

7.4 Beam Models

Beam models are similar to rod models in the sense that through the assumptions of Section 7.2, they quantify motion as a function of one spatial coordinate. However, beam dynamics are characterized by out-of-plane or transverse motion which necessitates balancing both moments and shear stresses to construct a strong formulation of the model. For homogeneous rods subject to uniform in-plane forces or stresses, any line suffices as 1-D reference line on which to represent dynamics. This is untrue for beams and one typically employs the *neutral line*, characterized by zero stress in pure bending regimes, as the reference line.

To provide prototypes that illustrate a number of the modeling issues associated with beams, plates and shells, we consider the structures depicted in Figure 7.13. The thin beam with surface-mounted patches exhibits effective or homogenized material parameters and piecewise inputs in the region covered by the patches but is simplified by the fact that the reference line and middle line coincide due to symmetry. This is not the case for the *asymmetric* polymer unimorph which motivates its use as a prototype for demonstrating the computation of the reference line as an initial step prior to moment computation.

In both cases, we let w and f respectively denote the transverse displacement and distributed out-of-plane force. The effective linear density (units of kg/m), Young's modulus, and Kelvin–Voigt damping coefficients for the composite structure are denoted by ρ, Y and c whereas material properties for constituent components are delineated by subscripts. Finally, we assume fixed-end conditions at $x = 0$ and free-end conditions at $x = \ell$.

As a point of notation, the thin beam model developed here is referred to as an Euler–Bernoulli model. The Timoshenko model which incorporates shear deformations and rotational inertia is developed in Section 7.8.

7.4.1 Unimorph Model

The unimorph model illustrates a number of issues associated with model development for beams so we consider it first. For simplicity, we frame the discussion in the context of the linear constitutive relations (7.1) and simply summarize the nonlinear input model resulting from (7.3) at the end of the section. Furthermore, while the in-plane and out-of-plane displacements are coupled due to the geometry, we will focus here on uncoupled out-of-plane displacements. The coupling will be discussed in Sections 7.6 and 7.7 in the context of shell, curved beam, and THUNDER models.

The geometric and material properties for the active PVDF layer and inactive polyimide layer are respectively delineated by the subscripts A and I. Both layers are assumed to have width b and the unimorph is assumed to have length ℓ.

Force and Moment Balancing

To establish equations of motion, we balance forces and moments associated with an infinitesimal beam element using the convention depicted in Figure 7.13.[30]

Force Balance

We first balance the forces associated with the shear resultants Q, distributed forces f, and viscous air damping which is assumed proportional to the transverse velocity with proportionality constant γ. Newton's second law then yields

$$\int_x^{x+\Delta x} \rho \frac{\partial^2 w}{\partial t^2}(t,s)ds = Q(t, x+\Delta x) - Q(t,x) + \int_x^{x+\Delta x} \left[f(t,s) - \gamma \frac{\partial w}{\partial t}(t,s) \right] ds$$

[30]We note that the moment and curvature conventions are opposite to those employed by some authors. The association of positive moments with negative curvature is consistent with the convention employed for general shells in Section 7.6 which in turn is consistent with 3-D elasticity relations. Both conventions yield the same final model as long as consistency is maintained.

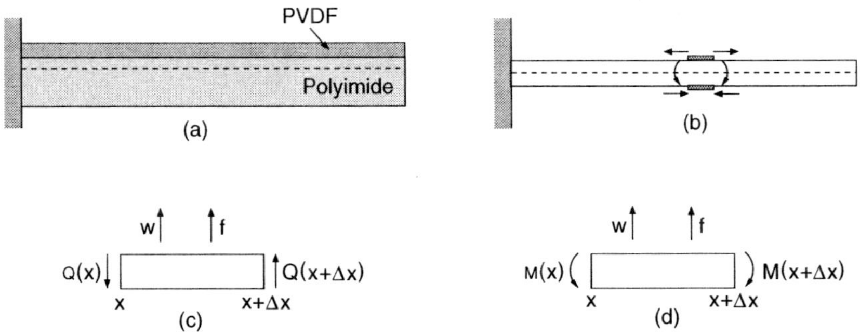

Figure 7.13. *(a) Asymmetric polymer unimorph comprised of an active PVDF layer and an inactive polyimide layer. (b) Cross-section of the beam from Figure 7.4 with symmetric, surface-mounted PZT patches. (c) and (d) Convention for the force and moment results employed when constructing the strong formulation of Euler-Bernoulli beam models.*

where the composite linear density is

$$\rho = h_A b \rho_A + h_I b \rho_I. \tag{7.26}$$

Dividing by Δx and taking $\Delta x \to 0$ yields

$$\rho \frac{\partial^2 w}{\partial t^2} + \gamma \frac{\partial w}{\partial t} = \frac{\partial Q}{\partial x} + f.$$

Moment Balance

We next balance moments about the left end of the element to obtain

$$M(t, x + \Delta x) - M(t, x) - Q(t, x + \Delta x)\Delta x + \int_x^{x+\Delta x} f(t, s)(s - x)dx = 0.$$

The retention of first-order terms after dividing by Δx and taking $\Delta x \to 0$ gives the relation

$$Q = \frac{\partial M}{\partial x} \tag{7.27}$$

relating the moment and shear resultant. This then yields

$$\rho \frac{\partial^2 w}{\partial t^2} + \gamma \frac{\partial w}{\partial t} - \frac{\partial^2 M}{\partial x^2} = f$$

as a strong formulation of the beam model.

Moment Evaluation

To complete the model, it is necessary to formulate the moment M in terms of geometric properties of the unimorph. To accomplish this, we must first determine the reference line which is defined to be the neutral line z_n that exhibits zero stress during bending — recall that through Assumptions 1–4 of Section 7.2, beam motion is defined in terms of the reference line dynamics — thus yielding 1-D models.

Neutral Line Specification

For linear inputs, (7.1) yields

$$\sigma = \begin{cases} Y_A\varepsilon + c_A\dot{\varepsilon} - Y_A \dfrac{d_{31}}{h_A} V & , \text{ Active layer} \\ Y_I\varepsilon + c_I\dot{\varepsilon} & , \text{ Inactive layer} \end{cases} \tag{7.28}$$

under the assumption of Kelvin–Voigt damping — the reader is referred to [122] for a formulation that employs more comprehensive viscoelastic Boltzmann damping relations. As illustrated for the stress profile depicted in Figure 7.14, the moment arm at height z in the unimorph has length $z - z_n$ so the total moment is given by

$$M = \int_{-h_I}^{h_A} b(z - z_n)\sigma \, dz. \tag{7.29}$$

To specify z_n, it is noted that at equilibrium the balance of forces, under Assumption 4 of Section 7.2 which posits a linear strain profile $\varepsilon(z) = \kappa(z - z_n)$ in the absence of in-plane strains, yields

$$\int_{-h_I}^{0} \kappa b Y_I (z - z_n) \, dz + \int_{0}^{h_A} \kappa b Y_A (z - z_n) \, dz = 0. \tag{7.30}$$

This gives the neutral line relation

$$z_n = \frac{Y_A h_A^2 - Y_I h_I^2}{2(Y_A h_A + Y_I h_I)}.$$

Analogous neutral surface representations for PZT-based unimorphs are determined in [295, 393].

Effective Parameters and Moment Components

The stress relation (7.28) has the form

$$\sigma = \sigma_e + \sigma_d + \sigma_{ext}$$

where σ_e, σ_d and σ_{ext} denote the elastic, damping and external components. Similarly, we can decompose the total moment into analogous components

$$M = M_e + M_d + M_{ext}.$$

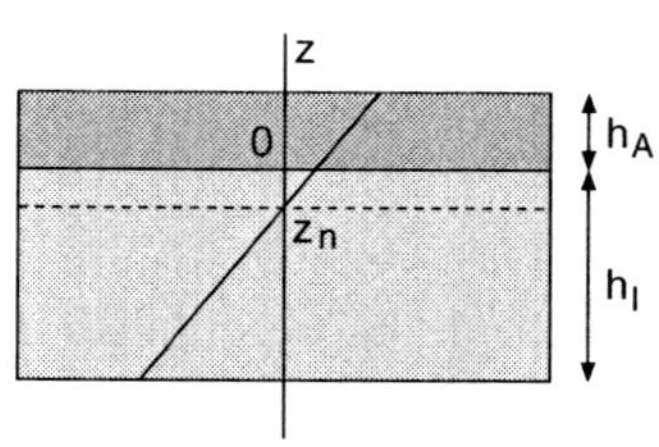

Figure 7.14. *Geometry used to compute the neutral line* z_n.

Since the strategy in thin beam theory is to represent all moments and forces through the thickness of the structure by resultants at the neutral line, it is necessary to specify these resultants either directly, in terms of the geometry and properties of constituent materials, or in terms of effective parameters for the combined structure. The latter approach provides the capability for incorporating material properties that are known (e.g., stiffness properties) while providing a general framework for the identification of unknown parameters (e.g., damping parameters).

We consider first the moment generated by the elastic component σ_e of the constitutive relation (7.28). To determine an effective Young's modulus Y for the composite structure, the general moment is equated to the components,

$$\int_{-h_I}^{h_A} bY\kappa(z-z_n)^2\,dz = \int_{-h_I}^{0} bY_I\kappa(z-z_n)^2\,dz + \int_{0}^{h_A} bY_A\kappa(z-z_n)^2\,dz\,,$$

to yield

$$Y = \frac{Y_I[(h_I+z_n)^3 - z_n^3] + Y_A[(h_A-z_n)^3 + z_n^3]}{(h_A-z_n)^3 + (h_I+z_n)^3}\,. \tag{7.31}$$

For thin beams, the relation

$$\kappa = -\frac{\partial^2 w}{\partial x^2} \tag{7.32}$$

provides a first-order approximation to the change in curvature — see Section 7.6 for details — so the elastic component of the moment is

$$\begin{aligned} M_e &= -\int_{-h_I}^{h_A} bY\frac{\partial^2 w}{\partial x^2}(z-z_n)^2\,dz \\ &= -YI\frac{\partial^2 w}{\partial x^2} \end{aligned} \tag{7.33}$$

where

$$I = \frac{b}{3}[(h_A-z_n)^3 + (h_I+z_n)^3]\,. \tag{7.34}$$

Through (7.31) and (7.34), the effective Young's modulus and generalized moment of inertia for the composite structure can be specified in terms of the geometry and Young's moduli for the constituent materials. Alternatively, the combined parameter YI can be treated as unknown and estimated through a least squares fit to data.

A similar analysis can be employed for the damping component of the moment. However, since values of the damping coefficients for the constituent materials are typically unavailable, we directly consider the moment relation

$$M_d = -cI\frac{\partial^3 w}{\partial x^2 \partial t} \tag{7.35}$$

where the parameter cI is considered unknown and is determined through inverse problem techniques.

Finally, the external moment is given by

$$\begin{aligned} M_{ext} &= -\int_0^{h_A} bY_A(z-z_n)\frac{d_{31}}{h_A}V(t)\,dz \\ &= k_p V(t) \end{aligned} \tag{7.36}$$

where

$$k_p = \frac{bY_A d_{31}}{2h_A}\left[z_n^2 - (h_A - z_n)^2\right]. \tag{7.37}$$

Strong Formulation of the Model with Boundary and Initial Conditions

The fixed-end condition at $x = 0$ enforces zero transverse displacement and slope which yields the boundary condition

$$w(t,0) = \frac{\partial w}{\partial x}(t,0) = 0.$$

Free-end conditions are characterized by the lack of a shear stress or moment; hence use of (7.27) to eliminate the former yields the boundary condition

$$M(t,\ell) = \frac{\partial M}{\partial x}(t,\ell) = 0.$$

Finally, the initial displacements and velocities are defined to be

$$w(0,x) = w_0(x) \quad , \quad \frac{\partial w}{\partial t}(0,x) = w_1(x).$$

The strong formulation of the Euler-Bernoulli model with linear inputs is thus

$$\begin{aligned} &\rho\frac{\partial^2 w}{\partial t^2} + \gamma\frac{\partial w}{\partial t} - \frac{\partial^2 M}{\partial x^2} = f(t,x) \\ &w(t,0) = \frac{\partial w}{\partial x}(t,0) = 0 \\ &M(t,\ell) = \frac{\partial M}{\partial x}(t,\ell) = 0 \\ &w(0,x) = w_0(x) \quad , \quad \frac{\partial w}{\partial t}(0,x) = w_1(x) \end{aligned} \tag{7.38}$$

where ρ is given by (7.26) and $M = M_e + M_d + M_{ext}$ has the elastic, damping and external components defined in (7.33), (7.35) and (7.36).

Weak Formulation of the Model — Linear Inputs

The elastic and damping components M_e and M_d yield fourth-order derivatives in (7.38) whereas differentiation of M_{ext} yields Dirac behavior at $x = \ell$. To avoid ensuing approximation difficulties, it is advantageous to consider a weak or variational formulation of the model developed either through integration by parts or Hamiltonian (energy) principles analogous to those detailed in Section 7.3.2 for the rod model. We summarize the former approach and refer the reader to [34] for details illustrating the construction of a beam model using variational principles.

We consider states $w(t,\cdot)$ in the state space

$$X = L^2(0,\ell)$$

and test functions ϕ in

$$V = H_0^2(0,\ell) = \left\{\phi \in H^2(0,\ell) \,|\, \phi(0) = \phi'(0) = 0\right\}.$$

The inner products

$$\langle \psi, \phi \rangle_X = \int_0^\ell \rho\psi\phi dx$$

$$\langle \psi, \phi \rangle_V = \int_0^\ell YI\psi''\phi'' dx$$

follow from the kinetic and strain (potential) energy components of the variational formulation — e.g., compare the inner products (7.16) and (7.17) for the rod model with the intermediate weak formulation (7.23) derived from the kinetic and potential energy relations (7.20).

Multiplication of (7.38) by test functions $\phi \in V$ and integration by parts yields the weak formulation

$$\int_0^\ell \rho \frac{\partial^2 w}{\partial t^2}\phi dx + \int_0^\ell \gamma \frac{\partial w}{\partial t}\phi dx - \int_0^\ell M\frac{d^2\phi}{dx^2} dx = \int_0^\ell f\phi dx$$

or

$$\begin{aligned}\int_0^\ell \rho \frac{\partial^2 w}{\partial t^2}\phi \, dx + \int_0^\ell \gamma \frac{\partial w}{\partial t}\phi \, dx + \int_0^\ell YI \frac{\partial^2 w}{\partial x^2}\frac{d^2\phi}{dx^2} dx \\ + \int_0^\ell cI \frac{\partial^3 w}{\partial x^2 \partial t}\frac{d^2\phi}{dx^2} dx = \int_0^\ell f\phi \, dx + \int_0^\ell k_p V(t) \frac{d^2\phi}{dx^2} dx\end{aligned} \tag{7.39}$$

of the beam model for the unimorph. Approximation techniques for the model in this form are discussed in Section 8.2.

Weak Formulation of the Model — Nonlinear Inputs

The development for nonlinear and hysteretic inputs is analogous and follows simply by employing the nonlinear constitutive (7.3) rather than (7.1) when computing the moment (7.36). This yields

$$\begin{aligned}\int_0^\ell \rho \frac{\partial^2 w}{\partial t^2}\phi \, dx + \int_0^\ell \gamma \frac{\partial w}{\partial t}\phi \, dx + \int_0^\ell YI \frac{\partial^2 w}{\partial x^2}\frac{d^2\phi}{dx^2} dx + \int_0^\ell cI \frac{\partial^3 w}{\partial x^2 \partial t}\frac{d^2\phi}{dx^2} dx \\ = \int_0^\ell f\phi \, dx + \left[k_1(P(E) - P_R) + k_2(P(E) - P_R)^2\right] \int_0^\ell \frac{d^2\phi}{dx^2} dx\end{aligned} \tag{7.40}$$

which must hold for all $\phi \in V$. The nonlinear E-P dependence is quantified by (7.3) or (2.114). The constants k_1 and k_2 have representations analogous to k_p in (7.37) but are treated as parameters to be estimated through a least squares fit since a_1 and a_2 from (7.3) are unknown.

Device Characterization

To illustrate attributes of the beam model when characterizing the PVDF-polyimide unimorph depicted in Figure 7.13, we summarize results from [122]. The experimental data consists of tip displacement measurements produced with 1 Hz peak input voltages of 25 V, 50 V, 75 V and 100 V as shown in Figure 7.15. Because these voltages are in a pre-switching range for PVDF, the linear input model was employed using the parameters summarized in Table 7.1. The relations (7.26), (7.31) and (7.34) were used to compute initial values for the effective parameters ρ and Y. Final values for all of the parameters were obtained through a least squares fit to the 100 V data and the resulting model was used to predict the tip displacement in response to 25 V, 50 V and 75 V inputs.

It is noted from Figure 7.15 that the model fit and predictions are very accurate in this linear regime. However, the resulting internal damping parameter

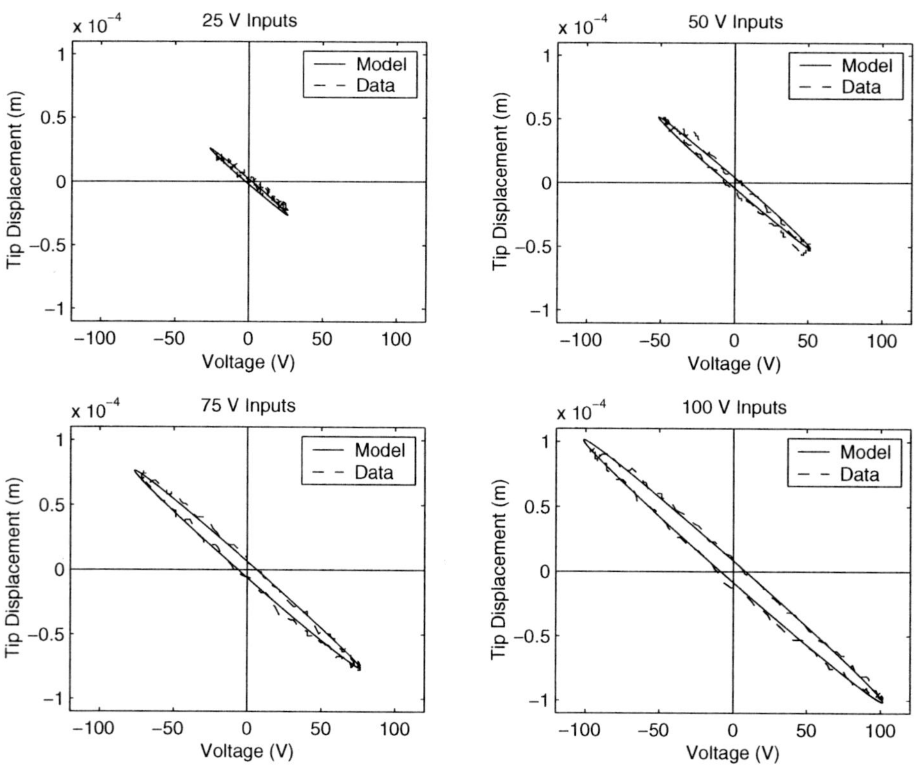

Figure 7.15. *Experimental data and model fit at 100 V, and model predictions at 25 V, 50 V and 75 V.*

Symbol	Units	Experimental Range	Employed in Model
ℓ	m	0.03	0.03
b	m	0.013	0.013
h_A	m	52×10^{-6}	52×10^{-6}
h_I	m	125×10^{-6}	137×10^{-6}
ρ_A	kg/m^3	1.78×10^3	1.78×10^3
ρ_I	kg/m^3	1.3×10^3	1.3×10^3
Y_A	N/m^2	2.0×10^9 – 2.6×10^9	2.0×10^9
Y_I	N/m^2	2.5×10^9 – 2.8×10^9	2.7×10^9
cI	N·s/m^2		2.2848×10^{-7}
γ	N·s/m^2		0.005
d_{31}	C/N	20×10^{-12} – 27×10^{-12}	20×10^{-12}

Table 7.1. *Experimental parameter ranges and values employed in the model.*

$cI = 2.2848 \times 10^{-7}$ is only two orders of magnitude smaller than the stiffness parameter $YI = 1.7250 \times 10^{-5}$. This is significantly larger than damping values estimated for elastic materials which are often five orders of magnitude less than corresponding stiffness parameters — e.g., see pages 134, 147 of [33]. These large damping coefficients reflect the viscoelastic nature of the unimorph, and the development of models and approximation techniques which incorporate Boltzmann damping constitute an active research area.

7.4.2 Uniform Beam with Surface-Mounted PZT Patches

Construction of the unimorph model illustrates issues associated with determination of the neutral line and effective density and stiffness parameters for a composite, asymmetric structure. To demonstrate some of the simplifications which result for symmetric beams and the quantification of piecewise inputs, we consider the thin beam with surface-mounted patches depicted in Figure 7.13(b). For simplicity, we consider a single patch pair but note that extension to multiple pairs is achieved in an analogous manner as detailed in Section 7.5 for a thin plate. We initially consider linear operating regimes for which application of diametrically opposite voltages generate pure bending moments and transverse motion. This is in contrast to equal voltages which generate in-plane motion, quantified using the techniques of Section 7.3, or general voltages which produce both in-plane and out-of-plane motion.[31]

We retain the notation convention established in Section 7.4.1 and let the subscript I denote beam material properties (e.g., properties of aluminum or steel) and let the subscript A denote PZT properties. The thickness coordinate z is configured so that $z = 0$ corresponds with the beam centerline as depicted in Figure 7.16.

[31]We note that in high drive regimes, opposite fields to the patch pairs produce both bending and in-plane motion due to the asymmetry of the E-ε relation about $E = 0$ as illustrated in Figure 2.10(b). These coupled effects are considered in Section 7.5.

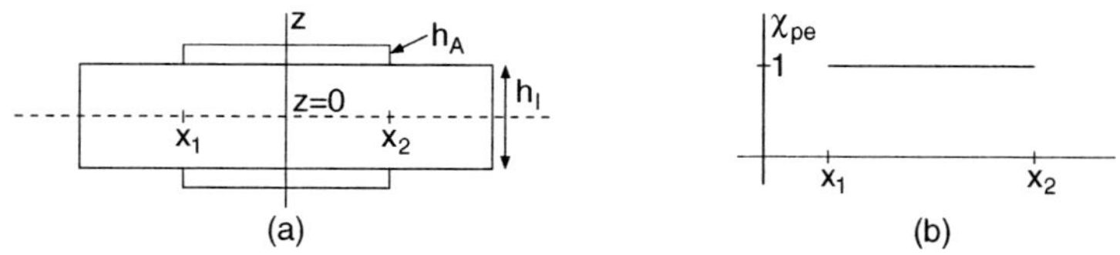

Figure 7.16. *(a) Coordinate system for moment computation and (b) characteristic function χ_{pe} which delineates the region with surface-mounted patches.*

Force and Moment Balancing

Forces and moments are balanced in a manner identical to that used to construct equations of motion for the unimorph. This yields

$$\rho\frac{\partial^2 w}{\partial t^2}+\gamma\frac{\partial w}{\partial t}-\frac{\partial^2 M}{\partial x^2}=f \tag{7.41}$$

where the linear density ρ is given by

$$\rho(x)=\begin{cases} 2h_A b\rho_A + h_I b\rho_I & , \quad x\in[x_1,x_2] \\ h_I b\rho_I & , \quad x\in[0,x_1)\cup(x_2,\ell] \end{cases}$$

and $[x_1, x_2]$ is the region covered by the patches. To consolidate notation, we employ the characteristic equation

$$\chi_{pe}(x)=\begin{cases} 1 & , \quad x\in[x_1,x_2] \\ 0 & , \quad x\in[0,x_1)\cup(x_2,\ell], \end{cases}$$

depicted in Figure 7.16(b), to formulate the density as

$$\rho(x)=h_I b\rho_I + 2\chi_{pe}(x)h_A b\rho_A. \tag{7.42}$$

Moment Evaluation

The conservation principles used to compute the neutral line z_n, effective stiffness YI, and external coupling parameter k_p are the same as those employed in Section 7.4.1 for the unimorph so we simply summarize here the final expressions for the thin beam geometry.

Force balancing in a manner analogous to (7.30) yields the centerline

$$z_n = 0$$

for the neutral line. This is consistent with the symmetry of the structure.

The elastic, damping and external moments

$$M_e=-YI(x)\frac{\partial^2 w}{\partial x^2} \quad , \quad M_d=-cI(x)\frac{\partial^3 w}{\partial x^2\partial t} \quad , \quad M_{ext}=k_p(x)V(t) \tag{7.43}$$

have the same form as the unimorph moments (7.33), (7.35) and (7.36). However, the geometry-dependent coefficients differ and are given by

$$\begin{aligned} YI(x) &= Y_I \frac{h_I^3 b}{12} + Y_A c_3 \chi_{pe}(x) \\ cI(x) &= c_I \frac{h_I^3 b}{12} + c_A c_3 \chi_{pe}(x) \\ k_p(x) &= \frac{2Y_A d_{31} c_2}{h_A} \chi_{pe}(x) \end{aligned} \tag{7.44}$$

where

$$\begin{aligned} c_2 &= b \int_{h_I/2}^{h_I/2+h_A} (z - z_n)\, dz = \frac{b}{2} \left[\left(\frac{h_I}{2} + h_A \right)^2 - \left(\frac{h_I}{2} \right)^2 \right] \\ c_3 &= b \int_{h_I/2}^{h_I/2+h_A} (z - z_n)^2\, dz = \frac{b}{3} \left[\left(\frac{h_I}{2} + h_A \right)^3 - \left(\frac{h_I}{2} \right)^3 \right]. \end{aligned} \tag{7.45}$$

Strong and Weak Forms of the Beam Model

Because the general equations of motion (7.41) and moment relations (7.43) are identical to those for the unimorph, the strong and weak forms of the models also agree, with geometry differences incorporated through the parameters ρ, YI, cI and k_p defined in (7.42) and (7.44). Hence the strong formulation of the model is given by (7.38) where it is noted that differentiation of the spatially-dependent parameters yields Dirac distributions at the patch edges. This is alleviated in the weak formulations (7.39) and (7.40) which simply involve differing material coefficients in the regions covered by and devoid of patches. When implementing the numerical methods of Section 8.2, one needs to ensure that the spline or finite element grid coincides with the patch edges to retain optimal convergence rates.

7.5 Plate Models

The rod and beam models developed in Sections 7.3 and 7.4 quantify the in-plane and out-of-plane motion of structures whose width is sufficiently small compared with the length that suitable accuracy is obtained by considering motion only as a function of length. In this section, we summarize the development of 2-D plate models quantifying the in-plane and out-of-plane motion in both the x and y-coordinates.

7.5.1 Rectangular Plate

We consider a plate of length ℓ, width a, and thickness h_I and let $\Omega = [0, \ell] \times [0, a]$ denote the support of the plate. We assume that N_A PZT patch pairs having thickness h_A are mounted on the surface of the plate with edges parallel to the x and

y-axes as depicted in Figure 7.17. The regions covered by the patches are denoted by $\Omega_1, \ldots, \Omega_{N_A}$. As in previous sections, the subscripts I and A on the density ρ, Young's modulus Y, and Kelvin–Voigt damping parameter c designate plate and patch values. The air damping coefficient is denoted by γ and the displacements of the reference surface in the x, y and z directions are respectively denoted by u, v and w. Finally, distributed forces are denoted by $\mathbf{f} = f_x \hat{\imath}_x + f_y \hat{\imath}_y + f_n \hat{\imath}_n$.

Force and Moment Balancing

When balancing forces and moments for an infinitesimal plate element, it is advantageous to employ the resultants in differential form and having the orientation depicted in Figure 7.18.[32] The differential notation is equivalent in the limit to the resultant convention employed in Sections 7.3 and 7.4 but simplifies both the 2-D balance of forces and moments and formulation of the deformed reference surface when constructing the nonlinear von Kármán plate model as summarized in Section 7.8.

Force Balancing

The balance of forces in the x-direction in combination with Newton's second law yields

$$\rho \frac{\partial^2 u}{\partial t^2} dx dy = \left(N_x + \frac{\partial N_x}{\partial x} dx \right) dy - N_x dy + \left(N_{yx} + \frac{\partial N_{yx}}{\partial y} dy \right) dx$$
$$- N_{yx} dx + \hat{f}_x dx dy$$

which implies that

$$\rho \frac{\partial^2 u}{\partial t^2} - \frac{\partial N_x}{\partial x} - \frac{\partial N_{yx}}{\partial y} = f_x. \tag{7.46}$$

The equilibrium equations

$$\rho \frac{\partial^2 v}{\partial t^2} - \frac{\partial N_y}{\partial y} - \frac{\partial N_{xy}}{\partial x} = f_y \quad , \quad \rho \frac{\partial^2 w}{\partial t^2} - \frac{\partial Q_x}{\partial x} - \frac{\partial Q_y}{\partial y} = f_n \tag{7.47}$$

[32]See Footnote 30 on page 316 for discussion regarding the moment convention.

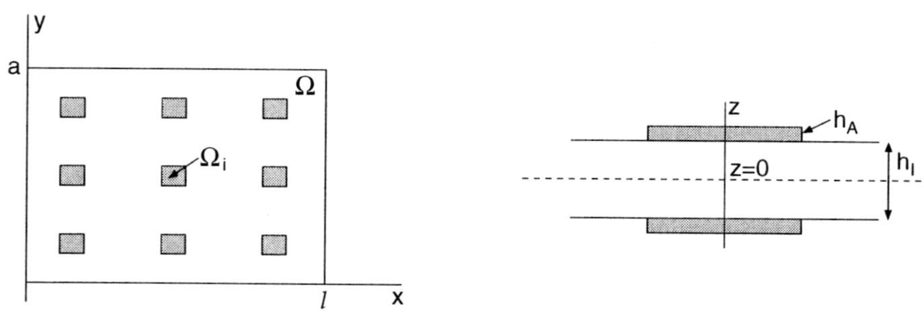

Figure 7.17. *Plate of length ℓ, width a, and thickness h_I with PZT actuators of thickness h_A covering the regions $\Omega_1, \ldots, \Omega_{N_A}$. Due to symmetry, the neutral surface z_n corresponds with the centerline $z = 0$.*

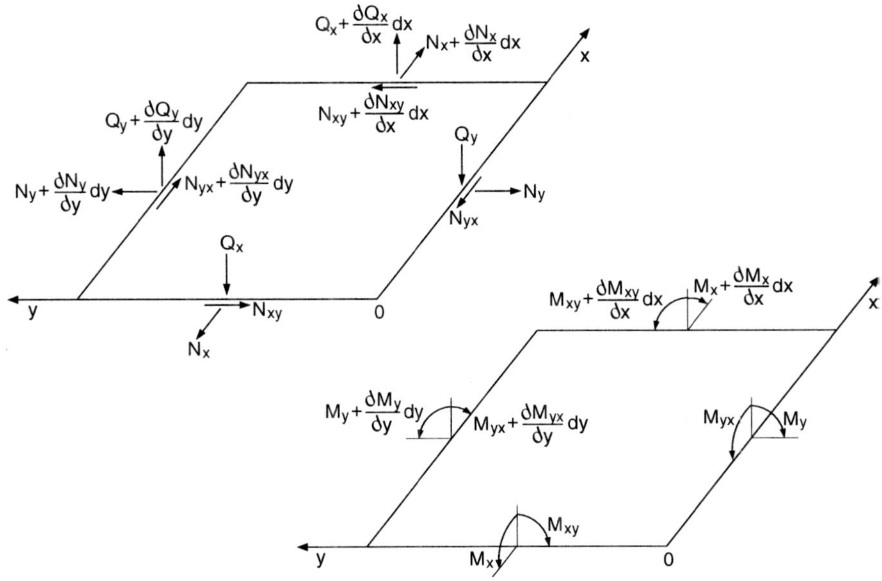

Figure 7.18. *Force and moment resultants for the infinitesimal plate element.*

in the y and z-directions are derived in a similar manner. In all of these relations, the composite density is given by

$$\rho(x,y) = \rho_I h_I + 2\sum_{i=1}^{N_A} \chi_{pe_i}(x,y)\rho_A h_A \tag{7.48}$$

where the characteristic function

$$\chi_{pe_i}(x,y) = \begin{cases} 1 & , \quad (x,y) \in \Omega_i \\ 0 & , \quad (x,y) \notin \Omega_i \end{cases} \tag{7.49}$$

isolates the region covered the the ith patch pair.

Moment Balancing

Moments are balanced with respect to a reference point which we choose as the point 0 in Figure 7.18. The balancing of moments with respect to y yields

$$\begin{aligned} &\left(M_x + \frac{\partial M_x}{\partial x}dx\right)dy - M_x dy - \left(Q_x + \frac{\partial Q_x}{\partial x}dx\right)dydx \\ &\quad + \left(M_{yx} + \frac{\partial M_{yx}}{\partial y}dy\right)dx - M_{yx}dx + Q_y\frac{dx}{2}dx \\ &\quad - \left(Q_y + \frac{\partial Q_y}{\partial y}dy\right)dx\frac{dx}{2} + \hat{f}_n dxdy\frac{dx}{2} = 0. \end{aligned}$$

Retention of first-order terms in accordance with Assumption 2 of Section 7.2 yields the equilibrium equation

$$\frac{\partial M_x}{\partial x} + \frac{\partial M_{yx}}{\partial y} - Q_x = 0. \tag{7.50}$$

In a similar manner, the relations

$$\frac{\partial M_y}{\partial y} + \frac{\partial M_{xy}}{\partial x} - Q_y = 0 \tag{7.51}$$

and

$$N_{xy} - N_{yx} = 0 \tag{7.52}$$

are determined by balancing moments with respect to x and z. It will be shown that due to the symmetry of the stress tensor, $N_{xy} = N_{yx}$ so (7.52) is automatically satisfied.

The uncoupled equations of motion can then be formulated as

$$\begin{aligned}
&\rho\frac{\partial^2 u}{\partial t^2} - \frac{\partial N_x}{\partial x} - \frac{\partial N_{yx}}{\partial y} = f_x \\
&\rho\frac{\partial^2 v}{\partial t^2} - \frac{\partial N_y}{\partial y} - \frac{\partial N_{xy}}{\partial x} = f_y \\
&\rho\frac{\partial^2 w}{\partial t^2} - \frac{\partial^2 M_x}{\partial x^2} - \frac{\partial^2 M_y}{\partial y^2} - \frac{\partial^2 M_{yx}}{\partial x \partial y} - \frac{\partial^2 M_{xy}}{\partial x \partial y} = f_n.
\end{aligned} \tag{7.53}$$

We next formulate the strain-displacement and stress-strain relations necessary to pose (7.53) in terms of the state variables u, v and w.

Resultant Formulation

The definitions of the force and moment resultants are the same as the 1-D definitions employed in Sections 7.3 and 7.4 when deriving rod and beam equations so we simply summarize here requisite 2-D relations. For the considered symmetric geometry, the reference surface z_n is the unperturbed middle surface so $z_n = 0$. Extension of the model to nonsymmetric structures is accomplished using theory analogous to that of Section 7.4.1.

Stress-Strain Relations

We summarize first constitutive relations which relate the normal strains $\varepsilon_x, \varepsilon_y$ and shear strains $\varepsilon_{xy}, \varepsilon_{yx}$ at arbitrary points in the plate to normal stresses σ_x, σ_y and shear stresses σ_{xy}, σ_{yx} having the orientation shown in Figure 7.19. This is accomplished using (7.2) or (7.4) with $\alpha = x$ and $\beta = y$. As detailed in [33, 291], symmetry of the stress tensor dictates that $\sigma_{xy} = \sigma_{yx}$ so we focus on relations for the first three pairs. Finally, we focus initially on the linear input relations (7.2), which provide suitable accuracy for a number of smart material applications, but note that identical analysis applies for the nonlinear input relations (7.4).

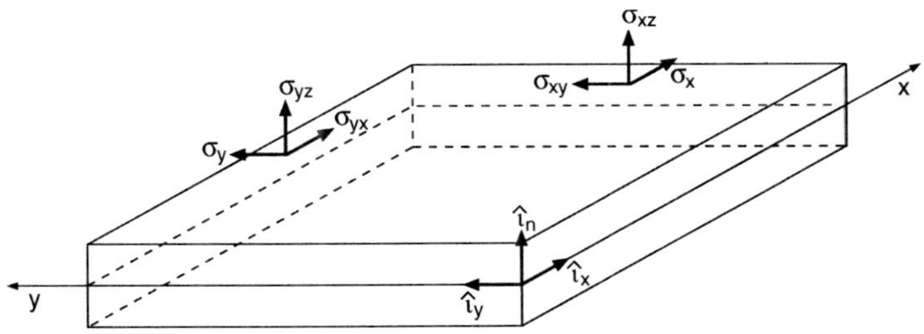

Figure 7.19. *Orientation of normal stresses* σ_x, σ_y *and shear stresses* σ_{xy}, σ_{yx}, σ_{xz}, σ_{yz}. *The convention for normal and shear strains is analogous.*

From the first relation in (7.2), it follows that

$$\sigma_x = \begin{cases} \sigma_{x_I} & , \text{ Plate } \left(|z| < \frac{h_I}{2}\right) \\ \sigma_{x_A} & , \text{ Patch } \left(\frac{h_I}{2} \le |z| \le \frac{h_I}{2} + h_A\right) \end{cases}$$

where

$$\begin{aligned} \sigma_{x_I} &= \frac{Y_I}{1-\nu_I^2}\left(\varepsilon_x + \nu_I \varepsilon_y\right) + \frac{c_I}{1-\nu_I^2}\left(\dot{\varepsilon}_x + \nu_I \dot{\varepsilon}_y\right) \\ \sigma_{x_A} &= \frac{Y_A}{1-\nu_A^2}\left(\varepsilon_x + \nu_A \varepsilon_y\right) + \frac{c_A}{1-\nu_A^2}\left(\dot{\varepsilon}_x + \nu_A \dot{\varepsilon}_y\right) - \frac{Y_A d_{31}}{h_A(1-\nu_A)} V. \end{aligned} \tag{7.54}$$

The relations for σ_y and $\sigma_{xy} = \sigma_{yx}$ follow in a similar manner. Nonlinear input relations are obtained through identical analysis using the polarization relation (7.4).

Strain-Displacement Relations

A fundamental tenet of thin beam, plate and shell theory is that motion is quantified in terms of displacements and rotation of the reference surface. To accomplish, we let e_x, e_y and e_{xy}, e_{yx} respectively denote normal and shear strains of the reference surface z_n. Moreover, κ_x, κ_y and κ_{xy} respectively denote changes in the curvature and twist of the reference surface.

By invoking Assumption 4 of Section 7.2, the strains $\varepsilon_x, \varepsilon_y, \varepsilon_{xy}$ at arbitrary positions z in the plate can be expressed as

$$\begin{aligned} \varepsilon_x &= e_x + \kappa_x z \\ \varepsilon_y &= e_y + \kappa_y z \\ \varepsilon_{xy} &= e_{xy} + \kappa_{xy} z. \end{aligned} \tag{7.55}$$

As depicted in Figure 7.20, the first term in each relation quantifies in-plane strains whereas the second characterizes strains due to bending.

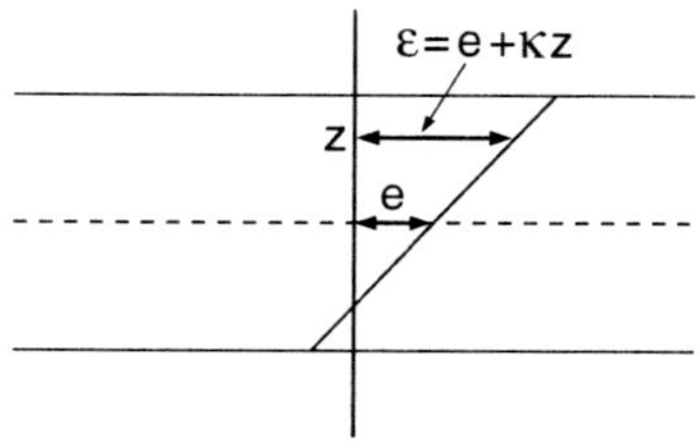

Figure 7.20. *Representative strain profile comprised of an in-plane component e and bending component κz.*

Extension of the strain definition (7.11) and curvature relation (7.32) to 2-D subsequently yields the kinematic relations

$$
\begin{aligned}
e_x &= \frac{\partial u}{\partial x} \quad , \quad e_y = \frac{\partial v}{\partial y} \quad , \quad e_{xy} = \frac{\partial v}{\partial x} + \frac{\partial u}{\partial y} \\
\kappa_x &= -\frac{\partial^2 w}{\partial x^2} \quad , \quad \kappa_y = -\frac{\partial^2 w}{\partial y^2} \quad , \quad \kappa_{xy} = -2\frac{\partial^2 w}{\partial x \partial y}.
\end{aligned}
\tag{7.56}
$$

The combination of (7.55) and (7.56) provides relations which quantify the general strains employed in stress-strain relations — e.g., (7.54) — in terms of displacement properties of the reference surface.

Force and Moment Resultants — General Relations

The force resultants $N_x, N_y, N_{xy} = N_{yx}$ and moment resultants $M_x, M_y, M_{xy} = M_{yx}$ are defined in a manner analogous to (7.9) and (7.29). Inclusion of the patch properties and inputs yields the general relations

$$
\begin{aligned}
\begin{bmatrix} N_x \\ N_y \\ N_{xy} \end{bmatrix} &= \int_{-h_I/2}^{h_I/2} \begin{bmatrix} \sigma_{x_I} \\ \sigma_{y_I} \\ \sigma_{xy_I} \end{bmatrix} dz + \sum_{i=1}^{N_A} \chi_{pe_i}(x,y) \left(\int_{h_I/2}^{h_I/2+h_A} \begin{bmatrix} \sigma_{x_A} \\ \sigma_{y_A} \\ \sigma_{xy_A} \end{bmatrix} dz \right. \\
&\qquad \left. + \int_{-h_I/2-h_A}^{-h_I/2} \begin{bmatrix} \sigma_{x_A} \\ \sigma_{y_A} \\ \sigma_{xy_A} \end{bmatrix} dz \right)
\end{aligned}
$$

$$
\begin{aligned}
\begin{bmatrix} M_x \\ M_y \\ M_{xy} \end{bmatrix} &= \int_{-h_I/2}^{h_I/2} \begin{bmatrix} \sigma_{x_I} \\ \sigma_{y_I} \\ \sigma_{xy_I} \end{bmatrix} z dz + \sum_{i=1}^{N_A} \chi_{pe_i}(x,y) \left(\int_{h_I/2}^{h_I/2+h_A} \begin{bmatrix} \sigma_{x_A} \\ \sigma_{y_A} \\ \sigma_{xy_A} \end{bmatrix} z dz \right. \\
&\qquad \left. + \int_{-h_I/2-h_A}^{-h_I/2} \begin{bmatrix} \sigma_{x_A} \\ \sigma_{y_A} \\ \sigma_{xy_A} \end{bmatrix} z dz \right)
\end{aligned}
$$

where the characteristic function is defined in (7.49). From (7.54), it is observed that the stresses have elastic, damping, and external components; hence the resultants can be expressed as

$$
\begin{aligned}
N_x &= N_{x_e} + N_{x_d} + N_{x_{ext}} \quad &, \quad M_x &= M_{x_e} + M_{x_d} + M_{x_{ext}} \\
N_y &= N_{y_e} + N_{y_d} + N_{y_{ext}} \quad &, \quad M_y &= M_{y_e} + M_{y_d} + M_{y_{ext}} \\
N_{xy} &= N_{xy_e} + N_{xy_d} + N_{xy_{ext}} \quad &, \quad M_{xy} &= M_{xy_e} + M_{xy_d} + M_{xy_{ext}}
\end{aligned}
\tag{7.57}
$$

where the subscripts e, d and ext respectively indicate elastic, damping and external components.

Force and Moment Resultants — Elastic Components

For the case under consideration, the symmetry of patch pairs simplifies the resultant formulation and yields

$$
\begin{aligned}
N_{x_e} &= \frac{Y_I h_I}{1-\nu_I^2}(e_x + \nu_I e_y) + \frac{2Y_A h_A}{1-\nu_A^2}(e_x + \nu_A e_y)\sum_{i=1}^{N_A}\chi_{pe_i}(x,y) \\
N_{y_e} &= \frac{Y_I h_I}{1-\nu_I^2}(e_y + \nu_I e_x) + \frac{2Y_A h_A}{1-\nu_A^2}(e_y + \nu_A e_x)\sum_{i=1}^{N_A}\chi_{pe_i}(x,y) \\
N_{xy_e} &= \frac{Y_I h_I}{2(1+\nu_I)} e_{xy} + \frac{Y_A h_A}{1+\nu_A} e_{xy}\sum_{i=1}^{N_A}\chi_{pe_i}(x,y) \\
M_{x_e} &= \frac{Y_I h_I^3}{12(1-\nu_I^2)}(\kappa_x + \nu_I \kappa_y) + \frac{2Y_A c_3}{1-\nu_A^2}(\kappa_x + \nu_A \kappa_y)\sum_{i=1}^{N_A}\chi_{pe_i}(x,y) \\
M_{y_e} &= \frac{Y_I h_I^3}{12(1-\nu_I^2)}(\kappa_y + \nu_I \kappa_x) + \frac{2Y_A c_3}{1-\nu_A^2}(\kappa_y + \nu_A \kappa_x)\sum_{i=1}^{N_A}\chi_{pe_i}(x,y) \\
M_{xy_e} &= \frac{Y_I h_I^3}{24(1+\nu_I)} \kappa_{xy} + \frac{Y_A c_3}{1+\nu_A} \kappa_{xy}\sum_{i=1}^{N_A}\chi_{pe_i}(x,y)
\end{aligned}
\tag{7.58}
$$

where $e_x, e_y, e_{xy}, \kappa_x, \kappa_y, \kappa_{xy}$ are defined in (7.56) and $c_3 = \int_{h_I/2}^{h_I/2+h_A}(z-z_n)^2\,dz$ is given in (7.45). For more general constructs, the same techniques are applied but the final expressions will reflect geometry-dependencies.

Force and Moment Resultants — Internal Damping Components

The resultant components that incorporate the Kelvin–Voigt damping have the same form as the elastic components but involve the temporal derivatives of strains and rotations; for example

$$
\begin{aligned}
N_{x_d} &= \frac{Y_I h_I}{1-\nu_I^2}(\dot{e}_x + \nu_I \dot{e}_y) + \frac{2Y_A h_A}{1-\nu_A^2}(\dot{e}_x + \nu_A \dot{e}_y)\sum_{i=1}^{N_A} \chi_{pe_i}(x,y) \\
M_{x_d} &= \frac{Y_I h_I^3}{12(1-\nu_I^2)}(\dot{\kappa}_x + \nu_I \dot{\kappa_y}) + \frac{2Y_A c_3}{1-\nu_A^2}(\dot{\kappa}_x + \nu_A \dot{\kappa}_y)\sum_{i=1}^{N_A} \chi_{pe_i}(x,y)
\end{aligned}
\tag{7.59}
$$

with analogous expressions for $N_{y_d}, N_{xy_d}, M_{y_d}$ and M_{xy_d}.

Force and Moment Resultants — External Components

Consider first the external components that result from the linear input relations (7.2) when voltages $V_{1i}(t)$ and $V_{2i}(t)$ are respectively applied to the inner and outer patches in the ith pair. Integration through the patch thickness yields

$$
\begin{aligned}
N_{x_{ext}} &= N_{y_{ext}} = \frac{-Y_A d_{31}}{1-\nu_A}\sum_{i=1}^{N_A}[V_{1i}(t) + V_{2i}(t)]\chi_{pe_i}(x,y) \\
N_{xy_{ext}} &= N_{yx_{ext}} = 0 \\
M_{x_{ext}} &= M_{y_{ext}} = \frac{-Y_A d_{31} c_2}{h_A(1-\nu_A)}\sum_{i=1}^{N_A}[V_{1i}(t) + V_{2i}(t)]\chi_{pe_i}(x,y) \\
M_{xy_{ext}} &= M_{yx_{ext}} = 0
\end{aligned}
\tag{7.60}
$$

where $c_2 = \int_{h_I/2}^{h_I/2+h_A}(z - z_n)dz$ is defined in (7.45).

It is observed that if equal voltages $V_i(t) = V_{1i}(t) = V_{2i}(t)$ are applied to the patches, then

$$
\begin{aligned}
N_{x_{ext}} &= N_{y_{ext}} = \frac{-2Y_A d_{31}}{1-\nu_A}\sum_{i=1}^{N_A} V_i(t)\chi_{pe_i}(x,y) \\
N_{xy_{ext}} &= N_{yx_{ext}} = M_{x_{ext}} = M_{y_{ext}} = M_{xy_{ext}} = M_{yx_{ext}} = 0
\end{aligned}
\tag{7.61}
$$

which produces solely in-plane motion. Alternatively, if $V_i(t) = V_{1i}(t) = -V_{2i}(t)$, only bending moments

$$
M_{x_{ext}} = M_{y_{ext}} = \frac{-2Y_A d_{31} c_2}{h_A(1-\nu_A)}\sum_{i=1}^{N_A} V_i(t)\chi_{pe_i}(x,y) \tag{7.62}
$$

are produced and the plate will exhibit transverse or out-of-plane motion. This is analogous to the drive regimes which provide in-plane and out-of-plane motion in the rod and beam models discussed in Sections 7.3 and 7.4.

The formulation of the external resultants for the nonlinear input relations (7.4) is analogous and yields

$$
\begin{aligned}
N_{x_{ext}} = N_{y_{ext}} = \frac{-h_A}{1-\nu_A} \sum_{i=1}^{N_A} & \left[a_1(P_{1i}(t) + P_{2i}(t) - 2P_R)\right. \\
& \left. + a_2\left((P_{1i}(t) - P_R)^2 + (P_{2i}(t) - P_R)^2\right)\right] \chi_{pe_i}(x,y) \\
M_{x_{ext}} = M_{y_{ext}} = \frac{-c_2}{1-\nu_A} \sum_{i=1}^{N_A} & \left[a_1(P_{1i}(t) - P_{2i}(t) - 2P_R)\right. \\
& \left. + a_2\left((P_{1i}(t) - P_R)^2 - (P_{2i}(t) - P_R)^2\right)\right] \chi_{pe_i}(x,y)
\end{aligned}
\tag{7.63}
$$

where P_{1i}, P_{2i} are the polarizations modeled by (7.4) or (2.114) in response to input fields E_{1i}, E_{2i} applied to the inner and outer patches in each pair. We note that in this case, $E_i = E_{1i} = E_{2i}$ and $E_i = E_{1i} = -E_{2i}$ do not produce solely in-plane force and out-or-plane bending due the asymmetry of the E-ε relation about $E = 0$ — e.g., see Figure 2.10(b). For low drive levels, however, the E-ε relation is approximately linear which leads to (7.61) and (7.62) resulting from the linear input model.

Boundary Conditions and Strong Model Formulation

Appropriate boundary conditions are determined by the requirement that no work is performed along the plate edge. To illustrate, consider the edge $x = 0$, $0 \leq y \leq a$. The work during deformation can be expressed as

$$
W = \int_0^a \left[N_x u + N_{xy} v + Q_x w + M_{xy}\theta_y + M_x \theta_x\right] dy \tag{7.64}
$$

where the rotations of the normal to the reference surface are approximated by

$$
\theta_x = \frac{\partial w}{\partial x} \quad , \quad \theta_y = \frac{\partial w}{\partial y}.
$$

Integration by parts gives

$$
\int_0^a \left[N_x u + N_{xy} v + \left(Q_x - \frac{\partial M_{xy}}{\partial y}\right) w + M_x \frac{\partial w}{\partial x}\right] dy + M_{xy} w \Big|_0^a = 0
$$

which yields the boundary conditions

$$
\begin{aligned}
u = 0 \quad & \text{or} \quad N_x = 0 \\
v = 0 \quad & \text{or} \quad N_{xy} = 0 \\
w = 0 \quad & \text{or} \quad Q_x - \frac{\partial M_{xy}}{\partial y} = 0 \\
\frac{\partial w}{\partial x} = 0 \quad & \text{or} \quad M_x = 0
\end{aligned}
$$

and $M_{xy} w|_0^a = 0$.

Analogous conditions hold for edges parallel to the y-axis. We point out that the first condition in each relation constitutes an *essential boundary condition* which must be enforced when constructing spaces of test functions V whereas the second is a *natural boundary condition* that is automatically satisfied by solutions to the weak formulation of the model.

Common boundary conditions employed when modeling smart material systems include the following.

(a) Clamped or fixed edge:

$$u = v = w = \frac{\partial w}{\partial x} = 0$$

(b) Free edge:

$$N_x = N_{xy} = \left(Q_x + \frac{\partial M_{xy}}{\partial y}\right) = M_x = 0$$

(c) Simply supported edge, not free to move:

$$u = v = w = M_x = 0$$

(d) Simply supported edge, free to move in x direction:

$$u = w = M_x = N_x = 0$$

The *shear diaphragm* condition (d) is popular from a theoretical perspective since it admits analytic solution for plates devoid of patches. For applications, however, the boundary conditions (a)–(c) typically provide a better approximation to physical conditions, thus necessitating the use of approximation techniques of the type discussed in Section 8.3.

For physical clamping conditions which dissipate energy, boundary conditions analogous to (7.13) can be developed through force balancing as summarized in Section 7.5.2 and detailed in [291].

The strong formulation of the model is then given by (7.53) with the general resultants specified by (7.57) and elastic, damping and external components specified by (7.58), (7.59) and (7.60) or (7.63).

Weak Model Formulation

From the perspective of approximation, the strong formulation of the model poses the same difficulties noted in Sections 7.3 and 7.4; namely, spatial differentiation of piecewise constant material parameters and inputs yields Dirac distributions and derivatives of Dirac distributions at actuator boundaries. This can severely impede the convergence of approximation techniques applied directly to the strong model formulation.

These difficulties are eliminated in weak formulations of the model obtained obtained either through energy principles analogous to those detailed in Section 7.3.2 or direct integration by parts. The state $\xi(t) = (u(t,\cdot,\cdot), v(t,\cdot,\cdot), w(t,\cdot,\cdot))$ is considered in the state space

$$X = L^2(\Omega) \times L^2(\Omega) \times L^2(\Omega)$$

where $\Omega = [0,\ell] \times [0,a]$ denotes the plate region. The space of test functions is taken to be

$$V = H_b^1(\Omega) \times H_b^1(\Omega) \times H_b^2(\Omega)$$

where H_b^1 and H_b^2 are subsets of H^1 and H^2 restricted to those functions which satisfy essential boundary conditions.

A weak formulation is

$$\begin{aligned}
&\int_\Omega \left\{ \rho \frac{\partial^2 u}{\partial t^2}\phi_1 + N_x \frac{\partial \phi_1}{\partial x} + N_{yx}\frac{\partial \phi_1}{\partial y} - f_x\phi_1 \right\} d\omega = 0 \\
&\int_\Omega \left\{ \rho \frac{\partial^2 v}{\partial t^2}\phi_2 + N_y \frac{\partial \phi_2}{\partial y} + N_{xy}\frac{\partial \phi_2}{\partial x} - f_y\phi_2 \right\} d\omega = 0 \\
&\int_\Omega \left\{ \rho \frac{\partial^2 w}{\partial t^2}\phi_3 - M_x \frac{\partial^2 \phi_3}{\partial x^2} - 2M_{xy}\frac{\partial^2 \phi_3}{\partial x \partial y} - M_y \frac{\partial^2 \phi_3}{\partial y^2} - f_n\phi_3 \right\} d\omega = 0
\end{aligned} \tag{7.65}$$

which must be satisfied for all $\Phi = (\phi_1, \phi_2, \phi_3) \in V$. The resultants are given by (7.57) with components defined in (7.58), (7.59) and (7.60) or (7.63).

As will be noted in Section 8.3, the approximation of u and v can be accomplished with linear finite elements whereas cubic Hermite elements or cubic B-splines are required to accommodate the second derivatives in the equation for w.

The differential equations are uncoupled, even for general voltages/fields and nonlinear and hysteretic input regimes. This is in contrast to the *nonlinear* von Kármán model summarized in Section 7.8 which incorporates coupling between in-plane and out-of-plane motion. As noted previously, only u and v vibrations are produced when equal voltages $V_i(t) = V_{1i}(t) = V_{2i}(t)$ are applied to the linear input relations whereas transverse motion modeled by the w relation is generated by diametrically out-of-phase voltages $V_i(t) = V_{1i}(t) = -V_{2i}(t)$. In high drive regimes, all three components of the motion are excited due to the asymmetry of the E-ε relation about $E = 0$ as manifested by the external resultant relations (7.63).

7.5.2 Circular Plate Model

Circular plates with circular or sectoral patches comprise a second common geometry in smart material applications. For modeling purposes, we consider a plate of radius a and thickness h_I with surface-mounted patches of thickness h_A placed in pairs as depicted in Figure 7.17(b). The region $\Omega = [0,a] \times [0,2\pi]$ delineates the plate region and the N_A regions covered by patch pairs are indicated by Ω_i.

The fundamental principles employed for model development are the same as those detailed in Section 7.5.1 for rectangular plates and we summarize here only the primary relations to illustrated geometry-induced differences. Details regarding the theory of circular plates can be found in [33, 291]. Finally, we consider only transverse vibrations since they comprise the primary response in many applications employing circular plates having fully clamped edges.

Force and Moment Balancing

The balance of moments with respect to r and θ yields

$$\frac{1}{r}\frac{\partial M_\theta}{\partial \theta} + \frac{\partial M_{r\theta}}{\partial r} + \frac{2}{r}M_{r\theta} - Q_\theta = 0$$

$$\frac{1}{r}M_r + \frac{\partial M_r}{\partial r} - \frac{1}{r}M_\theta + \frac{1}{r}\frac{\partial M_{\theta r}}{\partial \theta} - Q_r = 0$$

whereas force balancing yields

$$\rho\frac{\partial^2 w}{\partial t^2} - \frac{1}{r}Q_r - \frac{\partial Q_r}{\partial r} - \frac{1}{r}\frac{\partial Q_\theta}{\partial \theta} = f_n.$$

The synthesis of these relations yields the dynamic model

$$\rho\frac{\partial^2 w}{\partial t^2} - \frac{\partial^2 M_r}{\partial r^2} - \frac{2}{r}\frac{\partial M_r}{\partial r} + \frac{1}{r}\frac{M_\theta}{\partial r} - \frac{2}{r}\frac{\partial^2 M_{r\theta}}{\partial r \partial \theta} - \frac{2}{r^2}\frac{\partial M_{r\theta}}{\partial \theta} - \frac{1}{r^2}\frac{\partial^2 M_\theta}{\partial \theta^2} = f_n.$$

The density ρ has the form (7.48) to incorporate the differing material properties in regions covered by the patches.

Resultant Evaluation

The constitutive relations (7.2) and (7.4) and general strain relations (7.55) are independent of geometry so we employ them directly modulo a change of coordinates from (x, y) to (r, θ). Since we are considering only transverse vibrations, we have $e_r = e_\theta = e_{r\theta} = 0$ for the reference surface strains and hence only consider the curvature changes

$$\begin{aligned}\kappa_r &= -\frac{\partial^2 w}{\partial r^2}\\ \kappa_\theta &= -\frac{1}{r}\frac{\partial w}{\partial r} - \frac{1}{r^2}\frac{\partial^2 w}{\partial \theta^2}\\ \kappa_{r\theta} &= -\frac{\partial}{\partial r}\left(\frac{\partial w}{\partial \theta}\right)\end{aligned} \tag{7.66}$$

in the kinematic relations (7.55). The elastic, damping and external components in the general resultant relations

$$\begin{aligned}M_r &= M_{r_e} + M_{r_d} + M_{r_{ext}}\\ M_\theta &= M_{\theta_e} + M_{\theta_d} + M_{\theta_{ext}}\\ M_{r\theta} &= M_{r\theta_e} + M_{r\theta_d} + M_{r\theta_{ext}}\end{aligned} \tag{7.67}$$

are defined as follows.

Elastic Components

The elastic components of the bending resultants are

$$M_{r_e} = \frac{Y_I h_I^3}{12(1-\nu_I^2)}(\kappa_r + \nu_I \kappa_\theta) + \frac{2Y_A c_3}{1-\nu_A^2}(\kappa_r + \nu_A \kappa_\theta)\sum_{i=1}^{N_A} \chi_{pe_i}(r,\theta)$$

$$M_{y\theta_e} = \frac{Y_I h_I^3}{12(1-\nu_I^2)}(\kappa_\theta + \nu_I \kappa_r) + \frac{2Y_A c_3}{1-\nu_A^2}(\kappa_\theta + \nu_A \kappa_r)\sum_{i=1}^{N_A} \chi_{pe_i}(r,\theta)$$

$$M_{r\theta_e} = \frac{Y_I h_I^3}{24(1+\nu_I)}\kappa_{r\theta} + \frac{Y_A c_3}{1+\nu_A}\kappa_{r\theta}\sum_{i=1}^{N_A} \chi_{pe_i}(r,\theta)$$

where κ_r, κ_θ and $\kappa_{r\theta}$ are defined in (7.66) and $c_3 = \int_{h_I/2}^{h_I/2+h_A}(z-z_n)^2 dz$ is given by (7.45).

Damping Components

The damping components involve strain rates rather than strains and are

$$M_{r_d} = \frac{c_I h_I^3}{12(1-\nu_I^2)}(\dot\kappa_r + \nu_I \dot\kappa_\theta) + \frac{2c_A c_3}{1-\nu_A^2}(\dot\kappa_r + \nu_A \dot\kappa_\theta)\sum_{i=1}^{N_A} \chi_{pe_i}(r,\theta)$$

$$M_{\theta_d} = \frac{c_I h_I^3}{12(1-\nu_I^2)}(\dot\kappa_\theta + \nu_I \dot\kappa_r) + \frac{2c_A c_3}{1-\nu_A^2}(\dot\kappa_\theta + \nu_A \dot\kappa_r)\sum_{i=1}^{N_A} \chi_{pe_i}(r,\theta)$$

$$M_{r\theta_d} = \frac{c_I h_I^3}{24(1+\nu_I)}\dot\kappa_{r\theta} + \frac{c_A c_3}{1+\nu_A}\dot\kappa_{r\theta}\sum_{i=1}^{N_A} \chi_{pe_i}(r,\theta)$$

External Components

The external components are analogous to (7.60) and (7.63) for the rectangular plate. Hence for linear and nonlinear inputs they are

$$M_{r_{ext}} = M_{\theta_{ext}} = \frac{-Y_A d_{31} c_2}{h_A(1-\nu_A)}\sum_{i=1}^{N_A}[V_{1i}(t) + V_{2i}(t)]\chi_{pe_i}(r,\theta)$$

and

$$\begin{aligned} M_{r_{ext}} = M_{\theta_{ext}} = \frac{-c_2}{1-\nu_A}\sum_{i=1}^{N_A} & [a_1(P_{1i}(t) - P_{2i}(t) - 2P_R) \\ & + a_2\left((P_{1i}(t) - P_R)^2 - (P_{2i}(t) - P_R)^2\right)\right]\chi_{pe_i}(r,\vartheta). \end{aligned}$$

In both cases, the external twisting moments $M_{r\theta} = M_{\theta r}$ are zero.

Boundary Conditions

For physical devices with ideal clamps, zero slope and displacement are maintained around the plate perimeter yielding the fixed-edge condition

$$w(t,a,\theta) = \frac{\partial w}{\partial r}(t,a,\theta) = 0. \tag{7.68}$$

In applications, however, perfectly fixed-edge conditions are difficult to maintain and energy dissipation through the clamps often produces measured frequencies that are lower than predicted by (7.68). To incorporate dissipative edge motion, boundary deformations and rotations are considered to be governed by damped, elastic springs in a manner analogous to that employed when constructing the rod boundary condition (7.13). As detailed in [32,291], this yields the boundary moment conditions

$$\begin{aligned}
&\frac{1}{a}M_r(t,a,\theta) + \frac{\partial M_r}{\partial r}(t,a,\theta) \\
&\qquad = -k_a w(t,a,\theta) - c_a\frac{\partial w}{\partial t}(t,a,\theta) - \rho\frac{\partial^2 w}{\partial t^2}(t,a,\theta) \\
&M_r(t,a,\theta) = k_p\frac{\partial w}{\partial r}(t,a,\theta) + c_p\frac{\partial^2 w}{\partial r\partial t}(t,a,\theta).
\end{aligned} \tag{7.69}$$

It is observed that if one divides by the stiffness coefficients k_a and k_p and takes $k_a \to \infty$, $k_p \to \infty$, the dissipative boundary conditions (7.69) converge to the fixed-edge conditions (7.68). Alternatively, one obtains free-edge conditions in the absence of elastic, damping or inertial edge effects.

Weak Model Formulation

Consider the circular plate model with the fixed-edge conditions (7.68). The state space and space of test functions are taken to be

$$X = L^2(\Omega)$$

and

$$V = H_0^2(\Omega) = \left\{\phi \in H^2(\Omega) \,|\, \phi(a,\theta) = \frac{\partial \phi}{\partial r}(a,\theta) = 0\right\} \tag{7.70}$$

with the usual inner products.

The weak or variational formulation of the model is

$$\begin{aligned}
&\int_\Omega \rho\frac{\partial^2 w}{\partial t^2}\phi_3 d\omega + \int_\Omega M_r\frac{\partial^2\phi_3}{\partial r^2}d\omega + \int_\Omega \frac{1}{r^2}M_\theta\left(r\frac{\partial\phi_3}{\partial r} + \frac{\partial^2\phi_3}{\partial\theta^2}\right)d\omega \\
&\qquad +2\int_\Omega \frac{1}{r^2}M_{r\theta}\left(r\frac{\partial^2\phi_3}{\partial r\partial\theta} - \frac{\partial\phi_3}{\partial\theta}\right)d\omega = \int_\Omega f_n\phi_3 d\omega
\end{aligned} \tag{7.71}$$

which must be satisfied for all $\phi_3 \in V$. The differential is $d\omega = rd\theta dr$. Details regarding the weak model formulations for the dissipative boundary conditions (7.69) can be found in [32].

Model Validation

To illustrate the performance of the dynamic circular plate model (7.71), we consider the characterization of a circular aluminum plate with a single piezoceramic patch surface-mounted at the center of the plate as depicted in Figure 7.21. The plate had clamped boundary conditions, a radius of 9 in and a thickness of 0.05 in, and the PZT patch had a radius of 0.75 in and a thickness of 0.007 in (7 mils). Because the patch is small compared with the plate, in-plane motion due to the geometric asymmetry in the region covered by the patch is negligible and we consider only transverse vibrations generated by centered and noncentered strikes with an impact hammer. However, the patch contributions to the density (7.48) and resultant relations (7.67) are retained in the model, and it is illustrated in [33] that differing material properties are estimated in the region Ω_i covered by the patch during model identification. Details regarding this example can be found in [30,33] and we provide here only a summary of two dynamic responses.

Axisymmetric Response

We consider first the characterization of axisymmetric dynamics excited by a centered strike with a soft-headed impact hammer. The resulting time history and frequency response measured with the centered accelerometer $A_c = (0'', 0)$ are plotted in Figure 7.22. The measured force from the impact hammer was input to the discretized circular plate relation to obtain the modeled response. The frequency plot illustrates that four axisymmetric modes, having frequencies of 59.3, 227.8 516.4 and 917.7 Hz were excited in the experiment. The model accurately quantifies the low frequency dynamics but overdamps at high frequencies which is characteristic of the Kelvin–Voigt damping model.

For numerous applications, however, the high frequency dynamics typically have low magnitude and are highly damped, thus minimizing their impact on control design. Moreover, for structural acoustic applications, high frequency structural modes exhibit minimal coupling with acoustic modes and hence they provide

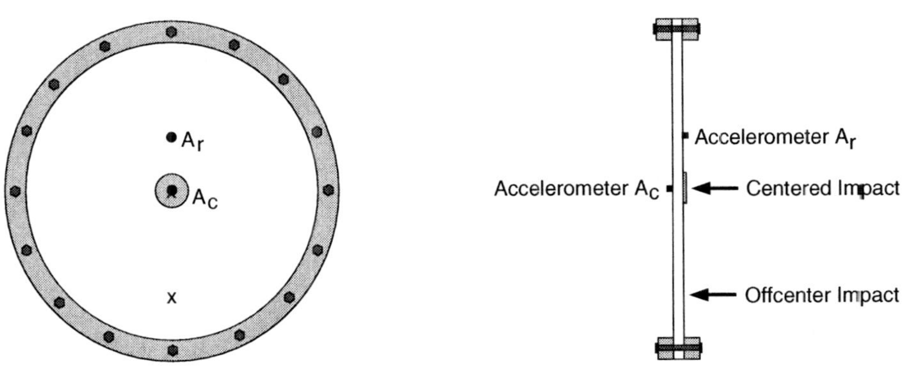

Figure 7.21. *Clamped circular plate with a single, centered, piezoceramic patch. Inputs were provided by centered and noncentered hammer impacts with acceleration measured at* $A_c = (0'', 0)$ *and* $A_r = (2'', 0)$.

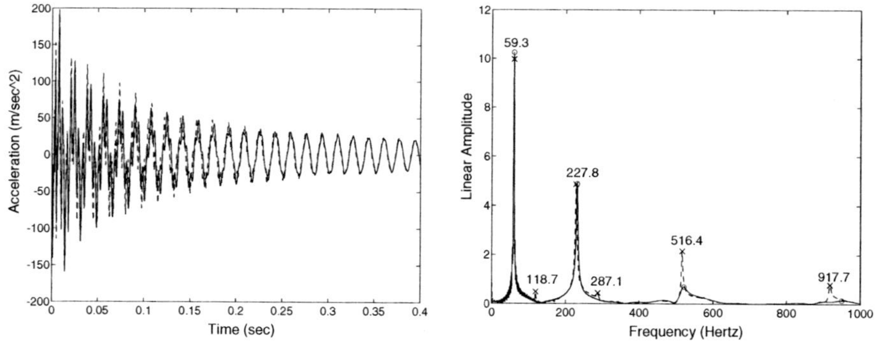

Figure 7.22. *Time history and frequency content at* $A_c = (0'', 0)$ *in response to a centered hammer impact: data (x – – –) and model (o ——).*

negligible contribution to structure-borne noise. Finally, feedback mechanisms can accommodate high frequency model limitations in model-based control designs. It is illustrated in [31, 33] that the circular plate model constructed in this manner can thus be employed for model-based LQG control design using the piezoceramic patch as an actuator.

Nonaxisymmetric Response

For axisymmetric regimes, the plate model (7.71) reduces to one spatial dimension. To demonstrate the 2-D nature of the model, we also illustrate the characterization of plate dynamics excited by a noncentered impact using a hard-tipped hammer at the point (7.27",0) depicted in Figure 7.21. The measured and modeled response at the point $A_r = (2'', 0)$ are plotted in Figure 7.23. It is observed that the model accurately characterizes the $(n, m) =$ (0,0), (0,2), (0,3), (1,1), (1,2), (2,0), (2,1), and (0,4) modes while underdamping the (1,0) and (0,1) modes and

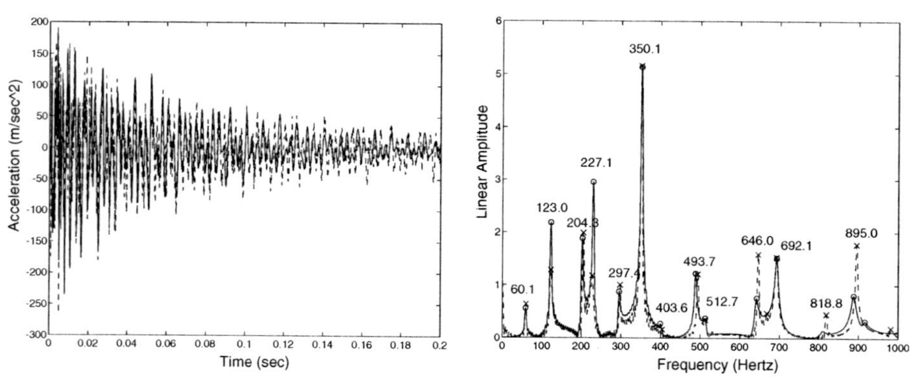

Figure 7.23. *Time history and frequency content at* $A_r = (2'', 0)$ *in response to a noncentered impact at (7.27",0): data (x – – –) and model (o ——).*

overdamping higher frequency modes. Despite the limitation of the Kelvin–Voigt damping relation, the model accurately characterizes eight modes which provides ample accuracy for model-based control design.

7.6 Shell Models – General Development

The rod, beam and plate models developed in previous sections comprise special cases of shell models. This class of structures also includes the cylindrical, bi-spherical and general shell configurations arising in the AFM, structural acoustic, THUNDER, and jet engine applications depicted in Figure 7.1. A comprehensive discussion of model development for shells transcends the scope of this chapter and we provide here only a summary of the theory with the goal of providing readers with a framework from which to start when constructing models for specific smart material applications. Details regarding general shell theory can be found in Dym [145], Flügge [164], Love [301], Markuš [318], Novozhilov [364], Soedel [453] and Timoshenko and Woinowsky-Krieger [480] whereas discussion focused on piezoelectric shells or shells with piezoelectric actuators is provided in [485–487].

The analysis in Section 7.5 of in-plane and out-of-plane motion for plate structures illustrates the moment and force balancing principles, constitutive stress-strain relations, and kinematic strain-displacement tenets used to construct models for 2-D composite structures comprised of both active and inactive components. The extensions required to incorporate curvature-induced coupling are geometric in nature and do not affect the fundamental constitutive behavior. Hence to simplify the discussion, we consider in this section the passive dynamics of undamped, homogeneous structures. Once the general geometric relations are established, the inclusion of damping and external inputs follows in a manner analogous to that detailed in Section 7.5 for plates. This will be further illustrated in Section 7.7 where the special cases of cylindrical shells and curved beams are considered.

7.6.1 Shell Coordinates

We consider a homogeneous thin structure of width h so that the reference surface z_n coincides with the unperturbed middle surface, $z_n = 0$, as depicted in Figure 7.24. From Assumption 4 of Section 7.2, it follows that the behavior at any point in the shell is quantified in terms of the motion of the reference surface so we begin there when specifying coordinates. Consider an orthogonal, curvilinear coordinate system on the reference surface chosen to coincide with lines of principle curvature. The third coordinate direction is chosen perpendicular to the reference surface through the shell thickness. The coordinates in the three directions are denoted α, β and z. If we designate the reference surface by

$$z_n = r(\alpha, \beta),$$

arbitrary points in the shell can be specified by

$$R(\alpha, \beta, z) = r(\alpha, \beta) + z\hat{\imath}_n$$

where $\hat{\imath}$ is the unit vector normal to the reference surface.

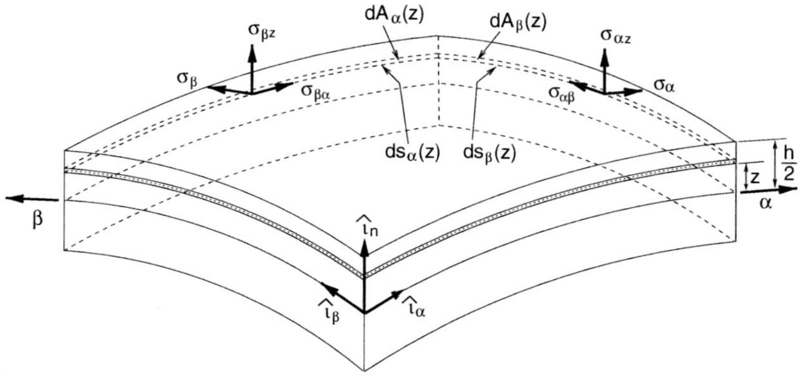

Figure 7.24. *Fundamental shell element in the* (α, β) *coordinate system.*

The radii of curvature in the α and β directions are denoted by R_α and R_β and Lamé constants A and B are defined by

$$A^2 = \frac{\partial r}{\partial \alpha} \cdot \frac{\partial r}{\partial \alpha} \quad , \quad B^2 = \frac{\partial r}{\partial \beta} \cdot \frac{\partial r}{\partial \beta}.$$

As detailed in [145, 292], the squared length of a differential length element is

$$\begin{aligned}(ds)^2 &= dR \cdot dR \\ &= A^2(1 + z/R_\alpha)^2(d\alpha)^2 + B^2(1 + z/R_\beta)^2(d\beta)^2 + (dz)^2.\end{aligned}$$

Hence a differential shell element at height z has edges of length

$$\begin{aligned}ds_\alpha(z) &= A(1 + z/R_\alpha)d\alpha \\ ds_\beta(z) &= B(1 + z/R_\beta)d\beta\end{aligned} \tag{7.72}$$

and faces of area

$$\begin{aligned}dA_\alpha(z) &= A(1 + z/R_\alpha)d\alpha dz \\ dA_\beta(z) &= B(1 + z/R_\beta)d\beta dz\end{aligned}$$

as depicted in Figure 7.24.

7.6.2 Force and Moment Balancing

Consider force and moment resultants having the orientation depicted in Figure 7.25 and let external forces be denoted by

$$\mathbf{f} = f_\alpha \hat{\boldsymbol{\imath}}_\alpha + f_\beta \hat{\boldsymbol{\imath}}_\beta + f_n \hat{\boldsymbol{\imath}}_n.$$

The displacements in the α, β and z directions are denoted by u, v and w.

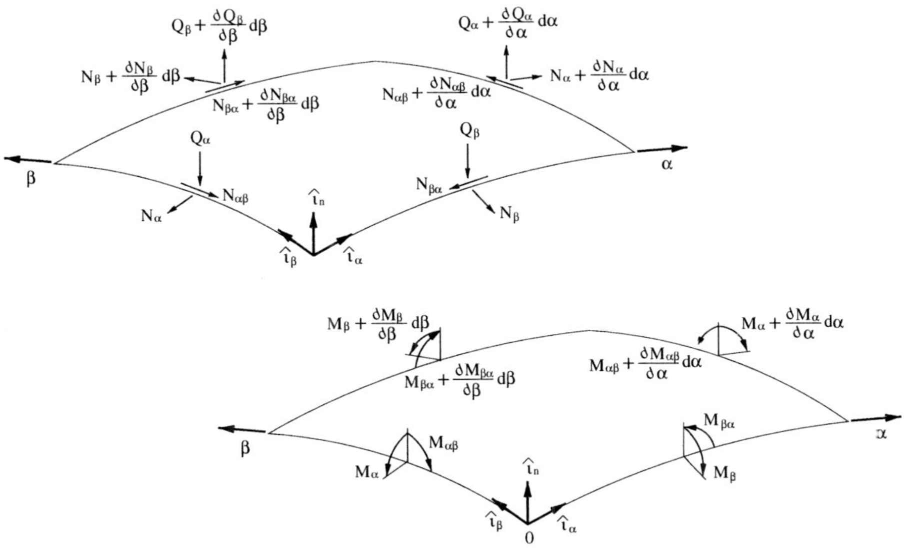

Figure 7.25. *Force and moment resultants in shell coordinates.*

Force balancing in a manner analogous to that detailed in Section 7.5.1 yields

$$\begin{aligned}
\rho h AB\frac{\partial^2 u}{\partial t^2} &= \frac{\partial}{\partial \alpha}(BN_\alpha) + \frac{\partial}{\partial \beta}(AN_{\beta\alpha}) + \frac{\partial A}{\partial \beta}N_{\alpha\beta} \\
&\quad - \frac{\partial B}{\partial \alpha}N_\beta + \frac{AB}{R_\alpha}Q_\alpha + ABf_\alpha \\
\rho h AB\frac{\partial^2 v}{\partial t^2} &= \frac{\partial}{\partial \beta}(AN_\beta) + \frac{\partial}{\partial \alpha}(BN_{\alpha\beta}) + \frac{\partial B}{\partial \alpha}N_{\beta\alpha} \\
&\quad - \frac{\partial A}{\partial \beta}N_\alpha + \frac{AB}{R_\beta}Q_\beta + ABf_\beta \\
\rho h AB\frac{\partial^2 w}{\partial t^2} &= -\frac{AB}{R_\alpha}N_\alpha - \frac{AB}{R_\beta}N_\beta + \frac{\partial}{\partial \alpha}(BQ_\alpha) + \frac{\partial}{\partial \beta}(AQ_\beta) + ABf_n
\end{aligned} \tag{7.73}$$

whereas moment balancing yields

$$\begin{aligned}
&\frac{\partial}{\partial \alpha}(BM_\alpha) + \frac{\partial}{\partial \beta}(AM_{\beta\alpha}) + \frac{\partial A}{\partial \beta}M_{\alpha\beta} - \frac{\partial B}{\partial \alpha}M_\beta - ABQ_\alpha + AB\hat{m}_\beta = 0 \\
&\frac{\partial}{\partial \beta}(AM_\beta) + \frac{\partial}{\partial \alpha}(BM_{\alpha\beta}) + \frac{\partial B}{\partial \alpha}M_{\beta\alpha} - \frac{\partial A}{\partial \beta}M_\alpha - ABQ_\beta + AB\hat{m}_\alpha = 0 \\
&N_{\alpha\beta} - N_{\beta\alpha} + \frac{M_{\alpha\beta}}{R_\alpha} - \frac{M_{\beta\alpha}}{R_\beta} = 0.
\end{aligned} \tag{7.74}$$

The relations (7.73) and (7.74) combine to form a strong formulation for shell models. We next specify the resultants in terms of reference surface strains and rotations.

7.6.3 Strain-Displacement Relations

Following the convention established in previous sections, we let $\sigma_\alpha, \sigma_\beta$ denote normal forces and $\sigma_{\alpha\beta} = \sigma_{\beta\alpha}$, $\sigma_{\alpha z}$, $\sigma_{\beta z}$ denote shear forces having the orientation shown in Figure 7.24. Similarly, we let $\varepsilon_\alpha, \varepsilon_\beta, \varepsilon_z$ and $\varepsilon_{\alpha\beta}$, $\varepsilon_{\alpha z}$, $\varepsilon_{\beta z}$ denote normal and shear strains at arbitrary points in the shell. Finally, $e_\alpha, e_\beta, e_{\alpha\beta}$ denote normal and shear strains in the reference surface and $\kappa_\alpha, \kappa_\beta, \kappa_{\alpha\beta}$ designate reference surface changes in curvature.

Strain-displacement relations for shells are derived from 3-D elasticity relations with various simplifications resulting from the hypotheses of Section 7.2. The simplifications typically involve when and where terms $\frac{z}{R_\alpha}$ and $\frac{z}{R_\beta}$ are neglected and various assumptions in this regard have led to a number of shell theories. The reader is referred to [33, 292] for a more comprehensive treatment and comparison of these theories and we focus solely on the simplified Donnell–Mushtari model and more accurate Byrne–Flügge-Lur'ye theory. The former is a subset of the latter and throughout the discussion, we underline terms in the Byrne–Flügge-Lur'ye theory that are neglected in the Donnell–Mushtari model.

It is an interesting commentary on the evolution of scientific investigations that several of the theories (e.g., Donnell–Mushtari and Byrne–Flügge-Lur'ye) advanced in parallel despite being investigated in isolation. Hence a number of the prominent theories have hyphenated designations.

We begin by posing 3-D elasticity relations in terms of shell coordinates. Letting U, V and W denote displacements in the α, β and z directions at arbitrary points in the shell, this yields the general strain-displacement relations

$$\begin{aligned}
\varepsilon_\alpha &= \frac{1}{1+z/R_\alpha}\left(\frac{1}{A}\frac{\partial U}{\partial \alpha} + \frac{V}{AB}\frac{\partial A}{\partial \beta} + \frac{W}{R_\alpha}\right) \\
\varepsilon_\beta &= \frac{1}{1+z/R_\beta}\left(\frac{1}{B}\frac{\partial V}{\partial \beta} + \frac{U}{AB}\frac{\partial B}{\partial \alpha} + \frac{W}{R_\beta}\right) \\
\varepsilon_z &= \frac{\partial W}{\partial z} \\
\varepsilon_{\alpha\beta} &= \frac{A(1+z/R_\alpha)}{B(1+z/R_\beta)}\frac{\partial}{\partial \beta}\left[\frac{U}{A(1+z/R_\alpha)}\right] \\
&\quad + \frac{B(1+z/R_\beta)}{A(1+z/R_\alpha)}\frac{\partial}{\partial \alpha}\left[\frac{V}{B(1+z/R_\beta)}\right] \\
\varepsilon_{\alpha z} &= \frac{1}{A(1+z/R_\alpha)}\frac{\partial W}{\partial \alpha} + A(1+z/R_\alpha)\frac{\partial}{\partial z}\left[\frac{U}{A(1+z/R_\alpha)}\right] \\
\varepsilon_{\beta z} &= \frac{1}{B(1+z/R_\beta)}\frac{\partial W}{\partial \beta} + B(1+z/R_\beta)\frac{\partial}{\partial z}\left[\frac{V}{B(1+z/R_\beta)}\right]
\end{aligned} \tag{7.75}$$

— see [145, 292] for details.

To simplify these relations, we now invoke Hypothesis 4 of Section 2.2 which posits that lines originally normal to the reference surface remain straight and normal during deformation. We first employ the assumption that deformations are linear in the thickness direction to pose displacements at arbitrary points

$$
\begin{aligned}
U(\alpha,\beta,z) &= u(\alpha,\beta) + z\theta_\alpha(\alpha,\beta)\\
V(\alpha,\beta,z) &= v(\alpha,\beta) + z\theta_\beta(\alpha,\beta)\\
W(\alpha,\beta,z) &= w(\alpha,\beta)
\end{aligned}
\tag{7.76}
$$

in terms of the displacements u, v, w and rotations $\theta_\alpha, \theta_\beta$ of the reference surface — e.g., see Figure 7.26. Secondly, the assumption that fibers remain normal and unextended implies that transverse shear strains $\varepsilon_{\alpha z}, \varepsilon_{\beta z}$ and normal strains ε_z are negligible; hence

$$
\varepsilon_{\alpha z} = \varepsilon_{\beta z} = \varepsilon_z = 0. \tag{7.77}
$$

The substitution of (7.76) into (7.75) and enforcement of (7.77) yields the relations

$$
\theta_\alpha = \frac{u}{R_\alpha} - \frac{1}{A}\frac{\partial w}{\partial \alpha} \quad , \quad \theta_\beta = \frac{v}{R_\beta} - \frac{1}{B}\frac{\partial w}{\partial \beta} \tag{7.78}
$$

for the rotations. Note that for thick structures with significant rotation, the Kirchhoff assumption, and hence (7.77), are relaxed and strain-displacement relations are formulated in terms of $\theta_\alpha, \theta_\beta$ as detailed for shells in [453] and plates in Section 7.8.1.

For thin structures, employment of (7.78) and (7.76) in (7.75) yields the expressions

$$
\begin{aligned}
\varepsilon_\alpha &= \frac{1}{(1+\underline{z/R_\alpha})}(e_\alpha + z\kappa_\alpha)\\
\varepsilon_\beta &= \frac{1}{(1+\underline{z/R_\beta})}(e_\beta + z\kappa_\beta)\\
\varepsilon_{\alpha\beta} &= \frac{1}{(1+\underline{z/R_\alpha})(1+\underline{z/R_\beta})}\left[\left(1 - \underline{\frac{z^2}{R_\alpha R_\beta}}\right)e_{\alpha\beta}\right.\\
&\qquad \left. + z\left(1 + \underline{\frac{z}{2R_\alpha} + \frac{z}{2R_\beta}}\right)\kappa_{\alpha\beta}\right].
\end{aligned}
\tag{7.79}
$$

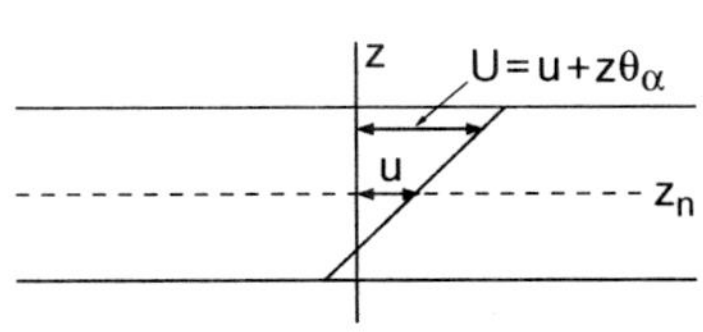

Figure 7.26. *Formulation of the displacement U in terms of the reference surface displacement u and rotation θ_α when $R_\alpha = \infty$.*

relating strains at an arbitrary point in the shell to reference surface strains

$$e_\alpha = \frac{1}{A}\frac{\partial u}{\partial \alpha} + \frac{v}{AB}\frac{\partial A}{\partial \beta} + \frac{w}{R_\alpha}$$

$$e_\beta = \frac{1}{B}\frac{\partial v}{\partial \beta} + \frac{u}{AB}\frac{\partial B}{\partial \alpha} + \frac{w}{R_\beta}$$

$$e_{\alpha\beta} = \frac{A}{B}\frac{\partial}{\partial \beta}\left(\frac{u}{A}\right) + \frac{B}{A}\frac{\partial}{\partial \alpha}\left(\frac{v}{B}\right)$$

and changes in curvature

$$\kappa_\alpha = \frac{1}{A}\frac{\partial \theta_\alpha}{\partial \alpha} + \frac{\theta_\beta}{AB}\frac{\partial A}{\partial \beta}$$

$$\kappa_\beta = \frac{1}{B}\frac{\partial \theta_\beta}{\partial \beta} + \frac{\theta_\alpha}{AB}\frac{\partial B}{\partial \alpha}$$

$$\kappa_{\alpha\beta} = \frac{A}{B}\frac{\partial}{\partial \beta}\left(\frac{\theta_\alpha}{A}\right) + \frac{B}{A}\frac{\partial}{\partial \alpha}\left(\frac{\theta_\beta}{B}\right) + \underline{\frac{1}{R_\alpha}\left(\frac{1}{B}\frac{\partial u}{\partial \beta} - \frac{v}{AB}\frac{\partial B}{\partial \alpha}\right)} \qquad (7.80)$$

$$\underline{+\frac{1}{R_\beta}\left(\frac{1}{A}\frac{\partial v}{\partial \alpha} - \frac{u}{AB}\frac{\partial A}{\partial \beta}\right)}.$$

In combination, relations (7.78)–(7.80) quantify the strain-displacement behavior in the Byrne–Flügge–Lur'ye model whereas the underlined terms are neglected in the Donnell–Mushtari model.

7.6.4 Stress-Strain Behavior

The constitutive relations (7.2) and (7.4) are independent of geometry and hence are directly applicable to general shell models. When combined with (7.78)–(7.80), this quantifies the stress-strain behavior for linear and nonlinear inputs. To clarify the discussion, we will neglect the damping and external components in the subsequent general shell development. Their inclusion is straight-forward as illustrated in Section 7.5 for the rectangular plate model.

7.6.5 Force and Moment Resultants

Force resultants are computed by equating the total force on the face of the differential element depicted in Figure 7.24 with an equivalent resultant acting on the reference surface. To illustrate, consider the force resultant due to the normal stress σ_α. Because the force acting on the area $dA_\beta(z) = ds_\beta(z)dz$ of the element is $\sigma_\alpha dA_\beta(z)$, equating the total force with a resultant acting along the arclength $ds_\beta = Bd\beta$ of the middle surface yields

$$N_\alpha ds_\beta = \int_{-h/2}^{h/2} \sigma_\alpha ds_\beta(z)dz.$$

From the relation (7.72) for $ds_\beta(z)$, it follows that

$$N_\alpha = \int_{-h/2}^{h/2} \sigma_\alpha \left(1 + \frac{z}{R_\beta}\right) dz$$

where N_α has units of force per unit length of middle or reference surface.

The full set of resultants, corresponding to stresses acting on faces perpendicular to the α-axis, can be expressed as

$$\begin{bmatrix} N_\alpha \\ N_{\alpha\beta} \\ Q_\alpha \end{bmatrix} = \int_{-h/2}^{h/2} \begin{bmatrix} \sigma_\alpha \\ \sigma_{\alpha\beta} \\ \sigma_{\alpha z} \end{bmatrix} \left(1 + \frac{z}{R_\beta}\right) dz \ .$$

Similarly, force resultants accommodating stresses perpendicular to the β-axis are

$$\begin{bmatrix} N_\beta \\ N_{\beta\alpha} \\ Q_\beta \end{bmatrix} = \int_{-h/2}^{h/2} \begin{bmatrix} \sigma_\beta \\ \sigma_{\beta\alpha} \\ \sigma_{\beta z} \end{bmatrix} \left(1 + \frac{z}{R_\alpha}\right) dz \ . \tag{7.81}$$

Moment resultants having units of moment per unit length of reference surface include the moment arm z and have the general form

$$\begin{aligned} \begin{bmatrix} M_\alpha \\ M_{\alpha\beta} \end{bmatrix} &= \int_{-h/2}^{h/2} \begin{bmatrix} \sigma_\alpha \\ \sigma_{\alpha\beta} \end{bmatrix} \left(1 + \frac{z}{R_\beta}\right) z dz \\ \begin{bmatrix} M_\beta \\ M_{\beta\alpha} \end{bmatrix} &= \int_{-h/2}^{h/2} \begin{bmatrix} \sigma_\beta \\ \sigma_{\beta\alpha} \end{bmatrix} \left(1 + \frac{z}{R_\alpha}\right) z dz. \end{aligned} \tag{7.82}$$

When evaluating the expressions, various geometric series approximations to the term $\frac{1}{1+z/R_i}, i = \alpha, \beta$, in (7.79) are invoked before integration. Based on the assumption that $\frac{z}{R_i} < 1$, the term is neglected in the Donnell–Mushtari theory whereas terms of degree greater than three are neglected in the Byrne–Flügge–Lur'ye model. In the absence of damping or external forces or moments, the latter case yields

$$\begin{aligned} N_\alpha &= \frac{Yh}{1-\nu^2}\left[e_\alpha + \nu e_\beta - \underline{\frac{h^2}{12}\left(\frac{1}{R_\alpha} - \frac{1}{R_\beta}\right)\left(\kappa_\alpha - \frac{e_\alpha}{R_\alpha}\right)}\right] \\ N_\beta &= \frac{Yh}{1-\nu^2}\left[e_\beta + \nu e_\alpha - \underline{\frac{h^2}{12}\left(\frac{1}{R_\beta} - \frac{1}{R_\alpha}\right)\left(\kappa_\beta - \frac{e_\beta}{R_\beta}\right)}\right] \\ N_{\alpha\beta} &= \frac{Yh}{2(1+\nu)}\left[e_{\alpha\beta} - \underline{\frac{h^2}{12}\left(\frac{1}{R_\alpha} - \frac{1}{R_\beta}\right)\left(\frac{\kappa_{\alpha\beta}}{2} - \frac{e_{\alpha\beta}}{R_\alpha}\right)}\right] \\ N_{\beta\alpha} &= \frac{Yh}{2(1+\nu)}\left[e_{\alpha\beta} - \underline{\frac{h^2}{12}\left(\frac{1}{R_\beta} - \frac{1}{R_\alpha}\right)\left(\frac{\kappa_{\alpha\beta}}{2} - \frac{e_{\alpha\beta}}{R_\beta}\right)}\right] \end{aligned} \tag{7.83}$$

and

$$
\begin{aligned}
M_\alpha &= \frac{Yh^3}{12(1-\nu^2)}\left[\kappa_\alpha + \nu\kappa_\beta - \underline{\left(\frac{1}{R_\alpha} - \frac{1}{R_\beta}\right)e_\alpha}\right]\\
M_\beta &= \frac{Yh^3}{12(1-\nu^2)}\left[\kappa_\beta + \nu\kappa_\alpha - \underline{\left(\frac{1}{R_\beta} - \frac{1}{R_\alpha}\right)e_\beta}\right]\\
M_{\alpha\beta} &= \frac{Yh^3}{24(1+\nu)}\left[\kappa_{\alpha\beta} - \underline{\frac{e_{\alpha\beta}}{R_\alpha}}\right]\\
M_{\beta\alpha} &= \frac{Yh^3}{24(1+\nu)}\left[\kappa_{\alpha\beta} - \underline{\frac{e_{\alpha\beta}}{R_\beta}}\right]
\end{aligned}
\tag{7.84}
$$

where underlined terms are neglected in the Donnell–Mushtari model. We point out that even though the symmetry of the stress tensor dictates that $\sigma_{\alpha\beta} = \sigma_{\beta\alpha}$, $N_{\alpha\beta} \neq N_{\beta\alpha}$ and $M_{\alpha\beta} \neq M_{\beta\alpha}$ unless $R_\alpha = R_\beta$ or higher-order terms are neglected.

7.6.6 Boundary Conditions

Boundary conditions can be specified using either Newtonian (force and moment balancing) principles analogous to those employed for rods in Section 7.3 or energy (work) relations similar to (7.64). As detailed on page 27 of [292], appropriate boundary conditions along edges α_1 and α_2 are

$$
\begin{aligned}
u = 0 \quad &\text{or} \quad N_\alpha = 0\\
v = 0 \quad &\text{or} \quad N_{\alpha\beta} + \frac{M_{\alpha\beta}}{R_\beta} = 0\\
w = 0 \quad &\text{or} \quad Q_\alpha + \frac{1}{B}\frac{\partial M_{\alpha\beta}}{\partial \beta} = 0\\
\theta_\alpha = 0 \quad &\text{or} \quad M_\alpha = 0
\end{aligned}
\tag{7.85}
$$

and $M_{\alpha\beta} w|_{\beta_1}^{\beta_2} = 0$. Note that if β is a closed curve (as will be the case with a cylindrical shell), then this last condition is satisfied identically. Analogous conditions along β_1 and β_2 edges are obtained by reversing the roles of α and β in (7.85).

7.7 Shell Models – Special Cases

The relations (7.73), with stress-strain behavior quantified by (7.2) or (7.4), resultants given by (7.83) and (7.84), and strain-displacement relations (7.79), provide a strong model formulation for general shell geometries. As noted in the introduction to the chapter, these relations are very general and include rod, beam and plate models as subsets. In this section, we illustrate the manner through which specific choices of the radii R_α, R_β and Lamé parameters A, B yield plate, cylindrical shell, and curved beam models. For clarity, we consider the low-order Donnell–Mushtari relations but note that similar analysis applies with the more accurate Byrne–Flügge–Lur'ye model.

7.7.1 Plate Model

To obtain a rectangular plate model, we take

$$\begin{aligned} \alpha &= x & \beta &= y \\ A &= 1 & B &= 1 \\ R_\alpha &= \infty & R_\beta &= \infty \end{aligned} \tag{7.86}$$

in the relations of Section 7.6. It is observed that (7.73) and (7.74), obtained through force and moment balancing, reduce to (7.46), (7.47) and (7.50)–(7.52) whereas the resultant expressions (7.83) and (7.84) reduce to the elastic plate components of (7.58). In a similar manner, the strain-displacement relations for the two geometries are made equivalent by the parameter choices (7.86). Hence the plate is simply a thin shell with no curvature in the undeformed state.

7.7.2 Cylindrical Shell Model — AFM Stage

As illustrated in Figure 7.1, cylindrical shells arise in smart material applications ranging from nanopositioning in an atomic force microscope (AFM) to control of structure-borne noise in structural acoustic cavities. For clarity, we illustrate the development of a cylindrical shell model in the context of the piezoceramic AFM stage depicted in Figure 7.1(a). The extension of the theory to composite shells with surface-mounted patches is analogous to that provided in Section 7.5 for plates.

We focus on characterizing the component of the actuator employed for transverse or axial placement of the sample relative to the microcantilever. For modeling purposes, we consider a shell of radius R, length ℓ, and thickness h with clamped boundary conditions at one end and dissipating elastic conditions at the other end. Because the shell is solely comprised of PZT, we omit subscripts on the material properties. For simplicity, we summarize the Donnell-Mushtari model but note that the Byrne–Flügge–Lur'ye relations are derived in an analogous manner.

We consider the axial direction to be along the x-axis and employ the parameters

$$\begin{aligned} \alpha &= x & \beta &= \theta \\ A &= 1 & B &= R \\ R_\alpha &= \infty & R_\beta &= R \end{aligned}$$

in the general shell relations summarized in Section 7.6.

Strong Model Formulation

Combination of the relations (7.73) and (7.74) with these parameter choices yields the dynamic model

$$
\begin{aligned}
&R\rho h \frac{\partial^2 u}{\partial t^2} - R\frac{\partial N_x}{\partial x} - \frac{\partial N_{x\theta}}{\partial \theta} = Rf_x \\
&R\rho h \frac{\partial^2 v}{\partial t^2} - \frac{\partial N_\theta}{\partial \theta} - R\frac{\partial N_{x\theta}}{\partial x} = Rf_\theta \\
&R\rho h \frac{\partial^2 w}{\partial t^2} - R\frac{\partial^2 M_x}{\partial x^2} - \frac{1}{R}\frac{\partial^2 M_\theta}{\partial \theta^2} - 2\frac{M_{x\theta}}{\partial x \partial \theta} + N_\theta = Rf_n.
\end{aligned}
\tag{7.87}
$$

The force and moment resultants

$$
\begin{aligned}
N_x &= \frac{Yh}{1-\nu^2}(e_x + \nu e_\theta) + \frac{ch}{1-\nu^2}(\dot{e}_x + \nu \dot{e}_\theta) - \frac{h}{1-\nu}[a_1(P-P_R) + a_2(P-P_R)^2] \\
N_\theta &= \frac{Yh}{1-\nu^2}(e_\theta + \nu e_x) + \frac{ch}{1-\nu^2}(\dot{e}_\theta + \nu \dot{e}_x) - \frac{h}{1-\nu}[a_1(P-P_R) + a_2(P-P_R)^2] \\
N_{x\theta} &= \frac{Yh}{2(1+\nu)}e_{x\theta} + \frac{ch}{2(1+\nu)}\dot{e}_{x\theta}
\end{aligned}
\tag{7.88}
$$

and

$$
\begin{aligned}
M_x &= \frac{Yh^3}{12(1-\nu^2)}(\kappa_x + \nu\kappa_\theta) + \frac{ch^3}{12(1-\nu^2)}(\dot{\kappa}_x + \nu\dot{\kappa}_\theta) \\
M_\theta &= \frac{Yh^3}{12(1-\nu^2)}(\kappa_\theta + \nu\kappa_x) + \frac{ch^3}{12(1-\nu^2)}(\dot{\kappa}_\theta + \nu\dot{\kappa}_x) \\
M_{x\theta} &= \frac{Yh^3}{24(1+\nu)}\kappa_{x\theta} + \frac{ch^3}{24(1+\nu)}\dot{\kappa}_{x\theta}
\end{aligned}
\tag{7.89}
$$

include elastic, damping and external components analogous to (7.58)–(7.63) for the plate. Note that we have employed the nonlinear input relations (7.4) when quantifying the external component since the characterization of hysteresis and constitutive nonlinearities can prove crucial when specifying nanoscale displacements. It is observed that in the Donnell–Mushtari theory, poling in a d_{31} manner to produce transverse or axial motion yields null moments since the terms $\frac{z}{R}$ are considered negligible compared with unity. For low drive regimes where the behavior is approximately linear, one can alternatively employ the linear input relation (7.2). Finally, the midsurface strains and changes in curvature are

$$
\begin{aligned}
&e_x = \frac{\partial u}{\partial x} \quad , \quad e_\theta = \frac{1}{R}\frac{\partial v}{\partial \theta} + \frac{w}{R} \quad , \quad e_{x\theta} = \frac{\partial v}{\partial x} + \frac{1}{R}\frac{\partial u}{\partial \theta} \\
&\kappa_x = -\frac{\partial^2 w}{\partial x^2} \quad , \quad \kappa_\theta = -\frac{1}{R^2}\frac{\partial^2 w}{\partial \theta^2} \quad , \quad \kappa_{x\theta} = -\frac{2}{R}\frac{\partial^2 w}{\partial x \partial \theta}.
\end{aligned}
\tag{7.90}
$$

A comparison between (7.87)–(7.90) and the corresponding plate relations reveals that a number of the terms are equivalent if one equates dy and $Rd\theta$. However, the presence of the term $\frac{w}{R}$ in the strain relation for e_θ produces a curvature-induced coupling between displacements not found in in models for flat structures. This impacts both the dynamics quantified by the model and associated approximate techniques for model simulation. It is also noted that if one equates the differentials dy and $Rd\theta$ and takes $R \to \infty$, the Donnell–Mushtari shell model reduces to the plate model.

The boundary conditions for the fixed-end at $x = 0$ are taken to be

$$u = v = w = \frac{\partial w}{\partial x} = 0$$

whereas the conditions

$$N_x = -m\frac{\partial^2 u}{\partial t^2} \quad , \quad N_{x\theta} + \frac{M_{x\theta}}{R} = 0$$

$$Q_x + \frac{1}{R}\frac{\partial M_{x\theta}}{\partial \theta} = 0 \quad , \quad M_x = 0$$

are employed at $x = \ell$. The first resultant condition incorporates the inertial force due to the mass m of the piezoceramic actuator employed for lateral translation along with the mass of the sample.

Weak Model Formulation

To reduce smoothness requirements for approximation and eliminate the Dirac behavior of external inputs at $x = \ell$, we also consider a weak formulation of the model. The state is taken to be $\xi(t) = (u(t,\cdot,\cdot), v(t,\cdot,\cdot), w(t,\cdot,\cdot), u(t,\ell,\cdot))$ in the state space

$$X = L^2(\Omega) \times L^2(\Omega) \times L^2(\Omega) \times L^2(0, 2\pi)$$

where

$$\Omega = [0,\ell] \times [0, 2\pi]$$

denotes the shell region. The space of test functions is specified as

$$V = \left\{\Phi = (\phi_1, \phi_2, \phi_3, \eta) \in X \,|\, \phi_1 \in H_0^1(\Omega), \phi_2 \in H_0^1(\Omega), \phi_3 \in H_0^2(\Omega)\right\}$$

where $\eta(\theta) = \phi_1(\ell, \theta)$ and

$$\begin{aligned} H_0^1(\Omega) &= \left\{\phi \in H^1(\Omega) \,|\, \phi(0,\theta) = 0\right\} \\ H_0^2(\Omega) &= \left\{\phi \in H^2(\Omega) \,|\, \phi(0,\theta) = \phi'(0,\theta) = 0\right\}. \end{aligned} \tag{7.91}$$

Through either variation principles analogous to those in Section 7 3.2 — e.g., see [33] — or integration by parts, one obtains the weak formulation of the thin shell model,

$$
\begin{aligned}
&\int_\Omega \left\{ R\rho h \frac{\partial^2 u}{\partial t^2}\phi_1 + RN_x \frac{\partial \phi_1}{\partial x} + N_{x\theta}\frac{\partial \phi_1}{\partial \theta} - Rf_x\phi_1 \right\} d\omega = 0 \\
&\int_\Omega \left\{ R\rho h \frac{\partial^2 v}{\partial t^2}\phi_2 + N_\theta \frac{\partial \phi_2}{\partial \theta} + RN_{x\theta}\frac{\partial \phi_2}{\partial x} - Rf_\theta\phi_2 \right\} d\omega = 0 \\
&\int_\Omega \left\{ R\rho h \frac{\partial^2 w}{\partial t^2}\phi_3 - RM_x \frac{\partial^2 \phi_3}{\partial x^2} - 2M_{x\theta}\frac{\partial^2 \phi_3}{\partial x \partial \theta} - \frac{1}{R}M_\theta \frac{\partial^2 \phi_3}{\partial \theta^2} \right. \\
&\qquad\qquad\qquad\qquad\qquad\qquad \left. + N_\theta \phi_3 - Rf_n\phi_3 \right\} d\omega = 0,
\end{aligned}
\tag{7.92}
$$

which must be satisfied for all $\Phi \in V$. The resultants are given by (7.88) and (7.89) with midsurface strains and changes in curvature designated in (7.90). Numerical methods for approximating solutions to (7.92) are discussed in Section 8.5.

7.7.3 Curved Beam Model

The narrow THUNDER transducer shown in Figure 7.4(c) is curved in the region covered by PZT and hence exhibits curvature-induced coupling between in-plane and out-of-plane motion. Moreover, it is sufficiently narrow that motion in the width (longitudinal) direction is negligible. Hence it behaves as a thin beam with coupled circumferential and transverse dynamics.

Thin beam models for such curved geometries can be obtained directly from previous shell models. As detailed in [491], the geometry in the patch region has been experimentally verified to have nearly constant radius of curvature in θ with negligible curvature in x, as predicted by thermomechanical theory, so we start with the Donnell–Mushtari cylindrical shell model summarized in Section 7.7.2. More accurate Byrne–Flügge–Lur'ye relations can be constructed by retaining higher-order terms in the manner indicated in previous sections.

We initially consider a homogeneous thin beam having width b, thickness h, and constant radius of curvature R as depicted in Figure 7.27. The circumferential and transverse displacements are denoted by v and w.

From (7.87), it follows that a strong formulation of the curved beam model for this geometry is

$$
\begin{aligned}
&R\rho h \frac{\partial^2 v}{\partial t^2} - \frac{\partial N_\theta}{\partial \theta} = Rf_\theta \\
&R\rho h \frac{\partial^2 w}{\partial t^2} - \frac{1}{R}\frac{\partial^2 M_\theta}{\partial \theta^2} + N_\theta = Rf_n
\end{aligned}
\tag{7.93}
$$

where the resultants are defined in (7.88) and (7.89). To illustrate the manner through which curvature-induced coupling between v and w components is introduced, consider the undamped case ($c = 0$) in the absence of voltage or field inputs.

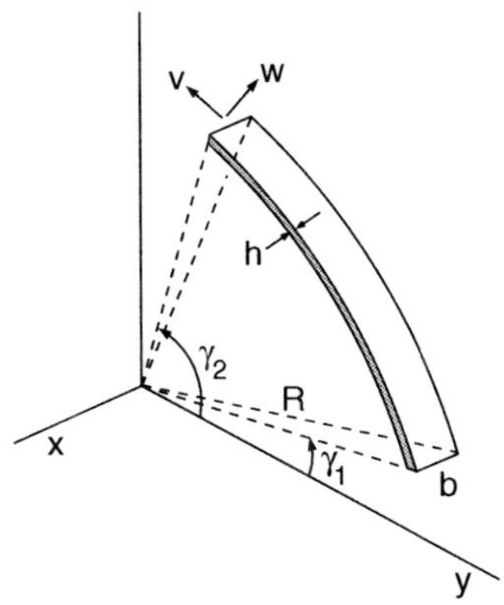

Figure 7.27. *Curved beam in which circumferential and transverse motion are coupled due to curvature.*

In this case, the resultants are

$$N_\theta = Yh\left(\frac{1}{R}\frac{\partial v}{\partial \theta} + \frac{w}{R}\right)$$

$$M_\theta = -\frac{Yh^3}{12R^2}\frac{\partial^2 w}{\partial \theta^2}.$$

Hence w-dependence is introduced in the equation of motion for v through the term $\frac{w}{R}$ whereas v-dependence in the second relation of (7.93) is introduced through the strain $\frac{\partial v}{\partial \theta}$.

To construct a corresponding variational or weak formulation, we consider states $\xi(t) = (v(t,\cdot), w(t,\cdot))$ in the state space

$$X = L^2(\Omega) \times L^2(\Omega)$$

where $\Omega = [\gamma_1, \gamma_2]$. The space of test functions is

$$V = H_b^2(\Omega) \times H_b^2(\Omega)$$

where the subscript b indicates subsets of the spaces $H^1(\Omega)$ and $H^2(\Omega)$ comprised of functions that satisfy essential boundary conditions. A weak formulation is then

$$\int_{\gamma_1}^{\gamma_2} \left\{ R\rho h \frac{\partial^2 v}{\partial t^2}\phi_1 + N_\theta \frac{d\phi_1}{d\theta} - Rf_\theta \phi_1 \right\} d\omega = 0$$

$$\int_{\gamma_1}^{\gamma_2} \left\{ R\rho h \frac{\partial^2 w}{\partial t^2}\phi_2 + N_\theta \phi_2 - \frac{1}{R} M_\theta \frac{d^2\phi_2}{d\theta^2} - Rf_n \phi_2 \right\} = 0$$

which must be satisfied for all $(\phi_1, \phi_2) \in V$.

7.7.4 Flat Beam Model

The uncoupled rod equations quantifying in-plane motion and flat beam equations characterizing out-of-plane motion were derived in Sections 7.3 and 7.4. They also follow directly from the general shell models with $\beta = y$, $B = 1$ and $R_\beta = \infty$ which implies that differentials $Rd\theta$ in the curved beam model are replaced by dy and $R \to \infty$ to yield the *uncoupled* relations

$$\rho h \frac{\partial^2 v}{\partial t^2} - \frac{\partial N_y}{\partial y} = f_y$$

$$\rho h \frac{\partial^2 w}{\partial t^2} - \frac{\partial^2 M_y}{\partial y^2} = f_n.$$

In the absence of damping or inputs, the resultants are

$$N_y = Yh\frac{\partial v}{\partial y} \quad , \quad M_y = -\frac{Yh^3}{12}\frac{\partial^2 w}{\partial y^2}$$

which are also uncoupled. The in-plane relation is exactly the undamped rod model (7.14) whereas the transverse expression is the undamped beam model (7.38). The inclusion of damping and input components yields the complete rod and beam models developed in Sections 7.3 and 7.4.

7.8 Timoshenko, Mindlin–Reissner, and von Kármán Models

The rod, beam, plate and shell models, developed in previous sections, are based on Assumptions 1–4 of Section 7.2. In this section, we relax various assumptions to obtain the linear Mindlin–Reissner and Timoshenko models, which incorporate shear deformations and rotational effects, and the nonlinear von Kármán relations.

7.8.1 Mindlin–Reissner Plate and Timoshenko Beam Models

The fourth hypothesis of Section 7.2 asserts that normal lines to the reference surface remain normal during bending. As illustrated in Figure 7.28, this is reasonable for thin structures with moderate rotational effects but fails in thick structures with significant rotation due to nonnegligible transverse shear deformations.[33] The Mindlin–Reissner and Timoshenko models result when Assumption 4 is relaxed to allow transverse shear strains while retaining the assumption that filaments remain straight and unstrained during deformation.

Mindlin–Reissner Plate Model

For simplicity, we consider a homogeneous plate of thickness h in the absence of damping ($c = 0$) and inputs ($V = E = 0$). The extensions to include these effects are analogous to those detailed in Sections 7.5.1.

[33]This is easily illustrated by noting how a thick paperback book bends as compared with bending of a thin book.

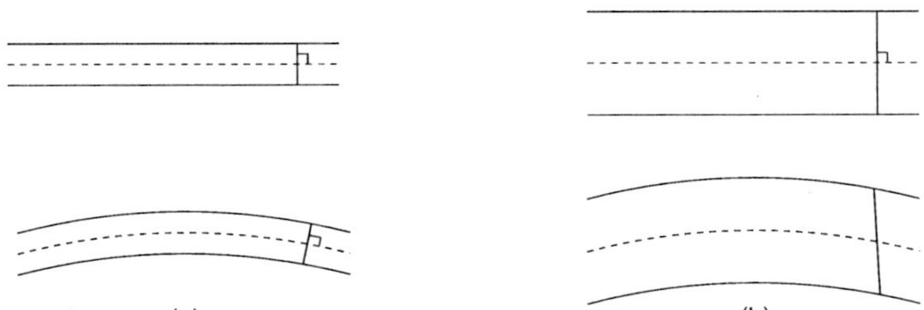

Figure 7.28. *Behavior of normal lines to the reference surface during bending. (a) Lines remain normal in thin structures with moderate rotation and (b) non-normal response in thick structures due to transverse shear strains.*

To formulate the model, we take $\alpha = x, A = 1, R_\alpha = \infty$, $\beta = y, B = 1$ and $R_\beta = \infty$ in the general shell relations (7.75)–(7.80) to obtain the strain-displacement relations

$$\begin{aligned} &\varepsilon_x = e_x + z\kappa_x \quad , \quad \varepsilon_y = e_y + z\kappa_y \\ &\varepsilon_{xy} = e_{xy} + z\kappa_{xy} \quad , \quad \varepsilon_{xz} = \frac{\partial w}{\partial x} + \theta_x \quad , \quad \varepsilon_{yz} = \frac{\partial w}{\partial y} + \theta_y \end{aligned} \tag{7.94}$$

where θ_x, θ_y are rotations of the reference surface and

$$\begin{aligned} &e_x = \frac{\partial u}{\partial x} \quad , \quad e_y = \frac{\partial v}{\partial y} \quad , \quad e_{xy} = \frac{\partial u}{\partial y} + \frac{\partial v}{\partial x} \\ &\kappa_x = \frac{\partial \theta_x}{\partial x} \quad , \quad \kappa_y = \frac{\partial \theta_y}{\partial y} \quad , \quad \kappa_{xy} = \frac{\partial \theta_x}{\partial y} + \frac{\partial \theta_y}{\partial x}. \end{aligned} \tag{7.95}$$

Note that if Kirchhoff's hypothesis is invoked so that $\varepsilon_{xz} = \varepsilon_{yz} = 0$ in (7.94), then the kinematic relations (7.95) are the same as the thin plate relations (7.56). However, retention of these terms eliminates one of the contradictions arising from the assumption of all four postulates [292].

The force and moment resultants

$$\begin{aligned} N_x &= \frac{Yh}{1-\nu^2}\left(\frac{\partial u}{\partial x} + \nu\frac{\partial v}{\partial y}\right) \quad , \quad M_x = \frac{Yh^3}{12(1-\nu^2)}\left(\frac{\partial \theta_x}{\partial x} + \nu\frac{\partial \theta_y}{\partial y}\right) \\ N_y &= \frac{Yh}{1-\nu^2}\left(\frac{\partial v}{\partial y} + \nu\frac{\partial u}{\partial x}\right) \quad , \quad M_y = \frac{Yh^3}{12(1-\nu^2)}\left(\frac{\partial \theta_y}{\partial y} + \nu\frac{\partial \theta_x}{\partial x}\right) \\ N_{xy} &= \frac{Yh}{2(1+\nu)}\left(\frac{\partial u}{\partial y} + \frac{\partial v}{\partial x}\right) \quad , \quad M_{xy} = \frac{Yh^3}{24(1+\nu)}\left(\frac{\partial \theta_y}{\partial x} + \frac{\partial \theta_x}{\partial y}\right) \end{aligned} \tag{7.96}$$

follow from (7.83) and (7.84) whereas the shear resultants

$$Q_x = \frac{K^2 Yh}{2(1+\nu)}\left(\frac{\partial w}{\partial x} + \theta_x\right) \quad , \quad Q_y = \frac{K^2 Yh}{2(1+\nu)}\left(\frac{\partial w}{\partial y} + \theta_y\right)$$

are provided by (7.81) and (7.82). The constant K^2 compensates for the fact that the outer surface of the plate cannot support a shear stress. Whereas averaging values can be used to compute theoretical values for K^2, in applications it is typically treated as a parameter to be estimated.

Force balancing in the manner detailed in Section 7.5.1 yields the dynamic equations

$$\rho h \frac{\partial^2 u}{\partial t^2} - \frac{\partial N_x}{\partial x} - \frac{\partial N_{yx}}{\partial y} = f_x$$

$$\rho h \frac{\partial^2 v}{\partial t^2} - \frac{\partial N_y}{\partial y} - \frac{N_{xy}}{\partial x} = f_y$$

$$\rho h \frac{\partial^2 w}{\partial t^2} - \frac{\partial Q_x}{\partial x} - \frac{\partial Q_y}{\partial y} = f_n$$

while inclusion of rotational inertia when balancing moments yields

$$\begin{aligned} \frac{\rho h^3}{12} \frac{\partial^2 \theta_x}{\partial t^2} &= \frac{\partial M_x}{\partial x} + \frac{\partial M_{xy}}{\partial y} - Q_x \\ \frac{\rho h^3}{12} \frac{\partial^2 \theta_y}{\partial t^2} &= \frac{\partial M_y}{\partial y} + \frac{\partial M_{xy}}{\partial x} - Q_y. \end{aligned} \tag{7.97}$$

It is observed that the relaxation of the Kirchhoff hypothesis and inclusion of rotational inertia affects only the transverse relation. It is detailed in [291,453] that inclusion of shear deformations decreases the stiffness whereas the rotational inertia increases mass effects. Both serve to decrease modeled frequencies and in a number of applications, including those with multiple frequencies, the Reissner–Mindlin plate model provides better accuracy than the Kirchhoff plate model developed in Section 7.5.

Timoshenko Beam Model

The 1-D analogue of the Mindlin–Reissner plate model is the Timoshenko beam model. Since it follows directly from (7.94)–(7.97) when one considers transverse displacements in addition to longitudinal displacements in either x or y, we do not repeat the relations. The advantages that the Timoshenko model provide over the Euler–Bernoulli model developed in Section 7.4 are the same as those provided by the Mindlin–Reissner plate model.

7.8.2 von Kármán Plate Model

As a result of Assumption 2 of Section 7.2, kinematic and equilibrium relations for the rod, beam, plate and shell models developed in previous sections, were considered with respect to the *unperturbed* reference surface. Furthermore, high-order strain-displacement terms were neglected in accordance with the assumption of small displacements. The results of Assumption 2 are twofold: (i) the models exhibit *linear* state-dependence, and (ii) the modeling relations for in-plane and out-of-plane motion are *decoupled* for flat structures ($R_\alpha = R_\beta = \infty$).

In this section, we relax this assumption to accommodate large displacements of the type often exhibited by THUNDER transducers and MEMs of the type depicted in Figure 7.4. This yields the nonlinear von Kármán plate model in which longitudinal and transverse displacements are coupled.

To clarify the discussion, we again consider a homogeneous plate of thickness h for which damping and external voltages or fields are neglected; hence the reference surface coincides with the middle surface so $z_n = 0$ and moments contain only elastic components. Extension of the model to incorporate damping, linear and nonlinear inputs, and geometric nonhomogeneities follow in the manner detailed in Section 7.5.1.

The first extension to accommodate displacements that are large compared with the thickness is to balance forces and moments with regard to the *deformed* reference surface depicted in Figure 7.29. When defining the deformation, it is typical to approximate the sine of rotation angles by changes in the slope.

Force and Moment Balancing

As detailed in [291, 480], balancing of transverse forces yields the nonlinear relations

$$\rho h \frac{\partial^2 w}{\partial t^2} = \frac{\partial Q_x}{\partial x} + \frac{\partial Q_y}{\partial y} + \frac{\partial}{\partial x}\left(N_x \frac{\partial w}{\partial x}\right) + \frac{\partial}{\partial y}\left(N_y \frac{\partial w}{\partial y}\right)$$

$$+ \frac{\partial}{\partial x}\left(N_{xy} \frac{\partial w}{\partial y}\right) + \frac{\partial}{\partial y}\left(N_{xy} \frac{\partial w}{\partial x}\right) + f_n$$

when third-order differential terms are neglected. Similarly, balancing forces in the

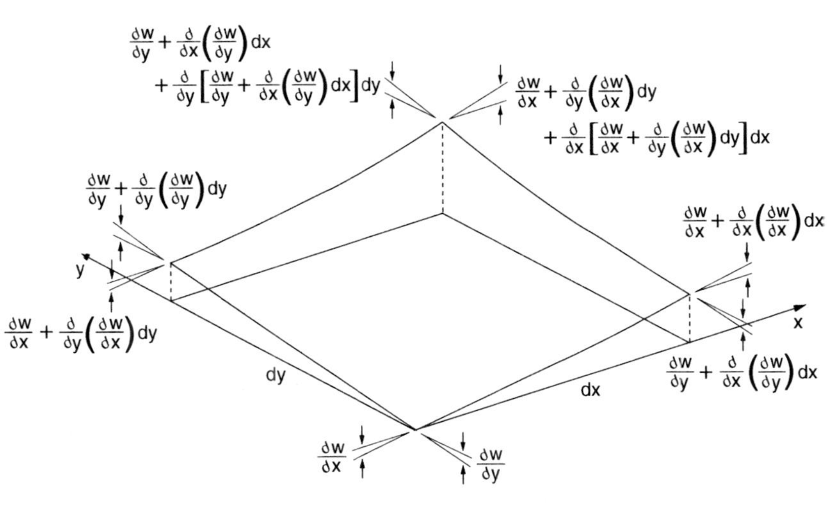

Figure 7.29. *Deformed reference surface for a thin plate.*

x and y directions yields the in-plane relations

$$\rho h \frac{\partial^2 u}{\partial t^2} = \frac{\partial N_x}{\partial x} + \frac{\partial N_{xy}}{\partial y} - \frac{\partial}{\partial x}\left(Q_x \frac{\partial w}{\partial x}\right) - \frac{\partial}{\partial y}\left(Q_y \frac{\partial w}{\partial x}\right) + f_x$$

$$\rho h \frac{\partial^2 v}{\partial t^2} = \frac{\partial N_y}{\partial y} + \frac{\partial N_{xy}}{\partial x} - \frac{\partial}{\partial y}\left(Q_y \frac{\partial w}{\partial y}\right) - \frac{\partial}{\partial x}\left(Q_x \frac{\partial w}{\partial y}\right) + f_y.$$

It is observed that these nonlinear equations reduce to the linear model (7.46) and (7.47) if high-order terms are neglected.

The incorporation of rotational inertia but neglect of high-order terms yields

$$\begin{aligned} \frac{\rho h^3}{12} \frac{\partial^3 w}{\partial x \partial t^2} &= Q_x - \frac{\partial M_x}{\partial x} - \frac{\partial M_{yx}}{\partial y} \\ \frac{\rho h^3}{12} \frac{\partial^3 w}{\partial y \partial t^2} &= Q_y - \frac{\partial M_y}{\partial y} - \frac{\partial M_{xy}}{\partial x} \end{aligned} \tag{7.98}$$

when moments are balanced. These are a simplification of the Mindlin–Reissner relations (7.97) based on the assumption that $\varepsilon_{xz} = \varepsilon_{yz} = 0$, and hence $\theta_x = -\frac{\partial w}{\partial x}, \theta_y = -\frac{\partial w}{\partial y}$, in accordance with Assumption 2 of Section 7.2. It is observed that (7.98) reduces to (7.50) and (7.51) if rotational inertia is neglected.

Force and Moment Resultants

To accommodate large displacements, quadratic terms are retained in the strain-displacement relations

$$\varepsilon_x = \frac{\partial u}{\partial x} + \frac{1}{2}\left(\frac{\partial w}{\partial x}\right)^2, \quad \varepsilon_y = \frac{\partial v}{\partial y} + \frac{1}{2}\left(\frac{\partial w}{\partial y}\right)^2, \quad \varepsilon_{xy} = \frac{\partial v}{\partial x} + \frac{\partial u}{\partial y} + \frac{\partial w}{\partial x}\frac{\partial w}{\partial y}$$

which yields the force resultants

$$N_x = \frac{Eh}{1-\nu^2}\left[\frac{\partial u}{\partial x} + \nu \frac{\partial v}{\partial y} + \frac{1}{2}\left(\frac{\partial w}{\partial x}\right)^2 + \frac{\nu}{2}\left(\frac{\partial w}{\partial y}\right)^2\right]$$

$$N_y = \frac{Eh}{1-\nu^2}\left[\frac{\partial v}{\partial y} + \nu \frac{\partial u}{\partial x} + \frac{1}{2}\left(\frac{\partial w}{\partial y}\right)^2 + \frac{\nu}{2}\left(\frac{\partial w}{\partial x}\right)^2\right]$$

$$N_{xy} = \frac{Eh}{2(1+\nu)}\left[\frac{\partial u}{\partial y} + \frac{\partial v}{\partial x} + \frac{\partial w}{\partial x}\frac{\partial w}{\partial y}\right].$$

The moment resultants remain the same as those defined in linear theory — e.g., see (7.57) and (7.58) or (7.96) with $\theta_x = -\frac{\partial w}{\partial x}, \theta_y = -\frac{\partial w}{\partial y}$.

We note that in this nonlinear von Kármán model, the longitudinal and transverse displacements are coupled due to the curvature of the deformed reference surface and the retention of high-order terms in the kinematic relations. As expected, the model reduces to the linear Kirchhoff model developed in Section 7.5 under the assumption of small displacements.

The reader is directed to [291] for further discussion regarding properties of the von Kármán model and to [277] for a derivation of the model using energy principles.

7.9 THUNDER Models

To illustrate principles developed throughout this chapter, we discuss issues pertaining to characterization of initial shapes and displacements generated by THUNDER transducers. We note that this is an active research topic and the model discussed here should be interpreted as initial formulations to illustrate issues rather than final frameworks which fully characterize the complex behavior of the devices. Limitations and open research questions will be indicated at various points in the discussion.

We consider a narrow THUNDER device of the type illustrated in Figure 7.4 and 7.30(a) which exhibits negligible curvature and motion in the width direction. One end is clamped while the other is constrained to slide freely in the horizontal direction as depicted in Figure 7.30(d). For specificity, we consider a physical transducer comprised of a stainless steel backing strip, an adhesive Layer of LaRC Si, a PZT layer, and a protective top coating of LaRC Si. The steel has dimensions 0.5 in × 2.5 in × 0.015 in and the centered PZT is 0.5 in × 1.5 in × 0.008 in. The mean thickness of the LaRC Si is 0.001 in. We include the specific constituent materials and dimensions to indicate a prototypical size but note that the modeling principles are generic and hence apply to a range of compounds and dimensions.

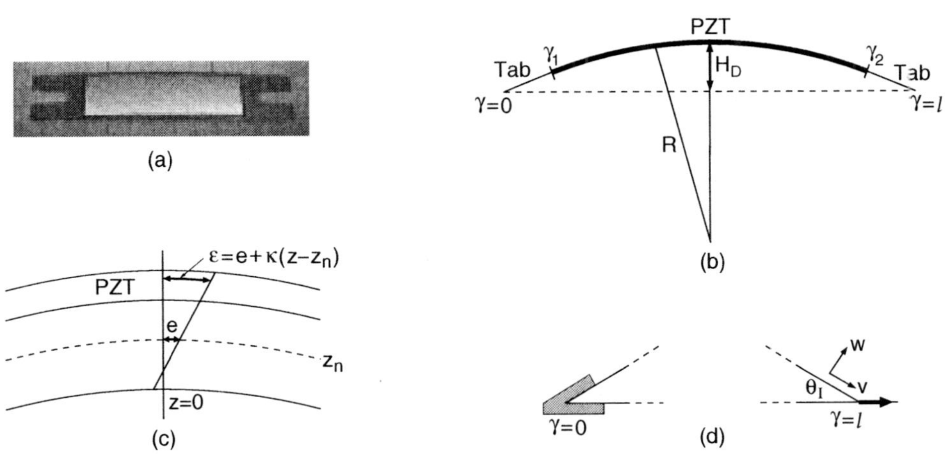

Figure 7.30. *(a) Narrow THUNDER transducer and (b) geometry comprised of flat tabs and a circular arc having radius of curvature R in the region $[\gamma_1, \gamma_2]$ covered by PZT. (c) Reference surface and decomposition of strains ε into an in-plane component e and a bending component $\kappa(z - z_n)$. (d) Fixed-end condition at $\gamma = 0$ and sliding end at $\gamma = \ell$.*

The geometry employed for model development is established in Figure 7.29(b) and (c). The coordinate for arclength is denoted by γ where $\gamma = x$ in the flat tabs and $\gamma = R\theta$ in the curved region covered by PZT. The ends of the PZT are delineated by γ_1 and γ_2, and the entire transducer has length ℓ and width b. Material properties and dimensions of the backing layer, LaRC Si adhesive, and PZT are respectively indicated by the subscripts I, S and A. We orient the thickness coordinate so that $z = 0$ corresponds to the outer edge of the backing material.

There are two related but distinct phases of model development. In the first, thermal, elastic and electromechanical forces are balanced to quantify the shape of the device as a function of constituent materials and manufacturing conditions. This is important both for device characterization and the inverse problem of constructing transducers having prescribed geometries and attributes. Secondly, curved and flat beam relations are coupled to construct dynamic models which quantify in-plane and out-of-plane displacements due to input fields.

To simplify the discussion, we focus primarily on models having linear state dependence and linear or nonlinear inputs. However, we caution the reader that the linear theory has limited applicability for large displacements and high drive regimes. Furthermore, the nonlinear input relations rely on the assumption that stresses do no exceed the critical stress σ_c which delineates the initiation of stress-induced switching. These relations must also be extended to incorporate the stress-dependent electromechanical behavior shown in Figure 1.6(c).

7.9.1 Linear State Dependence

Actuator Geometry

During the manufacturing process, the constituent materials are heated in a vacuum to approximately 325 ^{o}C under a pressure of 241.3 kPa. During the cooling process, the LaRC Si solidifies at approximately 270 ^{o}C and subsequent cooling produces curvature in the composite structure due to the differing thermal coefficients of the constituent materials. Because the Curie point of PZT5A (350 ^{o}C) is in the proximity of the peak manufacturing temperature, the device is subsequently repoled after cooling. This increases the radius of curvature R and hence decreases the dome height H_D of the physical device.

To consolidate notation, we let the indices $i = 1, \ldots, 4$ respectively correspond to the ordered subscripts I, S, A, S indicating the constituent materials as shown in Figure 7.31. The thermal coefficients are generically denoted by α, and e and κ respectively denote the reference surface strain and change in curvature. Finally, we define

$$H_i = \sum_{j=1}^{i} h_j \tag{7.99}$$

to indicate the z values that delineate the various layers — i.e., $H_0 = 0$, $H_1 = h_I, H_2 = h_I + h_S$, $H_3 = h_I + h_S + H_A$ and $H_4 = h_I + h_S + H_A + H_S$.

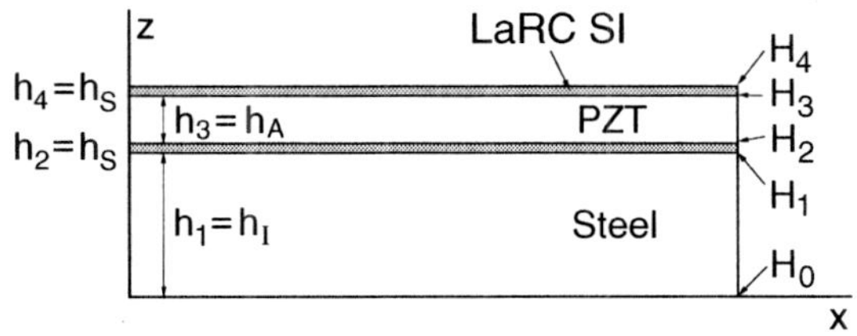

Figure 7.31. *Orientation employed when quantifying the transducer geometry.*

Under the assumption that strains are proportion to ΔT during cooling, the strain $\varepsilon(z)$ at a height z in the composite can be expressed as

$$\varepsilon(z) = \frac{\sigma(z)}{Y(z)} + \alpha(z)\Delta T - \frac{3}{2}\nu\lambda_s\delta(z) \tag{7.100}$$

where $Y(z) = Y_i$ and $\alpha(z) = \alpha_i$ for z in the ith layer, ν is the Poisson ratio and λ_s is the saturation electrostriction. The third term on the right hand side quantifies strains due to dipole rotation during repoling using analysis similar to that employed for magnetostrictive materials in Section 4.1.8. The Kronecker delta

$$\delta = \begin{cases} 1\,, & \text{if } z \text{ in PZT layer} \\ 0\,, & \text{otherwise} \end{cases}$$

isolates the electromechanical strains due to repoling to the PZT layer.

Additionally, the assumption that strains are linear through the thickness yields the relation

$$\varepsilon(z) = e + \kappa(z - z_n)$$

which is illustrated in Figure 7.30(c). The neutral or reference surface is specified through the force balance

$$\sum_{i=1}^{4} \int_{h_i} Y(z)\kappa(z - z_n)\,dz = 0 \tag{7.101}$$

which is analogous to (7.30) employed when computing the neutral surface for the unimorph. Evaluation of (7.101) yields

$$z_n = \frac{\sum_{i=1}^4 h_i^2 Y_i + 2\sum_{i=2}^4 h_i Y_i H_i}{2\sum_{i=1}^4 h_i Y_i}$$

where H_i is defined in (7.99).

To determine the neutral surface strain e and curvature change κ, forces and moments are balanced through the layers to provide the constraints

$$b\sum_{i=1}^{4} \int_{h_j} \sigma\,dz = 0 \quad , \quad b\sum_{i=1}^{4} \int_{h_j} z\sigma\,dz = 0$$

where

$$\sigma(z) = \left[e + \kappa(z - z_n) - \alpha(z)\Delta T + \frac{3}{2}\nu\lambda_s\delta(z)\right] Y(z).$$

This yields the 2×2 system

$$A\mathcal{E} = f \tag{7.102}$$

where $\mathcal{E} = [e, \kappa]^T$ and

$$A = \begin{bmatrix} \sum_{i=1}^4 Y_i(H_i - H_{i-1}) & \sum_{i=1}^4 Y_i\left[\frac{1}{2}(H_i^2 - H_{i-1}^2) - z_n(H_i - H_{i-1})\right] \\ \sum_{i=1}^4 \frac{1}{2}Y_i(H_i^2 - H_{i-1}^2) & \sum_{i=1}^4 Y_i\left[\frac{1}{3}(H_i^3 - H_{i-1}^3) - \frac{z_n}{2}(H_i^2 - H_{i-1}^2)\right] \end{bmatrix}$$

$$f = \begin{bmatrix} \sum_{i=1}^4 Y_i(\alpha_i\Delta T - 3/2\delta\nu\lambda_s)(H_i - H_{i-1}) \\ \sum_{i=1}^4 \frac{1}{2}Y_i(\alpha_i\Delta T - 3/2\delta\nu\lambda_s)(H_i^2 - H_{i-1}^2) \end{bmatrix}.$$

We note that when constructing A and f, some properties such as layer thicknesses and Young's moduli for steel and PZT can be directly measured or obtained from manufacturer specifications whereas other parameters — e.g., thermal coefficients and moduli for LaRC Si and the saturation electrostriction λ_s — are estimated through a least squares fit to the data.

For a given set of material properties and dimensions, solution of (7.102) yields e and κ and hence provides the radius of curvature

$$R = \frac{1}{\kappa}.$$

In experiments, however, one typically measures the dome height H_D depicted in Figure 7.29(b). For a transducer having flat tabs of length γ_t and PZT-covered region with arclength γ_s, it is shown in [77] that the dome height and radius of curvature are related by the expression

$$H_D = R\left[1 - \cos\left(\frac{\gamma_s}{2R}\right)\right] + \gamma_t \sin\left(\frac{\gamma_s}{2R}\right).$$

The performance of the model when predicting dome heights associated with various constituent materials is illustrated in [25, 77, 509].

It is noted that the previous analysis predicts a constant radius of curvature R through the region covered by PZT and flat tabs outside that region. These predictions have been experimentally validated in [491].

Displacement Model

The previous component of the model predicts the radius of curvature R and dome height H_D as a function of material properties and manufacturing conditions. Here we construct a dynamic model by combining the relations for a curved beam having radius of curvature R and flat beam expressions for the tabs. To simplify the discussion, we make the assumption that the LaRC Si layers have negligible effect on the dynamics and neglect their contribution. The extension of the model to include these adhesive layers is straight-forward.

To delineate the region covered by the patch, we define the characteristic function

$$\chi_{pe}(\gamma) = \begin{cases} 1 & , \quad \gamma_1 \leq \gamma \leq \gamma_2 \\ 0 & , \quad \text{otherwise} \end{cases}$$

where $\gamma = x$ in the tabs and $\gamma = R\theta$ in the patch region.

Under the assumption that rotational inertia and shear deformations are negligible, the longitudinal and transverse displacements v and w are quantified by the dynamic equations

$$\begin{aligned} &\rho(\gamma)\frac{\partial^2 v}{\partial t^2} - \frac{\partial N_\gamma}{\partial \gamma} = 0 \\ &\rho(\gamma)\frac{\partial^2 w}{\partial t^2} - \frac{\partial^2 M_\gamma}{\partial \gamma^2} + \frac{1}{R}N_\gamma \chi_{pe}(\gamma) = f_n \end{aligned} \tag{7.103}$$

as specified in (7.93). Here f_n denotes applied normal loads and

$$\rho(\gamma) = \rho_I h_I + \rho_A h_A \chi_{pe}(\gamma).$$

The resultant expressions

$$\begin{aligned} N_\gamma &= \int_0^{h_I} \sigma(z)\,dz + \int_{h_I}^{h_I+h_A} \sigma(z)\,dz \\ M_\gamma &= \int_0^{h_I} z\sigma(z)\,dz + \int_{h_I}^{h_I+h_A} z\sigma(z)\,dz \end{aligned}$$

with the constitutive relations

$$\begin{aligned} &\sigma(z) = Y(z)\varepsilon + c(z)\dot{\varepsilon} - \left[a_1 P + a_2 P^2\right]\chi_{pe}(\gamma)\delta(z) \\ &\varepsilon(z) = e + \kappa(z - z_n) \end{aligned}$$

yields

$$\begin{aligned} N_\gamma &= [Y_I h_I + Y_A h_A \chi_{pe}(\gamma)]\, e_\gamma + [Y_I c_1 + Y_A c_2 \chi_{pe}(\gamma)]\,\kappa \\ &+ [c_I h_I + c_A h_A \chi_{pe}(\gamma)]\,\dot{e}_\gamma + [c_I c_1 + c_A c_2 \chi_{pe}(\gamma)]\,\dot{\kappa} \\ &- h_A \chi_{pe}(\gamma)\left[a_1(P(E) - P_R) + a_2(P(E) - P_R)^2\right] \\ M_\gamma &= \left[Y_I h_I^2/2 + Y_A c_3 \chi_{pe}(\gamma)\right] e_\gamma + [Y_I c_4 + Y_A c_5 \chi_{pe}(\gamma)]\,\kappa \\ &+ \left[c_I h_I^2/2 + c_A c_3 \chi_{pe}(\gamma)\right] \dot{e}_\gamma + [c_I c_4 + c_A c_5 \chi_{pe}(\gamma)]\,\dot{\kappa} \\ &- c_2 \chi_{pe}(\gamma)\left[a_1(P(E) - P_R) + a_2(P(E) - P_R)^2\right] \end{aligned} \tag{7.104}$$

where

$$c_1 = \frac{H_1^2}{2} - H_1 z_n \quad , \quad c_2 = \left(\frac{H_2^2}{2} - \frac{H_1^2}{2}\right) - (H_2 - H_1)\, z_n$$

$$c_3 = \frac{H_2^2}{2} - \frac{H_1^2}{2} \quad , \quad c_4 = \frac{H_1^3}{3} - \frac{H_1^2}{2} z_n \quad , \quad c_5 = \left(\frac{H_2^3}{3} - \frac{H_1^3}{3}\right) - \left(\frac{H_2^2}{2} - \frac{H_1^2}{2}\right) z_n$$

and

$$H_1 = h_I \quad , \quad H_2 = h_I + h_A.$$

The nonlinear and hysteretic E-P relation is quantified by (7.3) or (2.114) and the reference surface strains and curvature changes are given by

$$e_\gamma = \frac{\partial u}{\partial \gamma} + \frac{w}{R}\chi_{pe}(\gamma) \quad , \quad \kappa_\gamma = -\frac{\partial^2 w}{\partial \gamma^2}. \tag{7.105}$$

In combination, (7.103)–(7.105) provide a strong formulation of the model.

Boundary Conditions

Recall that the transducer is assumed to have a fixed or clamped-end condition at $\gamma = 0$ and a sliding-end condition at $\gamma = \ell$. This yields the boundary conditions

$$v(t,0) = 0 \quad , \quad w(t,0) = \frac{\partial w}{\partial \gamma}(t,0) = 0$$

$$w(t,\ell) = v(t,\ell)\tan(\phi_I) \quad , \quad M_\gamma(t,\ell) = N_\gamma(t,\ell) = 0$$

where ϕ_I denotes the initial angle at $\gamma = \ell$ as depicted in Figure 7.30(d). As detailed in [25,509], the condition $N_\gamma(t,\ell)$ results from simplification of the physical constraint $N_\gamma(t,\ell) = -Q_\gamma(t,\ell)\tan(\theta_C)$, $\theta_C = \theta_I + \frac{\partial w}{\partial \gamma}$, based upon the assumption that Q_γ is negligible.

Weak Model Formulation

Consider states $(v(t,\cdot), w(t,\cdot))$ in the state space

$$X = L^2(0,\ell) \times L^2(0,\ell).$$

The space of test functions is

$$V = \big\{(\phi_1,\phi_2) \in H^1(0,\ell) \times H^2(0,\ell) \,\big|\, \phi_1(0) = 0, \phi_2(0) = \phi_2'(0) = 0, \\ \phi_2(\ell) = \phi_1(\ell)\tan(\theta_I)\big\}.$$

For all $(\phi_1,\phi_2) \in V$, a weak formulation of the model is

$$\begin{aligned} &\int_0^\ell \left\{\rho(\gamma)\frac{\partial^2 v}{\partial t^2}\phi_1 + N_\gamma \frac{d\phi_1}{d\gamma}\right\} d\gamma = 0 \\ &\int_0^\ell \left\{\rho(\gamma)\frac{\partial^2 w}{\partial t^2}\phi_2 + \frac{1}{R}N_\theta \phi_2 \chi_{pe}(\gamma) - M_\gamma \frac{d^2\phi_2}{d\gamma^2} - f_n\phi_2\right\} d\gamma = 0 \end{aligned} \tag{7.106}$$

where N_γ and M_γ are specified in (7.104). We note that the weak formulation requires continuity of v, w and $\frac{\partial w}{\partial \gamma}$ at the junctions γ_1 and γ_2 but accommodates discontinuities in higher derivatives.

7.9.2 Nonlinear State Dependence

The linear models developed in Section 7.9.1 should be used with caution in high drive regimes since they are based on the assumption of small displacements. To extend the framework to accommodate large displacements, which are a hallmark of the transducer, one can employ the nonlinear von Kármán theory summarized in Section 7.8.2. This includes two nonlinear effects: (i) formulation of the balance laws in terms of the deformed reference surface, and (ii) retention of quadratic terms in the strain-displacement relations.

Balancing forces and moments on the reference surface yields

$$\begin{aligned}\rho(\gamma)\frac{\partial^2 v}{\partial t^2} &= \frac{\partial N_\gamma}{\partial \gamma} - \frac{\partial}{\partial \gamma}\left[\left(k_1\frac{\partial^3 w}{\partial \gamma \partial t^2} + \frac{\partial M_\gamma}{\partial \gamma}\right)\frac{\partial w}{\partial \gamma}\right] \\ \rho(\gamma)\frac{\partial^2 w}{\partial t^2} &= k_1\frac{\partial^4 w}{\partial \gamma^2 \partial t^2} - \frac{\partial^2 M_\gamma}{\partial \gamma^2} + \frac{\partial}{\partial \gamma}\left(N_\gamma \frac{\partial w}{\partial \gamma}\right) + f_n\end{aligned} \tag{7.107}$$

when relations analogous to (7.98) are used to eliminate the shear force resultant Q_γ. The retention of quadratic strain-displacement terms yields the reference surface strain relation

$$e_\gamma = \frac{\partial u}{\partial \gamma} + \frac{1}{2}\left(\frac{\partial w}{\partial \gamma}\right)^2 + \frac{w}{R}\chi_{pe}(\gamma) \tag{7.108}$$

which is employed in the resultant expressions (7.104).

Nonlinear models employing strain-displacement relations of the form (7.108) have been constructed in [231–233] to characterize aspects of THUNDER and RAINBOW behavior. These models, which assume uniform curvature throughout the device and linear input behavior, illustrate that inclusion of geometric nonlinearities produces a flattening in the modeled shape as compared with the linear case. The experimental validation of (7.107) with nonlinear inputs and extension of the hysteresis models to incorporate stress-induced dipole switching constitutes and active research area.

7.10 Abstract Model Formulation

To facilitate well-posedness analysis, convergence analysis of approximation techniques, and infinite-dimensional control design, it is advantageous to pose models in an abstract Hilbert space formulation. We illustrate this for the beam model developed in Section 7.4 and cylindrical shell model from Section 7.7.2. Detailed analysis regarding well-posedness, convergence and control criteria can be found in [33] and included references.

7.10.1 Beam Model

Consider the state and test function spaces

$$X = L^2(0,\ell)$$
$$V = H_0^2(0,\ell) = \{\phi \in H^2(0,\ell) \,|\, \phi(0) = \phi'(0) = 0\}$$

with the inner products

$$\langle \psi, \phi \rangle_X = \int_0^\ell \rho \psi \phi \, dx$$

$$\langle (YI)\psi, \phi \rangle_V = \int_0^\ell YI \psi'' \phi'' \, dx.$$

It is observed that V is densely and continuously embedded in X with $|\phi|_X \leq c|\phi|_V$; this is expressed by $V \hookrightarrow X$. Moreover, when one defines conjugate dual spaces X^* and V^* — e.g., V^* denotes the linear space of all conjugate linear continuous functionals on V — two observations are important: (i) X^* can be identified with X through the Riesz map, and (ii) $X^* \hookrightarrow V^*$. Hence the two spaces comprise what is termed a Gelfand triple $V \to X \cong X^* \hookrightarrow V^*$ with pivot space X and duality pairing (duality product) $\langle \cdot, \cdot \rangle_{V^*,V}$. The latter is defined as the extension by continuity of the inner product $\langle \cdot, \cdot \rangle_X$ from $V \times X$ to $V^* \times X$. Hence elements $v^* \in V^*$ have the representation $v^*(v) = \langle v^*, v \rangle_{V^*,V}$.

Consider the weak formulation of the model (7.39),

$$\int_0^\ell \rho \frac{\partial^2 w}{\partial t^2} \phi \, dx + \int_0^\ell \gamma \frac{\partial w}{\partial t} \phi \, dx + \int_0^\ell YI \frac{\partial^2 w}{\partial x^2} \frac{d^2 \phi}{dx^2} \, dx + \int_0^\ell cI \frac{\partial^3 w}{\partial x^2 \partial t} \frac{d^2 \phi}{dx^2} \, dx$$
$$= \int_0^\ell f \phi \, dx + \left[a_1(P(E) - P_R) + a_2(P(E) - P_R)^2\right] \int_0^\ell \frac{d^2 \phi}{dx^2} \, dx \tag{7.109}$$

which must hold for all $\phi \in V$.

Abstract Second-Order Formulation

We begin by defining stiffness and damping sesquilinear forms $\sigma_i : V \times V \to \mathbb{C}$, $i = 1, 2$, by

$$\sigma_1(\psi, \phi) = \langle (YI)\psi, \phi \rangle_V$$
$$\sigma_2(\psi, \phi) = \langle (cI)\psi, \phi \rangle_V + \frac{1}{\rho} \langle \gamma \psi, \phi \rangle_X \tag{7.110}$$

where $\langle (cI)\psi, \phi \rangle_V$ differs from $\langle (YI)\psi, \phi \rangle_V$ only in the sense that the Young's modulus is replaced by the Kelvin–Voigt damping parameter. It can be directly verified that the stiffness form satisfies

(H1)	$\lvert\sigma_1(\psi, \phi)\rvert \leq c_1 \lvert\psi\rvert_V \lvert\phi\rvert_V$	for some $c_1 \in \mathbb{R}$	(Bounded)
(H2)	$\mathrm{Re}\, \sigma_1(\psi, \psi) \geq c_2 \lvert\psi\rvert_V^2$	for some $c_2 > 0$	(V-Elliptic)
(H3)	$\sigma_1(\psi, \phi) = \overline{\sigma_1(\phi, \psi)}$		(Symmetric)

for all $\psi, \phi \in V$. Moreover, the damping term σ_2 satisfies

$$\begin{aligned} &\text{(H4)} \qquad |\sigma_2(\psi,\phi)| \le c_3|\psi|_V|\phi|_V \ , \ \text{for some } c_3 \in \mathbb{R} \qquad \text{(Bounded)} \\ &\text{(H5)} \qquad \mathrm{Re}\,\sigma_2(\psi,\psi) \ge c_4|\psi|_V^2 \ , \ \text{for some } c_4 > 0 \qquad (V\text{-Elliptic}). \end{aligned} \tag{7.111}$$

The input space is taken to be the Hilbert space $U = \mathbb{R}$ and the input operator $B : U \to V^*$ is defined by

$$\langle [B(E)](t), \phi \rangle_{V^*,V} = \left[a_1[P(E(t)) - P_R] + a_2[P(E(t)) - P_R]^2\right] \int_0^\ell \phi''\, dx. \tag{7.112}$$

It is observed that B can be expressed as

$$[B(E)](t) = [b(E)](t) \cdot g \quad , \ g \in V^* \tag{7.113}$$

where

$$\begin{aligned} [b(E)](t) &= [P(E(t)) - P_R] + a_2[P(E(t)) - P_R]^2 \\ g(\phi) &= \int_0^\ell \phi''\, dx. \end{aligned} \tag{7.114}$$

If we let $\widetilde{f} = \frac{f}{\rho}$, the model (7.109) can be written in the abstract variational formulation

$$\begin{aligned} &\langle \ddot{w}(t), \phi \rangle_{V^*,V} + \sigma_2(\dot{w}(t), \phi) + \sigma_1(w(t), \phi) = \langle [B(E)](t) + \widetilde{f}, \phi \rangle_{V^*,V} \\ &w(0) = w_0 \quad , \quad \dot{w}(0) = w_1 \end{aligned} \tag{7.115}$$

for all $\phi \in V$.

Alternatively, one can define the operators $A_i \in \mathcal{L}(V, V^*), i = 1, 2$, by

$$\langle A_i\phi_1, \phi_2 \rangle_{V^*,V} = \sigma_i(\phi_1, \phi_2) \tag{7.116}$$

and formulate the model in operator form as

$$\begin{aligned} &\ddot{w}(t) + A_2\dot{w}(t) + A_1 w(t) = [B(E)](t) + \widetilde{f}(t) \\ &w(0) = w_0 \quad , \quad \dot{w}(0) = w_1 \end{aligned} \tag{7.117}$$

in the dual space V^*. This formulation illustrates the analogy between the infinite-dimensional, strongly damped elastic model and familiar finite-dimensional relations.

Model Well-Posedness

As a prelude to establishing the well-posedness of the beam model with hysteretic E-P relations, we provide a lemma which quantifies the smoothness of the input operator. In the next section, this lemma is also employed when establishing the equivalence of solutions.

Lemma 7.10.1. *Consider continuous field inputs $E \in C[0,T]$. The input operator B defined by (7.112) then satisfies*

$$B(E) \in L^2(0,T;V^*). \tag{7.118}$$

Proof. In Appendix B.3, we establish that for continuous input fields, $P \in C[0,T]$ which implies that b defined by (7.114) satisfies $b(\cdot) : C[0,T] \to C[0,T]$. Hence the norm

$$\|[B(E)](t)\|_{V^*} = \sup_{v \in V} \frac{|[b(E)](t) \cdot g(v)|}{\|v\|_V}$$

exists for each $t \in [0,T]$. Since $\|[B(E)](t)\|_{V^*} = |[b(E)](t)| \cdot \|g\|_{V^*}$, it follows that

$$\|B(E)\|^2_{L^2(0,T;V^*)} \leq \max_{t \in [0,T]} \left\{ |[b(E)](t)|^2 \right\} \cdot T \cdot \|g\|^2_{V^*}$$

which implies that

$$B(E) \in L^2(0,T;V^*).$$

□

The well-posedness of the model is established by the following theorem whose proof follows directly from Theorem 4.1 of [33] or Theorem 2.1 and Remark 2.1 of [26].

Theorem 7.10.2. *Let σ_1 and σ_2 be given by (7.110) and consider continuous field inputs $E \in C[0,T]$ and exogenous inputs $f \in L^2(0,T;V^*)$. There then exists a unique solution w to (7.115), or equivalently (7.117), which satisfies*

$$w \in C(0,T;V)$$
$$\dot{w} \in C(0,T;X).$$

Abstract First-Order Formulation

We also consider an abstract first-order formulation of the model which has mild solutions defined in terms of an analytic C_0-semigroup. As detailed in Chapter 7 of [33], this provides a framework which facilitates infinite-dimensional control design and subsequent approximation.

Define the product spaces $\mathcal{X} = V \times X$ and $\mathcal{V} = V \times V$ with the norms

$$|(\phi_1, \phi_2)|^2_{\mathcal{X}} = |\phi_1|^2_V + |\phi_2|^2_X$$
$$|(\phi_1, \phi_2)|^2_{\mathcal{V}} = |\phi_1|^2_V + |\phi_2|^2_V$$

so that $\mathcal{V} \hookrightarrow \mathcal{X} \cong \mathcal{X}^* \hookrightarrow \mathcal{V}^*$ again forms a Gelfand triple with $\mathcal{V}^* = V \times V^*$. The state is $z(t) = (w(t,\cdot), \dot{w}(t,\cdot)) \in \mathcal{X}$. Letting $\Phi = (\phi_1, \phi_2)$ and $\Psi = (\psi_1, \psi_2)$, the sesquilinear form $\sigma : \mathcal{V} \times \mathcal{V} \to \mathbb{C}$ is defined by

$$\sigma(\Psi, \Phi) = -\langle \psi_2, \phi_1 \rangle_V + \sigma_1(\psi_1, \phi_2) + \sigma_2(\psi_2, \phi_2)$$

and the product space forcing terms are formulated as

$$\mathcal{F}(t) = \begin{bmatrix} 0 \\ \widetilde{f}(t) \end{bmatrix} \quad , \quad [\mathcal{B}(E)](t) = \begin{bmatrix} 0 \\ [(B(E)](t) \end{bmatrix}.$$

The weak model formulation (7.117) can then be written as the first-order relation

$$\langle \dot{z}(t), \Phi \rangle_{\mathcal{V}^*,\mathcal{V}} + \sigma(z(t), \Phi) = \langle [\mathcal{B}(E)](t) + \mathcal{F}(t), \Phi \rangle_{\mathcal{V}^*,\mathcal{V}}$$

for $\Phi \in \mathcal{V}$. Formally, this equivalent to the strong formulation

$$\begin{aligned} \dot{z}(t) &= \mathcal{A}z(t) + [\mathcal{B}(E)](t) + \mathcal{F}(t) \quad \text{in } \mathcal{V}^* \\ z(0) &= z_0 = \begin{bmatrix} w_0 \\ w_1 \end{bmatrix} \end{aligned} \tag{7.119}$$

where the operator $\mathcal{A}$ is given by

$$\begin{aligned} \text{dom}\mathcal{A} &= \{(\phi_1, \phi_2) \in \mathcal{X} | \phi_2 \in V, A_1\phi_1 + A_2\phi_2 \in X\} \\ \mathcal{A} &= \begin{bmatrix} 0 & I \\ -A_1 & -A_2 \end{bmatrix}. \end{aligned} \tag{7.120}$$

In a manner analogous to (7.116), $\mathcal{A}$ can be related to the operator $\widetilde{\mathcal{A}} \in \mathcal{L}(\mathcal{V}, \mathcal{V}^*)$ defined by

$$\langle \widetilde{\mathcal{A}}\Psi, \Phi \rangle_{\mathcal{V}^*,\mathcal{V}} = \sigma(\Psi, \Phi).$$

Specifically, $\mathcal{A}$ is the negative of the restriction to dom$\mathcal{A}$ of $\widetilde{\mathcal{A}}$ so that $\sigma(\Psi, \Phi) = \langle -\mathcal{A}\Psi, \Phi \rangle_{\mathcal{X}}$ for $\Psi \in \text{dom}\mathcal{A}$, $\Phi \in \mathcal{V} \subset \mathcal{X}$.

The formulation (7.119) with $\mathcal{A}$ defined by (7.120) is formally analogous to the first-order formulation of finite-dimensional second-order systems.

Due to the presence of Kelvin–Voigt damping which causes σ_2 to satisfy the V-ellipticity and V-continuity conditions (H4) and (H5) of (7.111), it is established in Chapter 4 of [33] that σ is $\mathcal{V}$-elliptic and $\mathcal{A}$ generates an analytic semigroup $\mathcal{T}(t)$ on $\mathcal{V}, \mathcal{X}$ and $\mathcal{V}^*$.[34] From Lemma 7.10.1, it follows that $B(E) \in L^2(0, T; V^*)$ and hence $\mathcal{B} \in L^2(0, T; \mathcal{V}^*)$. Under the assumptions that $z_0 \in \mathcal{V}^*$ and $\mathcal{F} \in L^2(0, T; \mathcal{V}^*)$, a mild solution to (7.119) is given by

$$z(t) = \mathcal{T}(t)z_0 + \int_0^t \mathcal{T}(t-s)\left([\mathcal{B}(E)](s) + \mathcal{F}(s)\right) ds. \tag{7.121}$$

It is illustrated in Section 4.4 of [33] that under these conditions, the mild and weak solutions are equivalent.

[34]The domain defined in (7.120) is actually $\text{dom}_{\mathcal{X}}\mathcal{A}$. However, the use of one symbol when denoting semigroups or infinitesimal generators defined on each of the spaces in a Gelfand triple is common in the literature and does not cause ambiguity.

Remark 7.10.3. *In the case of weaker damping (e.g., air damping), weakened conditions (H4') and (H5') must be considered which leads to the generation of a C_0-semigroup $\mathcal{T}(t)$ on $\mathcal{X}$ that is not analytic. To accommodate inputs in $\mathcal{V}^*$, it is necessary to extend the semigroup to a larger space $\mathcal{W} = [\text{dom}\mathcal{A}^*]^* \supset \mathcal{V}^*$ using extrapolation space techniques similar to those used by DaPrato and Grisvard [121], Haraux [208] and Weissler [508]. Details regarding this extension and resulting equivalence of solutions can be found in [27, 33].*

7.10.2 Shell Model

To illustrate the generality of this approach, we also summarize the abstract formulation of the cylindrical shell model developed in Section 7.7.2. We consider fixed boundary conditions

$$u = v = w = \frac{\partial w}{\partial x} = 0$$

at $x = 0$ and free end conditions

$$N_x = N_{x\theta} + \frac{M_{x\theta}}{R} = M_x = Q_x + \frac{1}{R}\frac{\partial M_{x\theta}}{\partial \theta} = 0$$

at $x = \ell$.

We consider the state $\xi(t) = (u(t,\cdot,\cdot), v(t,\cdot,\cdot), w(t,\cdot,\cdot))$ in the state space

$$\begin{aligned} X &= L^2(\Omega) \times L^2(\Omega) \times L^2(\Omega) \\ \langle \psi, \phi \rangle_X &= \int_\Omega \rho h \psi_1 \overline{\phi}_1 \, d\omega + \int_\Omega \rho h \psi_2 \overline{\phi}_2 \, d\omega + \int_\Omega \rho h \psi_3 \overline{\phi}_3 \, d\omega \end{aligned}$$

where $\Omega = [0,\ell] \times [0, 2\pi]$ and $\psi = (\psi_1, \psi_2, \psi_3)$, $\phi = (\phi_1, \phi_2, \phi_3)$.[35] The space of test functions is

$$V = H_0^1(\Omega) \times H_0^1(\Omega) \times H_0^2(\Omega)$$

$$\begin{aligned} \langle (Y)\psi, \phi \rangle_V &= \int_\Omega \frac{Yh}{1-\nu^2} \left[(e_x + \nu e_\theta) \overline{\frac{\partial \phi_1}{\partial x}} + \frac{1}{2R}(1-\nu) e_{x\theta} \overline{\frac{\partial \phi_1}{\partial \theta}} \right] d\omega \\ &+ \int_\Omega \frac{Yh}{1-\nu^2} \left[(e_\theta + \nu e_x) \frac{1}{R} \overline{\frac{\partial \phi_2}{\partial \theta}} + \frac{1}{2}(1-\nu) e_{x\theta} \overline{\frac{\partial \phi_2}{\partial x}} \right] d\omega \\ &+ \int_\Omega \frac{Yh}{1-\nu^2} \left[\frac{1}{R}(e_\theta + \nu e_x)\overline{\phi}_3 - \frac{h^2}{12}(\kappa_x + \nu\kappa_\theta) \overline{\frac{\partial^2 \phi_3}{\partial x^2}} \right. \\ &\qquad \left. - \frac{h^2}{12R^2}(\kappa_\theta + \nu\kappa_x) \overline{\frac{\partial^2 \phi_3}{\partial \theta^2}} - \frac{h^2}{12R}(1-\nu)\kappa_{x\theta} \overline{\frac{\partial^2 \phi_3}{\partial x \partial \theta}} \right] d\omega . \end{aligned}$$

[35] As detailed in [130], retention of the complex conjugate in the inner product is necessary when implementing approximation techniques employing complex Fourier bases.

where $H_0^1(\Omega)$ and $H_0^2(\Omega)$ are defined in (7.91) and the reference surface strains and changes in curvature are defined in (7.90). We now summarize the abstract formulation of the model (7.92).

As with the beam model, we define sesquilinear forms

$$\sigma_1(\psi,\phi) = \langle (Y)\psi, \phi\rangle_V$$
$$\sigma_2(\psi,\phi) = \langle (c)\psi, \phi\rangle_V$$

which incorporate the stiffness and damping components. The input operator $B: U \to V^*$ is defined by

$$\begin{aligned}\langle [B(E)](t), \phi\rangle_{V^*,V} &= a_1[P(E(T)) - P_R]\frac{h}{1-\nu}\int_\Omega \left\{\frac{\partial\overline{\phi_1}}{\partial x} + \frac{1}{R}\frac{\partial\overline{\phi_2}}{\partial\theta}\right\} d\omega \\ &\quad + a_2[P(E(t)) - P_R]^2\frac{h}{1-\nu}\int_\Omega \left\{\frac{\partial\overline{\phi_1}}{\partial x} + \frac{1}{R}\frac{\partial\overline{\phi_2}}{\partial\theta}\right\} d\omega.\end{aligned}$$

The weak formulation can subsequently be posed as

$$\left\langle \ddot{\xi}(t), \phi\right\rangle_{V^*,V} + \sigma_2(\dot{\xi}(t),\phi) + \sigma_1(\xi(t),\phi) = \langle [B(E)](t) + \widehat{f}, \phi\rangle_{V^*,V}$$

where $\widehat{f} = \frac{1}{\rho h}(f_x, f_y, f_n)$. This is the same general abstract variational formulation for second-order systems that was considered in (7.115) for the beam and the remainder of the formulation follows that described in the context of the beam model.

Chapter 8

Numerical Techniques

The rod, beam, plate and shell models developed in Chapter 7 generally preclude analytic solution due to the boundary conditions and piecewise nature of material parameters and exogenous inputs. This necessitates the development of approximation techniques appropriate not only for simulations and transducer characterization but also for optimization and control design. Whereas we focus primarily on the first objective in this chapter, the use of these numerical models for subsequent transducer optimization and control design dictates that attention be paid to additional criteria, such as adjoint convergence, that arise in the context of constrained optimization — the reader is referred to [67] and references therein for details regarding approximation techniques pertaining to optimization and control formulations.

For all of the models, we first employ Galerkin approximations in space to obtain semi-discrete, vector-valued ODE that evolve in time — see pages 417–419 of Appendix A for a general discussion regarding the relation between Galerkin and finite element methods. This provides a natural setting for linear and nonlinear finite-dimensional control design and direct simulations. Since Galerkin or finite element approximation in space typically yield moderately stiff ODE systems, A-stable or stiff algorithms are advised when approximating solutions in time; this includes trapezoidal-based approximations or routines such as `ode15s.m` in MATLAB.

In all cases, we consider approximation in the context of the weak model formulations since this reduces smoothness requirements and accommodates in a natural manner discontinuous material parameters and inputs. This necessarily involves the integration of polynomial or trigonometric basis elements which we accomplish using Gaussian quadrature routines chosen to ensure exact integration for linear and cubic basis functions. We summarize these numerical integration algorithms in Section 8.1. Approximation techniques for rods, beams, plates and shells are subsequently described in Section 8.2–8.5.

Numerical approximation of distributed structural models is an extremely broad topic and includes issues such as shear locking and approximation techniques for control design which constitute active research areas. Rather than attempt to provide a comprehensive description of numerical techniques, we instead summarize certain fundamental methods appropriate for smart structures and indicate pitfalls and directions required to extend the algorithms to more complex settings. The reader is referred to [54, 479] for finite element theory, [390, 462, 527] for finite element implementation techniques, and [163, 383, 406] for the theory of splines, variational methods and Galerkin techniques. A detailed discussion illustrating finite element implementation in MATLAB is provided in [276] and m-files for implementing various models discussed in this chapter are provided at the website `http://www.siam.org/books/fr32`. Additional references specific to the various structures will be indicated in relevant sections.

8.1 Quadrature Techniques

Consider integrals of the form $I = \int_a^b f(x)dx$ where (a, b) is either finite or infinite and the value I is finite. The goal in numerical integration or quadrature is to approximate I by finite sums of the form

$$I_n = \sum_{i=0}^{n} w_i f(x_i)$$

where w_i and x_i respectively denote quadrature weights and nodes. Various approximation theories — e.g., based on Taylor expansions or the theory of orthogonal polynomials — yield choices for w_i and x_i that determine the rate at which I_n converges to I as $n \to \infty$.

8.1.1 Newton–Cotes Formulae

To illustrate issues associated with the choice of weights and nodes, we consider first closed and open Newton–Cotes formulae for finite $[a, b]$. The nodes for the two cases are $x_i = x_0 + ih$ where $i = 0, \ldots, n$ and

$$h = \frac{b-a}{n}, \quad x_0 = a, \quad \text{closed Newton–Cotes formulae}$$

$$h = \frac{b-a}{n+2}, \quad x_0 = a + h, \quad \text{open Newton–Cotes formulae.}$$

Hence nodes lie in the closed interval $[a, b]$ in the first case and the open interval (a, b) in the second. In all of the following formulae, ξ is a point in (a, b) and f is assumed sufficiently smooth so that requisite derivatives exist and are continuous. Details regarding the derivation of these relations can be found in [19]. The trapezoid and midpoint rules are illustrated in Figure 8.1.

Closed Newton–Cotes Formulae

1. $(n = 1)$ Trapezoidal rule

$$\int_a^b f(x)dx = \frac{h}{2}[f(x_0) + f(x_1)] - \frac{h^3}{12} f''(\xi) \tag{8.1}$$

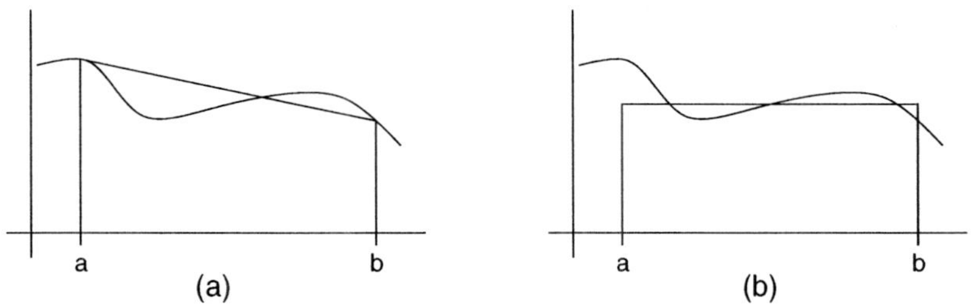

Figure 8.1. *(a) Trapezoidal rule and (b) midpoint rule for the interval* $[a, b]$.

2. $(n = 2)$ Simpson's rule

$$\int_a^b f(x)dx = \frac{h}{3}[f(x_0) + 4f(x_1) + f(x_2)] - \frac{h^5}{90}f^{(4)}(\xi) \tag{8.2}$$

3. $(n = 3)$ Simpson's three-eighths rule

$$\int_a^b f(x)dx = \frac{3h}{8}[f(x_0) + 3f(x_1) + 3f(x_2) + f(x_3)] - \frac{3h^5}{80}f^{(4)}(\xi) \tag{8.3}$$

4. $(n = 4)$ Milne's rule

$$\int_a^b f(x)dx = \frac{2h}{45}[7f(x_0)+32f(x_1)+12f(x_2)+32f(x_3)+7f(x_4)] - \frac{8h^7}{945}f^{(6)}(\xi) \tag{8.4}$$

Open Newton–Cotes Formulae

1. $(n = 0)$ Midpoint rule

$$\int_a^b f(x)dx = 2hf(x_0) + \frac{h^3}{3}f''(\xi) \tag{8.5}$$

2. $(n = 1)$

$$\int_a^b f(x)dx = \frac{3h}{2}[f(x_0) + f(x_1)] + \frac{3h^3}{4}f''(\xi) \tag{8.6}$$

3. $(n = 2)$

$$\int_a^b f(x)dx = \frac{4h}{3}[2f(x_0) - f(x_1) + 2f(x_2)] + \frac{14h^5}{45}f^{(4)}(\xi) \tag{8.7}$$

4. $(n = 3)$

$$\int_a^b f(x)dx = \frac{5h}{24}[11f(x_0) + f(x_1) + f(x_2) + 11f(x_3)] + \frac{95h^5}{144}f^{(4)}(\xi) \tag{8.8}$$

Accuracy of the Newton–Cotes Formulae

A numerical quadrature formula I_n is said to have *degree of precision* m if $I_n = I$ for all polynomials f such that $\deg(f) \leq m$ and $I_n \neq I$ for $\deg(f) = m+1$. Hence the trapezoid rule has degree of precision 1 which corroborates the observation that it integrates linear polynomials exactly. It is observed that formulae with an even index gain an extra degree of precision when compared with odd formulae which often makes them preferable. For later comparison with Gaussian quadrature routines, we note that the errors E_n and degrees of precision DP_n for the Newton–Cotes formulae are

$$|E_n| = \begin{cases} K_{1n} h^{n+2} f^{(n+1)}(\xi) & , \quad n \text{ odd} \\ K_{2n} h^{n+3} f^{(n+2)}(\xi) & , \quad n \text{ even} \end{cases}$$

and

$$DP_n = \begin{cases} n & , \quad n \text{ odd} \\ n+1 & , \quad n \text{ even} \end{cases}$$

where K_{1n} and K_{2n} are constants and it is assumed that $f \in C^{n+2}[a, b]$ if n is even and $f \in C^{n+1}[a, b]$ for odd n.

Whereas increasing n leads to improved accuracy, Newton–Cotes formulae are typically restricted to $n \leq 8$ to avoid stability issues. To achieve the requisite accuracy on large intervals $[a, b]$, including infinite intervals, alternatives are required. These include Romberg integration techniques which improve accuracy through Richardson extrapolation, composite rules, and Gaussian quadrature techniques. We summarize next the latter two options.

Composite Quadrature Techniques

An obvious technique to improve accuracy is to partition finite domains $[a, b]$ into subintervals and then apply the quadrature rules on each subinterval. The manner through which partitions are constructed depends on the choice of open versus closed Newton–Cotes formulae as illustrated in Figure 8.2 for the composite trapezoid and midpoint formulae. In both cases, is is assumed that $f \in C^2[a, b]$ and ξ is a point in (a, b).

Composite Trapezoidal Rule

The composite trapezoid rule for n subintervals is

$$\int_a^b f(x)dx = \frac{h}{2}\left[f(a) + \sum_{i=1}^{n-1} f(x_i) + f(b)\right] - \frac{b-a}{12} h^2 f''(\xi)$$

where $h = \frac{b-a}{n}$ and $x_i = a + ih$ for $i = 0, \ldots, n$. It is observed from Figure 8.2 that if n is doubled, present values of $f(x_i)$ are re-used, thus contributing to the efficiency of the method.

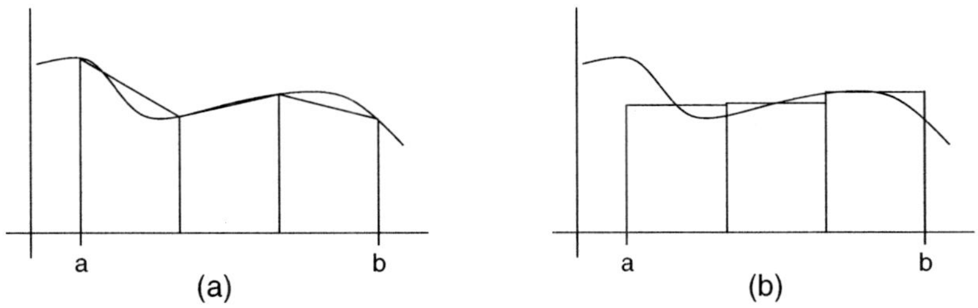

Figure 8.2. *(a) Composite trapezoidal rule and (b) composite midpoint rule with three subintervals.*

Composite Midpoint Rule

The composite midpoint formula for even n and $\frac{n}{2}+1$ subintervals is

$$\int_a^b f(x)dx = 2h\sum_{i=0}^{n/2} f(x_i) + \frac{b-a}{6}h^2 f''(\xi).$$

Here $h = \frac{b-a}{n+2}$ and $x_i = a + (2i+1)h$ for $i = 0, \ldots, n$. The approximation obtained with $n = 4$, and hence 3 subintervals is illustrated in Figure 8.2(b).

8.1.2 Gaussian Quadrature Techniques

It is observed that for the open and closed Newton–Cotes formulae, and corresponding composite rules, the quadrature points x_i are fixed *a priori* and quadrature weights w_i are determined to achieve a specified level of accuracy. Hence both the accuracy and degree of precision for the methods are roughly equivalent to the degrees of freedom associated with the weights. Alternatively, one can let both x_i and w_i be free parameters to achieve a maximal order of accuracy. This is the basis for Gaussian quadrature routines which provides the capability for exactly integrating polynomials up to degree $2n-1$ using n-point expansions.

To provide intuition, we initially consider the expansion

$$\int_{-1}^1 f(x)dx \approx \sum_{i=1}^n f(x_i)w_i$$

with $n = 1$ and $n = 2$. Defining the error as

$$E_n(f) = \int_{-1}^1 f(x)dx - \sum_{i=1}^n f(x_i)w_i,$$

we note that for polynomials $p_m = a_0 + a_1 x + \cdots + a_m x^m$,

$$E_n(p_m) = a_0 E_n(1) + a_1 E_n(x) + \cdots + a_m E_n(x^m).$$

It thus follows that the integration rule has degree of precision m if

$$E_n(x^i) = 0 \quad , \quad i = 0, \ldots, m.$$

1-Point Gaussian Quadrature

To determine the two parameters x_1 and w_1, we consider the constraints

$$E_1(1) = 0 \; , \; E_1(x) = 0$$

or

$$\int_{-1}^{1} 1dx - w_1 = 0 \; , \; \int_{-1}^{1} xdx - w_1 x_1 = 0.$$

This yields $x_1 = 0$ and $w_1 = 2$ and the general quadrature rule

$$\int_{-1}^{1} f(x)dx \approx 2f(0).$$

Note that this is simply the midpoint formula (8.5) which is illustrated for the interval $[a, b]$ in Figure 8.1(b).

2-Point Gaussian Quadrature

Here there are four parameters x_1, x_2, w_1, w_2 and four constraints

$$E_2(x^i) = \int_{-1}^{1} x^i dx - (w_1 x_1^i + w_2 x_2^i) = 0 \quad , \quad i = 0, 1, 2, 3.$$

This yields the nonlinear system of equations

$$\begin{aligned} w_1 + w_2 &= 2 \\ w_1 x_1 + w_2 x_2 &= 0 \\ w_1 x_1^2 + w_2 x_2^2 &= \frac{2}{3} \\ w_1 x_1^3 + w_2 x_2^3 &= 0 \end{aligned}$$

which has the unique solution

$$\begin{aligned} x_1 &= -\tfrac{1}{\sqrt{3}} \quad , \quad w_1 = 1 \\ x_2 &= \tfrac{1}{\sqrt{3}} \quad , \quad w_2 = 1. \end{aligned}$$

It is noted that the general quadrature rule

$$\int_{-1}^{1} f(x)dx \approx f\left(\frac{-1}{\sqrt{3}}\right) + f\left(\frac{1}{\sqrt{3}}\right) \tag{8.9}$$

has the same degree of precision as Simpson's rule (8.2) which required three nodes.

n-Point Gaussian Quadrature

For $n > 2$, solving the nonlinear systems of equations becomes prohibitive and Gaussian quadrature rules are typically formulated using interpolation theory for orthogonal polynomials. By considering families of orthogonal polynomials defined on the intervals $[-1,1]$, $[0,\infty)$ and $(-\infty,\infty)$, in addition to weighted integrands, this provides substantial flexibility for approximating a broad range of integrals. A complete discussion of this theory is beyond the scope of this chapter and we refer the reader to [19, 125, 528] for details regarding Gaussian quadrature routines for both single and multivariate approximation.

Gauss–Legendre Quadrature

Gauss–Legendre quadrature formulae are typically defined in terms of degree n Legendre polynomials

$$P_n(x) = \frac{1}{2^n n!} \cdot \frac{d^n}{dx^n}\left(x^2-1\right)^n \quad , \; n = 0,1,2,\ldots$$

on the interval $[-1,1]$ — see [123, 465] for a derivation of the Legendre polynomials through application of the Gram–Schmidt process to the sequence $1, x, x^2, \cdots$. Note that the first five Legendre polynomials are

$$\begin{aligned} P_0(x) &= 1 \\ P_1(x) &= x \\ P_2(x) &= \frac{1}{2}(3x^2-1) \\ P_3(x) &= \frac{1}{2}(5x^3-3x) \\ P_4(x) &= \frac{1}{8}(35x^4-30x^2+3). \end{aligned}$$

The quadrature relation is

$$\int_{-1}^{1} f(x)dx \approx \sum_{i=1}^{n} w_i f(x_i) \tag{8.10}$$

where the nodes x_i are zeroes of $P_n(x)$ and the weights are

$$w_i = \frac{-2}{(n+1)P_n'(x_i)P_{n+1}(x_i)} \quad , \qquad i = 1,\ldots,n,$$

as summarized in Table 8.1. For $f \in C^{2n}[-1,1]$, the errors are given by

$$E_n(f) = e_n \frac{f^{(2n)}(\xi)}{(2n)!}$$

where ξ is a point in $[-1,1]$ and $e_n \approx \frac{\pi}{4^n}$ as $n \to \infty$ [19]. As noted previously, this implies that the quadrature formula (8.10) is *exact* for polynomials having degree less than or equal to $2n-1$. Hence when approximating the solution to weak formulations for structural models, the degree n is chosen to be commensurate with finite element or spline basis and test functions.

n	Nodes x_i	Weights w_i
1	0	2
2	$\pm\frac{1}{\sqrt{3}}$	1
3	0	$\frac{8}{9}$
	$\pm\sqrt{\frac{3}{5}}$	$\frac{5}{9}$
4	$\pm\frac{\sqrt{15+2\sqrt{30}}}{\sqrt{35}}$	$\frac{49}{6(18+\sqrt{30})}$
	$\pm\frac{\sqrt{15-2\sqrt{30}}}{\sqrt{35}}$	$\frac{49}{6(18-\sqrt{30})}$

Table 8.1. *Nodes and weights for Gauss–Legendre quadrature on* $[-1,1]$.

Gauss–Laguerre and Gauss–Hermite Quadrature

The integrals in the polarization model (2.114) involve the domains $[0,\infty)$ and $(-\infty,\infty)$. One technique for approximating the integrals is to exploit decay exhibited by the integrand to truncate to finite domains. Alternatively, one can directly approximate the integrals using orthogonal polynomials defined on the half line and real line. This yields the Gauss–Laguerre quadrature relation

$$\int_0^\infty e^{-x} f(x)dx \approx \sum_{i=1}^n w_i f(x_i)$$

and Gauss–Hermite relation

$$\int_{-\infty}^\infty e^{-x^2} f(x)dx \approx \sum_{i=1}^n w_i f(x_i).$$

The weights and nodes for these formulae can be found in [528].

Gauss–Legendre Quadrature on $[a,b]$ and Composite Quadrature

To evaluate integrals on arbitrary domains using the Gauss–Legendre quadrature relation (8.10), one can employ the linear change of variables

$$\int_a^b f(x)dx = \frac{b-a}{2}\int_{-1}^1 f\left(\frac{a+b+\xi(b-a)}{2}\right)d\xi \tag{8.11}$$

to map to the interval $[-1,1]$. It follows that the nodes and weights x_i and w_i for $[a,b]$ are related to ξ_i and η_i for $[-1,1]$ by the relations

$$x_i = \frac{a+b}{2} + \frac{b-a}{2}\xi_i \quad , \quad w_i = \frac{b-a}{2}\eta_i.$$

This mapping can also be used to construct composite Gaussian quadrature routines and formulae appropriate for finite element and spline meshes. To illustrate, consider the partition of $[a,b]$ given by $x_j = a + jh$, $h = \frac{b-a}{n}$, for $j = 0, \ldots, n$ as illustrated in Figure 8.3. The nodes and weights for the 4-point Gauss–Legendre rule on the subinterval $[x_{j-1}, x_j]$ are then

$$\begin{aligned}
x_1 &= x_{j-1} + h\left[\frac{1}{2} - \frac{\sqrt{15+2\sqrt{30}}}{2\sqrt{35}}\right], & w_1 &= \frac{49h}{12(18+\sqrt{30})} \\
x_2 &= x_{j-1} + h\left[\frac{1}{2} - \frac{\sqrt{15-2\sqrt{30}}}{2\sqrt{35}}\right], & w_2 &= \frac{49h}{12(18-\sqrt{30})} \\
x_3 &= x_{j-1} + h\left[\frac{1}{2} + \frac{\sqrt{15-2\sqrt{30}}}{2\sqrt{35}}\right], & w_3 &= \frac{49h}{12(18-\sqrt{30})} \\
x_4 &= x_{j-1} + h\left[\frac{1}{2} + \frac{\sqrt{15+2\sqrt{30}}}{2\sqrt{35}}\right], & w_4 &= \frac{49h}{12(18+\sqrt{30})}.
\end{aligned} \tag{8.12}$$

We note that this will integrate exactly piecewise polynomials of order up to 7.

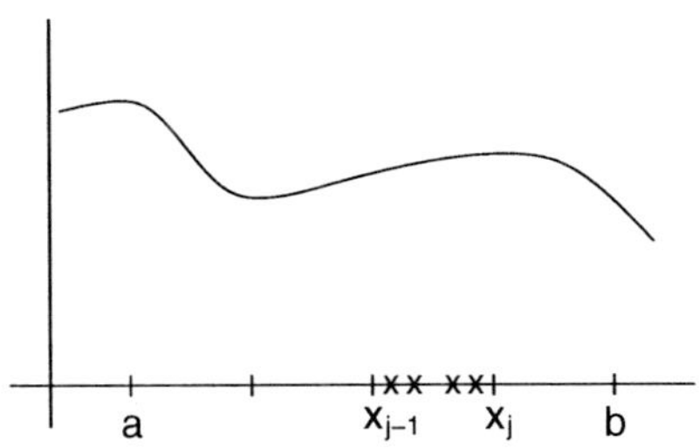

Figure 8.3. *Partition of $[a,b]$ into n subintervals and position of quadrature points for the 4-point Gauss–Legendre rule on $[x_{j-1}, x_j]$.*

8.1.3 2-D Quadrature Formulae

Approximation of weak formulations for plate and shell models requires the numerical evaluation of double integrals using quadrature rules of the form

$$\int_a^b \int_c^d f(x,y)\,dydx \approx \sum_{i=1}^{n_x} \sum_{j=1}^{n_y} f(x_i, y_j) w_i w_j.$$

Through the change of variables

$$\begin{aligned}
&\int_a^b \int_c^d f(x,y)\,dydx \\
&= \left(\frac{b-a}{2}\right)\left(\frac{d-c}{2}\right) \int_{-1}^1 \int_{-1}^1 f\left(\frac{a+b+\xi(b-a)}{2}, \frac{c+d+\upsilon(d-c)}{2}\right) d\upsilon d\xi,
\end{aligned}$$

these integrals can be mapped to the rectangular domain $[-1,1] \times [-1,1]$ so we consider first this case. This also provides the framework necessary for numerical integration using rectangular elements. Finally, we summarize formulae for triangular elements as required for general finite element analysis.

Rectangular Domains

The formulae for rectangular domains are obtained by tensoring 1-D relations. For Gaussian formulae, this yields the nodal placement depicted in Figure 8.4.

4-Point Gauss–Legendre Quadrature

The tensor product of the 1-D formula (8.9) yields the 4-point quadrature relation

$$\int_{-1}^{1}\int_{-1}^{1} f(\xi,\upsilon)d\upsilon d\xi \approx f(a,a)+f(a,b)+f(b,a)+f(b,b)$$

where $a = b = \frac{1}{\sqrt{3}}$. This relation is exact for polynomials of degree 2. Hence this algorithm would be employed when integrating linear quadrilateral elements.

9-Point Gauss–Legendre Quadrature

The tensor product of the 3-point formula from Table 8.1 yields

$$\begin{aligned}\int_{-1}^{1}\int_{-1}^{1} f(\xi,\upsilon)d\upsilon d\xi \quad &\approx \quad \frac{25}{81}\left[f(a,a), f(a,c)+f(c,a)+f(c,c)\right] \\ &+\frac{40}{81}\left[f(a,b)+f(c,b)+f(b,a)+f(b,c)\right]+\frac{64}{81}f(b,b)\end{aligned}$$

where $a = -\sqrt{\frac{3}{5}}$, $b = 0$ and $c = \sqrt{\frac{3}{5}}$. This relation is exact for degree 4 polynomials so it would be used to integrate quadratic elements [390].

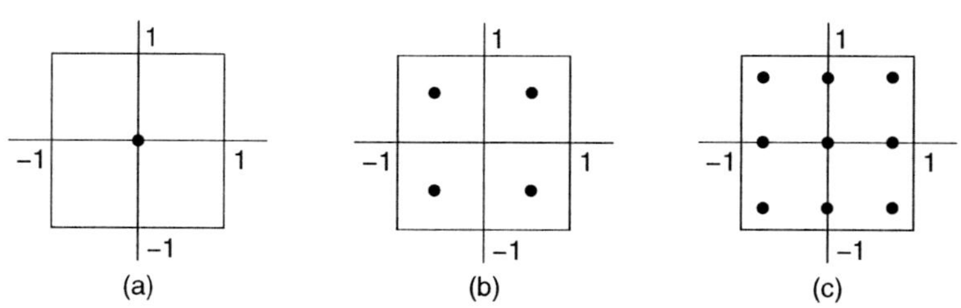

Figure 8.4. *Quadrature points in a 2-D rectangular domain: (a) 1-point rule, (b) 4-point rule, and (c) 9-point rule.*

Triangular Domains

For general finite element analysis, it is also necessary to consider quadrature formulae for triangular domains. This is often accomplished by considering transformations between physical space (x, y–coordinates) and computational space

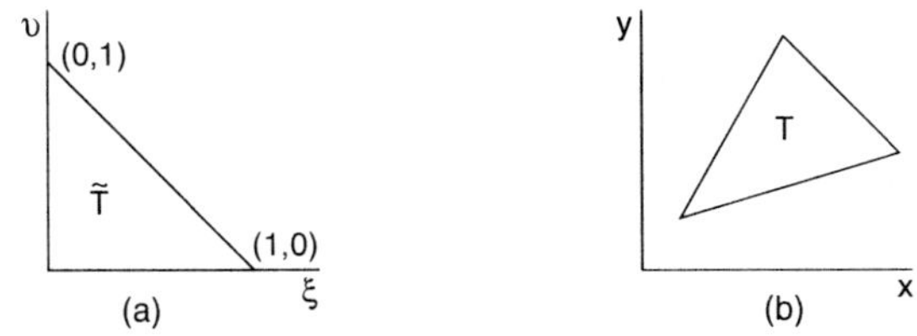

Figure 8.5. *(a) Master triangular element and (b) global element in physical space.*

(ξ, υ–coordinates), as shown in Figure 8.5, so we summarize first the construction of local coordinates that are independent of orientation.

The local coordinates L_1, L_2 and L_3 are defined as the ratio between the perpendicular distance s to a side and the altitude h of the side as depicted in Figure 8.6(a). This implies that $0 \leq L_i \leq 1$. To elucidate a second property of the elements, consider the triangle T_1, delineated by L_1 as shown in Figure 8.6(b), and the complete triangle T. From the area relations

$$A_{T_1} = \frac{sb}{2} \ , \ A_T = \frac{hb}{2},$$

it follows that $L_1 = \frac{A_{T_1}}{A_T}$. This motivates the designation of L_1, L_2, L_3 as *area coordinates* and establishes the relation

$$L_1 + L_2 + L_3 = 1.$$

From the definition of the local coordinate L_1, it follows that it satisfies the property

$$L_1 = \begin{cases} 1 \text{ at node } i \\ 0 \text{ at nodes } i \text{ and } k \end{cases}$$

with similar properties for L_2 and L_3. When combined with the linearity of the definition, this implies that local coordinates also provide the simplex linear elements N_i, N_j, N_k depicted in Figure 8.7; that is,

$$N_i = L_1 \ , \ N_j = L_2 \ , \ N_k = L_3.$$

This proves crucial when defining quadrature properties for the finite element method.

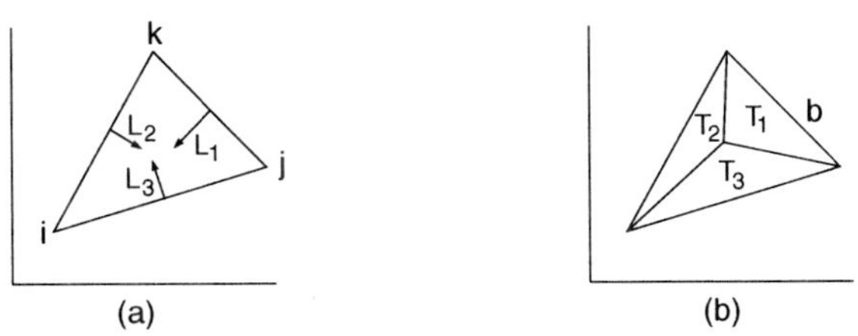

Figure 8.6. *(a) Local coordinates L_1, L_2, L_3 and (b) triangles T_1, T_2, T_3 having the areas $A_{T_1}, A_{T_2}, A_{T_3}$.*

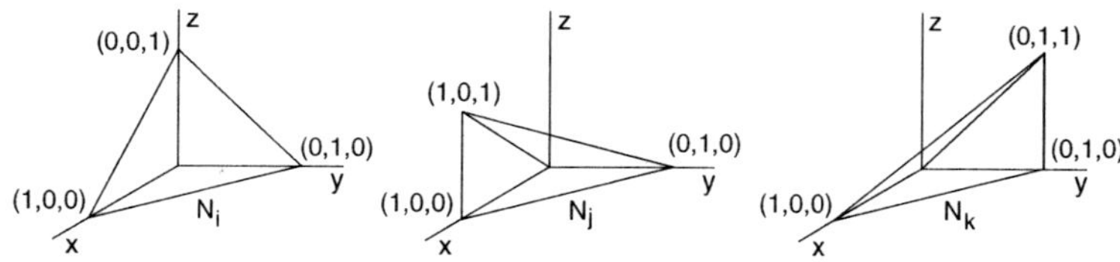

Figure 8.7. *Linear elements N_i, N_j and N_k.*

Letting J denote the Jacobian for the transformation between physical and master triangular elements, it follows that

$$\begin{aligned}\int_T f(x,y)dxdy &= \int_{\tilde{T}} f(L_1, L_2, L_3)\,|J|\,dL_1 dL_2 \\ &\approx \frac{1}{2}\sum_{i=1}^{n} w_i f(L_{1_i}, L_{2_i}, L_{3_i})|J_i|.\end{aligned}$$

The quadrature points, weights, and degrees of precision for $n = 1, 2, 3$ are summarized in Table 8.2 and higher-order formulae can be found in [205, 462].

To illustrate, the 1-point quadrature relation

$$\int_{\tilde{T}} f(\xi, \upsilon) d\xi d\upsilon \approx \frac{1}{2} f(1/3, 1/3)$$

integrates linear functions exactly whereas quadratic polynomials are integrated exactly by the formula

$$\int_{\tilde{T}} f(\xi, \upsilon) d\xi d\upsilon \approx \frac{1}{6}\left[f(1/2, 0) + f(0, 1/2) + f(1/2, 1/2)\right].$$

Note that all of the formulae yield the triangle area $A = \frac{1}{2}$ with $f(\xi, \upsilon) = 1$.

n	Degree of Precision	Local Coordinates L_1	L_2	L_3	Weights w		Geometric Location
1	1	$\frac{1}{3}$	$\frac{1}{3}$	$\frac{1}{3}$	1	a	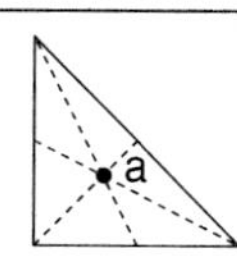
3	2	$\frac{1}{2}$	0	$\frac{1}{2}$	$\frac{1}{3}$	a	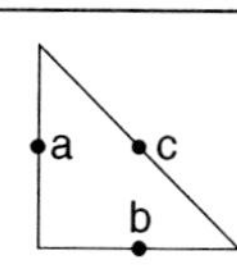
		$\frac{1}{2}$	$\frac{1}{2}$	0	$\frac{1}{3}$	b	
		0	$\frac{1}{2}$	$\frac{1}{2}$	$\frac{1}{3}$	c	
4	3	$\frac{1}{3}$	$\frac{1}{3}$	$\frac{1}{3}$	$-\frac{27}{48}$	a	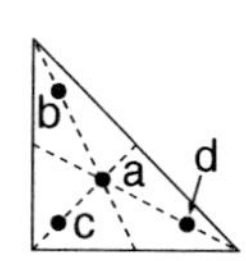
		$\frac{2}{15}$	$\frac{2}{15}$	$\frac{11}{15}$	$\frac{25}{48}$	b	
		$\frac{11}{15}$	$\frac{2}{15}$	$\frac{2}{15}$	$\frac{25}{48}$	c	
		$\frac{2}{15}$	$\frac{11}{15}$	$\frac{2}{15}$	$\frac{25}{48}$	d	

Table 8.2. *Quadrature points and weights for triangular elements.*

8.2 Numerical Approximation of the Rod Model

In this section, we illustrate approximation techniques for the distributed rod models developed in Section 7.3. For the general model (7.18), we first consider a direct Galerkin discretization in space followed by a finite difference discretization in time. This yields *global* mass, stiffness and damping matrices, a semi-discrete system appropriate for control design, and a fully discrete system feasible for simulations. Secondly, we consider an *elemental* analysis to demonstrate aspects of finite element assembly often employed for 2-D and 3-D characterization.

8.2.1 Global Discretization in Space

Consider the weak model formulation (7.18),

$$\begin{aligned}\int_0^\ell \rho A \frac{\partial^2 u}{\partial t^2}\phi dx + \int_0^\ell \left[YA\frac{\partial u}{\partial x} + cA\frac{\partial^2 u}{\partial x \partial t}\right]\frac{d\phi}{dx}dx = \int_0^\ell f\phi dx \\ +A\left[a_1\widetilde{P} + a_2\widetilde{P}^2\right]\int_0^\ell \frac{d\phi}{dx}dx - \left[k_\ell u(t,\ell) + c_\ell \frac{\partial u}{\partial t}(t,\ell) + m_\ell\frac{\partial^2 u}{\partial t^2}(t,\ell)\right]\phi(\ell),\end{aligned} \tag{8.13}$$

where $\widetilde{P} = P - P_R$, which must hold for ϕ in the space of test functions

$$V = \left\{\Phi = (\phi,\varphi) \in L^2(0,\ell)\times\mathbb{R}\,|\, \phi\in H^1(0,\ell), \phi(0) = 0, \phi(\ell) = \varphi\right\}.$$

The goal when constructing Galerkin solutions to (8.13) is to determine approximate solutions in finite dimensional subspaces V^N of V.

To construct V^N, we consider a uniform partition of the interval $[0,\ell]$ with points $x_j = jh, j = 0,\ldots,N$ and a uniform stepsize $h = \frac{\ell}{N}$ where N denotes the number of subintervals. The spatial basis $\{\phi_j\}_{j=1}^N$ used to construct V^N is comprised of linear splines

$$\begin{aligned}\phi_j(x) &= \frac{1}{h}\begin{cases} x - x_{j-1}\,, & x_{j-1}\le x < x_j \\ x_{j+1} - x\,, & x_j \le x \le x_{j+1} \\ 0\,, & \text{otherwise}\end{cases}\quad,\quad j = 1,\ldots,N-1 \\ \phi_N(x) &= \frac{1}{h}\begin{cases} x - x_{N-1}\,, & x_{N-1}\le x\le x_N \\ 0\,, & \text{otherwise}\end{cases}\end{aligned} \tag{8.14}$$

as depicted in Figure 8.8. It is observed that the basis functions satisfy the essential boundary condition $\phi_j(0) = 0$ for $j = 1,\ldots,N$. Furthermore, $\{\phi_j\}_{j=1}^N$ are differentiable throughout $(0,\ell)$, except at the countable set of gridpoints, and hence they

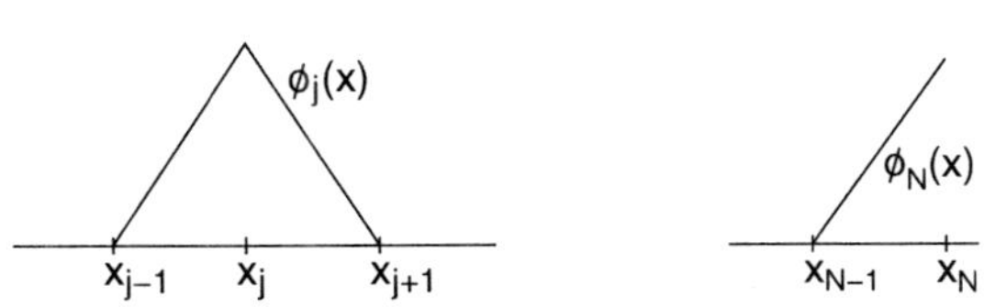

Figure 8.8. *Piecewise linear basis functions (a) $\phi_j(x)$ and (b) $\phi_N(x)$.*

are elements in $H^1(0,\ell)$. Letting $\varphi_j = \phi_j(\ell)$ and $\Phi_j = (\phi_j, \varphi_j)$, the approximating subspace is defined to be

$$V^N = \text{span}\,\{\Phi_j\}\,.$$

The solution to (8.13) is approximated by the expansion

$$u^N(t,x) = \sum_{j=1}^N u_j(t)\phi_j(x) \tag{8.15}$$

which satisfies $u^N(t,0) = 0$ and can achieve arbitrary displacements at $x = \ell$.

A semi-discrete second-order matrix system is obtained by considering the approximate solution $u^N(t,x)$ in (8.13) with the basis functions $\{\phi_i\}_{i=1}^N$ employed as test functions — this is equivalent to projecting the system (8.13) onto the finite-dimensional subspace V^N. The interchange of integration and summation yields the system

$$\begin{aligned}\sum_{j=1}^N \ddot{u}_j(t)\int_0^\ell \rho A\phi_i\phi_j dx + \sum_{j=1}^N \dot{u}_j(t)\int_0^\ell cA\phi_i'\phi_j' dx + \sum_{j=1}^N u_j(t)\int_0^\ell YA\phi_i'\phi_j' dx \\ = \int_0^\ell f\phi_i dx + A\left[a_1(P-P_R) + a_2(P-P_R)^2\right]\int_0^\ell \phi_i' dx \\ -\Big(k_\ell u_N(t)\phi_N(\ell) + c_\ell \dot{u}_N(t)\phi_N(\ell) + m_\ell \ddot{u}_N(t)\phi_N(\ell)\Big)\phi_N(\ell)\end{aligned}$$

which holds for $i = 1,\ldots,N$. This can be written as the second-order vector-valued system

$$\mathbb{M}\,\ddot{\mathbf{u}}(t) + \mathbb{Q}\,\dot{\mathbf{u}} + \mathbb{K}\,\mathbf{u}(t) = \mathbf{f}(t) + A\left[a_1(P(E(t)) - P_R) + a_2(P(E(t)) - P_R)^2\right]\mathbf{b} \tag{8.16}$$

where

$$\mathbf{u}(t) = [u_1(t),\ldots,u_N(t)]^T\,.$$

The *global* mass, stiffness and damping matrices have the components

$$[\mathbb{M}]_{ij} = \begin{cases} \displaystyle\int_0^\ell \rho A\phi_i\phi_j\,dx & , \quad i\neq N \text{ or } j\neq N \\ \displaystyle\int_0^\ell \rho A\phi_i\phi_j\,dx + m_\ell & , \quad i = N \text{ and } j = N \end{cases}$$

$$[\mathbb{K}]_{ij} = \begin{cases} \displaystyle\int_0^\ell YA\phi_i'\phi_j'\,dx & , \quad i\neq N \text{ or } j\neq N \\ \displaystyle\int_0^\ell YA\phi_i'\phi_j'\,dx + k_\ell & , \quad i = N \text{ and } j = N \end{cases} \tag{8.17}$$

$$[\mathbb{Q}]_{ij} = \begin{cases} \displaystyle\int_0^\ell cA\phi_i'\phi_j'\,dx & , \quad i\neq N \text{ or } j\neq N \\ \displaystyle\int_0^\ell cA\phi_i'\phi_j'\,dx + c_\ell & , \quad i = N \text{ and } j = N \end{cases}$$

whereas the force vectors are defined by

$$[\mathbf{f}\,]_i = \int_0^\ell f\phi_i dx \quad , \quad [\mathbf{b}]_i = \int_0^\ell \phi_i' dx. \tag{8.18}$$

To evaluate the integrals, it is observed that when ρ and Y are constant, the maximal degree occurs in the mass matrix which is comprised of quadratic polynomials on each subinterval $[x_{j-1}, x_j]$. In this case, 2-point composite Gaussian quadrature with nodes and weights

$$x_1 = x_{j-1} + \frac{h}{2}\left(1 - 1/\sqrt{3}\right) \quad , \quad w_1 = \frac{h}{2}$$

$$x_2 = x_{j-1} + \frac{h}{2}\left(1 + 1/\sqrt{3}\right) \quad , \quad w_2 = \frac{h}{2}$$

will provide *exact* integration on $[x_{j-1}, x_j]$ which yields the tridiagonal matrices

$$\mathbb{M} = \rho A h \begin{bmatrix} \frac{2}{3} & \frac{1}{6} & 0 & \cdots & & 0 \\ \frac{1}{6} & \frac{2}{3} & \frac{1}{6} & & & \\ & \ddots & \ddots & \ddots & & \\ & & & \frac{1}{6} & \frac{2}{3} & \frac{1}{6} \\ 0 & \cdots & & 0 & \frac{1}{6} & \frac{1}{3} + \frac{m_\ell}{h} \end{bmatrix} \quad , \quad \mathbb{K} = \frac{YA}{h} \begin{bmatrix} 2 & -1 & 0 & \cdots & & 0 \\ -1 & 2 & -1 & & & \\ & \ddots & \ddots & \ddots & & \\ & & & -1 & 2 & -1 \\ 0 & \cdots & & 0 & -1 & 1 + hk_\ell \end{bmatrix} \tag{8.19}$$

and vector

$$\mathbf{b} = [0\,, \,\ldots\,, \,1]^T.$$

The damping matrix has the representation $\mathbb{Q} = \frac{c}{Y}\mathbb{K}$ when c is constant.

To formulate a first-order semi-discrete system appropriate for finite-dimensional control design, we let $\mathbf{z} = [\mathbf{u}, \dot{\mathbf{u}}]^T$ and define the system matrix $\mathbb{A}$ and vectors $\mathbf{F}(t)$ and $\mathbf{B}$ by

$$\mathbb{A} = \begin{bmatrix} 0 & \mathbb{I} \\ -\mathbb{M}^{-1}\mathbb{K} & -\mathbb{M}^{-1}\mathbb{Q} \end{bmatrix},$$

$$\mathbf{F}(t) = \begin{bmatrix} 0 \\ \mathbb{M}^{-1}\mathbf{f}(t) \end{bmatrix} \quad , \quad \mathbf{B} = \begin{bmatrix} 0 \\ \mathbb{M}^{-1}\mathbf{b} \end{bmatrix}. \tag{8.20}$$

The second-order system (8.16) can subsequently be reformulated as

$$\begin{aligned} \dot{\mathbf{z}}(t) &= \mathbb{A}\,\mathbf{z}(t) + \mathbf{F}(t) + A\left[a_1(P(E(t)) - P_R) + a_2(P(E(t)) - P_R)^2\right]\mathbf{B} \\ \mathbf{z}(0) &= \mathbf{z}_0 \end{aligned} \tag{8.21}$$

where the $2N \times 1$ vector $\mathbf{z}_0$ denotes the projection of the initial conditions into the approximating subspace.

Temporal Discretization

The system (8.21) must be discretized in time to permit numerical or experimental implementation. The choice of approximation method is dictated by accuracy and stability requirements, storage capabilities, and sample rates. This can be accomplished using MATLAB routines such as `ode15s.m` which accommodate the moderate stiffness inherent to Galerkin approximation in space. Alternatively, a trapezoidal method can be advantageous for experimental implementation since it is moderately accurate, is A-stable, and requires minimal storage when implemented as a single step method. For temporal stepsizes Δt, a standard trapezoidal discretization yields the iteration

$$\begin{aligned}
\mathbf{z}_{k+1} &= \mathbb{W}\mathbf{z}_k + \frac{1}{2}\mathbb{V}\left[\mathbf{F}(t_k) + \mathbf{F}(t_{k+1})\right] \\
&\quad + \frac{1}{2}\left[a_1\widetilde{P}(E(t_k)) + a_1\widetilde{P}(E(t_{k+1})) + a_2\widetilde{P}^2(E(t_k)) + a_2\widetilde{P}^2(E(t_{k+1}))\right]\mathbb{V}\mathbf{B} \\
\mathbf{z}_0 &= \mathbf{z}(0)
\end{aligned}$$

where $\widetilde{P}(E) = P(E) - P_R$, $t_k = k\Delta t$, and $\mathbf{z}_k$ approximates $\mathbf{z}(t_k)$. The matrices

$$\mathbb{W} = \left(\mathbb{I} - \frac{\Delta t}{2}\mathbb{A}\right)^{-1}\left(\mathbb{I} + \frac{\Delta t}{2}\mathbb{A}\right) \quad , \quad \mathbb{V} = \Delta t\left(\mathbb{I} - \frac{\Delta t}{2}\mathbb{A}\right)^{-1}$$

need only be created once when numerically or experimentally implementing the method. This yields approximate solutions having $\mathcal{O}(h^2, (\Delta t)^2)$ accuracy. For applications in which data at future times t_{k+1} is unavailable, the modified trapezoidal algorithm

$$\begin{aligned}
\mathbf{z}_{k+1} &= \mathbb{W}\mathbf{z}_k + \mathbb{V}\mathbf{F}(t_k) + \left[a_1(P(E(t_k)) - P_R) + a_2(P(E(t_k)) - P_R)^2\right]\mathbb{V}\mathbf{B} \\
\mathbf{z}_0 &= \mathbf{z}(0)
\end{aligned}$$

can be employed. Whereas this decreases slightly the temporal accuracy, for large sample rates with correspondingly small stepsizes Δt, the accuracy is still commensurate with that of the data.

Remark 8.2.1. *The approximation of the eigenvalue problem associated with the undamped rod model yields the generalized eigenvalue problem*

$$\mathbb{K}\boldsymbol{\zeta} = \omega^2\mathbb{M}\boldsymbol{\zeta} \tag{8.22}$$

where the stiffness matrix $\mathbb{K}$ and mass matrix $\mathbb{M}$ are defined in (8.19). Hence (8.22) can be used to approximate the natural frequencies and modes for the undamped rod.

Remark 8.2.2. *Due to the presence of both internal damping and damping in the boundary condition at $x = \ell$, the eigenvalues of the system matrix $\mathbb{A}$ defined in (8.20) will all have negative real part. This property can be used to check the validity of the signs in the boundary condition (7.13) and weak formulation (8.13). For example, an incorrect formulation which added rather than subtracted the final boundary contribution will produce eigenvalues of $\mathbb{A}$ having positive real part which is inconsistent with the damping in the model.*

8.2.2 Elemental Analysis

The spatial discretization technique detailed in Section 8.2.1 illustrates the general philosophy of Galerkin approximation including the definition of global basis functions and their use for defining an approximating subspace $V^N \subset V$. In this section, we summarize a local, elemental approach to the problem. Whereas the two techniques yield equivalent mass, stiffness and damping matrices, the latter is significantly more efficient for general 2-D and 3-D geometries and hence is employed for general finite element analysis of complex structures.

To simplify the discussion and facilitate energy analysis, we consider the regime employed in Section 7.3.2 and take $c = f = P = m_\ell = c_\ell = k_\ell = 0$ in the weak formulation (8.13). Additionally, we assume that ρ and Y are constant. This model quantifies the dynamics of an undamped and unforced rod that is fixed at $x = 0$ and free at $x = \ell$. The space of test functions in this case is

$$V = H_0^1(0,\ell) = \left\{\phi \in H^1(0,\ell) \,|\, \phi(0) = 0\right\}.$$

Local Basis Elements

To illustrate the construction of local mass and stiffness matrices, we initially consider the approximation of rod dynamics on a local interval $[0, h]$ as depicted in Figure 8.9(a). In accordance with the assumption that u is differentiable in x at all but a countable number of points, we express displacements as

$$\begin{aligned} u(t,x) &= a_0(t) + a_1(t)x \\ &= \varphi^T(x)\mathbf{a}(t) \end{aligned}$$

where $\mathbf{a}(t) = [a_0(t)\,,\, a_1(t)]^T$ and $\varphi(x) = [1\,,\, x]^T$. To formulate u in terms of the nodal values $u_\ell(t)$ and $u_r(t)$ at $x = 0$ and $x = \ell$, we note that

$$\begin{bmatrix} u_\ell(t) \\ u_r(t) \end{bmatrix} = \begin{bmatrix} 1 & 0 \\ 1 & h \end{bmatrix} \begin{bmatrix} a_0(t) \\ a_1(t) \end{bmatrix}$$

or

$$\mathbf{u}(t) = \mathbb{T}\mathbf{a}(t)$$

where

$$\mathbb{T} = \begin{bmatrix} 1 & 0 \\ 1 & h \end{bmatrix} \quad , \quad \mathbf{u}(t) = \begin{bmatrix} u_\ell(t) \\ u_r(t) \end{bmatrix}.$$

Figure 8.9. *(a) Rod displacements $u_\ell(t)$ and $u_r(t)$ at the left and right endpoints of the local interval $[0,h]$. (b) Local linear basis functions $\widetilde{\phi}_1(x)$ and $\widetilde{\phi}_2(x)$.*

By observing that $\mathbf{a}(t) = \mathbb{S}\mathbf{u}(t)$, where

$$\mathbb{S} = \mathbb{T}^{-1} = \begin{bmatrix} 1 & 0 \\ -\frac{1}{h} & \frac{1}{h} \end{bmatrix},$$

it follows that displacements can be represented as

$$u(t,x) = \boldsymbol{\phi}^T(x)\mathbf{u}(t). \tag{8.23}$$

Here $\boldsymbol{\phi}(x) = \left[\widetilde{\phi}_1(x), \widetilde{\phi}_2(x)\right]^T$ contains the *local basis elements*

$$\widetilde{\phi}_1(x) = 1 - \frac{x}{h} \quad , \quad \widetilde{\phi}_2(x) = \frac{x}{h}$$

depicted in Figure 8.9(b).

Note that (8.23) is a local version of the expansion (8.15) employed when constructing approximate solutions using the global basis functions $\{\phi_j\}_{j=1}^N$ defined in (8.14). The local region $[0,h]$ and global interval $[x_{j-1}, x_j]$ are analogous to the master and physical triangles depicted in Figure 8.5 whereas the local basis set $\{\widetilde{\phi}_1, \widetilde{\phi}_2\}$ is the 1-D analogue of the 2-D elements $\{N_i, N_j, N_k\}$ shown in Figure 8.7.

Hamiltonian Formulation

To specify a dynamic model quantifying the displacements u, we employ the Hamiltonian framework detailed in Section 7.3.2 for the infinite dimensional problem. Here we consider displacements $u \in V^N = \text{span}\{\widetilde{\phi}_1, \widetilde{\phi}_2\}$ as dictated by our approximation framework.

We first note that for this class of displacements, the squared strains and velocities can be expressed as

$$u_x^2(t,x) = \mathbf{u}^T(t)\mathbb{S}^T\mathbb{D}(x)\mathbb{S}\mathbf{u}(t)$$
$$u_t^2(t,x) = \dot{\mathbf{u}}^T(t)\mathbb{S}^T\mathbb{F}(x)\mathbb{S}\dot{\mathbf{u}}(t)$$

where

$$\mathbb{D}(x) = \boldsymbol{\varphi}_x(x)\boldsymbol{\varphi}_x^T(x)^T = \begin{bmatrix} 0 & 0 \\ 0 & 1 \end{bmatrix} \quad , \quad \mathbb{F}(x) = \boldsymbol{\varphi}(x)\boldsymbol{\varphi}^T(x) = \begin{bmatrix} 1 & x \\ x & x^2 \end{bmatrix}.$$

The potential and kinetic energies (7.20) can thus be expressed as

$$U = \frac{YA}{2}\mathbf{u}^T(t)\mathbb{S}^T \cdot \int_0^h \mathbb{D}(x)dx \cdot \mathbb{S}\mathbf{u}(t)$$

$$K = \frac{\rho A}{2}\dot{\mathbf{u}}^T(t)\mathbb{S}^T \cdot \int_0^h \mathbb{F}(x)dx \cdot \mathbb{S}\dot{\mathbf{u}}(t).$$

Application of Hamilton's principle in the manner detailed in Section 7.3.2 yields the relation

$$\rho A\mathbb{S}^T \cdot \int_0^h \mathbb{F}(x)dx \cdot \mathbb{S}\ddot{\mathbf{u}}(t) + YA\mathbb{S}^T \cdot \int_0^h \mathbb{D}(x)dx \cdot \mathbb{S}\mathbf{u}(t) = 0. \tag{8.24}$$

Local Mass and Stiffness Matrices

The weak formulation (8.24) can be written as

$$\mathbb{M}_e \ddot{\mathbf{u}}(t) + \mathbb{K}_e \mathbf{u} = 0$$

where

$$\mathbb{M}_e = \rho A \mathbb{S}^T \cdot \int_0^h \mathbb{F}(x)dx \cdot \mathbb{S} \quad , \quad \mathbb{K}_e = YA\mathbb{S}^T \cdot \int_0^h \mathbb{D}(x)dx \cdot \mathbb{S}$$

are the *local* mass and stiffness matrices. Evaluation of the integrals yields the analytic formulations

$$\mathbb{M}_e = \rho Ah \begin{bmatrix} \frac{1}{3} & \frac{1}{6} \\ \frac{1}{6} & \frac{1}{3} \end{bmatrix} \quad , \quad \mathbb{K}_e = \frac{YA}{h} \begin{bmatrix} 1 & -1 \\ -1 & 1 \end{bmatrix}.$$

Global Mass and Stiffness Matrices

To motivate the techniques used to construct global mass and stiffness matrices, we first partition the rod interval $[0,\ell]$ into two subregions as shown in Figure 8.10(a) — hence $h = \frac{\ell}{2}$. With the requirement that $u_{1r}(t) = u_{2\ell}(t)$, the nodal unknowns in this case are $\mathbf{u}(t) = [u_{1\ell}(t), u_{2\ell}(t), u_{2r}(t)]^T$. Combination of the local relations subsequently yields the global system

$$\mathbb{M}\ddot{\mathbf{u}} + \mathbb{K}\mathbf{u} = 0$$

where the *global* mass and stiffness matrices are given by

$$\mathbb{M} = \rho Ah \begin{bmatrix} \frac{1}{3} & \frac{1}{6} & 0 \\ \frac{1}{6} & \frac{2}{3} & \frac{1}{6} \\ 0 & \frac{1}{6} & \frac{1}{3} \end{bmatrix} \quad , \quad \mathbb{K} = \frac{YA}{h} \begin{bmatrix} 1 & -1 & 0 \\ -1 & 2 & -1 \\ 0 & -1 & 1 \end{bmatrix}.$$

We note that the second row in the mass matrix is obtained by summing the element relations

$$\frac{\rho A\ell}{2}\left(\frac{1}{6}\ddot{u}_{1\ell} + \frac{1}{3}\ddot{u}_{2\ell}\right) = 0$$

$$\frac{\rho A\ell}{2}\left(\frac{1}{3}\ddot{u}_{2\ell} + \frac{1}{6}\ddot{u}_{2r}\right) = 0$$

after enforcing $\ddot{u}_{1r} = \ddot{u}_{2\ell}$. A similar strategy is employed when constructing the stiffness matrix and the general procedure is illustrated in Figure 8.11(a).

Figure 8.10. *Partition of the rod domain $[0,\ell]$ into (a) 2 subregions and (b) n subregions.*

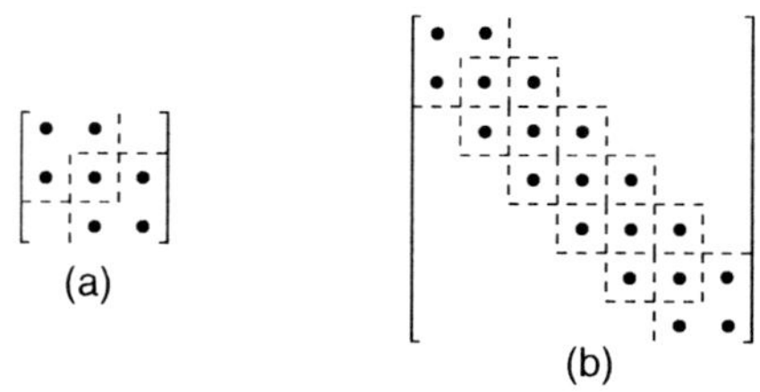

Figure 8.11. *Combination of elemental mass and stiffness matrices* $\mathbb{M}_e$ *and* $\mathbb{K}_e$ *to construct global matrices* $\mathbb{M}$ *and* $\mathbb{K}$*: (a) 2 subregions, and (b) n subregions.*

The process for general partitions $x_j = jh, j = 0, 1, \ldots, N$ with $h = \frac{\ell}{N}$ is identical and leads to a similar summation process as shown in Figure 8.11(b). Enforcing the boundary condition $u_{1\ell} = 0$ subsequently yields the global mass and stiffness matrices

$$\mathbb{M} = \rho A h \begin{bmatrix} \frac{2}{3} & \frac{1}{6} & 0 & \cdots & 0 \\ \frac{1}{6} & \frac{2}{3} & \frac{1}{6} & & \\ & \ddots & \ddots & \ddots & \\ & & \frac{1}{6} & \frac{2}{3} & \frac{1}{6} \\ 0 & \cdots & 0 & \frac{1}{6} & \frac{1}{3} \end{bmatrix} \quad , \quad \mathbb{K} = \frac{YA}{h} \begin{bmatrix} 2 & -1 & 0 & \cdots & 0 \\ -1 & 2 & -1 & & \\ & \ddots & \ddots & \ddots & \\ & & -1 & 2 & -1 \\ 0 & \cdots & 0 & -1 & 1 \end{bmatrix} . \tag{8.25}$$

A comparison between (8.25) and (8.19) obtained through global analysis illustrates that the matrices are identical when one takes $m_\ell = k_\ell = 0$ in the latter set.

The technique can be extended to incorporate linear and nonlinear inputs by employing the augmented action integral (7.25) and an extended Hamilton's principle. Internal and boundary damping is incorporated by employing extended constitutive relations as detailed in the previous section.

Whereas the global discretization techniques and elemental analysis provide the same semi-discrete systems, the latter technique facilitates implementation for general 2-D and 3-D geometries discretized using triangular meshes. Additional details regarding finite element techniques for rod models can be found in [36, 276].

8.2.3 Examples and Software

The performance of the discretized rod model is illustrated in Section 7.3.3 in the context of characterizing displacements generated by a stacked PZT actuator from an AFM stage. This example illustrates the effects of variable drive levels and the incorporation of frequency-dependent hysteresis mechanisms via the nonlinear constitutive relations developed in Chapter 2.

The performance of the approximation techniques when used to characterize the hysteretic dynamics of a Terfenol-D transducer is reported in [119, 120]. In this case, the domain wall model was used to provide the constitutive relations which provide the basis for constructing the distributed rod model. In both cases, the mass, stiffness and damping matrices have the general representations (8.17) —

$\mathbb{M}$ and $\mathbb{K}$ have the specific representations (8.19) when ρ and Y are constant — and the input vectors are given by (8.18). The only differences in the models occur in the scalar-valued input relations used to characterize the nonlinear hysteresis.

MATLAB m-files for implementing both the rod model and the model for the stacked PZT actuator employed in the AFM stage can be found at the website `http://www.siam.org/books/fr32`.

8.3 Numerical Approximation of the Beam Model

The strategy for approximating the beam models developed in Section 7.4 is analogous to that employed for the rod models — spline or finite element discretizations in space are used to construct semi-discrete systems appropriate for control design or subsequent temporal discretization for simulations. Unlike rod models, quantification of the bending moments or strain energy associated with bending yields second derivatives in weak beam formulations which must be accommodated by basis functions. In Section 8.3.1 we summarize the use of cubic B-splines to construct approximating subspaces whereas a cubic Hermite basis is constructed in Section 8.3.2 through techniques analogous to those of Section 8.2.2.

8.3.1 Cubic Spline Basis

We illustrate beam approximation in the context of a cantilever beam, having a fixed-end at $x = 0$ and free-end at $x = \ell$, with surface-mounted patches. For test functions ϕ in the space

$$V = H_0^2(0,\ell) = \left\{\phi \in H^2(0,\ell) \,|\, \phi(0) = \phi'(0) = 0\right\},$$

the weak formulation of the model for linear inputs is

$$\begin{aligned}\int_0^\ell \rho\frac{\partial^2 w}{\partial t^2}\phi dx + \gamma\int_0^\ell \frac{\partial w}{\partial t}\phi dx + \int_0^\ell YI\frac{\partial^2 w}{\partial x^2}\frac{d^2\phi}{dx^2}dx \\ + \int_0^\ell cI\frac{\partial^3 w}{\partial x^2\partial t}\frac{d^2\phi}{dx^2}dx = \int_0^\ell f\phi dx + V(t)\int_0^\ell k_p\frac{d^2\phi}{dx^2}dx\end{aligned} \tag{8.26}$$

where the piecewise constants ρ, YI, cI and k_p are defined in (7.42) and (7.44).

To formulate approximate solutions based on cubic B-Splines, consider the partition $x_j = jh$, where $h = \frac{\ell}{N}$ and $j = 0, \ldots, N$. For $j = -1, 0, 1, \ldots, N+1$, it is shown in [383] that standard cubic B-splines are defined by

$$\widehat{\phi}_j(x) = \frac{1}{h^3}\begin{cases}(x - x_{j-2})^3\,, & x \in [x_{j-2}, x_{j-1}) \\ h^3 + 3h^2(x - x_{j-1}) + 3h(x - x_{j-1})^2 - 3(x - x_{j-1})^3\,, & x \in [x_{j-1}, x_j) \\ h^3 + 3h^2(x_{j+1} - x) + 3h(x_{j+1} - x)^2 - 3(x_{j+1} - x)^3\,, & x \in [x_j, x_{j+1}) \\ (x_{j+2} - x)^3\,, & x \in [x_{j+1}, x_{j+2}) \\ 0\,, & \text{otherwise}\end{cases} \tag{8.27}$$

as illustrated in Figure 8.12. To accommodate the essential boundary conditions $w(t,0) = \frac{\partial w}{\partial x}(t,0) = 0$ and corresponding restrictions $\phi(0) = \phi'(0) = 0$ required of functions in V, we employ the basis functions

$$\phi_j(x) = \begin{cases} \widehat{\phi}_0(x) - 2\widehat{\phi}_{-1}(x) - 2\widehat{\phi}_1(x) \quad , \quad j = 1 \\ \widehat{\phi}_j(x) \quad , \quad j = 2, \dots, N+1 \end{cases} \tag{8.28}$$

which are modified to ensure that $\phi_j \in V$ for $j = 1, \dots, N+1$. Finally, we consider approximate solutions

$$w^N(t,x) = \sum_{j=1}^{N+1} w_j(t)\phi_j(x)$$

in the space $V^N = \text{span}\{\phi_j\} \subset V$.

In a manner similar to that detailed in Section 8.2.1, consideration of w^N in (8.26) with basis functions employed as test functions yields the semi-discrete system

$$\mathbb{M}\,\ddot{\mathbf{w}} + \mathbb{Q}\,\dot{\mathbf{w}} + \mathbb{K}\,\mathbf{w} = \mathbf{f} + V(t)\mathbf{b} \tag{8.29}$$

where

$$\mathbf{w}(t) = [w_1(t)\,, \ \dots \ , \, w_{N+1}(t)]^T\,. \tag{8.30}$$

The *global* mass, damping and stiffness matrices are defined by

$$\begin{aligned} [\mathbb{M}]_{ij} &= \int_0^\ell \rho\phi_i\phi_j dx \\ [\mathbb{Q}]_{ij} &= \int_0^\ell \left[\gamma\phi_i\phi_j + cI\phi_i''\phi_j''\right] dx \\ [\mathbb{K}]_{ij} &= \int_0^\ell YI\phi_i''\phi_j'' dx \end{aligned} \tag{8.31}$$

whereas the force vectors have the components

$$[\mathbf{f}]_i = \int_0^\ell f\phi_i dx \quad , \quad [\mathbf{b}]_i = \int_0^\ell \phi_i'' dx. \tag{8.32}$$

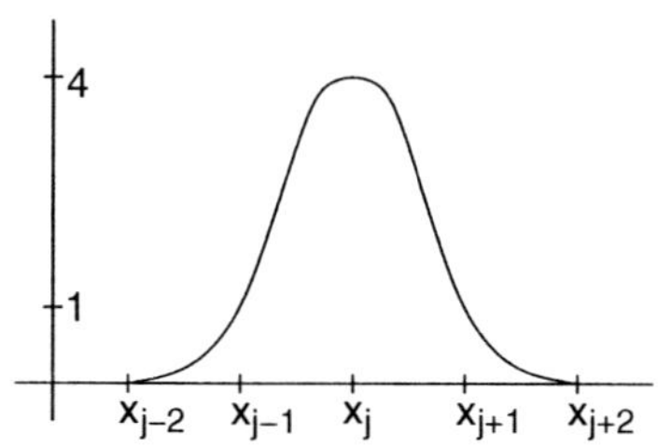

Figure 8.12. *Cubic B-spline* $\phi_j(x)$.

The second-order vector-value system (8.29) has the same form as (8.16) which was developed for the rod model and the techniques of Section 8.2.1 can be used to construct a first-order system appropriate for design and control.

Properties of Cubic Spline Approximates

We summarize here properties of the cubic spline approximates — additional comparison between this class of approximate solutions and the cubic Hermite approximates developed in Section 8.3.2 can be found in Section 8.3.3.

Integral Evaluation

From (8.31) it is observed that the construction of the mass, damping and stiffness matrices requires the evaluation of integrals whose integrands are piecewise polynomials of order 2 or 6 when ρ, c and Y are constant. To ensure exact integration, we employ the composite 4-point Gauss–Legendre rule (8.12), which exhibits degree of precision 7 accuracy, on each subinterval $[x_{j-1}, x_j]$. For the fixed-free end boundary conditions considered here, and constant Young's modulus, this yields the stiffness matrix

$$\mathbb{K} = \frac{YI}{h^3}\begin{bmatrix} 648 & 96 & 6 & -12 & & & & & & & \\ 96 & 96 & -54 & 0 & 6 & & & & & & \\ 6 & -54 & 96 & -54 & 0 & 6 & & & & & \\ -12 & 0 & -54 & 96 & -54 & 0 & 6 & & & & \\ & 6 & 0 & -54 & 96 & -54 & 0 & 6 & & & \\ & & \ddots & \ddots & \ddots & \ddots & \ddots & \ddots & \ddots & & \\ & & & 6 & 0 & -54 & 96 & -54 & 0 & 6 \\ & & & & 6 & 0 & -54 & 84 & -36 & 0 \\ & & & & & 6 & 0 & -36 & 48 & -18 \\ & & & & & & 6 & 0 & -18 & 12 \end{bmatrix}.$$

For the beam with surface-mounted patches, the parameters will be piecewise constant with discontinuities at the gridpoints aligned with patch ends.

Partition Construction and Accuracy of the Method

For constant parameters ρ, YI and cI, the accuracy of the cubic spline approximate is $\mathcal{O}(h^4)$. This implies that errors will diminish by a factor of approximately 16 when stepsizes h are halved. The same asymptotic accuracy is achieved for piecewise constant parameters associated with the surface-mounted patches if the partition is aligned with the patch edges; that is, gridpoints x_j must correspond with the patch endpoints x_1 and x_2. The accuracy degradation which occurs if this condition is neglected depends in part on the magnitude of the discontinuity of ρ, YI, cI and k_p at x_1 and x_2.

Alignment of the partition with the patch ends is easily addressed in simulation and control designs for which patches are bonded at predefined locations.

A more pertinent issue arises when addressing the problem of *optimal patch location* for which x_1 and x_2 are parameters determined through an optimization routine [132, 155, 202]. This necessitates consideration of variable or adaptive meshes and constitutes an active research area.

Projection Method

It is important to note that approximation using the modified B-spline basis (8.28) constitutes a projection rather than interpolation method as is the case for the cubic Hermite methods summarized in Section 8.3.2. Hence the coefficients (8.30) do not approximate nodal values of the true solution and modified basis functions are required to accommodate essential boundary conditions. In this sense, B-spline approximation shares a projective kinship with globally defined spectral methods — e.g., Legendre or Fourier — while retaining the sparsity associated with locally defined polynomial elements. As will be detailed in Section 8.3.3, the advantage of cubic B-splines over cubic Hermite elements lies in the fact that half as many coefficients are required in the former case. The non-interpolatory nature of B-splines constitutes the primary disadvantage which is especially pertinent when accommodating boundary or interface conditions between components of the structure — e.g., curved and flat portions of THUNDER transducers.

8.3.2 Cubic Hermite Basis

To illustrate the construction of a finite element basis which interpolates displacements and slopes at the partition points, we initially consider the model (8.26) in the absence of internal or air damping and external forces — hence $c = \gamma = f = V = 0$. Additionally, we take ρ and Y to be constant to highlight the structure of constituent matrices.

Local Mass and Stiffness Matrices

It was detailed in Section 7.4 that thin beam models quantify both the transverse displacement and rotation of the neutral line so we begin the elemental analysis by quantifying the displacements w_ℓ, w_r and slopes θ_ℓ, θ_r on an arbitrary interval $[0, h]$ as depicted in Figure 8.13. Once we have constructed local mass and stiffness matrices, we will extend the analysis to partitions of the full beam interval $[0, \ell]$ to construct global system matrices.

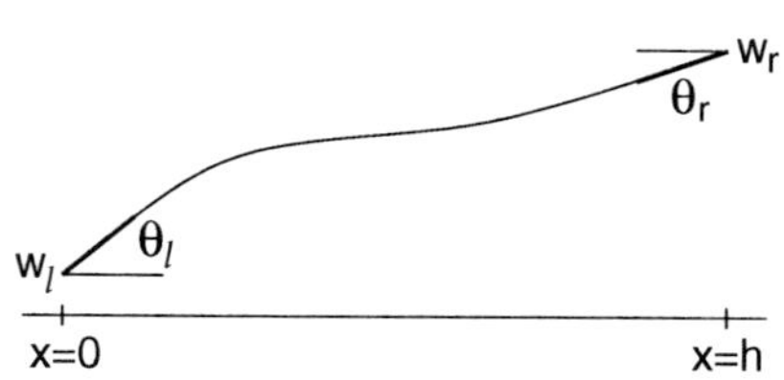

Figure 8.13. *Displacements w_ℓ, w_r and slopes θ_ℓ, θ_r at the ends of a cubic element on the interval $[0, h]$.*

Because the characterization of $\mathbf{w}(t) = [w_\ell(t), \theta_\ell(t), w_r(t), \theta_r(t)]^T$ involves 4 degrees of freedom , we consider cubic representations

$$\begin{aligned} w(t,x) &= a_0(t) + a_1(t)x + a_2(t)x^2 + a_3(t)x^3 \\ &= \boldsymbol{\varphi}^T(x)\mathbf{a}(t) \end{aligned} \tag{8.33}$$

where $\mathbf{a}(t) = [a_0(t), a_1(t), a_2(t), a_3(t)]^T$ and $\boldsymbol{\varphi}(x) = [1, x, x^2, x^3]^T$. By noting that $\theta(t,x) = \frac{\partial w}{\partial x}(t,x)$ and enforcing the interpolation conditions at $x = 0$ and h, the nodal coefficients can be represented as

$$\mathbf{w}(t) = \mathbb{T}\mathbf{a}(t)$$

where

$$\mathbb{T} = \begin{bmatrix} 1 & 0 & 0 & 0 \\ 0 & 1 & 0 & 0 \\ 1 & h & h^2 & h^3 \\ 0 & 1 & 2h & 3h^2 \end{bmatrix}.$$

Substitution of $\mathbf{a}(t) = \mathbb{T}^{-1}\mathbf{w}(t)$ into (8.33) yields the expansion

$$w(t,x) = \boldsymbol{\phi}^T(x)\mathbf{w}(t)$$

where $\boldsymbol{\phi} = [\widetilde{\phi}_1, \widetilde{\phi}_2, \widetilde{\phi}_3, \widetilde{\phi}_4]^T$ comprises the local cubic Hermite basis functions

$$\begin{aligned} \widetilde{\phi}_1(x) &= \frac{1}{h^3}(h-x)^2[2x+h] \quad , \quad & \widetilde{\phi}_3(x) &= \frac{1}{h^3}x^2[2(h-x)+h] \\ \widetilde{\phi}_2(x) &= \frac{1}{h^2}x(x-h)^2 \quad , \quad & \widetilde{\phi}_4(x) &= \frac{1}{h^2}x^2(x-h). \end{aligned} \tag{8.34}$$

As shown in Figure 8.14, the elements $\widetilde{\phi}_j(x)$ have displacement or slope values of 0 or 1 at $x = 0, h$. This ensures that the coefficients $\mathbf{w}(t) = [w_\ell(t), \theta_\ell(t), w_r(t), \theta_r(t)]^T$ interpolate the beam displacements and slopes at $x = 0, h$.

From the relations

$$U = \frac{1}{2}YI\int_0^h \left(\frac{\partial^2 w}{\partial x^2}\right)^2 dx \quad , \quad K = \frac{1}{2}\rho\int_0^h \left(\frac{\partial^2 w}{\partial t^2}\right)^2 dx$$

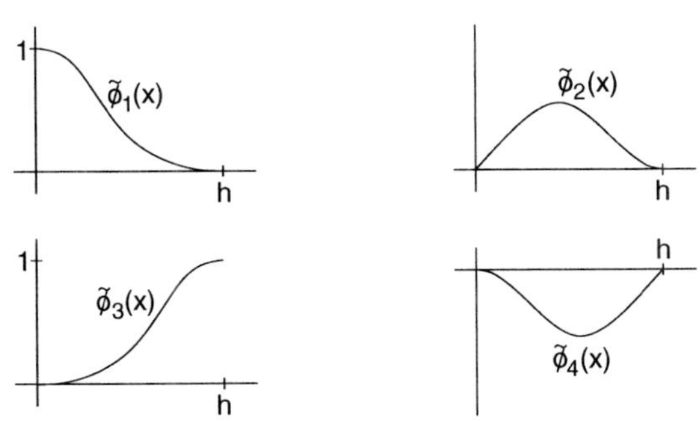

Figure 8.14. *Cubic Hermite basis functions $\widetilde{\phi}_1, \ldots, \widetilde{\phi}_4$.*

for the potential energy due to bending and kinetic energy, it follows that for the class of approximate displacements (8.33),

$$U = \frac{YI}{2}\mathbf{w}^T(t)\mathbb{S}^T \cdot \int_0^h \mathbb{D}(x)dx \cdot \mathbb{S}\mathbf{w}(t)$$

$$K = \frac{\rho}{2}\dot{\mathbf{w}}^T(t)\mathbb{S}^T \cdot \int_0^h \mathbb{F}(x)dx \cdot \mathbb{S}\dot{\mathbf{w}}(t)$$

where $\mathbb{S} = \mathbb{T}^{-1}$ and

$$\mathbb{D}(x) = \boldsymbol{\varphi}_{xx}(x)\boldsymbol{\varphi}_{xx}{}^T(x) = \begin{bmatrix} 0 & 0 & 0 & 0 \\ 0 & 0 & 0 & 0 \\ 0 & 0 & 4 & 12x \\ 0 & 0 & 12x & 36x^2 \end{bmatrix}$$

$$\mathbb{F}(x) = \boldsymbol{\varphi}(x)\boldsymbol{\varphi}^T(x) = \begin{bmatrix} 1 & x & x^2 & x^3 \\ x & x^2 & x^3 & x^5 \\ x^2 & x^3 & x^4 & x^5 \\ x^3 & x^4 & x^5 & x^6 \end{bmatrix}.$$

Application of Hamilton's principle in the manner detailed in Section 7.3.2 yields the second-order vector system

$$\mathbb{M}_e\ddot{\mathbf{w}} + \mathbb{K}_e\mathbf{w} = 0$$

where the local mass and stiffness matrices are

$$\mathbb{M}_e = \frac{\rho h}{420}\begin{bmatrix} 156 & 22h & 54 & -13h \\ 22h & 4h^2 & 13h & -3h^2 \\ 54 & 13h & 156 & -22h \\ -13h & -3h^2 & -22h & 4h^2 \end{bmatrix} \quad , \quad \mathbb{K}_e = \frac{YI}{h^3}\begin{bmatrix} 12 & 6h & -12 & 6h \\ 6h & 4h^2 & -6h & 2h^2 \\ -12 & -6h & 12 & -6h \\ 6h & 2h^2 & -6h & 4h^2 \end{bmatrix}.$$

Global Mass and Stiffness Matrices

Global mass and stiffness matrices are constructed by combining local relations subject to the constraint that displacements and slopes match at the interfaces. To illustrate, we first subdivide the beam support $[0, \ell]$ into two subregions as depicted in Figure 8.10(a). By enforcing the interface conditions

$$w_{1r}(t) = w_{2\ell}(t) \quad , \quad \theta_{1r}(t) = \theta_{2\ell}(t),$$

we obtain the global matrices

$$\mathbb{M} = \frac{\rho h}{420}\begin{bmatrix} 156 & 22h & 54 & -13h & 0 & 0 \\ 22h & 4h^2 & 13h & -3h^2 & 0 & 0 \\ 54 & 13h & 312 & 0 & 54 & -13h \\ -13h & -3h^2 & 0 & 8h^2 & 13h & -3h^2 \\ 0 & 0 & 54 & 13h & 156 & -22h \\ 0 & 0 & -13h & -3h^2 & -22h & 4h^2 \end{bmatrix}$$

and

$$\mathbb{K} = \frac{YI}{h^3}\begin{bmatrix} 12 & 6h & -12 & 6h & 0 & 0 \\ 6h & 4h^2 & -6h & 2h^2 & 0 & 0 \\ -12 & -6h & 24 & 0 & -12 & 6h \\ 6h & 2h^2 & 0 & 8h^2 & -6h & 2h^2 \\ 0 & 0 & -12 & -6h & 12 & -6h \\ 0 & 0 & 6h & 2h^2 & -6h & 4h^2 \end{bmatrix}$$

where $h = \frac{\ell}{2}$.

Due to the fixed-end conditions at $x = 0$, it follows that $w_{1\ell} = \theta_{1\ell} = 0$. After re-ordering the vector of nodal values as

$$\mathbf{w}(t) = [w_{1r}(t), w_{2r}(t), \theta_{1r}(t), \theta_{2r}(t)]^T,$$

the dynamics are quantified by the system

$$\mathbb{M}\ddot{\mathbf{w}} + \mathbb{K}\mathbf{w} = 0$$

where

$$\mathbb{M} = \frac{\rho h}{420}\begin{bmatrix} 312 & 54 & 0 & -13h \\ 54 & 156 & 13h & -22h \\ 0 & 13h & 8h^2 & -3h^2 \\ -13h & -22h & -3h^2 & 4h^2 \end{bmatrix}, \quad \mathbb{K} = \frac{YI}{h^3}\begin{bmatrix} 24 & -12 & 0 & 6h \\ -12 & 12 & -6h & -6h \\ 0 & -6h & 8h^2 & 2h^2 \\ 6h & -6h & 2h^2 & 4h^2 \end{bmatrix}. \tag{8.35}$$

The process for general partitions $x_j = jh$ with $h = \frac{\ell}{N}$ is analogous and leads to a similar summation process when constructing global system matrices — see Figure 8.15. As detailed in previous sections, internal damping can be incorporated by employing more general constitutive relations based on the assumption that stress is proportional to a linear combination of strain and strain rate.

Global Discretization

We illustrated in Section 8.2 that either global Galerkin techniques or local elemental analysis could be employed when implementing linear finite element methods. The same is true with cubic Hermite elements. Whereas the local elemental analysis illustrates the implementation philosophy for general 2-D and 3-D

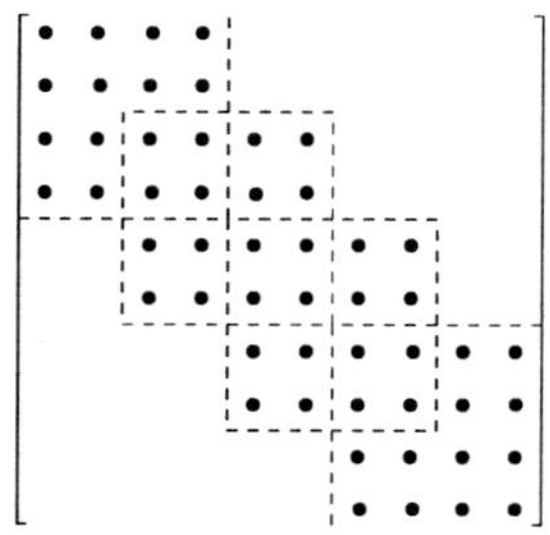

Figure 8.15. *Combination of elemental mass and stiffness matrices $\mathbb{M}_e$ and $\mathbb{K}_e$ to construct global matrices $\mathbb{M}$ and $\mathbb{K}$.*

structures, global Galerkin techniques analogous to those in Section 8.2 can be more efficient to implement for beam discretization. The global discretization also demonstrates similarities and differences between the interpolatory cubic Hermite approximates and the projective cubic spline technique discussed in Section 8.3.1.

For the partition $x_j = jh$, $h = \frac{\ell}{N}$, $j = 0, \ldots, N$, the global Hermite basis functions are taken to be

$$\phi_{w_j}(x) = \frac{1}{h^3} \begin{cases} (x - x_{j-1})^2[2(x_j - x) + h] & , \quad x \in [x_{j-1}, x_j] \\ (x_{j+1} - x)^2[2(x - x_j) + h] & , \quad x \in (x_j, x_{j+1}] \\ 0 \ , \ \text{otherwise} \end{cases}$$

$$\phi_{\theta_j}(x) = \frac{1}{h^2} \begin{cases} (x - x_{j-1})^2(x - x_j) & , \quad x \in [x_{j-1}, x_j] \\ (x_{j+1} - x)^2(x - x_j) & , \quad x \in (x_j, x_{j+1}] \\ 0 \ , \ \text{otherwise} \end{cases}$$

for $j = 1, \ldots, N-1$. The definitions of ϕ_{w_N} and ϕ_{θ_N} are analogous but involves only the interval $[x_{N-1}, x_N]$. As illustrated in Figure 8.16, the global basis functions ϕ_{w_j} and ϕ_{θ_j} are the concatenation of the local displacement elements $\widetilde{\phi}_3, \widetilde{\phi}_1$ and slope elements $\widetilde{\phi}_4, \widetilde{\phi}_2$ defined in (8.34) and shown in Figure 8.12.

The approximating subspace is taken to be

$$V^N = \text{span}\left\{\phi_{w_j}, \phi_{\theta_j}\right\}_{j=1}^N$$

and the approximate solution is

$$w^N(t,x) = \sum_{j=1}^N w_j(t)\phi_{w_j}(x) + \sum_{j=1}^N \theta_j(t)\phi_{\theta_j}(x). \tag{8.36}$$

We note that by omitting ϕ_{w_0} and ϕ_{θ_0}, elements $v \in V^N$ are guaranteed to satisfy $v(0) = v'(0) = 0$ which ensures that $V^N \subset V = H_0^2(0,\ell)$. The ordering of nodal coefficients provided by (8.36) yields banded, tridiagonal mass, damping and stiffness matrices that are Toeplitz along all but the last row and column.

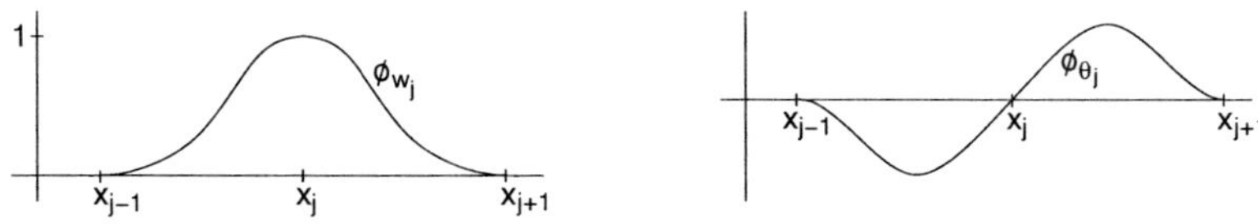

Figure 8.16. *Global Hermite basis functions ϕ_{w_j} and ϕ_{θ_j}.*

The projection of (8.26) onto V^N and use of basis functions as test functions yields the second-order system

$$\mathbb{M}\ddot{\mathbf{w}} + \mathbb{Q}\dot{\mathbf{w}} + \mathbb{K}\mathbf{w} = \mathbf{f} + V(t)\mathbf{b}$$

where the $2N \times 1$ vector $\mathbf{w}(t)$ has the nodal ordering

$$\mathbf{w}(t) = [w_1(t), \ldots, w_N(t), \theta_1(t), \ldots, \theta_N(t)]^T.$$

The global system matrices $\mathbb{M}, \mathbb{Q}, \mathbb{K}$ and vectors $\mathbf{f}, \mathbf{b}$ have definitions analogous to (8.31) and (8.32) where ϕ_i and ϕ_j now represent the combined basis $\{\phi_{w_i}, \phi_{\theta_i}\}$ and $\{\phi_{w_j}, \phi_{\theta_j}\}$. As with the matrices arising from the cubic B-spline discretization discussed in Section 8.3.1, the integrals involve piecewise polynomials of degree less than or equal to 6 so exact integration is achieved using the composite 4-point Gauss–Legendre rule (8.12) on each subinterval $[x_{j-1}, x_j]$.

For constant stiffness YI, this yields the stiffness matrix

$$\mathbb{K} = \frac{YI}{h^3}\left[\begin{array}{ccccc|ccccc}
24 & -12 & & & & 0 & 6h & & & \\
-12 & 24 & -12 & & & -6h & 0 & 6h & & \\
 & \ddots & \ddots & \ddots & & & \ddots & \ddots & \ddots & \\
 & & -12 & 24 & -12 & & & -6h & 0 & 6h \\
 & & & 12 & 12 & & & & -6h & -6h \\
\hline
0 & -6h & & & & 8h^2 & 2h^2 & & & \\
6h & 0 & -6h & & & 2h^2 & 8h^2 & 2h^2 & & \\
 & \ddots & \ddots & \ddots & & & \ddots & \ddots & \ddots & \\
 & & 6h & 0 & -6h & & & 2h^2 & 8h^2 & 2h^2 \\
 & & & 6h & -6h & & & & 2h^2 & 4h^2
\end{array}\right]. \qquad (8.37)$$

It is observed that the stiffness matrix $\mathbb{K}$ in (8.35), which was derived through elemental analysis, is a special case of (8.37) when $N = 2$. For the beam with surface-mounted patches, the differing values of YI in the patch region are simply incorporated in those regions of the partition which coincide with the patch. The mass and damping matrices are constructed in an analogous manner.

We note that the parameter ordering $\mathbf{w}(t) = [w_1(t), \theta_1(t), \ldots, w_N(t), \theta_N(t)]^T$ eliminates the block tridiagonal structure and yields block diagonal matrices where each block has a support of 6 diagonals. The disadvantage of this ordering scheme is that the Toeplitz nature of the matrices is destroyed which complicates matrix construction. Additional details regarding elemental and global approximation using cubic Hermite elements can be found in [36, 203, 422].

8.3.3 Comparison between Cubic Spline and Cubic Hermite Approximates

The cubic B-spline and cubic Hermite techniques detailed in Sections 8.3.1 and 8.3.2 illustrate two commonly employed Galerkin techniques for approximating beam models. Both provide $\mathcal{O}(h^4)$ spatial convergence rates as long as partitions are aligned with patches to accommodate discontinuities in mass, damping, stiffness and input parameters. As noted in Section 8.3.1, the cubic spline discretization is a projection method whereas the cubic Hermite method is interpolatory in the sense that the coefficients $\mathbf{w}(t)$ are nodal values of the displacement and slope at the partition points. Hence the cubic spline technique is more closely related to general Galerkin expansions — e.g., employing Legendre or Fourier bases — whereas the cubic Hermite expansion employs the finite element philosophy which, as detailed on pages 417–419 of Appendix A, is subsumed in the Galerkin framework.

The primary advantage of the cubic Hermite method lies in its interpolatory nature. This simplifies the enforcement of essential boundary conditions and facilitates characterization of complex structures which require nodal matching at the junction of differing geometries. For example, the transition from flat tabs to the curved patch region in THUNDER transducers — see the models (7.103), (7.106), or (7.107) in Section 7.9 — is easily accommodated by matching nodal values with Hermite elements whereas it is difficult to implement with cubic splines.

The disadvantage of the Hermite approximate is that it requires roughly twice as many coefficients as the spline expansion since both displacements and slopes are discretized. The increased dimensionality of system matrices must be accommodated when employing the model for control designs which require real-time implementation.

8.3.4 Examples and Software

Attributes of the discretized beam model, when used to characterize the PVDF–polyimide unimorph depicted in Figure 7.13, are illustrated in Section 7.4.1. Experimental validation of the discretized model for a beam with surface-mounted piezoceramic patches is addressed in Chapter 5 of [33].

MATLAB m-files for implementing the unimorph and beam models are located at the website `http://www.siam.org/books/fr32`.

8.4 Numerical Approximation of the Plate Model

In this section, we summarize approximation techniques for the rectangular and circular plate models developed in Section 7.5. We consider regimes in which transverse and longitudinal displacements can be decoupled and focus on approximating the former using Galerkin expansions employing spline or Fourier bases in space. As illustrated in Section 8.5 when discussing thin shell approximation, linear elements can be employed to discretize longitudinal displacements if warranted by the application. A full discussion regarding finite element methods for plates is beyond the scope of this discussion and the reader is referred to [276, 390, 422, 527] for details about this topic.

8.4.1 Rectangular Plate Approximation

We consider a rectangular plate with $x \in [0, \ell]$ and $y \in [0, a]$ as depicted in Figure 7.17. As in Section 7.5.1, we assume that the plate has thickness h and has N_A surface-mounted piezoceramic patches of thickness h_I whose edges are parallel with the x and y-axes. The plate is assumed to have fixed-edge conditions at $x = 0$, $y = 0$ and free boundary conditions for the remaining two edges. To simplify notation, we let $\Omega = [0, \ell] \times [0, a]$ denote the plate region. Finally, we consider linear input regimes with voltages $V_i(t) = V_{1i}(t) = -V_{2i}(t)$. Extension to nonlinear input regimes follows immediately when voltage inputs are replaced by the nonlinear polarization relations.

From (7.65), the transverse displacements are quantified by the weak model formulation

$$\int_\Omega \left\{ \rho \frac{\partial^2 w}{\partial t^2}\phi - M_x \frac{\partial^2 \phi}{\partial x^2} - 2M_{xy}\frac{\partial^2 \phi}{\partial x \partial y} - M_y \frac{\partial^2 \phi}{\partial y^2} - f_n \phi \right\} d\omega = 0 \tag{8.38}$$

for ϕ in the space of test functions

$$V = H_0^2(\Omega) = \{\phi \in H^2(\Omega) \,|\, \phi(0,y) = \phi_x(0,y) = 0 \text{ for } 0 \le y \le a,$$
$$\phi(x,0) = \phi_y(x,0) = 0 \text{ for } 0 \le x \le \ell\}.$$

The density ρ is specified by (7.48) and the moments are defined in (7.57) with components specified in (7.58)–(7.63). We consider the case of linear patch inputs but note that nonlinear inputs are accommodated in an identical manner.

The philosophy when approximating the dynamics of (8.38) is identical to that employed in Sections 8.2 and 8.3 for the rod and beam models. The relation is projected onto a spline-based finite-dimensional subspace $V^N \subset V$ to obtain a semi-discrete system appropriate for finite-dimensional control design. This vector-valued system can subsequently be simulated by employing finite difference discretizations in time or standard software for moderately stiff systems.

Consider a partition $\{(x_m, y_n)\}$ of Ω where $x_m = mh_x$, $y_n = nh_y$, with $h_x = \frac{\ell}{N_x}$, $h_y = \frac{a}{N_y}$ and $m = 0, \ldots, N_x$, $n = 0, \ldots, N_y$. Using the definition (8.28), we define modified cubic spline basis functions $\phi_m(x)$ and $\phi_n(y)$ on the intervals $[0, \ell]$ and $[0, a]$. The product space basis is then taken to be

$$\Phi_k(x,y) = \phi_m(x)\phi_n(y) \tag{8.39}$$

and the approximating subspace is

$$V^N = \text{span}\,\{\Phi_k\}_{k=1}^{N_w}$$

where $N_w = (N_x+1)(N_y+1)$. Approximate displacements have the representation

$$\begin{aligned} w^N(t,x,y) &= \sum_{k=1}^{N_w} w_k(t)\Phi_k(x,y) \\ &= \sum_{m=1}^{N_x+1}\sum_{n=1}^{N_y+1} w_{mn}(t)\phi_m(x)\phi_n(y). \end{aligned}$$

The restriction of the infinite-dimensional model (8.38) to the finite-dimensional subspace $V^N \subset V$ yields the vector-valued system

$$\mathbb{M}\ddot{\mathbf{w}} + \mathbb{Q}\dot{\mathbf{w}} + \mathbb{K}\mathbf{w} = \mathbf{f} + \frac{2Y_A d_{31} c_2}{h_A(1-\nu_A)}\sum_{i=1}^{N_A} V_i(t)\mathbf{b}_i \tag{8.40}$$

where $\mathbf{w}(t) = [w_1(t),\ldots,w_{N_w}(t)]^T$. The mass and stiffness matrices are defined componentwise by

$$\begin{aligned} [\mathbb{M}]_{jk} &= \int_\Omega \rho\Phi_j\Phi_k d\omega \\ [\mathbb{K}]_{jk} &= [\mathbb{K}_1]_{jk} + [\mathbb{K}_2]_{jk} + [\mathbb{K}_3]_{jk} + [\mathbb{K}_4]_{jk} + [\mathbb{K}_5]_{jk} \end{aligned}$$

where

$$[\mathbb{K}_1]_{jk} = \int_\Omega \left(\frac{Y_I h_I^3}{12(1-\nu_I^2)} + \frac{2Y_A c_3}{1-\nu_A^2}\sum_{i=1}^{N_A}\chi_{pe_i}(x,y)\right)\frac{\partial^2\Phi_j}{\partial x^2}\frac{\partial^2\Phi_k}{\partial x^2}d\omega$$

$$[\mathbb{K}_2]_{jk} = \int_\Omega \left(\frac{Y_I h_I^3\nu_I}{12(1-\nu_I^2)} + \frac{2Y_A c_3\nu_A}{1-\nu_A^2}\sum_{i=1}^{N_A}\chi_{pe_i}(x,y)\right)\frac{\partial^2\Phi_j}{\partial x^2}\frac{\partial^2\Phi_k}{\partial y^2}d\omega$$

$$[\mathbb{K}_3]_{jk} = \int_\Omega \left(\frac{Y_I h_I^3}{12(1-\nu_I^2)} + \frac{2Y_A c_3}{1-\nu_A^2}\sum_{i=1}^{N_A}\chi_{pe_i}(x,y)\right)\frac{\partial^2\Phi_j}{\partial y^2}\frac{\partial^2\Phi_k}{\partial y^2}d\omega$$

$$[\mathbb{K}_4]_{jk} = \int_\Omega \left(\frac{Y_I h_I^3\nu_I}{12(1-\nu_I^2)} + \frac{2Y_A c_3\nu_A}{1-\nu_A^2}\sum_{i=1}^{N_A}\chi_{pe_i}(x,y)\right)\frac{\partial^2\Phi_j}{\partial y^2}\frac{\partial^2\Phi_k}{\partial x^2}d\omega$$

$$[\mathbb{K}_5]_{jk} = \int_\Omega \left(\frac{Y_I h_I^3}{12(1+\nu_I)} + \frac{Y_A c_3}{1+\nu_A}\sum_{i=1}^{N_A}\chi_{pe_i}(x,y)\right)\frac{\partial^2\Phi_j}{\partial x\partial y}\frac{\partial^2\Phi_k}{\partial x\partial y}d\omega.$$

The damping matrix $\mathbb{Q}$ is constructed in a manner analogous to $\mathbb{K}$. Finally, the vectors are defined by

$$\begin{aligned} [\mathbf{f}]_j &= \int_\Omega f_n\Phi_j d\omega \\ [\mathbf{b}_i]_j &= \int_\Omega \chi_{pe_i}(x,y)\left(\frac{\partial^2\Phi_j}{\partial x^2} + \frac{\partial^2\Phi_j}{\partial y^2}\right)d\omega. \end{aligned}$$

The second-order system (8.40) has the same form as (8.16) which was developed for the rod in Section 8.2.1. Hence the techniques in that section can be used to construct a corresponding first-order system appropriate for control design or simulation.

8.4.2 Circular Plate Approximation

The circular plate model (7.71) with space of test functions (7.70) represents a setting in which the geometry and differential operator are not the tensor product of 1-D components. When combined with the inherent periodicity in θ, this motivates consideration of basis elements comprised of cubic B-splines in r and Fourier components in θ. We provide here an outline of the approximate system construction and refer the reader to [430] for details regarding this development.

The circumferential component of the basis is taken to be $\phi_m(\theta) = e^{im\theta}$ where $m = -M, \ldots, M$. The form of the radial component is motivated by the analytic Bessel behavior of the undamped plate that is devoid of patches. Let $\phi_n^m(r)$ denote the nth cubic spline modified to satisfy $\phi_n^m(a) = \frac{d\phi_n^m(a)}{dr} = 0$ — e.g., see (8.28) — along with the condition $\frac{d\phi_n^m(0)}{dr} = 0$ which is required to ensure differentiability at the origin.

The plate basis is then taken to be

$$\Phi_k(r,\theta) = r^{|\widehat{m}|}\phi_n^m(r)e^{im\theta}$$

where

$$\widehat{m} = \begin{cases} 0 & , \quad m = 0 \\ 1 & , \quad m \neq 0. \end{cases}$$

As detailed in [430], the inclusion of the weighting term $r^{|\widehat{m}|}$ is motivated by the asymptotic behavior of the Bessel functions as $r \to 0$ and ensures the uniqueness of the solution at the origin. The approximating subspace is

$$V^N = \text{span}\{\Phi_k\}$$

and approximate solutions have the representation

$$\begin{aligned} w^N(t,r,\theta) &= \sum_{k=1}^{N_w} w_k(t)\Phi_k(r,\theta) \\ &= \sum_{m=-M}^{M}\sum_{n=1}^{N^m} w_{mn}(t)r^{|\widehat{m}|}\phi_n^m(r)e^{im\theta}. \end{aligned}$$

Here

$$N^m = \begin{cases} N & , \quad m = 0 \\ N+1 & , \quad m \neq 0. \end{cases}$$

where N denotes the number of modified cubic splines.

Details regarding the construction of component matrices and vectors are provided in [430]. The performance of the discretized model when characterizing the dynamics of a circular plate is summarized in Section 7.5 and detailed in [430].

8.4.3 Examples and Software

The accuracy and limitations of the discretized circular plate model for characterizing both axisymmetric and nonaxisymmetric plate vibrations are illustrated in Section 7.5.2. In the axisymmetric case, the model accurately quantifies low to moderate frequency dynamics but overdamps high frequency modes which is characteristic of the Kelvin–Voigt damping model. It is illustrated that in the nonaxisymmetric regime, which is truly 2-D, the model accurately characterizes the dynamics associated with 8 of the 11 measured modes.

The reader can obtain MATLAB m-files for the approximation of the rectangular plate model at the website `http://www.siam.org/books/fr32`.

8.5 Numerical Approximation of the Shell Model

The final structure under consideration is the cylindrical shell model (7.92) developed in Section 7.7. To simplify the discussion, we consider fixed-edge conditions $u = v = w = \frac{\partial w}{\partial x} = 0$ at $x = 0, \ell$ and hence the space of test functions is

$$V = H_0^1(\Omega) \times H_0^1(\Omega) \times H_0^2(\Omega)$$

where $\Omega = [0, \ell] \times [0, 2\pi]$ denotes the shell region and

$$H_0^1(\Omega) = \left\{\phi \in H^1(\Omega) \,|\, \phi(0, \theta) = \phi(\ell, \theta) = 0\right\}$$
$$H_0^2(\Omega) = \left\{\phi \in H^2(\Omega) \,|\, \phi(0, \theta) = \phi(\ell, \theta) = \phi_x(0, \theta) = \phi_x(\ell, \theta) = 0\right\}.$$

We summarize here the cubic spline–Fourier approximation method developed in [130] and illustrated for control design in [131]. As noted in the first citation, two phenomena which plague the approximation of shell models are *shear locking* and *membrane* or *shear-membrane locking*. Shear locking, which has also been studied extensively in the context of Reissner-Mindlin plate models, is due to element incompatibility when enforcing the Kirchhoff-Love constraint of vanishing transverse shear strains as the shell thickness h tends to zero [14]. Membrane locking occurs when the total deformation energy is bending-dominated and is due to smoothness and asymptotic constraints in the shell model which are not appropriately represented by the approximation method — e.g., [21, 22, 290, 380]. If these constraints are not satisfied by approximating elements, the numerical solution is often overly stiff in the sense that the model exhibits bending dynamics which the approximate solution cannot match. As detailed in [290], mesh sizes must be chosen significantly smaller than the shell thickness to ensure accurate approximations with high-order finite elements in such bending-dominated regimes. It is noted that even with such mesh size restrictions, low-order finite element methods often fail in such regimes.

The discussion in this section is meant to provide the reader with an overview of issues associated with shell approximation and a brief summary of one discretization technique. Details and subtleties associated with this topic can found in [39,192] and previously cited references.

Basis Construction and Approximating Subspaces

To approximate the longitudinal, circumferential and transverse displacements u, v and w, it is necessary to construct bases for finite-dimensional subspaces of $H_0^1(\Omega)$ and $H_0^2(\Omega)$. This is accomplished using linear and cubic splines modified to accommodate the fixed boundary conditions.

We consider a uniform partition along the x-axis with gridpoints $x_n = nh$, $h = \frac{\ell}{N}$ and $n = 0, \ldots, N$. For $n = 1, \ldots, N-1$, we employ the linear splines

$$\phi_{u_n}(x) = \phi_{v_n}(x) = \frac{1}{h}\begin{cases} x - x_{n-1}\,, & x_{n-1} \le x < x_n \\ x_{n+1} - x\,, & x_n \le x \le x_{n+1} \\ 0\,, & \text{otherwise} \end{cases}$$

and modified cubic splines

$$\phi_{w_n}(x) = \begin{cases} \widehat{\phi}_0(x) - 2\widehat{\phi}_{-1}(x) - 2\widehat{\phi}_1(x)\,, & n = 1 \\ \widehat{\phi}_n(x)\,, \quad n = 2, \ldots, N-2 \\ \widehat{\phi}_N(x) - 2\widehat{\phi}_{N-1}(x) - 2\widehat{\phi}_{N+1}(x)\,, & n = N-1 \end{cases}$$

where the standard B-splines are defined in (8.27). For $n = 1, \ldots, N-1$ and $m = -M, \ldots, M$, the product space bases, in complex form, are taken to be

$$\begin{aligned} \Phi_{u_k}(\theta, x) &= e^{im\theta}\phi_{u_n}(x) \\ \Phi_{v_k}(\theta, x) &= e^{im\theta}\phi_{v_n}(x) \\ \Phi_{w_k}(\theta, x) &= e^{im\theta}\phi_{w_n}(x). \end{aligned}$$

To provide an equivalent real form, one can employ the representation

$$\Phi_{u_k}(\theta, x) = \left\{\begin{matrix} \cos(m\theta) \\ 1 \\ \sin(m\theta) \end{matrix}\right\}\phi_{u_n}(x)\,, \quad m = 1, \ldots, M$$

with similar definitions for Φ_{v_k} and Φ_{w_k}. The approximating subspaces are

$$\begin{aligned} V_u^N &= \text{span}\,\{\Phi_{u_k}\}_{k=1}^{N_u} \\ V_v^N &= \text{span}\,\{\Phi_{v_k}\}_{k=1}^{N_v} \\ V_w^N &= \text{span}\,\{\Phi_{w_k}\}_{k=1}^{N_w} \end{aligned}$$

and the approximate displacements are represented by the expansions

$$\begin{aligned} u^N(t, \theta, x) &= \sum_{k=1}^{N_u} u_k(t)\Phi_{u_k}(\theta, x) \\ v^N(t, \theta, x) &= \sum_{k=1}^{N_v} v_k(t)\Phi_{v_k}(\theta, x) \\ w^N(t, \theta, x) &= \sum_{k=1}^{N_w} w_k(t)\Phi_{w_k}(\theta, x). \end{aligned} \tag{8.41}$$

For a single partition, it follows that $N_u = N_v = N_w = (2N-1)(2M+1)$. Through the construction of ϕ_{u_n}, ϕ_{v_n} and ϕ_{w_n}, the approximate displacements u^N, v^N and w^N satisfy the fixed-end conditions and $V^N = V_u^N \times V_v^N \times V_w^N \subset V$. We note that consideration of real components yields the equivalent representation

$$\begin{aligned} u^N(t,\theta,x) &= \sum_{n=1}^{N-1} u_{0n}(t)\phi_{u_n}(x) \\ &+ \sum_{m=1}^{M}\sum_{n=1}^{N-1} u_{mn}(t)\cos(m\theta)\phi_{u_n}(x) + \sum_{m=1}^{M}\sum_{n=1}^{N-1} \widetilde{u}_{mn}(t)\sin(m\theta)\phi_{u_n}(x) \end{aligned}$$

with similar expressions for $v^N(t,\theta,x)$ and $w^N(t,\theta,x)$.

System Matrices

To determine the generalized Fourier coefficients $u_k(t), v_k(t)$ and $w_k(t)$, we employ the same approach as in previous sections and orthogonalize the residual with respect to the linearly independent test functions used to construct the approximating subspaces. To construct the resulting vector-valued system, we consolidate the coefficients in the vectors

$$\mathbf{u}(t) = \begin{bmatrix} u_1(t) \\ \vdots \\ u_{N_u}(t) \end{bmatrix}, \quad \mathbf{v}(t) = \begin{bmatrix} v_1(t) \\ \vdots \\ v_{N_v}(t) \end{bmatrix}, \quad \mathbf{w}(t) = \begin{bmatrix} w_1(t) \\ \vdots \\ w_{N_w}(t) \end{bmatrix}.$$

The full set of coefficients is then represented as $\boldsymbol{\vartheta}(t) = [\mathbf{u}(t), \mathbf{v}(t), \mathbf{w}(t)]^T$ where $N = N_u + N_v + N_w$.

The mass, stiffness and damping matrices have the form

$$\mathbb{M} = \left[\begin{array}{c|c|c} \mathbb{U}_m & & \\ \hline & \mathbb{V}_m & \\ \hline & & \mathbb{W}_m \end{array}\right],$$

$$\mathbb{K} = \left[\begin{array}{c|c|c} \mathbb{U}_{11} + \mathbb{U}_{12} & \mathbb{V}_{11} + \mathbb{V}_{12} & \mathbb{W}_{11} \\ \hline \mathbb{U}_{21} + \mathbb{U}_{22} & \mathbb{V}_{21} + \mathbb{V}_{22} & \mathbb{W}_{21} \\ \hline \mathbb{U}_{31} & \mathbb{V}_{31} & \sum_{k=1}^{6} \mathbb{W}_{3k} \end{array}\right],$$

$$\mathbb{Q} = \left[\begin{array}{c|c|c} \widetilde{\mathbb{U}}_{11} + \widetilde{\mathbb{U}}_{12} & \widetilde{\mathbb{V}}_{11} + \widetilde{\mathbb{V}}_{12} & \widetilde{\mathbb{W}}_{11} \\ \hline \widetilde{\mathbb{U}}_{21} + \widetilde{\mathbb{U}}_{22} & \widetilde{\mathbb{V}}_{21} + \widetilde{\mathbb{V}}_{22} & \widetilde{\mathbb{W}}_{21} \\ \hline \widetilde{\mathbb{U}}_{31} & \widetilde{\mathbb{V}}_{31} & \sum_{k=1}^{6} \widetilde{\mathbb{W}}_{3k} \end{array}\right]$$

and the exogenous vectors are

$$\mathbf{f}(t) = [\mathbf{f}_u\,,\, \mathbf{f}_v\,,\, \mathbf{f}_w]^T \quad , \quad \mathbf{b} = [\mathbf{b}_u\,,\, \mathbf{b}_v\,,\, ,0]^T.$$

The various submatrices contain individual components which arise when the weak formulation (7.92) is restricted to V^N. For example, the approximation of the mass and stiffness components in the longitudinal equation of (7.92) yields

$$[\mathbb{U}_m]_{jk} = \int_\Omega \rho h R \Phi_{u_k} \Phi_{u_j} d\omega,$$

$$[\mathbb{U}_{11}]_{jk} = \int_\Omega \frac{YhR}{1-\nu^2} \frac{\partial \Phi_{u_k}}{\partial x} \frac{\partial \Phi_{u_j}}{\partial x} d\omega,$$

$$[\mathbb{V}_{11}]_{jk} = \int_\Omega \frac{Yh\nu}{1-\nu^2} \frac{\partial \Phi_{v_k}}{\partial \theta} \frac{\partial \Phi_{u_j}}{\partial x} d\omega,$$

$$[\mathbb{W}_{11}]_{jk} = \int_\Omega \frac{Yh\nu}{1-\nu^2} \Phi_{w_k} \frac{\partial \Phi_{u_j}}{\partial x} d\omega,$$

$$[\mathbb{V}_{12}]_{jk} = \int_\Omega \frac{Yh}{2(1+\nu)} \frac{\partial \Phi_{v_k}}{\partial x} \frac{\partial \Phi_{u_j}}{\partial \theta} d\omega,$$

$$[\mathbb{U}_{12}]_{jk} = \int_\Omega \frac{Yh}{2R(1+\nu)} \frac{\partial \Phi_{u_k}}{\partial \theta} \frac{\partial \Phi_{u_j}}{\partial \theta} d\omega,$$

$$[\mathbf{f}_u]_j = \int_\Omega R f_x \Phi_{u_j} d\omega,$$

$$[\mathbf{b}_u]_j = \int_\Omega \frac{Rh}{1-\nu} \frac{\partial \Phi_{u_j}}{\partial x} d\omega$$

with similar expressions for the remaining submatrices.

In the usual manner, the second-order system

$$\mathbb{M}\ddot{\boldsymbol{\vartheta}}(t) + \mathbb{Q}\dot{\boldsymbol{\vartheta}}(t) + \mathbb{K}\boldsymbol{\vartheta}(t) = \mathbf{f}(t) + \left[a_1(P(E(t)) - P_R) + a_2(P(E(t)) - P_R)^2\right]\mathbf{b}$$
$$\vartheta(0) = \vartheta_0 \ , \ \dot{\vartheta}(0) = \vartheta_1$$

can be reformulated as the first-order Cauchy equation

$$\dot{\mathbf{z}}^N(t) = \mathbb{A}\,\mathbf{z}(t) + \mathbf{F}(t) + \left[a_1(P(E(t)) - P_R) + a_2(P(E(t)) - P_R)^2\right)\mathbf{B}$$
$$\mathbf{z}(0) = \mathbf{z}_0 \ ,$$

where $\mathbf{z} = [\boldsymbol{\vartheta}(t), \dot{\boldsymbol{\vartheta}}(t)]^T$ and

$$\mathbb{A} = \begin{bmatrix} 0 & \mathbb{I} \\ -\mathbb{M}^{-1}\mathbb{K} & -\mathbb{M}^{-1}\mathbb{Q} \end{bmatrix} \ , \quad \mathbf{F}(t) = \begin{bmatrix} 0 \\ \mathbb{M}^{-1}(\mathbf{t}) \end{bmatrix} \ , \quad \mathbf{B} = \begin{bmatrix} 0 \\ \mathbb{M}^{-1}\mathbf{b} \end{bmatrix}.$$

The system in this form is anemable to simulation, parameter estimation and control design. Note that the system can be adapted to alternative boundary conditions through modifications of the first and last basis functions. Flexibility in this regard is also a hallmark of the method.

Appendix A

Glossary of Terms

Anhysteretic Magnetization, M_{an} — The anhysteretic magnetization M_{an} is the hysteresis-free magnetization curve obtained by superimposing decaying AC fields centered at a locus of DC field values. As illustrated in Figure 4.9, M_{an} is a single-valued sigmoid curve that is antisymmetric with respect to the field H. From a theoretical perspective, M_{an} represents the magnetization that would result in the absence of imperfections or pinning sites; hence it represents the *global* equilibrium configuration of the magnetization for a specified field value.

Anisotropy — Anisotropy indicates that material properties have different values when measured in different directions. Examples include magnetic, shape and crystalline anisotropies which have values that are dependent on the choice of axes. This is in contrast with isotropic properties which are independent of direction.

Banach Space — A Banach space is a normed vector space that is complete in the metric induced by the norm.

Bimorph — A bimorph generically designates a structure comprised of two active elements, either bonded to each other or to an intermediate substrate as depicted in Figure A.1. Opposing stresses generated in the active elements produce out-of-plane or transverse motion in the transducer. See also unimorph.

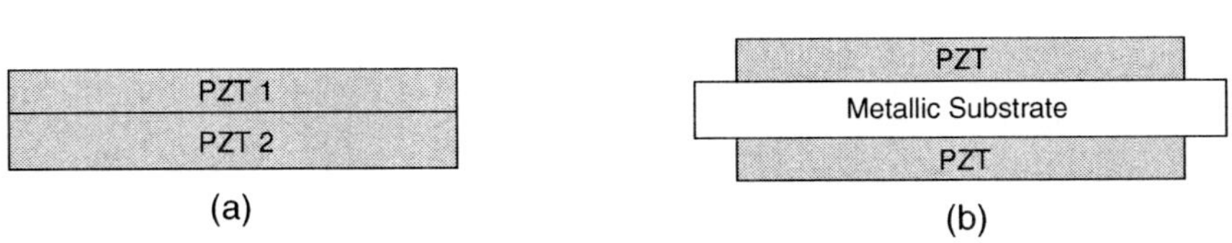

Figure A.1. *Bimorph comprised of (a) two PZT patches and (b) PZT patches bonded to a metallic substrate.*

Boltzmann's Relation — See Entropy.

$\boldsymbol{C^1[a,b]}$ — The space $C^1[a,b]$ is the normed space of all continuously differentiable functions on $I = [a,b]$ with the norm defined by

$$\|w\| = \max_{x \in I} |w(x)| + \max_{x \in I} |w'(x)|.$$

$\boldsymbol{C^n[a,b]}$ — The normed space $C^n[a,b]$ comprises the set of functions that are n-times differentiable on $I = [a,b]$. The norm in this case is

$$\|w\| = \sum_{i=0}^{n} \max_{x \in I} \left| w^{(i)}(x) \right| .$$

The space $C^0[a,b]$ is typically denoted $C[a,b]$.

$\boldsymbol{C(0,T;X)}$ — Let X be a Banach space. The function space $C(0,T;X)$ is defined as the set of X-valued continuous functions defined on $[0,T]$ having the norm

$$\|f\|_{C(0,T;X)} = \sup_{t \in [0,T]} \|f(t)\|_X .$$

Coercive Field (Electric), $\boldsymbol{E_c}$ — The coercive electric field is that which reduces the polarization to zero as depicted in Figure A.2.

Coercive Field (Magnetic), $\boldsymbol{H_c}$ — The magnetic coercive field is that required to drive the magnetization to zero starting at an arbitrary level as shown in Figure A.3. In many references, the coercivity is defined to be synonymous although some authors define it as the field required to reduce the magnetization to zero from negative saturation.

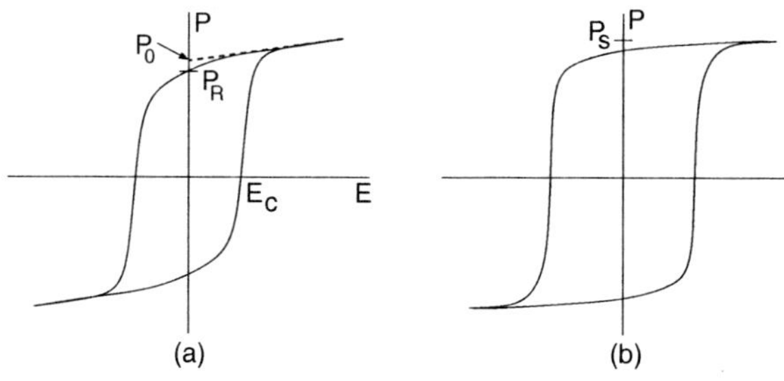

Figure A.2. *(a) Ferroelectric hysteresis, coercive field E_c, remanent polarization P_R, and spontaneous polarization P_0. (b) Saturation polarization P_s which results when $\frac{dP}{dE} \to 0$.*

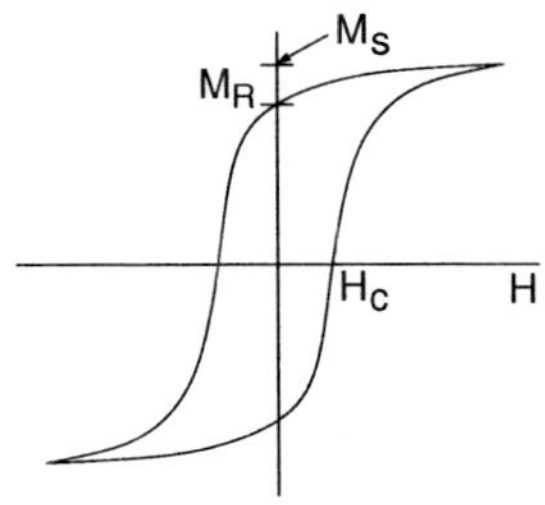

Figure A.3. *Saturation magnetization M_s, magnetic remanence M_R, and coercive field H_c.*

Coercive Stress, σ_c — The positive stress which reduces the strain to zero in ferroelastic materials is the coercive stress — e.g., see Figure A.4.

Compact Operator — Let X and Y be normed spaces and let $\mathcal{K} : X \to Y$ be a linear operator. The operator $\mathcal{K}$ is compact (completely continuous) if for every bounded subset M of X, the closure of the image, $\overline{\mathcal{K}(M)}$, is a compact subset of Y. The theory of compact operators originated with the investigation of integral equations arising in mathematical physics — e.g., see [104]. The theory subsequently motivated a significant body of functional analysis due to the property that compact operators exhibit nearly finite-dimensional structure while defined on infinite dimensional spaces — e.g., the theory of Section B.2 relies on the property that $\mathcal{K}$ in the polarization model can be represented in terms of a sequence of finite-dimensional operators.

Curie Point, T_c — The Curie point or transition temperature is the temperature which delineates disordered high-temperature phases and ordered low-temperature phases. Below T_c, the ferromagnetic, ferroelectric, and ferroelastic phases are absolutely stable although metastable disordered states can be induced by applied fields or stresses. In the context of the order parameter e, the Curie point is the temperature above which the order parameter vanishes — hence the magnetization

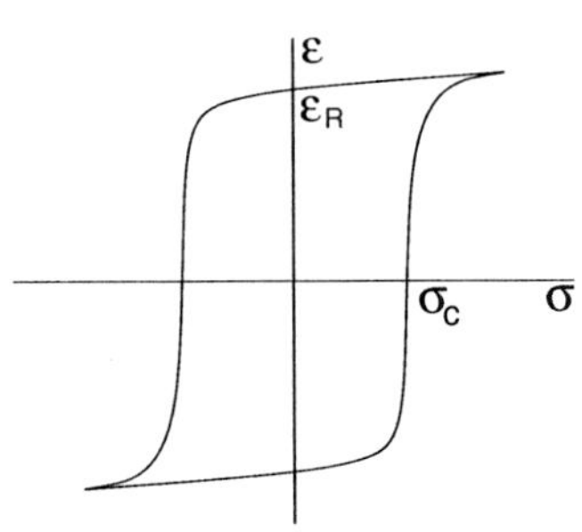

Figure A.4. *Ferroelastic stress-strain behavior, remanent strain ε_R, and coercive stress σ_c.*

M, polarization P and strains ε vanish above T_c in the absence of applied fields or stresses.

Curie Temperature, T_0 — The Curie temperature, or Curie–Weiss temperature, is typically extracted from the Curie–Weiss law

$$\epsilon = \frac{C}{T - T_0} + \epsilon_0$$

where C is termed the Curie constant and ϵ and ϵ_0 are the material permittivity and permittivity of free space. In materials exhibiting second-order phase transitions, the Curie temperature and Curie point are often approximately equal whereas in materials with first-order transitions, T_0 can be more than 10° C lower than T_c.

D Field — The electric $\mathbf{D}$ field is a vector-valued quantity specified by

$$\mathbf{D} = \epsilon_0 \mathbf{E} + \mathbf{P}$$

where $\epsilon_0 = 8.854 \times 10^{-12}$ F/m denotes the permittivity of a vacuum. Because the normal component of the D field equals the charge per unit area on an electrode, this quantity is often measured in experiments.

Degree of Precision — A numerical quadrature formula I_n is said to have degree of precision m if $I_n = I$ for all polynomials f such that $\deg(f) \leq m$ and $I_n \neq I$ for $\deg(f) = m + 1$.

Differential — Given a function $F(x, y)$, it is always possible to specify the differential

$$dF = \left[\frac{\partial F}{\partial x}\right]_y dx + \left[\frac{\partial F}{\partial y}\right]_x dy.$$

However, given an arbitrary differential $dF = A(x, y)dx + B(x, y)dy$, relating changes in a dependent variable F to changes dx and dy in the independent variables, a function F yielding this differential generally does not exist. If a function F does exist, dF is termed an exact, perfect or total differential whereas it is designated an inexact differential if F does not exist. As detailed in [3], a differential of two independent variables is exact if and only if A and B satisfy the conditions

$$\left[\frac{\partial A}{\partial y}\right]_x = \left[\frac{\partial B}{\partial x}\right]_y.$$

Equivalently, the differential F is exact if

$$\int_a^b dF = F(b) - F(a)$$

for all paths between the points $a = (x_1, y_1)$ and $b = (x_2, y_2)$.

In thermodynamics, the specification of whether or not a differential is exact is important since functions F associated with exact differentials represent state functions of the system. Examples of exact differentials are $dU, d\psi$ and dG and hence the internal energy U, Helmholtz energy ψ, and Gibbs energy G are state functions. The differentials dW and dQ for the work and heat are generally inexact — in some references, these inexact differentials are denoted by $đW$ and $đQ$ or $d'W$ and $d'Q$.

To illustrate the physical ramifications of inexact differentials, consider the force $\mathbf{F}(x,y) = xy\hat{\imath} + xy\hat{\jmath}$. Because $\mathbf{dr} = dx\hat{\imath} + dy\hat{\jmath}$, the work along a general path γ is

$$W = \int_\gamma xydx + xydy$$

so that $A(x,y) = B(x,y) = xy$. Because $\frac{\partial A}{\partial y} \neq \frac{\partial B}{\partial y}$, $dW = xydx + xydy$ is not exact which implies that the integral $\int_\gamma dW$ cannot be uniquely defined in terms of work values $W(x_1, y_1)$ and $W(x_2, y_2)$. To illustrate, consider two paths

$$\gamma_1 = \{(x,y)|x(t) = 0, y(t) = 2t, t \in [0, 1/2]; x(t) = 2(t - 1/2), y(t) = 0, t \in (1/2, 1]\}$$

$$\gamma_2 = \{(x,y)|x(t) = y(t) = t, t \in [0,1]\}$$

between $(0,0)$ and $(1,1)$ as depicted in Figure A.5. The corresponding work relations

$$W_1 = \int_{\gamma_1} \mathbf{F} \cdot \mathbf{dr} = \frac{1}{2}$$

$$W_2 = \int_{\gamma_2} \mathbf{F} \cdot \mathbf{dr} = \frac{2}{3}$$

illustrate that the work cannot be uniquely defined in terms of function values $W(0,0)$ and $W(1,1)$.

From a physical perspective, the property that dW is an inexact differential implies that it is meaningless to designate the "work in a system." This is in contrast to state functions, such as pressure or internal energy, that define properties of the system based solely on information regarding initial and final states.

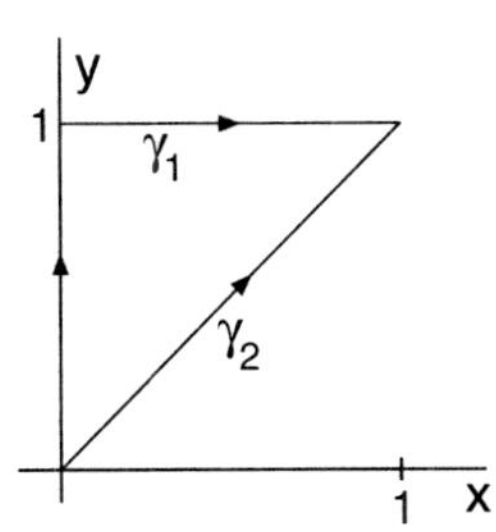

Figure A.5. *Paths γ_1 and γ_2 used when computing the work required to move from $(0,0)$ to $(1,1)$ in the force field $\mathbf{F}(x,y) = xy\hat{\imath} + xy\hat{\jmath}$.*

Domains and Domain Walls — See Ferroelastic, Ferroelectric and Ferromagnetic Domains and Domain Walls.

Einstein Summation — The Einstein summation convention is a notation in which repeated indices are summed over their range. Hence the expression $a_{ij}x_k$ does not indicate summation whereas the expression $a_{ij}x_{ijk}$ has the meaning

$$a_{ij}x_{ijk} = \sum_{i=1}^{N_i}\sum_{j=1}^{N_j} a_{ij}x_{ijk}.$$

Electrostriction, λ_s — The quadratic E-ε behavior common to all materials is termed electrostriction. In most materials, the effect is small and is dominated by linear behavior. Certain relaxor ferroelectric compounds provide an exception which necessitates the inclusion of quadratic field or polarization terms when modeling strain behavior. This phenomenon is analogous to magnetostriction in ferromagnetic materials.

Entropy, S — The entropy quantifies the amount of energy in a physical system that is unavailable to perform work. From the second law of thermodynamics, the change in entropy can be expressed as

$$dS = \frac{dQ}{T} + S_{irr}$$

where $S_{irr} \geq 0$ denotes the entropy production due to irreversible processes and dQ quantifies the change in heat. Hence $dS = \frac{dQ}{T}$ for reversible processes and $dS \geq \frac{dQ}{T}$ for irreversible processes.

Boltzmann's relation

$$S = k \ln W,$$

where $k = 1.38 \times 10^{-23}$ J/K denotes Boltzmann's constant, quantifies the entropy in terms of the probability W associated with microstate arrangements that are consistent with an observed macroscopic thermodynamic state — formulation in terms of the natural logarithm ensures that the products arising from the combinatorial arguments used to quantify W contributed linearly to S. This relation provides a connection between statistical mechanics and thermodynamics, and related analysis provides a framework for establishing that thermodynamic equilibria can be associated with most probable states. Since a perfectly ordered state constitutes a single microstate, Boltzmann's relation establishes that entropy is a state variable related to the amount of disorder in the system. When combined with the observation that equilibrium states are those associated with most probable microstate arrangements, this implies that systems evolve toward disorder.

The reader is referred to [142] for an introduction to the thermodynamic and statistical mechanics properties of entropy.

Essential Boundary Conditions — Essential boundary conditions are those which must be explicitly enforced in variational or weak model formulations. An example is the Dirichlet condition $v(t, 0) = 0$ in second-order rode models. Compare with natural boundary conditions.

Ferroelastic Domains and Domain Walls — Ferroelastic domains are regions in ferroelastic crystals distinguished by strain states having the same crystallographic orientation — e.g., martensite or austenite. The transition regions between ferroelastic domains are termed domain walls.

Ferroelastic Materials — Ferroelastic materials are defined as those which, at temperatures below the Curie point, exhibit two or more orientation states in the absence of applied mechanical stresses which can be reoriented by an applied stress. This produces hysteresis in the stress-strain relation which is an inherent property of ferroelastic compounds such as shape memory alloys. The terminology is motivated by analogous behavior originally observed in iron-based ferromagnetic compounds.

Ferroelectric Domains and Domain Walls — Ferroelectric domains are regions in ferroelectric materials in which dipoles — equivalently, polarization — are aligned. The transition regions delineating ferroelectric domains are termed domain walls. Two common orientations are 180° domains which minimize electrostatic energy and 90° domains which minimize ferroelastic energy — see Figure 2.4.

Ferroelectric Materials — A material is termed ferroelectric if, at temperatures below the Curie point, it exhibits a domain structure and spontaneous polarization which can be reoriented by an electric field.

Ferromagnetic Domains and Domain Walls — Ferromagnetic domains are regions in which magnetic moments are aligned and hence exhibit a spontaneous magnetization. As originally proposed by Pierre Weiss in 1906 and 1907, this moment alignment is due to interaction fields H_i which serve to align neighboring spins. The transition regions between domains are termed domain walls.

Ferromagnetic Materials — Ferromagnetic materials are characterized by the presence of a domain structure having nonzero spontaneous magnetization at temperatures below the Curie point. Examples include iron, steel, nickel, and Terfenol-D.

Galerkin Methods — Galerkin methods comprise a broad class of weighted residual techniques used to represent and approximate solutions to linear and nonlinear differential equations. This class of techniques includes finite element, Raleigh–Ritz, collocation and least squares methods as special cases.

To illustrate, consider the linear operator equation

$$Au = f$$

defined on the space X with inner product $\langle \cdot, \cdot \rangle$. For finite dimensional spaces $V^N = \text{span}\{\phi_1, \ldots, \phi_N\}$ and $Y^N = \text{span}\{\varphi_1, \ldots, \varphi_N\}$, the goal in general Galerkin methods is to find $u^N \in V^N$ which satisfies

$$\left\langle Au^N - f, \varphi_i \right\rangle = 0$$

for $i = 1, \ldots, N$. Since $\{\phi_i\}_{i=1}^N$ provides a basis for V^N, the approximate solution can be represented by

$$u^N = \sum_{j=1}^N u_j \phi_j$$

which yields the system

$$\sum_{j=1}^N \langle A\phi_j, \varphi_i \rangle u_j = \langle f, \varphi_i \rangle$$

for $i = 1, \ldots, N$. Here φ_i are typically referred to as weight or test functions, ϕ_j are commonly termed basis, trial, or shape functions, and u^N is denoted a trial or approximate solution. Methods for which the space of trial functions V^N differs from the space of test functions Y^N are often termed Petrov–Galerkin methods. We also point out that whereas V^N and Y^N are usually subspaces of X, this is not the case for certain techniques such as nonconforming finite element methods.

We summarize various choices for φ_k and conditions on A to illustrate the manner through which the Galerkin framework subsumes finite element, Rayleigh–Ritz, collocation and least squares methods. Details regarding the hierarchy of numerical methods can be found in [20, 91, 383, 389].

1. *Rayleigh–Ritz* — Consider the case when $\varphi_i = \phi_i$ so that

   $$\sum_{j=1}^N \langle A\phi_j, \phi_i \rangle u_j = \langle f, \phi_i \rangle$$

   for $i = 1, \ldots, N$. When A is symmetric and positive definite, this is the Rayleigh–Ritz method and the solution is equivalent to that obtained by minimizing

   $$J(u) = \langle Au, u \rangle - \langle f, u \rangle$$

   with respect to $u \in V^N$. By relaxing the symmetry and positivity criterion on A, the Galerkin method acquires significant flexibility for a broad range of applications.

2. *Finite Element* — Consider again the case when $\varphi_i = \phi_i$ so the spaces of test and trial functions coincide. In the most fundamental case, the finite element method is a Rayleigh–Ritz method in which basis functions are piecewise polynomials having compact support — e.g., piecewise linear or cubic

polynomials. However, the methodology can be applied to nonsymmetric operators A in which case the finite element method is a special case of general Galerkin methods with φ_i and ϕ_i taken to be piecewise polynomials.

3. *Method of Least Squares* — The choice $\varphi_i = A\phi_i$ yields the system

$$\sum_{j=1}^{N} \langle A\phi_j, A\phi_i\rangle\, u_j = \langle f, A\phi_i\rangle \quad , \quad i = 1, \ldots, N$$

and the method of least squares. This formulation offers theoretical and computational advantages for certain regimes.

4. *Collocation* — Collocation methods are obtained by specifying test functions as

$$\varphi_i(x) = \delta(x - x_i)$$

where δ denotes the Dirac distribution and x_i are nodal values.

Grains and Grain Boundaries — Grains are defined as regions of uniform anisotropy delineated by transition zones termed grain boundaries.

***H*^{1}(*a*, *b*)** — Consider the inner product

$$\langle v, w, \rangle_{H^1(a,b)} = \int_a^b \left[v(x)\overline{w(x)} + v'(x)\overline{w'(x)} \right] dx$$

and corresponding norm

$$\begin{aligned} \|w\|_{H^1(a,b)} &= \left(\int_a^b \left[|w(x)|^2 + |w'(x)|^2 \right] dx \right)^{1/2} \\ &= \langle w, w \rangle_{H^1(a,b)}^{1/2} \end{aligned} \qquad \text{(A.1)}$$

where the overbar denotes complex conjugation. The Hilbert space $H^1(a,b)$ is defined as the set of functions w having finite norm (A.1). Hence both the functions and their derivatives are square integrable or elements in $L^2(a,b)$.

***H*$_0^1$(*a*, *b*)** — The function space $H_0^1(a,b)$ denotes the subset of $H^1(a,b)$ whose elements satisfy essential boundary conditions. For example, the functions satisfying the condition $v(a) = 0$ yield the space

$$H_0^1(a,b) = \left\{ v \in H^1(a,b) \mid v(a) = 0 \right\}.$$

$H^2(a,b)$ — The definition of $H^2(a,b)$ is analogous to $H^1(a,b)$. The space consists of functions w having finite norm

$$\|w\|_{H^2(a,b)} = \langle w, w \rangle^{1/2}_{H^2(a,b)}$$

where

$$\langle v, w \rangle_{H^2(a,b)} = \sum_{i=0}^{2} \int_a^b v^{(i)} \overline{w^{(i)}} dx.$$

$H_0^2(a,b)$ — The space $H_0^2(a,b)$ consists of functions in $H^2(a,b)$ that satisfy essential boundary conditions. For example, the subset of functions satisfying $v(a) = v'(a) = 0$ yields the space

$$H_0^2(a,b) = \left\{ v \in H^2(a,b) \mid v(a) = v'(a) = 0 \right\}.$$

Hilbert Space — A complete inner product space is designated a Hilbert space.

Hysteresis — The term hysteresis connotes a lag effect between inputs and outputs in a system. In the context of smart materials, common inputs include electric and magnetic fields, stresses and heat, and outputs include polarization, magnetization and strains. Hysteretic processes are often said to have memory but this interpretation should be used with care. Whereas memory capabilities can often be exploited in hysteretic systems — e.g., magnetic memory relies on ferromagnetic hysteresis — the mechanisms which produce hysteresis are inherently related to the material and energy properties of a compound rather than specific memory mechanisms. The lag associated with hysteretic systems is fundamental in control applications since it produces delays or phase shifts which must be accommodated in control algorithms and designs. Contrary to the belief of those who have struggled with hysteretic systems, the term hysteresis appears to bear no etymological ties with the word hysteria.

Hysteron — We define a hysteron as a generic, multivalued kernel used when quantifying hysteresis in ferroelectric, ferromagnetic or ferroelastic materials. From a physical perspective, a hysteron can be interpreted as quantifying the irreversible changes in the polarization, magnetization or strain for a single lattice unit as fields or stresses are increased through critical threshold values. This can be due to dipole or moment switching or a local phase transformation. In Preisach models, hysterons are defined phenomenologically to have values of ± 1 with thresholds at arbitrary values s_1 and s_2 or interaction and coercive field values s_i and s_c. Energy principles are used to construct the hysterons in the homogenized energy framework which yields differing kernels depending on the assumptions made when constructing the Gibbs energy and quantifying equilibrium energy configurations — see Figure A.6.

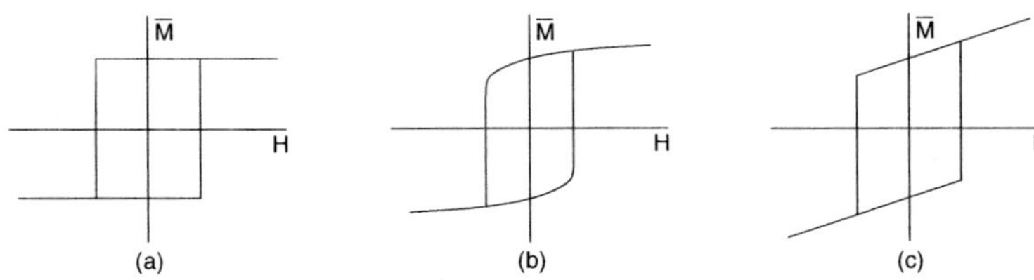

Figure A.6. *Hysterons employed in magnetic models: (a) Preisach hysteron, and energy-based hysterons derived from (b) statistical mechanics tenets and (c) constructed using a piecewise quadratic Gibbs energy relation.*

Infimum — Let S denote a set that is bounded below and let s_0 be a lower bound that is greater than or equal to all other lower bounds. Then s_0 is termed the greatest upper bound or infimum of S and is denoted by

$$s_0 \equiv \inf_S x = \inf S.$$

For example, the open interval $S = (0, 1)$ has the infimum $\inf S = 0$ even though $0 \notin S$. See also supremum.

$\boldsymbol{L^2(a,b)}$ — The function space $L^2(a, b)$ is defined as the set of all square integrable functions w having finite norm

$$\|w\|_2 = \left(\int_a^b |w(x)|^2 dx \right)^{1/2}. \tag{A.2}$$

Note that $L^2(a, b)$ is the completion of the vector space of continuous functions on $[a, b]$ with respect to the norm (A.2). When considered with the inner product

$$\langle v, w \rangle = \int_a^b v(x)\overline{w(x)} dx,$$

it is observed that $L^2(a, b)$ is a Hilbert space.

$\boldsymbol{L^2(0,T;X)}$ — Let X be a Banach space. The function space $L^2(0, T; X)$ is the class of Lebesgue measurable functions defined on $(0, T)$ with range X such that

$$\|f\|_{L^2(0,T;X)} = \left(\int_0^T \|f(t)\|_X^2 dt \right)^{1/2} < \infty.$$

Landau Notation — Let f and g be functions of the real variable x. The asymptotic $\mathcal{O}$ and o order statements are defined as follows:

$$f(x) = \mathcal{O}(g(x)) \text{ as } x \to 0 \text{ if } |f(x)/g(x)| < C \text{ for all small } x \text{ and some constant } C$$

$$f(x) = o(g(x)) \text{ as } x \to 0 \text{ if } \lim_{x \to 0} f(x)/g(x) = 0.$$

Similar definitions hold for the limit $x \to \infty$.

Magnetic Induction, B — The magnetic induction provides a measure of the magnetic flux density (Wb/m^2) due to an applied field $\mathbf{H}$. The relation

$$\mathbf{B} = \mu \mathbf{H}$$

in ferromagnetic materials is typically nonlinear and hysteretic so that the permeability μ must be interpreted as a multi-valued map rather than a function.

The flux density can be measured using a pickup coil with N loops and area A by employing the Faraday-Lenz law

$$V = -NA\frac{dB}{dt}$$

to compute $\frac{dB}{dt}$ given measurements of the induced voltage V. Integration subsequently yields the induction or flux density B.

Magnetic Remanence, M_R — This is the magnetization which results when the material is magnetized to saturation, M_s, and the field is then taken to zero — see Figure A.3. Some authors differentiate the remanent magnetization as that resulting when the field is applied to an arbitrary level and is then removed. In this case, the remanence provides an upper bound for all remanent values; however, we do not exploit this difference in our discussion.

Magnetization, M — Consider first a nonhomogeneous material having N_V magnetic dipoles $\mathbf{m}_i$ per unit volume. The magnetization is defined by

$$\mathbf{M} = \lim_{dV \to 0} \left[\frac{1}{dV} \sum_{i=1}^{N_V dV} \mathbf{m}_i \right] = N_V \mathbf{m}_{av}$$

where dV is a reference volume element and $\mathbf{m}_{av}$ is the average magnetic moment. Magnetization thus has units of amperes per meter (A/m) since $\mathbf{m}$ has units of Am^2. For additional details, see [23].

To develop hysteresis models appropriate for a number of applications, we also consider the case of homogeneous materials comprised of uniform moments with strength m_0 constrained to the direction of the applied field or diametrically opposite to it. The magnetization due to N moments can then be represented by the scalar relation

$$M = \frac{m_0}{V} \sum_{i=1}^{N} s_i$$

where $s_i = \pm 1$ designates the moment orientation.

For solenoid-driven transducers, the magnetization can be computed by invoking the Faraday-Lenz law

$$V = -NA\frac{dB}{dt}$$

to compute $\frac{dB}{dt}$ given measurements of the voltage V induced in a pickup coil having N loops and area A. Integration yields B, and the magnetization is subsequently given by $M = \frac{1}{\mu_0}B - H$.

Magnetostriction, $\lambda_0, \lambda_s, \lambda$ — The general term magnetostriction refers to strains generated in response to an applied magnetic field or during the ferromagnetic to paramagnetic phase transitions. The spontaneous magnetostriction λ_0 quantifies strains generated in the latter case whereas the saturation magnetostriction λ_s quantifies the maximal strain generated by a saturating applied field. The Joule magnetostriction λ characterizes intermediate strains due to nonsaturating fields. It is the magnetostrictive property of ferromagnetic materials that provides them with actuator capabilities.

Metastable States — Metastable states are those associated with relative rather than absolute energy minima.

Natural Boundary Conditions — Boundary conditions that are implicity satisfied in weak or variational model formulations are designated natural boundary conditions. An example is the Neumann condition $\frac{\partial v}{\partial x}(t,0) = 0$ in second-order rod models. Unlike essential boundary conditions, these do not have to be explicitly enforced in Galerkin or finite element bases and approximate solutions.

Order Parameter, e — The order parameter e is a scalar or vector-valued variable of state which characterizes the difference between two phases in the sense that the thermal average is nonzero below the Curie point T_c and zero above it in the absence of applied fields or stresses. For ferroelectric, ferromagnetic and ferroelastic materials, appropriate order parameters are the polarization P, magnetization M and strain ε.

Paraelectric Phase — The disordered, anhysteretic phase which occurs at temperatures above the Curie point is typically designated the paraelectric phase through analogy with its magnetic counterpart, the paramagnetic phase. The permittivity exhibits Curie–Weiss behavior in the paraelectric phase.

Paramagnetic Phase — The paramagnetic phase represents the disordered, anhysteretic state of ferromagnetic materials at temperatures above the Curie point.

Paramagnetism — Paramagnetism refers to a weak magnetism producing a weak susceptibility $\chi = M/H$ on the order of $\chi \approx 10^{-5}$ to 10^{-3} — for comparison, χ can range between 1 and 10 for Terfenol-D operating at room temperature. Unlike ferromagnetism, which is characterized by moment interactions which produce the domain structure, spontaneous magnetization, and interaction fields, spin interactions in paramagnetism are weak thus producing anhysteretic but nonlinear H-M behavior. Pierre Curie established that χ exhibits the temperature-dependence

$$\chi = \frac{C}{T}$$

in paramagnetic materials which is a special case of the Curie–Weiss law

$$\chi = \frac{C}{T - T_0}.$$

It should be noted that ferromagnetic materials exhibit paramagnetic behavior at temperatures above the Curie point.

Permeability of Free Space, μ_0 — In the SI system, the permeability of free space has the value $\mu_0 = 4\pi \times 10^{-7}$ H/m.

Permittivity of Free Space, ϵ_0 — The permittivity of free space has the value $\epsilon_0 = 8.854 \times 10^{-12}$ F/m.

Piezoelectric Effect — When used alone, the phrase "piezoelectric effect" typically refers to the direct piezoelectric effect which constitutes the change in polarity which results from an applied stress. This generates a voltage change and charge which can be measured and provides the materials with sensor capabilities. The converse piezoelectric effect constitutes the reversible strains generated in ferroelectric materials in response to an applied field. This provides actuator capabilities. It should be noted that in both cases, the designation piezoelectric has a linear connotation and hence should be used only when describing low to moderate E-ε or σ-P behavior where linear approximations provide reasonable accuracy.

Piezoelectric Materials — Materials exhibiting piezoelectric effects are often designated as piezoelectric materials. This can be a misnomer in the sense that compounds such as PZT are commonly referred to as piezoelectric materials even though they are inherently nonlinear and hysteretic. In this case, the designation carries the implicit assumption of low drive behavior where linear approximations provide reasonable accuracy. Piezoelectric materials which are not ferroelectric include bone and skin.

Piezomagnetic Relations — The linear constitutive relations used to approximate the direct and converse magnetomechanical effects at low to moderate drive levels are often termed piezomagnetic constitutive relations through analogy with the linear piezoelectric relations which quantify the direct and converse piezoelectric effects.

Pinning Sites — Pinning sites in ferroelectric, ferromagnetic and ferroelastic materials are generically defined as material nonhomogeneities that provide locations which energetically favor domain wall formation. They can be due to a number of factors including stress concentrations, impurities, defects such as voids or cracks, and second-phase regions.

Polarization, P — For a collection of general dipoles $\mathbf{p}_i$ in a nonhomogeneous medium, the polarization is defined by

$$\mathbf{P} = \lim_{dV \to 0} \left[\frac{1}{dV} \sum_{i=1}^{N_V dV} \mathbf{p}_i \right]$$

where dV is a reference volume and N_V denotes the number of dipoles per unit volume. The polarization thus designates the dipole moment density of the material and has units of coulombs per square meter (C/m^2). Additional details can be found in [23].

When developing hysteresis models for ferroelectric materials, we also consider the special case derived under the assumptions of homogeneous material properties in which dipoles having uniform strength p_0 are constrained to be oriented either in the direction of an applied electric field or diametrically opposite to it. The polarization resulting from N dipoles is scalar-valued and given by

$$P = \frac{p_0}{V} \sum_{i=1}^{N} s_i$$

where $s_i = \pm 1$ designates the dipole orientation.

Experimentally, polarization is measured with a Sawyer tower so it complements the electric D field in providing a measurable electric output quantity. The polarization is related to $\mathbf{D}$ and $\mathbf{E}$ by the relation

$$\mathbf{D} = \mathbf{P} + \epsilon_0 \mathbf{E}$$

where ϵ_0 is the permittivity of free space.

Polarization Switching — Polarization or dipole switching denotes the process in which applied field or mechanical stresses reorient the remanent polarization P_R to a new value P_r — generally opposite and equal in which case $P_r = -P_R$.

Poling — A ferroelectric material is poled by applying a DC field exceeding the coercive field, often at elevated temperatures, to align dipoles and produce a net remanent polarization.

Remanent Magnetization, M_R — See Magnetic Remanence.

Remanent Polarization, P_R — The remanent polarization is analogous to the remanent magnetization and strain, and consists of the polarization which remains when the applied field is reduced to zero following positive saturation — see Figure A.2.

Remanent Strain, ε_R — The remanent strain is that which remains in a ferroelastic crystal when stresses are reduced from a large positive value to zero as depicted in Figure A.4.

Resultants — The resultant of a system of forces is the simplest force representation which can replace the original forces without changing their effect on the rigid body. The definition of a moment resultant is similar. In the context of rod, beam, plate and shell models, this usually consists of replacing forces and moments through the structure's thickness by force and moment resultants acting at the neutral surface.

The static response of a body is determined by the equilibrium condition in which resultants sum to zero whereas the dynamic response is determined from Newton's second law by equating the resultants with corresponding inertial terms — e.g., mass times acceleration for linear motion.

Saturation Magnetization, M_s — The saturation magnetization M_s is that measured when all moments are aligned parallel. For homogeneous materials comprised of moments with strength m_0, M_s is given by

$$M_s = \frac{Nm_0}{V}$$

where N denotes the number of magnetic moments and V is a representative volume. As illustrated in Figure A.3, M_s represents the limiting magnetization achieved if the field H is increased indefinitely.

Saturation Polarization, P_s — For materials in which $\frac{dP}{dE} \to 0$ or becomes sufficiently small, the maximal polarization is termed the saturation polarization through analogy with the saturation magnetization — e.g., see Figure A.2. For homogeneous materials comprised of N dipoles of strength p_0, the saturation polarization is given by

$$P_s = \frac{Np_0}{V}.$$

Spontaneous Magnetization, M_0 — The spontaneous magnetization is that which forms as temperatures are cooled through the Curie point and domains form. It is clearly temperature-dependent and, due to thermal effects, is smaller than the saturation magnetization M_s. For a single domain material, M_0 limits to M_s at 0 K. As shown in Figure 4.8, a demagnetized material consists of randomly oriented domains each having nonzero spontaneous magnetization.

Spontaneous Polarization, P_0 — The spontaneous polarization is the polarization within a single ferroelectric domain at temperatures below the Curie point in the absence of an applied field. At the macroscopic level, it is measured by extrapolating the post-switching slope $\frac{dP}{dE}$ back to the polarization axis as depicted in Figure A.2.

Strain, ε — The strain is defined as the limiting change in length of an infinitesimal element divided by the length of the element; thus

$$\varepsilon = \lim_{\Delta x \to 0} \frac{\Delta u}{\Delta x} = \frac{du}{dx}.$$

Physically, strain represents a local, relative, nondimensional measure of displacement.

Stress, σ — Stress is analogous to pressure and is defined as force per unit area; thus

$$\sigma = \frac{F}{A}.$$

A compressive stress is negative whereas a tensile stress is positive.

Supremum — Let S denote a set that is bounded above and let S_0 be an upper bound that is smaller than or equal to all other upper bounds. Then S_0 is termed the supremum or least upper bound of S and is denoted by

$$S_0 \equiv \sup_S x = \sup S.$$

To illustrate, let S denote the open interval $S = (0, 1)$. Then $\sup S = 1$ even though $1 \notin S$. See also infimum.

Thermodynamic Potentials — Thermodynamics potentials are functionals whose minima yield equilibrium states of the system in the presence of constraints. In the context of material characterization, the internal energy, Helmholtz energy, and various Gibbs energy relations constitute thermodynamic potentials for different independent variable and constraint sets.

Transducer — A transducer is generically defined as any device which converts energy from one form to another. For example, electromechanical transducers convert current, field or voltage inputs to stress or strain outputs and vice versa. In many instances, the designation transducer is used to describe devices which serve as both actuators and sensors. Hence a magnetostrictive device which derives actuator and sensor capabilities from converse and direct magnetomechanical coupling is generically termed a magnetostrictive transducer.

Unimorph — In the context of smart materials, a unimorph generically refers to a transducer comprised of an active element bonded to an inactive substrate. Figure 7.4(b) on page 299 illustrates a unimorph comprised of an active PVDF layer bonded to an inactive polyimide layer. PZT bonded to metallic substrates comprises a second class of unimorphs that has been widely considered for applications. In actuators, transverse motion is generated when field or voltage-induced strains in the active layer respond against the inactive layer. For sensor applications, measured voltages are used to ascertain motion. See also bimorph.

Appendix B

Mathematical Theory

B.1 Dirac Sequences

We summarize here the attributes of a Dirac sequence and provide a theorem which establishes that the convolution of Dirac sequences with suitably smooth functions f will limit to a point evaluation of f. This theorem is employed in Chapters 2 and 4 to illustrate the convergence of Gaussian dipole and moment distributions to a Dirac distribution in the limit of small relative thermal energies kT/V. It also provides a framework for demonstrate that models which include thermal relaxation after-effects converge to models based on the assumption of negligible thermal energy as control volumes are taken to be arbitrarily large.

Theorem B.1. *Let $\{\phi_j\}$ be a sequence satisfying the following properties:*

(i) $\phi_j > 0$ *for all* j

(ii) $\int \phi_j(y)dy = 1$ *for all* j

(iii) Given $\varepsilon, \delta > 0$*, there exists* j_0 *such that*

$$\int_{|y|\geq\delta} \phi_j(y)dy < \varepsilon$$

for all $j \geq j_0$.

Let f be piecewise continuous on $\mathbb{R}$, continuous on the interval $[a, b]$, and satisfy the decay property:

(iv) Given $\varepsilon, \delta > 0$*, there exists* j_0 *such that*

$$\int_{|y|\geq\delta} \phi_j(y)f(x-y)dy < \varepsilon$$

for all $j \geq j_0$ *and* $x \in [a, b]$.

Then for $x \in [a,b]$, $\phi_k * f$ converges to f; that is

$$\int \phi_j(y) f(x-y)dy \to f(x)$$

or

$$\int \phi_j(x-y) f(y)dy \to f(x).$$

Proof. From (ii) it follows that

$$f(x) = f(x)\int \phi_j(y)dy = \int f(x)\phi_j(y)dy$$

so that for $x \in [a,b]$,

$$\begin{aligned} \phi_j * f(x) - f(x) &= \int \phi_j(y) f(x-y)dy - \int \phi_j(y) f(x)dy \\ &= \int \phi_j(y)[f(x-y) - f(x)]dy\,. \end{aligned}$$

From the continuity of f, it follows that for fixed ε, there exists δ such that

$$|f(x-y) - f(x)| < \varepsilon$$

for $|y| < \delta$. For this δ, we write

$$\begin{aligned} |\phi_j * f(x) - f(x)| \;\leq\; & \int_{|y|<\delta} \phi_j(y)|f(x-y) - f(x)|dy + \left| \int_{|y|\geq\delta} \phi_j(y) f(x-y)dy \right| \\ & + \int_{|y|\geq\delta} \phi_j(y)|f(x)|dy. \end{aligned}$$

For sufficiently large j, the integral over the region $|y| < \delta$ is bounded by ϵ due to the continuity of f on $[a,b]$ whereas the second integral is bounded by ε due to (iv). Finally the third integral is bounded by $\|f\|_\infty \varepsilon$, where $\|f\|_\infty \equiv \max_{x\in[a,b]} |f(x)|$, thus yielding the desired convergence. The convolution expression follows from a direct change of variables. □

We note that a sequence of functions $\{\phi_j\}$ satisfying the properties (i)-(iii) is termed a Dirac sequence on $\mathbb{R}^1$. Additionally, if we replace the assumption (iv) by the condition that f is bounded and measurable on $\mathbb{R}$, then Theorem 1 is a 1-D version of Theorem 3.1 from page 228 of [287].

B.2 Compactness of the Polarization Operator

It is demonstrated in (2.114) of Section 2.6.4 that the polarization model for moderate frequency operating regimes with low thermal activation can be formulated as

$$[P(E)](t) = \int_0^\infty \int_{-\infty}^\infty \nu(E_c, E_I)[\overline{P}(E + E_I; E_c, \xi)](t)\, dE_I\, dE_c. \tag{B.1}$$

One choice of kernel is $\overline{P}(E + E_I; E_c, \xi) = \frac{E}{\eta} + P_R\delta(E; E_c, E_I)$ where ξ delineates initial dipole orientations. The parameter δ has a value of 1 for positively oriented dipoles and is -1 for negative orientations. By invoking the physical decay criteria (2.113) for the density ν, (B.1) can be approximated to arbitrary accuracy by consideration of

$$[P(E)](t) = \iint_{\Omega_2} \nu(E_c, E_I)[\overline{P}(E + E_I; E_c, \xi)](t)\, dE_I\, dE_c \tag{B.2}$$

on the compact domain

$$\Omega_2 = \{(E_c, E_I) \in \mathbb{R}_+ \times \mathbb{R} \mid \nu(E_c, E_I) \geq \epsilon\}.$$

Furthermore, we let the minimum and maximum admissible input fields be denoted E_{min} and E_{max} and define

$$\Omega_1 = [E_{min}, E_{max}].$$

As in Section 2.6.6, we consider parameters $q = \nu$ in the parameter space

$$\mathcal{Q} = L^2(\Omega_2) \tag{B.3}$$

and define the observation operator $\mathcal{C}P = P(E)$ on the observation space

$$\mathcal{Y} = L^2(P_{min}, P_{max}). \tag{B.4}$$

The polarization model (B.2) can then be formulated as

$$y(E) = \mathcal{K}q(E)$$

where

$$E \in C[\Omega_1] \subset L^2(\Omega_1)$$

and the parameter-to-observation $\mathcal{K}$ is defined by

$$\mathcal{K}q = \mathcal{C} \iint_{\Omega_2} k(\cdot + E_I, E_c) q(E_c, E_I)\, dE_I\, dE_c. \tag{B.5}$$

It is readily observed that due to the affine construction of $k = \overline{P}$, $k \in L^1(\Omega)$ and $k \in L^2(\Omega)$ where

$$\Omega = \Omega_1 \times \Omega_2.$$

The property that $k \in L^1(\Omega)$ is typical for convolution operators whereas $k \in L^2(\Omega)$ facilitates construction of a generalized Fourier basis for the operator. We employ this latter property to establish that $\mathcal{K}$ is compact for the given choice of spaces. As a prelude, we state the following theorem which is Theorem 5.24.8 from [356].

Theorem B.2. *Let X and Y be Banach spaces and let $\mathcal{K}_N : X \to Y, N = 1, 2, \ldots,$ be a sequence of compact linear operators converging to a bounded linear operator $\mathcal{K} : X \to Y$; that is, $\|\mathcal{K}_N - \mathcal{K}\| \to 0$ as $N \to \infty$. Then $\mathcal{K}$ is a compact linear operator.*

Remark B.1. *Consider the parameter space $\mathcal{Q}$ and observation space $\mathcal{Y}$ defined in (B.3) and (B.4). The integral operator given by (B.5) is a compact operator. We establish this by demonstrating that $\mathcal{K}$ is the limit of a sequence of finite rank operators followed by the use of Theorem B.2.*

We first construct an orthonormal basis $\{\phi_i\}$ for $L^2(\Omega)$. It is illustrated in [356] that

$$\varphi_\ell(s) = \frac{1}{\sqrt{E_{max} - E_{min}}} \exp\left[2\pi i\ell \cdot \frac{s - E_{min}}{E_{max} - E_{min}}\right], \quad \ell = 0, \pm 1, \pm 2, \cdots$$

forms an orthonormal basis for $L^2(\Omega_1)$. With an analogous basis definition for $L^2(\Omega_2)$, it follows that an orthonormal basis for $L^2(\Omega)$ is

$$\phi_{\ell m}(s, t, v) = \varphi_\ell(s)\overline{\varphi_m(t, v)}$$

which we re-index as $\{\phi_i\}$.

It follows that every $f \in L^2(\Omega)$ has the generalized Fourier series representation

$$f = \sum_i \langle f, \phi_i \rangle \phi_i$$

where $\langle \cdot, \cdot \rangle$ denotes the usual L^2 inner product. The norm representation

$$\|f\|^2 = \sum_i |\langle f, \phi_i \rangle|^2$$

follows from Plancheral's theorem. Moreover, we can represent $\mathcal{K}$ and approximating finite-rank operators $\mathcal{K}_N$ by

$$\mathcal{K}f = \sum_i \langle f, \phi_i \rangle \psi_i$$

$$\mathcal{K}_N f = \sum_{i=1}^{N} \langle f, \phi_i \rangle \psi_i$$

where $\psi_i \equiv \mathcal{K}\phi_i$.

To establish the convergence $\mathcal{K} \to \mathcal{K}_N$, we note that

$$\begin{aligned}
\|\mathcal{K}f - \mathcal{K}_N f\| &= \left\| \sum_{i \geq N+1} \langle f, \phi_i \rangle \, \psi_i \right\| \\
&\leq \sum_{i \geq N+1} | \langle f, \phi_i \rangle | \, \|\psi_i\| \\
&\leq \left[\sum_{i \geq N+1} | \langle f, \phi_i \rangle |^2 \right]^{1/2} \left[\sum_{i \geq N+1} \|\psi_i\|^2 \right]^{1/2} \\
&\leq \|f\| \left[\sum_{i \geq N+1} \|\psi_i\|^2 \right]^{1/2}
\end{aligned}$$

where the third inequality follows from the Schwartz inequality. Furthermore, we observe that

$$\begin{aligned}
\sum_i \|\psi_i\|^2 &= \sum_i \left[\int_{\Omega_1} |\mathcal{K}\phi_i(E)|^2 \, dE \right] \\
&= \sum_i \int_{\Omega_1} \left| \iint_{\Omega_2} k(E + E_I, E_c) \phi_i(E_c, E_I) \, dE_I \, dE_c \right|^2 dE \\
&= \int_{\Omega_1} \left[\sum_i \left| \iint_{\Omega_2} k(E + E_I, E_c) \phi_i(E_c, E_I) \, dE_I \, dE_c \right|^2 \right] dE \\
&= \int_{\Omega_1} \left[\iint_{\Omega_2} |k(E + E_I, E_c)|^2 \, dE_I \, dE_c \right] dE < \infty
\end{aligned}$$

where the last step follows from Plancheral's theorem. The convergence of $\sum_i \|\psi_i\|^2$ implies that $\sum_{i \geq N+1} \|\psi_i\|^2 \to 0$ as $N \to \infty$. Thus for $\varepsilon > 0$, there exists N_ε such that for $N > N_\varepsilon$,

$$\|\mathcal{K} - \mathcal{K}_N\| = \sup_{f \neq 0} \frac{\|\mathcal{K}f - \mathcal{K}_N f\|}{\|f\|} < \varepsilon$$

which establishes that

$$\lim_{N \to \infty} \|\mathcal{K} - \mathcal{K}_N\| = 0.$$

Since the range of $\mathcal{K}_N$ is finite, it follows that $\mathcal{K}_N$ is a compact operator. The compactness of $\mathcal{K}$ follows from Theorem B.2 since it is the norm limit of a sequence of compact operators.

The compactness of the integral operator $\mathcal{K}$ given by (B.5) is to be expected since it is a special case of a Hilbert–Schmidt operator which, in general, can be characterized as having an L^2 kernel. This is evidenced by the fact that the proof given here is a modification of that in [217] for Hilbert–Schmidt operators with kernels in $L^2(\mathbb{R}^{2n})$. Details regarding the extension of this theory to more general measure spaces can be found in [287].

B.3 Continuity of the Homogenized Energy Model

We establish here the continuity of the homogenized free energy model (2.114),

$$[P(E)](t) = \int_0^\infty \int_{-\infty}^\infty [\overline{P}(E+E_I;E_c,\xi)](t)\nu_1(E_c)\nu_2(E_I)\,dE_I\,dE_c \tag{B.6}$$

as a function of both field and time. The densities ν_1 and ν_2 satisfy the conditions (2.113) and the kernel $\overline{P}$ has the form

$$\overline{P}(E) = \frac{E}{\eta} + P_R\delta(E;E_c,E_I)$$

where $\delta = 1$ for positively oriented dipoles and $\delta = -1$ for those with negative orientation — see (2.89) and (2.90). The continuity of the hysteresis model is important both for material characterization and for establishing the well-posedness of rod, beam, plate and shell models constructed using the nonlinear and hysteretic constitutive relations employing polarization, magnetization, and strain representations of the form (B.6).

We first note that there are at most three values at which δ can change sign: $-E_c, E_c$ and $-E_c \leq E_T \leq E_c$. The third is determined by the initial dipole distribution ξ, as depicted for the continuous model in Figure A.4(a) and discretized model in Figure 2.34, and is typically chosen so that $E_T = 0$ when $E + E_I = 0$.

We also note that the decay conditions (2.113) dictate that ν_1 and ν_2 satisfy the relations

$$|\nu_2(E_I)| \leq c_2,$$

$$\int_{-\infty}^\infty \nu_2(E_I)dE_I \leq b_2,$$

$$\int_0^\infty \nu_1(E_c)dE_c \leq b_1$$

where b_1, b_2 and c_2 are finite constants.

To establish the continuity of P with respect to E, we consider the behavior at field values E_0 and E_1 where, without loss of generality, we take $E_0 < E_1$. When

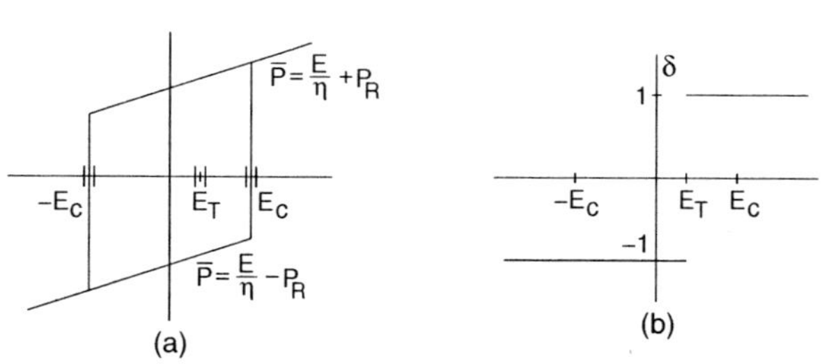

Figure B.1. *(a) Points $E + E_I = -E_c, E_c$ and E_T at which $\delta = \pm 1$ changes sign and (b) behavior of δ associated with the initial dipole distribution at $E + E_I = E_T$.*

integrating with respect to E_I, we decompose the interval $(-\infty, \infty)$ into seven regions delineated by the points $-E_c, E_c, E_T$ as shown in Figure B.1(a). For this decomposition, we note that

$$|\overline{P}(E_1 + E_I; E_c, \xi) - \overline{P}(E_0 + E_I; E_c, \xi)| = \begin{cases} \frac{1}{\eta}(E_1 - E_0)\ , & \text{region excludes } -E_c, E_c, E_T \\ \frac{1}{\eta}(E_1 - E_0) + 2P_R\ , & \text{region includes } -E_c, E_c, E_T\ . \end{cases}$$

To consolidate notation, we define the integrals

$$I(a,b) = \int_a^b \frac{1}{\eta}(E_1 - E_0)\nu_2(E_I)\, dE_I$$

$$I_{P_R}(a,b) = \int_a^b \left[\frac{1}{\eta}(E_1 - E_0) + 2P_R\right]\nu_2(E_I)\, dE_I.$$

It subsequently follows that

$$\begin{aligned}
|P(E_1) - P(E_0)| \leq & \int_0^\infty \{|I(-\infty, -E_c - E_1)| + |I_{P_R}(-E_c - E_1, -E_c - E_0)| \\
& + |I(-E_c - E_0, E_T - E_1)| + |I_{P_R}(E_T - E_1, E_T - E_0)| \\
& + |I(E_T - E_0, E_c - E_1)| + |I_{P_R}(E_c - E_1, E_c - E_0)| \\
& + |I(E_c - E_0, \infty)|\}\, \nu_1(E_c)\, dE_c \\
\leq & \ (E_1 - E_0)\int_0^\infty \left\{\frac{4c_2}{\eta} + 3b_2\left[\frac{1}{\eta}(E_1 - E_0) + 2P_R\right]\right\}\nu_1(E_c)\, dE_c \\
\leq & \ (E_1 - E_0)b_1\left(\frac{4c_2}{\eta} + 3b_2\left[\frac{1}{\eta}(E_1 - E_0) + 2P_R\right]\right).
\end{aligned}$$

For $\varepsilon > 0$, take

$$\delta = \min\left\{\frac{\varepsilon}{b_1\left(\frac{4c_2}{\eta} + 3b_2\left[\frac{1}{\eta}(E_1 - E_0) + 2P_R\right]\right)}, 1\right\}.$$

Under the assumption that E is continuous in time and $E_0 = E(t_0), E_1 = E(t_1)$, for every $\delta > 0$ there exists $\widetilde{\delta} > 0$ such that if $|t_1 - t_0| < \widetilde{\delta}$, we are guaranteed that $|E_1 - E_0| < \delta$. It follows that if $|t_1 - t_0| < \widetilde{\delta}$, the polarization values satisfy the bound

$$|[P(E)](t_1) - [P(E)](t_0)| \leq \varepsilon$$

thus establishing the continuity of the hysteresis model. This holds for all major and minor loops with analogous results for the magnetization model (4.86) and strain model (5.26). The model fits and predictions in Figures 2.40, 4.40, 4.41 and 5.23 illustrate the continuity achieved with the discretized model employing large discretization limits.

Appendix C

Legendre Transforms, Calculus of Varations, and Mechanics Principles

C.1 Legendre Transforms

Legendre transforms map functions in a vector space to functions in the dual space. From a theoretical perspective, they play a fundamental role in the construction of dual Banach spaces in functional analysis and the concepts of tangential coordinates and projective duality in algebraic geometry. Within the realm of model development for smart systems, Legendre transforms are employed when defining thermodynamic potentials and establishing the correspondence between Lagrangian and Hamiltonian frameworks for dynamic systems.

Definitions

A function f is termed *convex* if

$$f(\alpha x + (1-\alpha)y) \leq \alpha f(x) + (1-\alpha)f(y)$$

for all x and y within the domain and all α, $0 < \alpha < 1$ — see Figure C.1. The definition is geometric and does not require that f be differentiable. If sufficiently smooth, however, f will be convex if and only if $f''(x) \geq 0$.

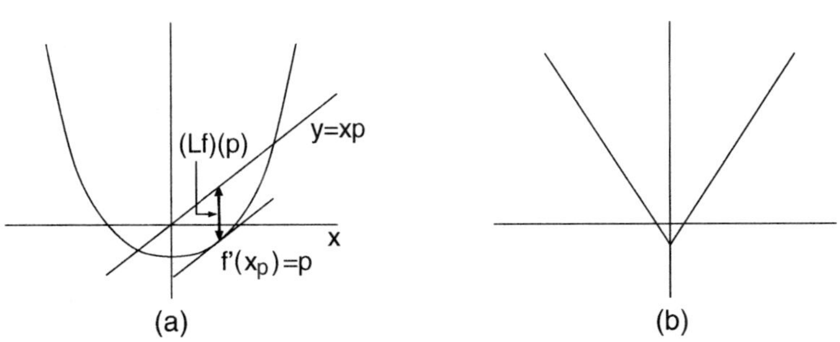

Figure C.1. *Convex functions f which are (a) differentiable throughout the domain and (b) nondifferentiable at points in the domain.*

Let $f : \mathbb{R}^1 \to \mathbb{R}^1$ be a convex function. The Legendre transform $g(p) = (Lf)(p)$ is defined by

$$(Lf)(p) = \sup_x [xp - f(x)].$$

Note that this specifies $x = g(p)$ as a function of p. If f is differentiable, the Legendre transform can be expressed as

$$\begin{aligned}(Lf)(p) &= \max_x [xp - f(x)] \\ &= x_p p - f(x_p)\end{aligned}$$

where x_p solves

$$p = f'(x_p). \tag{C.1}$$

The existence and uniqueness of x_p in this case is due to the convexity of f.

If we employ a dot product rather than the scalar product, the analogous definition

$$(Lf)(p) = \sup_x [x \cdot p - f(x)] \tag{C.2}$$

holds for convex $f : \mathbb{R}^n \to \mathbb{R}^1$.

Properties

Two properties of the Legendre transform are of fundamental importance for both analysis and applications: (i) it maps convex functions to convex functions, and (ii) the Legendre transform is self-dual or an involution. The first plays a fundamental role in the optimization of energy relations since it dictates the manner through which minimum energy states for one energy definition are related to those of its transform. The second property states that $L(Lf) = f$. For differentiable functions, this follows from the property that if p and x are related by $p = f'(x)$, then $x = g'(p)$. Proofs and ramifications of these properties can be found in [15].

Further attributes of the Legendre transform in 1-D are illustrated by the following examples.

Example 1. Let $f(x) = x^2$. From the condition $p = 2x$, it follows that

$$g(p) = (Lf)(p) = \frac{p^2}{2} - \frac{p^2}{4} = \frac{p^2}{4}.$$

Example 2. Let $f(v) = \frac{1}{2}mv^2$ denote the kinetic energy for a particle of mass m. The necessary condition (C.1) yields the momentum equation

$$p = mv.$$

The Legendre transform in this case is

$$g(p) = \frac{p^2}{2m}.$$

Further ramifications of this relation for Hamiltonian mechanics formulations will be provided in Section C.3.

Example 3. Consider the elastic Helmholtz energy relation

$$\psi(\varepsilon) = \frac{1}{2} Y \varepsilon^2$$

which results from (2.18) when polarization is neglected and strains ε are restricted to 1-D. The negative Legendre transform is

$$\begin{aligned} -(L\psi)(\sigma) &= -\left[\frac{\sigma^2}{Y} - \frac{1}{2}\left(\frac{\sigma}{Y}\right)^2\right] \\ &= -\frac{1}{2} s \sigma^2 \end{aligned}$$

where σ is an applied stress and $s = \frac{1}{Y}$ is the 1-D compliance. Comparison with (2.22) illustrates that

$$G(\sigma) = -(L\psi)(\sigma)$$

defines the elastic Gibbs energy for the system. Similarly, it is shown in Section 2.2.3 that $G(E) = -(L\psi)(E)$ for the Helmholtz energy $\psi(P) = \frac{1}{2}\alpha P^2$.

C.2 Principles from the Calculus of Variations

To provide fundamental relations used when establishing the Lagrange mechanics framework in Section C.3, we summarize selected principles pertaining to the calculus of variations. Additional details can be found in [15, 307, 505].

Gateaux and Fréchet Differentials

Calculus of variations is concerned with the extrema of functionals so we begin with a summary of differential theory for vector spaces. Throughout this discussion, X is a vector space, Y is a normed space, and $T : D \subset X \to Y$ is a (possibly nonlinear) transformation. For the case $Y = \mathbb{R}$, the transformation is a real-valued functional which we will denote by J.

Definition C.2.1. *Consider $x \in D \subset X$ and arbitrary $\eta \in X$. If the limit*

$$\delta T(x;\eta) = \lim_{\epsilon \to 0} \frac{1}{\epsilon}\left[T(x+\epsilon\eta) - T(x)\right]$$

exists for each $\eta \in X$, T is said to be Gateaux differentiable at x and $\delta T(x;\eta)$ is termed the Gateaux differential of T at x with increment or perturbation η.

For functionals J, the Gateaux differential, when it exists, is

$$\delta J(x;\eta) = \frac{d}{d\epsilon} J(x+\epsilon\eta)\Big|_{\epsilon=0}.$$

Note that for each fixed $x \in D$, $\delta J(x;\eta)$ is a functional with respect to $\eta \in X$.

Definition C.2.2. *T is said to be Fréchet differentiable at $x \in D$ in the normed space X if for each $\eta \in X$, there exists $\delta T(x;\eta) \in Y$ which is linear, continuous with respect to η, and satisfies*

$$\lim_{\|\eta\|\to 0} \frac{\|T(x+\eta) - T(x) - \delta T(x;\eta)\|}{\|\eta\|} = 0.$$

When it exists, $\delta T(x;\eta)$ is termed the Fréchet differential of T at x with increment η.

The Gateaux differential generalizes the concept of directional derivatives whereas the Fréchet differential generalizes the definition of differentiability. Because the Gateaux differential requires no norm on X, it cannot be directly used to establish continuity and is significantly weaker than the Fréchet differential. This is illustrated by the calculus example

$$f(x,y) = \begin{cases} 0 \quad , & (x,y) \in \Omega_1 = \{(x,y) \mid y \neq x^2 \text{ or } x = y = 0\} \\ 1 \quad , & (x,y) \in \Omega_2 = \{(x,y) \mid y = x^2 \text{ and } x \neq 0, y \neq 0\}. \end{cases}$$

All directional derivatives exist at $(0,0)$ but the function is both discontinuous and nondifferentiable at that point.

We note that the existence of the Fréchet differential implies the existence of the Gateaux differential in which case the two will be equal.

Extrema of Functionals

The following theorem establishes a necessary condition for a functional to have an extremum (minimum or maximum) at the point x_0.

Theorem C.1. *Let the functional $J : X \to \mathbb{R}$ have a Gateaux differential $\delta J(x;\eta)$. If J has an extremum at x_0, then $\delta J(x_0;\eta) = 0$ for all $\eta \in X$.*

Proof. If J has an extremum at x_0, it follows that the function $J(x_0 + \epsilon\eta)$ of the real variable ϵ has an extremum at $\epsilon = 0$. This implies that

$$\frac{d}{d\epsilon} J(x_0 + \epsilon\eta)\Big|_{\epsilon=0} = 0$$

and hence $\delta(x_0;\eta) = 0$. □

Points x_0 at which extrema occur are termed *stationary points.* We note that in the context of mechanics, Theorem C.1 is often referred to as *Hamilton's principle* or *Hamilton's principle of least motion.*

Euler–Lagrange Equations

Consider the problem of finding a function x which minimizes the functional

$$J = \int_{t_0}^{t_1} \mathcal{L}[x(t), \dot{x}(t), t]dt. \tag{C.3}$$

The function $\mathcal{L}$ is assumed to be continuous in $x, \dot{x}, t$ and have continuous partial derivatives with respect to x and $\dot{x}$. We also assume that the endpoints $x(t_0)$ and $x(t_1)$ are fixed.

To specify the admissible class of solutions, we consider variations of the form

$$\widehat{x}(t) = x(t) + \epsilon\eta(t)$$

where η satisfies

$$\begin{aligned} &(i) \quad \eta \in C^1[t_0, t_1] \\ &(ii) \quad \eta(t_0) = \eta(t_1) = 0. \end{aligned} \tag{C.4}$$

The first criterion guarantees the continuity of solutions and their temporal derivatives whereas the second guarantees that

$$\widehat{x}(t_0) = x(t_0) \quad , \quad \widehat{x}(t_1) = x(t_1)$$

in accordance with the condition of fixed endpoints. The second condition is depicted in Figure C.2.

The Gateaux differential is

$$\begin{aligned} \delta J(x;\eta) &= \frac{d}{d\epsilon}\int_{t_0}^{t_1} \mathcal{L}\left(x + \epsilon\eta, \dot{x} + \epsilon\dot{\eta}, t\right) dt \,\Big|_{\epsilon=0} \\ &= \int_{t_0}^{t_1} \frac{\partial \mathcal{L}}{\partial x}(x,\dot{x},t)\eta(t)dt + \int_{t_0}^{t_1} \frac{\partial \mathcal{L}}{\partial \dot{x}}(x,\dot{x},t)\dot{\eta}(t)dt \end{aligned}$$

which can be verified to be a Fréchet differential. Under the assumption that $\frac{d}{dt}\frac{\partial \mathcal{L}}{\partial \dot{x}}$ exists and is continuous, integration by parts and application of Theorem C.1 yields

$$\begin{aligned} \delta J(x;\eta) &= \int_{t_0}^{t_1} \left[\frac{\partial \mathcal{L}}{\partial x}(x,\dot{x},t) - \frac{d}{dt}\frac{\partial \mathcal{L}}{\partial \dot{x}}(x,\dot{x},t)\right]\eta(t)dt + \frac{\partial \mathcal{L}}{\partial \dot{x}}(x,\dot{x},t)\eta(t)\Big|_{t_0}^{t_1} \\ &= \int_{t_0}^{t_1} \left[\frac{\partial \mathcal{L}}{\partial x}(x,\dot{x},t) - \frac{d}{dt}\frac{\partial \mathcal{L}}{\partial \dot{x}}(x,\dot{x},t)\right]\eta(t)dt \\ &= 0 \end{aligned}$$

which must hold for all η satisfying the admissibility conditions (C.4). Because η is continuous, it follows that the optimal x must satisfy the Euler–Lagrange equation

$$\frac{d}{dt}\frac{\partial \mathcal{L}}{\partial \dot{x}}(x,\dot{x},t) - \frac{\partial \mathcal{L}}{\partial x}(x,\dot{x},t) = 0. \tag{C.5}$$

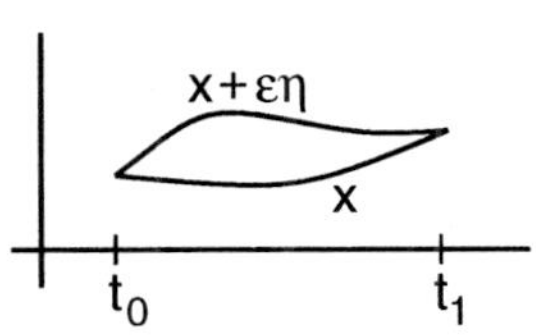

Figure C.2. *Admissible variations in the trajectory x.*

A derivation of (C.5) which relaxes the *a priori* continuity condition on $\frac{d}{dt}\frac{\partial \mathcal{L}}{\partial \dot{x}}$ is provided in [307].

Example 4. Let $x = \mathbb{R}$ and take $\mathcal{L}(x, \dot{x}, t) = \sqrt{1+\dot{x}^2}$ so that

$$J = \int_{t_0}^{t_1} \sqrt{1+\dot{x}^2} dt$$

defines the length of the curve between the endpoints $x(t_0)$ and $x(t_1)$. Here $\frac{\partial \mathcal{L}}{\partial x} = 0$ and $\frac{\partial \mathcal{L}}{\partial \dot{x}} = \frac{\dot{x}}{\sqrt{1+\dot{x}^2}}$ so the Euler–Lagrange equation is

$$\frac{d}{dt}\left(\frac{\dot{x}}{\sqrt{1+\dot{x}^2}}\right) = 0.$$

Integration yields $\dot{x} = c_1$ and hence $x(t) = c_1 t + c_2$. As expected, the length is minimized by a straight line with c_1 and c_2 determined by $x(t_0)$ and $x(t_1)$.

C.3 Classical, Lagrangian and Hamiltonian Mechanics

We summarize here basic tenets of classical, Lagrangian and Hamiltonian mechanics to provide aspects of the framework employed when constructing constitutive and structural models for smart material systems.

Classical Mechanics

Classical or Newtonian mechanics can be described as the physics of forces or moments acting on a body. To simplify the discussion, we consider only forces acting on a point particle of mass m and refer the reader to [15] for discussion regarding more complex systems. We let $\mathbf{r} = \mathbf{r}(x_1, x_2, x_3, t)$ denote the position of the particle and $\mathbf{v} = \dot{\mathbf{r}}$ denote its velocity.

One of the cornerstones of classical mechanics is Newton's second law

$$\mathbf{F} = \frac{d}{dt}(m\mathbf{v}) \tag{C.6}$$

which states that the change in momentum $\mathbf{p} = m\mathbf{v}$ is equal to the sum of all applied forces. When the mass is time invariant, this yields the familiar relation

$$\mathbf{F} = m\mathbf{a}.$$

Three scalar quantities which are fundamental for quantifying the static and dynamic response of the body are the work, kinetic energy and potential energy. The work dW caused by a force $\mathbf{F}$ acting for a distance $d\mathbf{r}$ is

$$dW = \mathbf{F} \cdot d\mathbf{r}$$

so the total work required to move a particle from point P_1 to point P_2 along a path γ is

$$W = \int_\gamma \mathbf{F} \cdot d\mathbf{r}.$$

The kinetic energy K is quantified by the quadratic form

$$K = \frac{1}{2}m|\mathbf{v}|^2 = \frac{1}{2}\mathbf{v} \cdot \mathbf{v}.$$

The change in potential energy is defined in terms of the work required to move the particle from P_1 to P_2 in a conservative force field — hence $\oint \mathbf{F} \cdot d\mathbf{r} = 0$ for any closed path. If we denote the potential energy at the endpoints by U_1 and U_2, then

$$U_2 - U_1 = -W.$$

For conservative forces $\mathbf{F}$, the potential energy is related to the force by the gradient relation

$$\mathbf{F} = -\nabla U. \tag{C.7}$$

In 1-D, one can integrate (C.7) to obtain

$$U(x) = -\int_{x_0}^{x} F(s)ds;$$

however, in 2-D and 3-D this is not always possible. The negative sign in these relations can be motivated by the observation that if $\mathbf{F}$ denotes the force due to gravity, the potential energy increases as the particle is lifted.

The *total energy* is the sum

$$\begin{aligned}\mathcal{H} &= K + U \\ &= \tfrac{1}{2}m\dot{\mathbf{r}} \cdot \dot{\mathbf{r}} + U\end{aligned}$$

of the kinetic and potential energies. For conservative forces in 3-D, it is observed that

$$\begin{aligned}\frac{\partial \mathcal{H}}{\partial t} &= m\ddot{\mathbf{r}} \cdot \dot{\mathbf{r}} + \sum_{i=1}^{3} \frac{\partial U}{\partial r_i}\frac{dr_i}{dt} \\ &= (m\ddot{\mathbf{r}} - \mathbf{F}) \cdot \dot{\mathbf{r}} \\ &= 0.\end{aligned}$$

This expresses the *law of energy conservation* for conservative systems.

Lagrangian Mechanics

The development of Lagrangian theory for mechanical systems is based on the observation that variational principles form the basis for several of the fundamental laws obtained from Newtonian principles — e.g., force and moment balancing. To illustrate, consider the functional (C.3),

$$J = \int_{t_0}^{t_1} \mathcal{L}(\mathbf{r}, \dot{\mathbf{r}}, t)dt,$$

where the Lagrangian

$$\mathcal{L} = K - U$$

is taken to be the difference between the kinetic and potential energies. It was shown in (C.5) that the extremum satisfies the Euler–Lagrange equations

$$\frac{d}{dt}\frac{\partial \mathcal{L}}{\partial \dot{r}_i} - \frac{\partial \mathcal{L}}{\partial r_i} = 0$$

for $i = 1, \ldots, 3$. Noting that $U = U(\mathbf{r})$ and $K = \frac{1}{2}m\sum_{i=1}^{3}\dot{r}_i^2$, it follows that

$$\frac{\partial \mathcal{L}}{\partial r_i} = -\frac{\partial U}{\partial r_i} \quad , \quad \frac{\partial \mathcal{L}}{\partial \dot{r}_i} = m\dot{r}_i$$

where the latter relation defines the momentum in terms of the Lagrangian. Hence the Euler–Lagrange equations yield

$$m\ddot{\mathbf{r}} = \mathbf{F}$$

which is precisely Newton's second law.

Until now, we have employed rectangular coordinates when summarizing fundamental physical principles. For many systems, however, other coordinates may be more natural — e.g., polar or spherical. Hence it is common to employ a minimal number of *generalized coordinates*

$$\mathbf{q} = (q_1, \ldots, q_n)$$

required to specify the motion of a particle, body, or system. The following definition generalizes several of the concepts previously discussed in the context of a point mass in rectangular coordinates.

Definition C.3.1. *All relations hold for $i = 1, \ldots, n$.*

$\mathbf{r} = \mathbf{r}(q_1, \ldots, q_n, t)$	*Position vector for the body*
$\dot{q}_i = \frac{dq_i}{dt}$	*Generalized velocities*
$\mathcal{L}(\mathbf{q}, \dot{\mathbf{q}}, t) = K - U$	*Lagrangian*
$\mathcal{A} = \int_{t_0}^{t_1} \mathcal{L}(\mathbf{q}, \dot{\mathbf{q}}, t)\, dt$	*Action integral*
$\frac{\partial \mathcal{L}}{\partial q_i}$	*Generalized forces*
$\frac{\partial \mathcal{L}}{\partial \dot{q}_i} = p_i$	*Generalized or conjugate momenta*
$\frac{d}{dt}\frac{\partial \mathcal{L}}{\partial \dot{q}_i} - \frac{\partial \mathcal{L}}{\partial q_i} = 0$	*Euler–Lagrange equations*

We note that in rectangular coordinates, the generalized momenta p_i are precisely the linear momenta $m\dot{x}_i$ whereas they are the angular momenta in polar coordinates. For arbitrary choices of generalized coordinates, the physical interpretation of p_i is less direct.

Details regarding the physics embodied in the Lagrangian framework can be found in [15] and additional theory is provided in [319].

Hamiltonian Mechanics

Whereas Lagrangian mechanics is based on a variational interpretation of physical principles, the Hamiltonian framework relies on total energy principles. In the former, the Lagrangian

$$\mathcal{L}(\mathbf{q}, \dot{\mathbf{q}}, t) = K(\dot{\mathbf{q}}) - U(\mathbf{q}, t),$$

defined in terms of generalized coordinates and their derivatives, provides the fundamental function used to quantify physical properties. In the Hamiltonian framework, the Hamiltonian

$$\mathcal{H}(\mathbf{q}, \mathbf{p}, t) = \dot{\mathbf{q}} \cdot \mathbf{p} - \mathcal{L}(\mathbf{q}, \dot{\mathbf{q}}, t), \tag{C.8}$$

where $p_i = \frac{\partial \mathcal{L}}{\partial \dot{q}_i}$ are conjugate or generalized momenta, is the fundamental quantity. From (C.2), it is observed that $\mathcal{H}$ is the Legendre transform of $\mathcal{L}$.

The differential of (C.8) is

$$\begin{aligned} d\mathcal{H} &= \sum_{i=1}^{n} \left[p_i d\dot{q}_i + \dot{q}_i dp_i - \frac{\partial \mathcal{L}}{\partial q_i} dq_i - \frac{\partial \mathcal{L}}{\partial \dot{q}_i} d\dot{q}_i \right] - \frac{\partial \mathcal{L}}{\partial t} dt \\ &= \sum_{i=1}^{n} [p_i d\dot{q}_i + \dot{q}_i dp_i - \dot{p}_i dq_i - p_i d\dot{q}_i] - \frac{\partial \mathcal{L}}{\partial t} dt \qquad \text{(C.9)} \\ &= \sum_{i=1}^{n} [\dot{q}_i dp_i - \dot{p}_i dq_i] - \frac{\partial \mathcal{L}}{\partial t} dt \end{aligned}$$

where the second step results from the definition $p_i = \frac{\partial \mathcal{L}}{\partial \dot{q}_i}$ and identity $\dot{p}_i = \frac{\partial \mathcal{L}}{\partial q_i}$ resulting from the Euler–Lagrange equations. Equating (C.9) with the total differential

$$d\mathcal{H} = \sum_{i=1}^{n} \left[\frac{\partial \mathcal{H}}{\partial q_i} dq_i + \frac{\partial \mathcal{H}}{\partial p_i} dp_i \right] + \frac{\partial \mathcal{H}}{\partial t} dt$$

yields Hamilton's equations

$$\dot{q}_i = \frac{\partial \mathcal{H}}{\partial p_i} \quad , \quad \dot{p}_i = -\frac{\partial \mathcal{H}}{\partial q_i} \quad , \quad \frac{\partial \mathcal{H}}{\partial t} = -\frac{\partial \mathcal{L}}{\partial t} \quad , \quad i = 1, \ldots, n.$$

Note that the first-order Hamilton's equations are equivalent to the second-order Euler–Lagrange equations.

Example 5. To illustrate properties of the Hamiltonian and Hamilton's equations of motion, we consider the 1-D motion of a mass m in response to a conservative force F. In this case

$$\mathcal{L}(x,\dot{x},t) = \frac{1}{2}m\dot{x}^2 - U(x)$$

and $p = \frac{\partial \mathcal{L}}{\partial \dot{x}} = m\dot{x}$ so that

$$\begin{aligned} \mathcal{H}(x,p) &= m\dot{x}^2 - \left(\frac{1}{2}m\dot{x}^2 - U(x)\right) \\ &= K(\dot{x}) + U(x) \\ &= \frac{p^2}{2m} + U(x). \end{aligned}$$

Hence it is observed that $\mathcal{H}$ is the sum of the kinetic and potential energies which is true in general for conservative systems. Furthermore, Hamilton's equations are

$$\frac{\partial \mathcal{H}}{\partial p} = \frac{p}{m} = \dot{x}$$

$$\frac{\partial \mathcal{H}}{\partial x} = \frac{\partial U}{\partial x} = -\dot{p}.$$

The first expression simply relates the generalized momentum to the velocity and the second is Newton's second law (C.6).

Whereas the Lagrangian framework has the advantage of a variational basis, the Hamiltonian perspective is advantageous for certain applications of perturbation theory (celestial mechanics) and the characterization of complex systems arising in statistical mechanics and ergotic theory. It has also proven fundamental in the development of theories for optics and quantum mechanics. Further details regarding the physics and theory of Hamiltonian mechanics can be found in [15, 319].

Appendix D

Inversion Algorithm

This appendix provides details regarding the algorithm used to invert the discretized polarization model (2.115). A similar algorithm can be used to invert the magnetization model (4.87). For the two cases, forward Algorithm 2.6.4 and Algorithm 4.7.3 form the nucleus for the inversion procedure.

Employing the notation of Section 2.6.5, we define the matrices

$$\Delta_{init} = \begin{bmatrix} -1 & \cdots & -1 & 1 & \cdots & 1 \\ \vdots & & \vdots & \vdots & & \vdots \\ -1 & \cdots & -1 & 1 & \cdots & 1 \end{bmatrix}_{N_i \times N_j}, \quad \mathcal{E}_c = \begin{bmatrix} E_{c_1} & \cdots & E_{c_1} \\ \vdots & & \vdots \\ E_{c_{N_i}} & \cdots & E_{c_{N_i}} \end{bmatrix}_{N_i \times N_j}$$

$$\mathcal{E}_k = \begin{bmatrix} E_k + E_{I_1} & \cdots & E_k + E_{I_{N_j}} \\ \vdots & & \vdots \\ E_k + E_{I_1} & \cdots & E_k + E_{I_{N_j}} \end{bmatrix}_{N_i \times N_j} \tag{D.1}$$

and vectors

$$V^T = \left[v_1 \nu_1(E_{c_1}), \; \cdots, v_{N_i} \nu_1(E_{c_{N_i}}) \right]_{1 \times N_i}$$

$$W^T = \left[w_1 \nu_2(E_{I_1}), \; \cdots, w_{N_j} \nu_2(E_{I_{N_j}}) \right]_{1 \times N_j}$$

where $E_k = E(t_k)$ is the kth value of the input field.

As noted in Section 2.6.7, the inverse algorithm can be summarized as follows. For given values of P_{k-1} and P_k, forward Algorithm 2.6.4 is used to increment the field until the predicted polarization P_{tmp} has advanced beyond the specified value P_k. At this point, the final field value E_k is determined through linear interpolation between the final two predicted field values. Algorithm D.1.1 augments Algorithm 2.6.8 by providing additional details regarding the manner in which ΔE is adaptively updated to improve efficiency.

Algorithm D.1.1.

$\Delta_{prev} = \Delta_{init}$

Set Scales

- $scale = (E_{max} - E_{min}) \cdot 4$
- $num\_steps = 1$

for $k = 2 : N_k$

$dP = P_k - P_{k-1}$

$E_{tmp} = E_{k-1}$, $P_{tmp} = P_{k-1}$, $ind = 1$

Specify ΔE

- if $num\_steps == 1$

  $step = \frac{1}{5} \cdot scale$

  elseif $num\_steps >= 10$

  $step = 2 \cdot step$

  else

  $step = step$

  end
- $\Delta E = step \cdot dP$

while $\mathrm{sgn}(dP) \cdot (P_k - P_{tmp}) >= 0$

$E_{tmp} = E_{tmp} + \Delta E$

Construct $\mathcal{E}_k$ given by $(D.1)$ using E_{tmp}

$\Delta = \mathrm{sgn}(\mathcal{E}_k + \mathcal{E}_c. * \Delta_{prev})$

$P_{tmp} = V^T \left[\frac{1}{\eta}\mathcal{E}_k + P_R\Delta\right] W$

$\Delta_{prev} = \Delta$

$ind = ind + 1$

end

E_k specified by linear interpolation

$slope = \Delta P_{tmp} / \Delta E_{tmp}$

$E_k = E_{tmp} - (P_{tmp} - P_k)/slope$

$num\_steps = ind$

end

Bibliography

[1] M. Achenbach, "A model for an alloy with shape memory," *International Journal of Plasticity,* 5, pp. 371–395, 1989.

[2] M. Achenbach and I. Müller, "Simulation of material behavior of alloys with shape memory," *Archives of Mechanics*, 37(6), pp. 573–585, 1985.

[3] C.J. Adkins, *Equilibrium Thermodynamics*, Cambridge University Press, Cambridge, UK, 1994.

[4] A. Aharoni, *Introduction to the Theory of Ferromagnetism*, Second Edition, Oxford University Press, Oxford, UK, 2000.

[5] A. Aharoni, "Micromagnetics: Past, present and future," *Physica B*, 306, pp. 1–9, 2001.

[6] R. Ahluwalia and W. Cao, "Influence of dipolar defects on switching behavior in ferroelectrics," *Physical Review B*, 63(1), article 012103, 2000.

[7] S. Aizawa, T. Kakizawa and M. Higasino, "Case studies of smart materials for civil structures," *Smart Materials and Structures*, 7, pp. 617–626, 1998.

[8] K. Aizu, "Possible species of ferromagnetic, ferroelectric, and ferroelastic crystals," *Physical Review B*, 2(3), pp. 754–772, 1970.

[9] M.A. Akbas and P.K. Davies, "Domain growth in $Pb(Mg_{1/3}\,Nb_{2/3})O_3$ perovskite relaxor ferroelectric oxides," *Journal of the American Ceramic Society*, 80(11), pp. 2933–2936, 1989.

[10] Ó. Alejos and E. Della Torre, "Magnetic aftereffect dependence on the moving parameter of the Preisach model," *Physica B*, 306, pp. 67–71, 2001.

[11] E.H. Anderson and N.W. Hagood, "Self-sensing piezoelectric actuation: Analysis and application to controlled structures," AIAA/ASME/ASCE/AHS/ASC Structures, Structural Dynamics and Materials Conference, Dallas, TX, pp. 2141–2155, 1992.

[12] J.C. Anderson, *Dielectrics*, Reinhold Publishing Corporation, New York, 1964.

[13] W.D. Armstrong, "Magnetization and magnetostriction processes in $Tb_{0.27-0.30}$ $Dy_{0.73-0.70}$ $Fe_{1.9-2.0}$," *Journal of Applied Physics*, 81(5), pp. 2321–2326, 1997.

[14] D.N. Arnold and R.S. Falk, "A uniformly accurate finite element for the Reissner-Mindlin plate," *SIAM Journal of Numerical Analysis*, 26, pp. 1276–1290, 1989.

[15] V.I. Arnold, *Mathematical Methods in Classical Mechanics*, Second Edition, Translated by K. Vogtmann and A. Weinstein, Springer–Verlag, New York, 1989.

[16] M. Ashhab, M.V. Salapaka, M. Dahleh and I. Mezic, "Dynamical analysis and control of micro-cantilevers," *Automatica*, 35(10), pp. 1663–1670, 1999.

[17] D.L. Atherton and J.R. Beattie, "A mean field Stoner–Wohlfarth hysteresis model," *IEEE Transactions on Magnetics*, 26(6), pp. 3059–3063, 1990.

[18] D.L. Atherton, B. Szpunar and J.A. Szpunar, "A new approach to Preisach diagrams," *IEEE Transactions on Magnetics*, 23(3), pp. 1856–1865, 1987.

[19] K.E. Atkinson, *An Introduction to Numerical Analysis*, John Wiley and Sons, New York, 1978.

[20] O. Axelsson and V.A. Barker, *Finite Element Solution of Boundary Value Problems*, SIAM Classics in Applied Mathematics, SIAM, Philadelphia, 2001.

[21] I. Babuška and M. Suri, "On locking and robustness in the finite element method," *SIAM Journal of Numerical Analysis*, 29(5), pp. 1261–1293, 1992.

[22] I. Babuška and M. Suri, "Locking effects in the finite element approximation of elasticity problems," *Numerische Mathematik*, 62, pp. 439–463, 1992.

[23] C.A. Balanis, *Advanced Engineering Electromagnetics*, John Wiley and Sons, New York, 1989.

[24] B.L. Ball and R.C. Smith, "A stress-dependent hysteresis model for PZT-based transducers," Smart Structures and Materials 2004, Proceedings of the SPIE, Volume 5383, pp. 23–30, 2004.

[25] B.L. Ball, R.C. Smith and Z. Ounaies, "A dynamic model for THUNDER transducers," Smart Structures and Materials 2003, Proceedings of the SPIE, Volume 5049, pp. 100–111, 2003.

[26] H.T. Banks, K. Ito and Y. Wang, "Well-posedness for damped second order systems with unbounded input operators," *Differential and Integral Equations*, 8, pp. 587–606, 1995.

[27] H.T. Banks, F. Kappel and C. Wang, "Weak solutions and differentiability for size structured population models," in *Distributed Parameter Systems: Control and Applications*, F. Kappel et al, Eds., ISNM Vol. 100, Birkhäuser, pp. 35–50, 1991.

[28] H.T. Banks, A.J. Kurdila and G. Webb, "Identification of hysteretic control influence operators representing smart actuators, Part I: formulation," *Mathematical Problems in Engineering*, 3, pp. 287–328, 1997.

[29] H.T. Banks, A.J. Kurdila and G. Webb, "Identification of hysteretic control influence operators representing smart actuators, Part II: convergent approximations," *Journal of Intelligent Material Systems and Structures*, 8(6), pp. 536–550, 1997.

[30] H.T. Banks, R.C. Smith, D.E. Brown, V.L. Metcalf and R.J. Silcox, "The estimation of material and patch parameters in a PDE-based circular plate model," *Journal of Sound and Vibration*, 199(5), pp. 777–799, 1997.

[31] H.T. Banks, R.C. Smith, D.E. Brown, R.J. Silcox, and V.L. Metcalf, "Experimental confirmation of a PDE-based approach to design of feedback controls," *SIAM Journal on Control and Optimization*, 35(4), pp. 1263-1296, 1997.

[32] H.T. Banks, R.C. Smith and Y. Wang, "Modeling and parameter estimation for an imperfectly clamped plate," in *Computation and Control IV,* K.L. Bowers and J. Lund, Eds., Birkhäuser, Boston, pp. 23–42, 1995.

[33] H.T. Banks, R.C. Smith and Y. Wang, *Smart Material Structures: Modeling, Estimation and Control*, Masson/John Wiley, Paris/ Chichester, 1996.

[34] H.T. Banks and H.T. Tran, *Mathematical and Experimental Modeling of Physical and Biological Processes*, SIAM, Philadelphia, PA, to appear.

[35] H. Barkhausen, "Two phenomena uncovered with help of the new amplifiers," *Berichte der Deutschen Physikalischen Gesellschaft* (later *Zeitschrift für Physik*), 20, pp. 401–403, 1919.

[36] E.B. Becker, G.F. Carey and J.T. Oden, *Finite Elements: An Introduction; Volume 1*, Prentice–Hall, Inc., Englewood Cliffs, NJ, 1981.

[37] D. Berlincourt, D.R. Curran and H. Jaffe, "Piezoelectric and piezomagnetic materials and their function in transducers," *Physical Acoustics*, Volume 1 (Part A), W.E. Mason, Ed., pp. 169–270, 1964.

[38] D. Bernardini and T.J. Pence, "Shape-memory materials, modeling," in *Encyclopedia of Smart Materials,* M. Schwartz, Ed., John Wiley and Sons, New York, pp. 964–980, 2002.

[39] M. Bernadou and J.M. Boisserie, *The Finite Element Method in Thin Shell Theory: Applications to Arch Dam Simulations*, Birkhäusser, Boston, 1982.

[40] G. Bertotti, *Hysteresis in Magnetism: For Physicists, Materials Scientists, and Engineers*, Academic Press, San Diego, CA, 1998.

[41] J.J. Binney, N.J. Dowrick, A.J. Fisher and M.E.J. Newman, *The Theory of Critical Phenomena: An Introduction to the Renormalization Group*, Clarendon Press, Oxford, UK, 1999.

[42] R.R. Birss, C.A. Faunce and E.D. Isaac, "Magnetomechanical effects in iron and iron-carbon alloys," *Journal of Applied Physics D: Applied Physics*, 4, pp. 1040–1048, 1971.

[43] F. Bitter, "On inhomogeneities in the magnetization of ferromagnetic materials," *Physical Review*, 38, pp. 1903–1905, 1931.

[44] Z. Bo and D.C. Lagoudas, "Thermomechanical modeling of polycrystalline SMAs under cyclic loading, Part I: theoretical derivations," *International Journal of Engineering Science*, 37, pp. 1089–1140, 1999.

[45] Z. Bo and D.C. Lagoudas, "Thermomechanical modeling of polycrystalline SMAs under cyclic loading, Part III: evolution of plastic strains and two-way shape memory effect," *International Journal of Engineering Science*, 37, pp. 1175–1203, 1999.

[46] Z. Bo and D.C. Lagoudas, "Thermomechanical modeling of polycrystalline SMAs under cyclic loading, Part IV: modeling of minor hysteresis loops," *International Journal of Engineering Science*, 37, pp. 1205–1249, 1999.

[47] C. Body, G. Reyne and G. Meunier, "Nonlinear finite element modelling of magneto-mechanical phenomena in giant magnetostrictive thin films," *IEEE Transactions on Magnetics*, 33(2), pp. 1620–1623, 1997.

[48] R.T. Bonnecaze and J.F. Brady, "Dynamic simulation of an electrorheological fluid," *Journal of Chemical Physics*, 96, pp. 2183–2203, 1992.

[49] O. Boser, "Statistical theory of hysteresis in ferroelectric materials," *Journal of Applied Physics*, 62(4), pp. 1344–1348, 1987.

[50] J.G. Boyd and D.C. Lagoudas, "Thermomechanical response of shape memory composites," *Journal of Intelligent Material Systems and Structures*, 5, pp. 333–346, 1994.

[51] R.M. Bozorth, *Ferromagnetism*, D. Van Nostrand Company, Inc., New York, 1951.

[52] W.L. Bragg and E.J. Williams, "The effect of thermal agitation on atomic arrangement in alloys," *Proceedings of the Royal Society of London*, A145, pp. 699–730, 1934.

[53] W.L. Bragg and E.J. Williams, "The effect of thermal agitation on atomic arrangement in alloys II," *Proceedings of the Royal Society of London*, A151, pp. 540–566, 1935.

[54] S.C. Brenner and L.R. Scott, *The Mathematical Theory of Finite Element Methods*, Springer–Verlag, New York, 1994.

[55] L.C. Brinson, "One-dimensional constitutive behavior of shape memory alloys: Thermomechanical derivation with non-constant material functions and redefined martensite internal variable," *Journal of Intelligent Material Systems and Structures*, 4, pp. 229–242, 1993.

[56] D. Brooks, "Applicability of simplified expressions for design with electrorheological fluids," *Journal of Intelligent Material Systems and Structures*, 4, pp. 409–414, 1993.

[57] M. Brokate and J. Sprekels, *Hysteresis and Phase Transitions*, Springer–Verlag, New York, 1996.

[58] S.A. Brown, C.L. Hom, M. Massuda, J. Prodey, K. Bridger, N. Shankar and S.R. Winzer, "The electro-mechanical behavior of a $Pb(Mg_{1/3},Nb_{2/3})O_3$-$PbTiO_3$-$BaTiO_3$ relaxor ferroelectric: An experimental and analytical study," *Journal of the American Ceramic Society*, 79, pp. 2271–2282, 1996.

[59] W.F. Brown, Jr., *Magnetostatic Principles in Ferromagnetism*, North–Holland Publishing Company, Amsterdam, 1962.

[60] W.F. Brown, Jr., *Magnetoelastic Interactions*, Springer–Verlag, 1966.

[61] W.F. Brown, Jr., "Domain, micromagnetics, and beyond, reminiscences and assessments, *Journal of Applied Physics*, 49, pp. 1937-1942, 1978.

[62] W.J. Buehler, J.V. Gilfrich and R.C. Wiley, "Effect of low-temperature phase changes on the mechanical properties of alloys near composition NiTi," *Journal of Applied Physics*, 34(5), pp. 1475–1477, 1963.

[63] W.J. Buehler and R.C. Wiley, *Naval Ordnance Laboratory Report 61–75*, 1961.

[64] W.J. Buehler and R.C. Wiley, "Nitinols are nonmagnetic, corrosion resistant, hardenable," *Materials in Design Engineering*, 55(2), pp. 82–83, 1962.

[65] B. Bundara, M. Tokuda, B. Kuselj, B. Ule and J.V. Tuma, "Superelastic tension and bending characteristics of shape memory alloys," *Metals and Materials*, 6(4), pp. 293–299, 2000.

[66] S.E. Burke and J.E. Hubbard, Jr., "Spatial filtering concepts in distributed parameter control," *ASME Journal of Dynamic Systems, Measurement, and Control*, 112, pp. 565–573, 1990.

[67] J.A. Burns, "Nonlinear distributed parameter control systems with non-normal linearizations: Applications and approximations," in *Research Directions in Distributed Parameter Systems*, R.C. Smith and M.A. Demetriou, Eds., pp. 17–53, SIAM, Philadelphia, PA, 2003.

[68] B. Burton, D. Nosse and M.J. Dapino, "High power density actuation through Terfenol-D resonant motion and magnetorheological flow control," Smart Structures and Materials 2004, Proceedings of the SPIE, Volume 5390, pp. 104–115, 2004.

[69] S.A. Burton, N. Makris, I. Konstantopoulos and P.J. Antsaklis, "Modeling the response of ER damper: Phenomenology and emulation," *ASCE Journal of Engineering Mechanics*, 122(9), pp. 897–906, 1996.

[70] J.L. Butler, S.C. Butler and A.E. Clark, "Unidirectional magnetostrictive/ piezoelectric hybrid transducer," *Journal of the Acoustical Society of America*, 88(1), pp. 7–11, 1990.

[71] S.C. Butler and F.A. Tito, "A broadband hybrid magnetostrictive/ piezoelectric transducer array," *OCEANS 2000 MTS/IEEE*, Volume 3, pp. 1469–1475, 2000.

[72] S. Büttgenbach, S. Bütefisch, M. Leester-Schädel and A. Wogersien, "Shape memory microactuators," *Microsystem Technologies*, 7, pp. 165–170, 2001.

[73] W.G. Cady, *Piezoelectricity*, McGraw-Hill Book Company, Inc., New York, 1946.

[74] F.T. Calkins and G.W. Butler, "Subsonic jet noise reduction variable geometry chevron," Proceedings of the 42nd AIAA Aerospace Sciences Meeting and Exhibit, 2004.

[75] F.T. Calkins, R.C. Smith and A.B. Flatau, "An energy-based hysteresis model for magnetostrictive transducers," *IEEE Transactions on Magnetics*, 36(2), pp. 429–439, 2000.

[76] W. Cao, S. Tavener and S. Xie, "Simulation of boundary condition influence in a second-order ferroelectric phase transition," *Journal of Applied Physics*, 86(10), pp. 5739–5746, 1999.

[77] M. Capozzoli, J. Gopalakrishnan, K. Hogan, J. Massad, T. Tokarchik, S. Wilmarth, H.T. Banks, K.M. Mossi and R.C. Smith, "Modeling aspects concerning THUNDER actuators," Smart Structures and Materials 1999, Proceedings of the SPIE, Volume 3667, pp. 719–727, 1999.

[78] J.D. Carlson, D.M. Catanzarite and K.A. St. Clair, "Commercial magnetorheological fluid devices," *International Journal of Modern Physics B*, 10(23 & 24), pp. 2857–2865, 1996.

[79] J.D. Carlson and M.R. Jolly, "MR fluid, foam and elastomer devices," *Mechatronics*, 10, pp. 555–569, 2000.

[80] J.D. Carlson and K.D. Weiss, "A growing attraction to magnetic fields," *Machine Design*, 66(15), pp. 61–64, 1994.

[81] M.E. Caspari and W.J. Merz, "The electromechanical behavior of $BaTiO_3$ single-domain crystals," *Physical Review*, 80(6), pp. 1082–1089, 1950.

[82] L.C. Chang and T.A. Read, "Plastic deformation and diffusionless phase changes in metals — the gold-cadmium beta phase," *Transactions of the American Institute of Mining and Metallurgical Engineers*, 191, pp. 47–52, 1951.

[83] I.-W. Chen and Y. Wang, "A domain wall model for relaxor ferroelectrics," *Ferroelectrics*, 206, pp. 245–263, 1998.

[84] J. Chen, H.M. Chan and M.P. Harmer, "Ordering structure and dielectric properties of undoped and La/Na-doped $Pb(Mg_{1/3}\ Nb_{2/3})O_3$," *Journal of the American Ceramic Society*, 72(4), pp. 593–598, 1989.

[85] W. Chen and C.S. Lynch, "A model for simulating polarization switching and AF-F phase changes in ferroelectric ceramics," *Journal of Intelligent Material Systems and Structures*, 9, pp. 427–432, 1998.

[86] W. Chen and C.S. Lynch, "A micro-electro-mechanical model for polarization switching of ferroelectric materials," *Acta Materialia*, 46(15), pp. 5303–5311, 1998.

[87] X. Chen, D.N. Fang and K.C. Hwang, "Micromechanics simulation of ferroelectric polarisation switching," *Acta Materialia*, 45(8), pp. 3181–3189, 1997.

[88] S. Chikazumi, *Physics of Ferromagnetism*, Second Edition, English edition prepared with the assistance of C.D. Graham, Jr., Clarendon Press, Oxford, 1997.

[89] E.K.P. Chong and S.H. Żak, *An Introduction to Optimization*, John Wiley and Sons, New York, 1996.

[90] H. Chopra, C. Ji and V. Kokorin, "Magnetic-field-induced twin boundary motion in magnetic shape-memory alloys," *Physical Review B*, 61, pp. 4913–4915, 2000.

[91] P.G. Ciarlet, *The Finite Element Method for Elliptic Problems*, SIAM Classics in Applied Mathematics, SIAM, Philadelphia, 2002.

[92] A.E. Clark and H.S. Belson, "Giant room-temperature magnetostrictions in $TbFe_2$ and $DyFe_2$," *Physical Review B*, 5, pp. 3642–3644, 1972.

[93] A.E. Clark, J.B. Restorff, M. Wun-Fogle, T.A. Lograsso and D.L. Schlagel, "Magnetostrictive properties of body-centered-cubic Fe-Ga and Fe-Ga-Al alloys," *IEEE Transactions on Magnetics*, 36(5), pp. 3238–3240, 2000.

[94] A.E. Clark, H.T. Savage and M.L. Spano, "Effect of stress on magnetostriction and magnetization of single crystal $Tb_{.27}Dy_{.73}Fe_2$," *IEEE Transactions on Magnetics*, 20, pp. 1443–1445, 1984.

[95] A.E. Clark, M. Wun-Fogle, J.B. Restorff, T.A. Lograsso, A.R. Ross and D.L. Schlagel, "Magnetostrictive Galfenol/Alfenol single crystal alloys under large compressive stresses," Proceedings of the 7th International Conference on New Actuators, H. Borgmann, Ed., Bremen, Germany, pp. 111, 2000.

[96] R.L. Clark, R.A. Burdisso and C.R. Fuller, "Design approaches for shaping polyvinylidene fluoride sensors in active structural acoustic control (ASAC)," *Journal of Intelligent Material Systems and Structures*, 4, pp. 354–365, 1993.

[97] R.L. Clark and S.E. Burke, "Practical limitations in achieving shaped modal sensors with induced strain materials," *Journal of Vibration and Acoustics*, 118, pp. 668–675, 1996.

[98] R.L. Clark and C.R. Fuller, "Modal sensing of efficient acoustic radiators with polyvinylidene fluoride distributed sensors in active structural acoustic control approaches," *Journal of the Acoustical Society of America*, 91(6), pp. 3321–3329, 1992.

[99] R.L. Clark, W.R. Saunders and G.P. Gibbs, *Adaptive Structures: Dynamics and Control*, John Wiley and Sons, New York, 1998.

[100] R.O. Claus, Ed., *Fiber Optic Sensor-Based Smart Materials and Structures*, Institute of Physics Publishing, Bristol, UK, 1992.

[101] A.C.F. Cocks and R.M. McMeeking, "A phenomenological constitutive law for the behavior of ferroelectric ceramics," *Ferroelectrics*, 228, pp. 219–228, 1999.

[102] R.E. Cohen, Ed., *First-Principles Calculations for Ferroelectrics*, American Institute of Physics, Woodbury, NY, 1998.

[103] D.G. Cole and R.L. Clark, "Adaptive compensation of piezoelectric sensoriactuators," *Journal of Intelligent Material Systems and Structures*, 5, pp. 665–672, 1994.

[104] R. Courant and D. Hilbert, *Methods of Mathematical Physics*, Volumes 1 and 2, Springer–Verlag, Berlin, 1937; reprinted by John Wiley and Sons, New York, 1989.

[105] D.J. Craik and R.S. Tebble, *Ferromagnetism and Ferromagnetic Domains*, North–Holland Publishing Company, Amsterdam, 1965.

[106] D.J. Craik and M.J. Wood, "Magnetization changes induced by stress in a constant applied field," *Journal of Applied Physics D: Applied Physics*, 3, pp. 1009–1016, 1970.

[107] D. Croft and S. Devasia, "Vibration compensation for high speed scanning tunneling microscopy," *Review of Scientific Instruments*, 70(12), pp. 4600–4605, 1999.

[108] D. Croft, G. Shed and S. Devasia, "Creep, hysteresis, and vibration compensation for piezoactuators: Atomic force microscopy application," *Journal of Dynamic Systems, Measurement, and Control*, 23, pp. 35–43, 2001.

[109] L.E. Cross, "Relaxor ferroelectrics," *Ferroelectrics,* 76, pp. 241–267, 1987.

[110] L.E. Cross, "Recent developments in piezoelectric ferroelectric materials and composites," Proceedings of the 4th European and 2nd MIMR Conference, G.R. Tomlinson and W.A. Bullough, Eds., Harrogate, UK, July 6-8, pp. 89–97, 1998.

[111] B.D. Cullity, *Introduction to Magnetic Materials*, Addison-Wesley, Reading, MA, 1972.

[112] B. Culshaw, *Optical Fibre Sensing and Signal Processing,* Peter Peregrinus, London 1984.

[113] B. Culshaw, *Smart Structures and Materials*, Artech House, Boston, 1996.

[114] D. Damjanovic and R.E. Newnham, "Electrostrictive and piezoelectric materials for actuator applications," *Journal of Intelligent Material Systems and Structures*, 3, pp. 190–208, 1992.

[115] A. Daniele, S. Salapaka, M.V. Salapaka and M. Dahleh, "Piezoelectric scanners for atomic force microscopes: Design of lateral sensors, identification and control," Proceedings of the America Control Conference, pp. 253–257, 1999.

[116] M.J. Dapino, "On magnetostrictive materials and their use in adaptive structures," *International Journal of Structural Engineering and Mechanics,* 17(3-4), pp. 303–329, 2004.

[117] M.J. Dapino, "Magnetostrictive materials," *Encyclopedia of Smart Materials*, M. Schwartz, Ed., John Wiley and Sons, New York, pp. 600–620, 2002.

[118] M.J. Dapino, F.T. Calkins and A.B. Flatau, "Magnetostrictive devices." *Wiley Encyclopedia of Electrical and Electronics Engineering*, John G. Webster, Ed., John Wiley and Sons, Inc., New York, Volume 12, pp. 278–305, 1999.

[119] M.J. Dapino, R.C. Smith, L.E. Faidley and A.B. Flatau, "A coupled structural-magnetic strain and stress model for magnetostrictive transducers," *Journal of Intelligent Material Systems and Structures*, 11(2), pp. 134–152, 2000.

[120] M.J. Dapino, R.C. Smith and A.B. Flatau, "A structural strain model for magnetostrictive transducers," *IEEE Transactions on Magnetics*, 36(3), pp. 545–556, 2000.

[121] G. DaPrato and P. Grisvard, Maximal regularity for evolution equations by interpolation and extrapolation, *Journal of Functional Analysis*, 58, 1984, pp. 107–124.

[122] G. Daspit, C. Martin, J.-H. Pyo, C. Smith, H. To, K.M. Furati, Z. Ounaies and R.C. Smith, "Model development for piezoelectric polymer unimorphs," Smart Structures and Materials 2002, Proceedings of the SPIE, Volume 4693, pp. 514–524, 2002.

[123] H.F. Davis, *Fourier Series and Orthogonal Functions*, Dover Publications, Inc., New York, 1963.

[124] L.C. Davis, "Model of magnetorheological elastomers," *Journal of Applied Physics*, 85(6), pp. 3348–3351, 1999.

[125] P.J. Davis and P. Rabinowitz, *Numerical Integration*, Blaisdell Publishing Company, Waltham, MA, 1967.

[126] E. Della Torre, *Magnetic Hysteresis*, IEEE Press, New York, 1999.

[127] E. Della Torre and L.H. Bennett, "A Preisach model for aftereffect," *IEEE Transactions on Magnetics*, 34(4), pp 1276–1278, 1998.

[128] E. Della Torre, L.H. Bennett, R.A. Fry and Ó. Alejos, "Preisach–Arrhenius model for thermal aftereffect," *IEEE Transactions on Magnetics*, 38(5), pp. 3409–3416, 2002.

[129] E. Della Torre and F. Vajda, "Parameter identification of the complete moving hysteresis model using major loop data," *IEEE Transactions on Magnetics*, 30(6), pp. 4987–5000, 1994.

[130] R.C.H. del Rosario and R.C. Smith, "Spline approximation of thin shell dynamics," *International Journal for Numerical Methods in Engineering*, 40, pp. 2807–2840, 1997.

[131] R.C.H. del Rosario and R.C. Smith, "LQR control of thin shell dynamics: Formulation and numerical implementation," *Journal of Intelligent Material Systems and Structures*, 9(4), pp. 301–320, 1998.

[132] M.A. Demetriou, "Integrated actuator-sensor placement and hybrid controller design of flexible structures under worst case spatiotemporal disturbance variations," *Journal of Intelligent Material Systems and Structures*, to appear.

[133] R. DesRoches and M. Delemont, "Seismic retrofit of simply supported bridges using shape memory alloys," *Engineering Structures*, 24, pp. 325–332, 2002.

[134] A.F. Devonshire, "Theory of ferroelectrics," *Philosophical Magazine*, 3(10), pp. 85–130, 1954.

[135] J. Dosch, D.J. Inman, and E. Garcia, "A self-sensing piezoelectric actuator for collocated control, *Journal of Intelligent Material Systems and Structures*, 3, pp. 166–185, 1992.

[136] P.R. Downey, M.J. Dapino and R.C. Smith, "Analysis of hybrid PMN/ Terfenol broadband transducers in mechanical series configuration," Smart Structures and Materials 2003, Proceedings of the SPIE, Volume 5049, pp. 168–179, 2003.

[137] M.E. Drougard, "Detailed study of switching current in barium titanate," *Journal of Applied Physics*, 31(2), pp. 352–355, 1960.

[138] T.A. Duenas and G.P. Carman, "Large magnetostrictive response of Terfenol-D resin composites," *Journal of Applied Physics*, 87(9), pp. 4696–4701, 2000.

[139] T.A. Duenas, L. Hsu and G.P. Carman, "Magnetostrictive composite material systems analytical/experimental," Materials Research Society Symposium Proceedings, 459, pp. 527–543, 1997.

[140] T.W. Duerig, "The use of superelasticity in modern medicine," *MRS Bulletin*, 27(2), pp. 101–104, 2002.

[141] T.W. Duerig, A. Pelton and D. Stöckel, "An overview of nitinol medical applications," *Materials Science and Engineering: A, Structural Materials: Properties, Microstructure and Processing*, Volumes 273–275, pp. 149–160, 1999.

[142] J.S. Dugdale, *Entropy and its Physical Meaning*, Taylor and Francis, London, UK, 1996.

[143] E. du Trémolet de Lacheisserie, *Magnetostriction: Theory and Applications of Magnetoelasticity*, CRSC Press, Boca Raton, FL, 1993.

[144] S.J. Dyke, B.F. Spencer, M.K. Sain and J.D. Carlson, "An experimental study of MR dampers for seismic protection," *Smart Materials and Structures*, 7, pp. 693–703, 1998.

[145] C.L. Dym, *Introduction to the Theory of Shells,* Pergamon Press, New York, 1974.

[146] M.A. Ealey and J.F. Washeba, "Continuous facesheet low voltage deformable mirrors," *Optical Engineering*, 29(10), pp. 1191–1198, 1990.

[147] B. Edmonds, Jr., J. Ernstberger, K. Ghosh, J. Malaugh, D. Nfodjo, W. Samyono, X. Xu, D. Dausch, S. Goodwin and R.C. Smith, "Electrostatic operation and curvature modeling for a MEMs flexible film actuator," Smart Structures and Materials 2004, Proceedings of the SPIE, Volume 5383, pp. 134–143, 2004.

[148] R.E. Edwards, *Functional Analysis: Theory and Applications,* Dover, New York, 1995.

[149] K. Elliot. "Titan vibroacoustics," Proceedings of the NASA-Industry Conference on Launch Environments of ELV Payloads, Elkridge, MD, pp. 189–215, 1990.

[150] H.W. Engl, M. Hanke and A. Neubauer, *Regularization of Inverse Problems,* Kluwer Academic Publishers, Dordrecht; Boston, 1996.

[151] D.H. Everett, "A general approach to hysteresis. Part 3: a formal treatment of the independent domain model of hysteresis," *Transactions of the Faraday Society*, 50, pp. 1077–1096, 1954.

[152] D.H. Everett, "A general approach to hysteresis. Part 4: an alternative formulation of the domain model," *Transactions of the Faraday Society*, 51, pp. 1551–1557, 1955.

[153] D.H. Everett and F.W. Smith, "A general approach to hysteresis. Part 2: development of the domain theory," *Transactions of the Faraday Society*, 50, pp. 187–197, 1954.

[154] D.H. Everett and W.I. Whitton, "A general approach to hysteresis," *Transactions of the Faraday Society*, 48, pp. 749–757, 1952.

[155] R. Fahroo and Y. Wang, "Optimal location of piezoceramic actuators for vibration suppression of a flexible structure," Proceedings of the 36th IEEE Conference on Decision and Control, pp. 1966–1971, 1997.

[156] L.E. Faidley, M.J. Dapino, G.N. Washington, R.C. Smith and T.A. Lograsso, "Analytical and experimental issues in Ni-Mn-Ga transducers," Smart Structures and Materials 2003, Proceedings of the SPIE, Volume 5049, pp. 1–12, 2003.

[157] F. Falk, "Model free energy, mechanics, and thermodynamics of shape memory alloys," *Acta Metallurgica*, 28, pp. 1773–1780, 1980.

[158] F. Falk, "Ginzburg–Landau theory of static domain walls in shape-memory alloys," *Zeitschrift für Physik, B, Condensed Matter*, 51, pp. 177–185, 1983.

[159] F. Falk, "One-dimensional model of shape memory alloys," *Archives of Mechanics*, 35(1), pp. 63–84, 1983.

[160] V.N. Fedosov and A.S. Sidorkin, "Quasielastic displacements of domain boundaries in ferroelectrics," *Soviet Physics Solid State*, 18(6), pp. 964–968, 1976.

[161] R.P. Feynman, R.B. Leighton and M. Sands, *Lectures on Physics*, Volume 2, Addison-Wesley Publishing Company, Reading, MA, 1964.

[162] F. Filisko, "Electrorheological materials," *Encyclopedia of Smart Materials*, M. Schwartz, Ed., John Wiley and Sons, New York, pp. 376–391, 2002.

[163] C.A.J. Fletcher, *Computational Galerkin Methods*, Springer–Verlag, New York, 1984.

[164] W. Flügge, *Stresses in Shells,* Second Edition, Springer-Verlag, New York, 1973.

[165] J. Fousek and V. Janoušek, "The contribution of domain-wall oscillations to the small-signal permittivity of triglycine sulphate," *Physica Status Solidi*, 13, pp. 195–206, 1966.

[166] A. Fröhlich, A. Brückner-Foit and S. Weyer, "Effective properties of piezoelectric polycrystals," Smart Structures and Materials 2000, Proceedings of the SPIE, Volume 3992, pp. 279–287, 2000.

[167] S. Fu, Y. Huo and I. Müller, "Thermodynamics of pseudoelasticiy – an analytic approach," *Acta Mechanica*, 99, pp. 1-19, 1993.

[168] M. Fujimoto, *The Physics of Structural Phase Transformations*, Springer–Verlag, New York, 1997.

[169] E. Fukada, "Piezoelectricity of bone and osteogenesis of piezoelectric films," *Mechanisms of Growth Control*, R.O. Becker, Ed., C.C. Thomas, Springfield, IL, 1981.

[170] E. Fukada, "Piezoelectricity and pyroelectricity of biopolymers," *Ferroelectric Polymers: Chemistry, Physics, and Applications,* H.S. Nalwa, Ed., Marcell Dekker, Inc., New York, 1995.

[171] H. Funakubo, Ed., *Shape Memory Alloys*, Translated from the Japanese by J.B. Kennedy, Gordon and Breach Science Publishers, New York, 1984.

[172] W.S. Galinaitis, D.S. Joseph and R.C. Rogers, "Parameter identification for Preisach operators with singular measures," *Physica B*, 306, pp. 149–154, 2001.

[173] W.S. Galinaitis and R.C. Rogers, "Compensation for hysteresis using bivariate Preisach models," Smart Structures and Materials 1997, Proceedings of the SPIE, Volume 3039, pp. 538–547, 1997.

[174] W.S. Galinaitis and R.C. Rogers, "Control of a hysteretic actuator using inverse hysteresis compensation," Smart Structures and Materials 1998, Proceedings of the SPIE, Volume 3323, pp. 267–277, 1998.

[175] K. Gall, H. Sehitoglu, R. Anderson, I. Karaman, Y.I. Chumlyakov and I.V. Kireeva, "On the mechanical behavior of single crystal NiTi shape memory alloys and related polycrystalline phenomenon," *Materials Science and Engineering A*, A317, pp. 85–92, 2001.

[176] D.R. Gamota and F.E. Filisko, "Dynamic mechanical studies of electrorheological materials: Moderate frequencies," *Journal of Rheology*, 35(3), pp. 399–426, 1991.

[177] M.V. Gandhi and B.S. Thompson, *Smart Materials and Structures*, Chapman and Hall, London, 1992.

[178] X. Gao, M. Huang and L.C. Brinson, "A multivariant micromechanical model for SMAs, Part 1: crystallographic issues for single crystal model," *International Journal of Plasticity*, 16, pp. 1345-1369, 2000.

[179] P. Ge and M. Jouaneh, "Modeling hysteresis in piezoceramic actuators," *Precision Engineering*, 17, pp. 211–221, 1995.

[180] P. Ge and M. Jouaneh, "Tracking control of a piezoceramic actuator," *IEEE Transactions on Control Systems Technology*, 4(3), pp. 209–216, 1996.

[181] P. deGennes, K. Okumura, M. Shahinpoor and K. Kim, "Mechanoelectric effects in ionic gels," *Europhysics Letters*, 50(4), pp. 513–518, 2000.

[182] F. Gilletta, P. Lauginie et L. Taurel, "Relaxation diéectrique dans les cristaux de sulfate de glycocolle multidomaines," *Comptes Rendus Hebdomadaires des Séances de L'Académie des Sciences, Série B*, 270, pp. 94–96, 1970.

[183] J.M. Ginder, M.E. Nichols, L.D. Elie and S.M. Clark, "Controllable-stiffness components based on magnetorheological elastomers," Smart Structures and Materials 2000, Proceedings of the SPIE, Volume 3985, pp. 418–425, 2000.

[184] V.L. Ginzburg, *Zhurnal Éksperimental'noĭ i Teoreticheskoĭ Fiziki (JETP)*, 15, p. 739, 1945.

[185] K.R.C. Gisser, M.J. Geselbracht, A. Cappellari, L. Hunsberger, A.B. Ellis, J. Perepezko and G.C. Lisensky, "Nickel-titanium memory metal," *Journal of Chemical Education*, 71(4), pp. 334–340, 1994.

[186] A.E. Glazounov, A.J. Bell and A.K. Tagantsev, "Relaxors as superparaelectrics with distributions of the local transition temperature," *Journal of Physics: Condensed Matter*, 7(21), pp. 4145–4168, 1995.

[187] J. Gonzalo, *Effective Field Approaches to Phase Transitions and Some Applications to Ferroelectrics*, World Scientific, Singapore, 1991.

[188] B.C. Goo and C. Lexcellent, "Micromechanics-based modeling of two-way memory effect of a single crystalline shape-memory alloy," *Acta Materialia*, 45(2), pp. 727–737, 1997.

[189] J.B. Goodenough, "Summary of losses in magnetic materials," *IEEE Transactions on Magnetics*, 38(5), pp. 3398–3408, 2002.

[190] R.B. Gorbet, D.W.L. Wang and K.A. Morris, "Preisach model identification of a two-wire SMA actuator," *IEEE International Conference on Robotics and Automation*, pp. 2161–2167, 1998.

[191] W. Gorsky, "Röntgenugraphische untersuchung von umwandlungen in der legierung CuAu," *Zeitschrift für Physik*, 50, pp. 64–81, 1928.

[192] P.L. Gould, *Finite Element Analysis of Shells of Revolution*, Pitman Advanced Publishing Program, Boston, 1985.

[193] S. Govindjee and G.J. Hall, "A computational model for shape memory alloys," *International Journal of Solids and Structures*, 37, pp. 735–760, 2000.

[194] E.J. Graesser and F.A. Cozzarelli, "Shape memory alloys as new materials for aseismic isolation," *Journal of Engineering Mechanics*, 117(11), pp. 2590–2608, 1991.

[195] A.J. Grodzinsky and J.R. Melcher, "Electromechanical transduction with charged polyelectrolyte membranes," *IEEE Transactions on Biomedical Engineering*, 23, pp. 421–433, 1976.

[196] C.W. Groetsch, *The Theory of Tikhonov Regularization for Fredholm Equations of the First Kind,* Pitman, Boston, 1984.

[197] Y. Gu, R.L. Clark, C.R. Fuller and A.C. Zander, "Experiments on active control of plate vibrations using piezoelectric actuators and polyvinylidene fluoride modal sensors," *ASME Journal of Vibration and Acoustics*, 116, pp. 303–308, 1994.

[198] D. ter Haar, *Elements of Statistical Mechanics,* 3rd Edition, Butterworth-Heinemann, Oxford, 1995.

[199] C.W. Haas and W.F. Jaep, "Domain wall model for ferroelastics," *Physics Letters A*, 49A(1), pp. 77–78, 1974.

[200] G.H. Haertling, "RAINBOW ceramics – a new type of ultra-high-displacement actuator," *American Ceramic Society Bulletin*, 73, pp. 93–96, 1994.

[201] G.H. Haertling, "Chemically reduced PLZT ceramics for ultra-high-displacement actuators, *Ferroelectrics*, 154, pp. 101–106, 1994.

[202] D. Halim and S.O.R. Moheimani, "An optimization approach to optimal placement of collocated piezoelectric actuators and sensors on a thin plate," *Mechatronics*, 13, pp. 27–47, 2003.

[203] C.A. Hall and T.A. Porsching, *Numerical Analysis of Partial Differential Equations*, Prentice Hall, Englewood Cliffs, NJ, 1990.

[204] W.R. Hamilton, "On a general method in dynamics; By which the study of the motions of all free systems of attracting or repelling points is reduced to the search and differentiation of one central relation or characteristic function," *Philosophical Transactions of the Royal Society of London*, 124, pp. 247–308, 1834.

[205] P.C. Hammer, O.P. Marlowe and A.H. Stroud, "Numerical integration over simplexes and cones," *Mathematical Tables and other Aids to Computation*, 10(55), pp. 130–137.

[206] P.K. Hansma, V.B. Elings, O. Marti and C.E. Bracker, "Scanning tunneling microscopy and atomic force microscopy: Application to biology and technology," *Science*, 242, pp. 209–242, 1988.

[207] T. Hao, "Electrorheological fluids," *Encyclopedia of Smart Materials*, M. Schwartz, Ed., John Wiley and Sons, New York, pp. 362–376, 2002.

[208] A. Haraux, "Linear semigroups in Banach spaces," in *Semigroups, Theory and Applications, II*, H. Brezis et al., Eds., Pitman Research Notes in Mathematics, Volume 152, Longman, London, pp. 93–135, 1986.

[209] J. Harrison and Z. Ounaies, "Piezoelectric polymers," *Encyclopedia of Smart Materials*, M. Schwartz, Ed., John Wiley and Sons, New York, pp. 860–873, 2002.

[210] A.G. Hatch, R.C. Smith and T. De, "Model development and control design for high speed atomic force microscopy," Smart Structures and Materials 2004, Proceedings of the SPIE, Volume 5383, pp. 457–468, 2004.

[211] M.J. Haun, E. Furman, S.J. Jang, H.A. McKinstry and L.E. Cross, "Thermodynamic theory of $PbTiO_3$," *Journal of Applied Physics*, 62(8), pp. 3331–3338, 1987.

[212] D. Hebda and S.R. White, "Effect of training conditions and extended thermal cycling on nitinol two-way shape memory behavior," *Smart Materials and Structures*, 4, pp. 298–304, 1995.

[213] C.E. Hecht, *Statistical Thermodynamics and Kinetic Theory*, Dover Publications, New York, 1998.

[214] W. Heisenberg, "Zur theorie des ferromagnetismus," *Zeitschrift für Physik*, 49, pp. 619–636, 1928.

[215] K.O. Hill and G. Meltz, "Fiber Bragg grating technology fundamentals and overview," *Journal of Lightwave Technology*, 15(8), pp. 1263–1276.

[216] A.D. Hilton, D.J. Barber, C.A. Randall and T.R. Shrout, "On short range ordering in the perovskite lead magnesium niobate," *Journal of Materials Science*, 25, pp. 3461–3466, 1990.

[217] P.D. Hislop and I.M Sigal, *Introduction to Spectral Theory with Applications to Schrödinger Operators*, Springer–Verlag, New York 1996.

[218] C.L. Hom, "Simulating electrostrictive deformable mirrors: II. Nonlinear dynamic analysis," *Smart Materials and Structures* 8, pp. 700–708, 1999.

[219] C.L. Hom, P.D. Dean and S.R. Winzer, "Simulating electrostrictive deformable mirrors: I. Nonlinear static analysis," *Smart Materials and Structures* 8, pp. 691–699, 1999.

[220] C.L. Hom and N. Shankar, "A fully coupled constitutive model for electrostrictive ceramic materials," *Journal of Intelligent Material Systems and Structures*, 5, pp. 795–801, 1994.

[221] C.L. Hom and N. Shankar, "Modeling nonlinearity in electrostrictive sonar transducers," *Journal of the Acoustical Society of America*, 104(4), pp. 1903–1913, 1998.

[222] C.L. Hom and N. Shankar, "A constitutive model for relaxor ferroelectrics," Smart Structures and Materials 1999, Proceedings of the SPIE, Volume 3667, pp. 134–144, 1999.

[223] L. Hou and D.S. Grummon, "Transformational superelasticity in sputtered titanium-nickel thin films," *Scripta Metallurgica*, 33(6), pp. 989-995, 1995.

[224] H.-C. Huang and A.S. Usmani, *Finite Element Analysis for Heat Transfer*, Springer–Verlag, Berlin, 1994.

[225] M. Huang and L.C. Brinson, "A multivariant model for single crystal shape memory alloy behavior," *Journal of the Mechanics and Physics of Solids*, 46, pp. 1379–1409, 1998.

[226] J.E. Huber, N.A. Fleck, C.M. Landis and R.M. McMeeking, "A constitutive model for ferroelectric polycrystals," *Journal of the Mechanics and Physics of Solids*, 47, pp. 1663–1697, 1999.

[227] D. Hughes and J.T. Wen, "Preisach modeling of piezoceramic and shape memory alloy hysteresis," *Smart Materials and Structures*, 6, pp. 287–300, 1997.

[228] F.V. Hunt, *Electroacoustics: The Analysis of Transduction, and Its Historical Background*, Published by the American Institute of Physics for the Acoustical Society of America, New York, 1982.

[229] Y. Huo, "A mathematical model for the hysteresis in shape memory alloys," *Continuum Mechanics and Thermodynamics*, 1, pp. 283–303, 1989.

[230] S.C. Hwang, C.S. Lynch and R.M. McMeeking, "Ferroelectric/ferroelastic interactions and a polarization switching model," *Acta Metallurgica et Materialia,* 43(5), pp. 2073–2084, 1995.

[231] M.W. Hyer and A. Jilani, "Predicting the deformation characteristics of rectangular unsymmetrically laminated piezoelectric materials," *Smart Materials and Structures*, 7, pp. 1–8, 1998.

[232] M.W. Hyer and A. Jilani, "Predicting the axisymmetric manufacturing deformations of disk-style benders," *Journal of Intelligent Material Systems and Structures*, 11, pp. 370–381, 2000.

[233] M.W. Hyer and A. Jilani, "Deformation characteristics of circular RAINBOW actuators," *Smart Materials and Structures*, 11, pp. 175–195, 2002.

[234] T. Ikeda, *Fundamentals of Piezoelectricity*, Oxford University Press, Oxford, UK, 1990.

[235] D.J. Inman and S.H.S. Carneiro, "Smart structures, structural health monitoring and crack detection," in *Research Directions in Distributed Parameter Systems*, R.C. Smith and M.A. Demetriou, Eds., pp. 169–186, SIAM, Philadelphia, PA, 2003.

[236] A. Ishida and V. Martynov, "Sputter-deposited shape-memory alloy thin films: Properties and applications," *MRS Bulletin*, 27(2), pp. 111–114, 2002.

[237] E. Ising, "Beitrag zur theorie des ferromagnetismus," *Zeitschrift für Physik*, 31, pp. 253–258, 1925.

[238] Y. Ivshin and T.J. Pence, "A thermomechanical model for a one variant shape memory material," *Journal of Intelligent Material Systems and Structures*, 5, pp. 455–473, 1994.

[239] R.V. Iyer, "Recursive estimation of the Preisach density function for a smart actuator," Smart Structures and Materials 2004, Proceedings of the SPIE, Volume 5383, pp. 202–210, 2004.

[240] R.V. Iyer and P.S. Krishnaprasad, "On a low-dimensional model for ferromagnetism," *Nonlinear Analysis: Theory, Methods and Application,* to appear.

[241] R.V. Iyer and M.E. Shirley, "Hysteresis parameter identification with limited experimental data," *IEEE Transactions on Magnetics*, 40(5), pp. 3227–3239, 2004.

[242] R.V. Iyer, X. Tan, P.S. Krishnaprasad, "Approximate inversion of a hysteresis operator with application to control of smart actuators," *IEEE Transactions on Automatic Control*, to appear.

[243] B. Jaffe, W.R. Cook, Jr. and H. Jaffe, *Piezoelectric Ceramics*, Academic Press, New York, 1971.

[244] R.D. James and M. Wuttig, "Magnetostriction of martensite," *Philosophical Magazine A*, 77(5), pp. 1273–1299, 1998.

[245] H. Janocha, Ed., *Adaptronics and Smart Structures*, Springer–Verlag, Berlin, 1999.

[246] H. Janocha and B. Clephas, "Hybrid actuator with piezoelectric and magnetostrictive material," *Actuator 96*, Proc. 5th International Conference on New Actuators, pp. 304–307, 1996.

[247] D.C. Jiles, *Introduction to Magnetism and Magnetic Materials*, Chapman and Hall, New York, 1991.

[248] D.C. Jiles, "A self-consistent generalized model for the calculation of minor loop excursions in the theory of hysteresis," *IEEE Transactions on Magnetics*, 28(5), pp. 2602–2604, 1992.

[249] D.C. Jiles, "Theory of the magnetomechanical effect," *Journal of Physics D: Applied Physics*, 28, pp. 1537–1546, 1995.

[250] D.C. Jiles and D.L. Atherton, "Theory of the magnetisation process in ferromagnetics and its application to the magnetomechanical effect," *Journal of Physics D: Applied Physics*, 17, pp. 1265–1281, 1984.

[251] D.C. Jiles and D.L. Atherton, "Theory of ferromagnetic hysteresis," *Magnetism and Magnetic Materials*, 61, pp. 48–60, 1986.

[252] D.C. Jiles and J.B. Thoelke, "Theoretical modelling of the effects of anisotropy and stress on the magnetization and magnetostriction of $Tb_{0.3}Dy_{0.7}Fe_2$," *Journal of Magnetism and Magnetic Materials*, 134, pp. 143–160, 1994.

[253] D.C. Jiles, J.B. Thoelke and M.K. Devine, "Numerical determination of hysteresis parameters for the modeling of magnetic properties using the theory of ferromagnetic hysteresis," *IEEE Transactions on Magnetics*, 28(1), pp. 27–35, 1992.

[254] M.R. Jolly, J.W. Bender and J.D. Carlson, "Properties and applications of commercial magnetorheological fluids," Smart Structures and Materials 1998, Proceedings of the SPIE, pp. 262–275, 1998.

[255] M.R. Jolly, J.D. Carlson and B.C. Munoz, "A model of the behavior of magnetorheological materials," *Smart Materials and Structures*, 5, pp. 607–614, 1996.

[256] F. Jona and G. Shirane, *Ferroelectric Crystals*, Dover Publications, Inc., New York, 1993.

[257] T. Kakeshita and K. Ullakko, "Giant magnetostriction in ferromagnetic shape-memory alloys," *MRS Bulletin*, 27(2), pp. 105–109, 2002.

[258] G.M. Kamath and N.M. Wereley, "Nonlinear viscoelastic plastic model for electrorheological fluids," *Smart Materials and Structures,* 6(3), pp. 351–359, 1997.

[259] M. Kamlah and Q. Jiang, "A constitutive model for ferroelectric PZT ceramics under uniaxial loading, *Smart Materials and Structures*, 8, pp. 441–459, 1999.

[260] A. Katchalsky, "Rapid swelling and deswelling of reversible gels of polymeric acids by ionization," *Experientia*, 5, pp. 318–319, 1949.

[261] H. Kawai, "The piezoelectricity of poly(vinylidene fluoride)," *Japanese Journal of Applied Physics*, 8, pp. 975–976, 1969.

[262] A.D. Kersey, "A review of recent developments in fiber optic sensor technology," *Optical Fiber Technology*, 2, pp. 291–317, 1996.

[263] C. Kittel, *Introduction to Solid State Physics*, Fifth Edition, John Wiley and Sons, New York, 1976.

[264] U.F. Kocks, A.S. Argon and M.F. Ashby, *Thermodynamics and Kinetics of Slip*, Volume 19 in *Progress in Materials Science*, B. Chalmers, J.W. Christian and T.B. Massalsk, Eds., Pergamon Press, Oxford, 1975.

[265] M. Kohl, D. Dittmann, E. Quandt and B. Winzek, "Thin film shape memory microvalves with adjustable operation temperature," *Sensors and Actuators A: Physical*, 83(1-3), pp. 214–219, 2000.

[266] N.C. Koon, A.I. Schindler and F.L. Carter, "Giant magnetostriction in cubic rare earth-iron compounds of the type RFe_2," *Physics Letters A*, 37, pp. 412–414, 1971.

[267] N.A. Koratkar and I. Chopra, "Testing and validation of a Froude scaled helicopter rotor model with piezo-bimorph actuated trailing edge flaps," Smart Structures and Materials 1997, Proceedings of the SPIE, Volume 3041, pp. 183–205, 1997.

[268] M.A. Krasnosel'skiĭ and A.V. Pokrovskiĭ, *Systems with Hysteresis*, Nauka, Moscow, 1983; Translated by M. Niezgódka, Springer–Verlag, Berlin, 1989.

[269] P. Krejčí and J. Sprekels, "On a system of nonlinear PDEs with temperature-dependent hysteresis in one-dimensional thermoplasticity," *Journal of Mathematical Analysis and Applications,* 209, pp. 25–46, 1997.

[270] H. Kronmüller and M. Fähnle, *Micromagnetism and the Microstructure of Ferromagnetic Solids*, Cambridge University Press, Cambridge, UK, 2003.

[271] P. Krulevitch, A.P. Lee, P.B. Ramsey, J.C. Trevino, J. Hamilton and M.A. Northrup, "Thin film shape memory alloy microactuators," *Journal of Microelectromechanical Systems*, 5(4), pp. 270–282, 1996.

[272] A. Ktena, D.I. Fotiadis, P.D. Spanos and C.V. Massalas, "A Preisach model identification procedure and simulation of hysteresis in ferromagnets and shape-memory alloys, *Physica B: Condensed Matter*, 306(1–4), pp. 84–90, 2001.

[273] J.N. Kudva, "Overview of the DARPA smart wing project," *Journal of Intelligent Material Systems and Structures*, 15(4), pp. 261–267, 2004.

[274] J.N. Kudva, B. Sanders, J. Pinkerton-Florance and E. Garcia, "The DARPA/AFRL/NASA Smart Wing Program — Final overview," Smart Structures and Materials 2002, Proceedings of the SPIE, Volume 4698, pp. 37–43, 2002.

[275] W. Kuhn, B. Hargitay, A. Katchalsky and H. Eisenberg, "Reversible dilation and contraction by change the state of ionization of high-polyer acid networks," *Nature*, 165, pp. 514–516, 1950.

[276] Y.W. Kwon and H. Bang, *The Finite Element Method Using Matlab*, CRSC Press, Boca Raton, FL, 1997.

[277] J.E. Lagnese and J.-L. Lions, *Modelling Analysis and Control of Thin Plates*, Collection *Recherches en Mathématiques Appliquées*, Masson, Paris, 1989.

[278] D.C. Lagoudas and A. Bhattacharyya, "On the correspondence between micromechanical models for isothermal pseudoelastic response of shape memory alloys and the Preisach model for hysteresis," *Mathematics and Mechanics of Solids*, 2(4), pp. 405–440, 1997.

[279] D.C. Lagoudas and Z. Bo, "Thermomechanical modeling of polycrystalline SMAs under cyclic loading, Part II: material characterization and experimental results for a stable transformation cycle," *International Journal of Engineering Science*, 37, pp. 1141–1173, 1999.

[280] B.D. Laikhtman, "Flexural vibrations of domain walls and dielectric dispersion of ferroelectrics," *Soviet Physics Solid State*, 15(1), pp. 62–68, 1973.

[281] L.D. Landau, "Zur theorie der phasenumwandlungen I, II," *Physikalische Zeitschrift der Sowjetunion*, 11, p. 26 and p. 545, 1937; *Zhurnal Éksperimental'noĭ i Teoreticheskoĭ Fiziki (JETP)*, 7, p. 19 and p. 627, 1937. English translation: "On the theory of phase transitions," pp. 193–216, *Collected Papers of L.D. Landau*, D. ter Haar, Ed., Published jointly by Gordan and Breach and Pergamon Press, New York, 1965.

[282] L.D. Landau and E.M. Lifshitz, "On the theory of the dispersion of magnetic permeability in ferromagnetic bodies," *Physikalische Zeitschrift der Sowjetunion*, 8, pp. 153–169, 1935.

[283] L.D. Landau and E.M. Lifshitz, *Statistical Physics, Part 1*, 3rd Edition, Translated by J.B. Sykes and M.J. Kearsley, Butterworth and Heinemann, Oxford, UK, 1982.

[284] L.D. Landau, E.M Lifshitz and L.P. Pitaevskii, *Electrodynamics of Continuous Media,* 2nd Edition, Translated by J.B. Sykes, J.S. Bell and M.J. Kearsley, Butterworth and Heinemann, Oxford, UK, 1982.

[285] C.M. Landis, "Non-linear constitutive modeling of ferroelectrics," *Current Opinion in Solid State and Materials Science*, 8, pp. 59–69, 2004.

[286] C.M. Landis, "On the strain saturation conditions for polycrystalline ferroelastic materials," *Journal of Applied Mechanics*, 70, pp. 470–478, 2003.

[287] S. Lang, *Real and Functional Analysis,* Springer–Verlag, New York, 1993.

[288] C.K. Lee and F.C. Moon, "Modal sensors/actuators," *Journal of Applied Mechanics*, 56, pp. 434–441, 1990.

[289] E.W. Lee and J.E.L. Bishop, "Magnetic behavior of single-domain particles," *Proceedings of the Physical Society*, 89, pp. 661–675, 1966.

[290] Y. Leino and J. Pitkäranta, "On the membrane locking of h–p finite elements in a cylindrical shell problem," *International Journal for Numerical Methods in Engineering*, 37, pp. 1053–1070, 1994.

[291] A.W. Leissa, *Vibration of Plates*, NASA SP-160, 1969, Reprinted by the Acoustical Society of America through the American Institute of Physics, 1993.

[292] A.W. Leissa, *Vibration of Shells*, NASA SP-288, 1973, Reprinted by the Acoustical Society of America through the American Institute of Physics, 1993.

[293] A.P. Levanyuk, "Interaction between ferroelastic domain walls mediated by thermal fluctuations and defects," *Phase Transitions*, B55(1–4), pp. 127–133, 1995.

[294] G. Li, E. Furman and G.H. Haertling, "Stress-enhanced displacements in PLZT rainbow actuators, *Journal of the American Ceramic Society*, 80, pp. 1382–1388, 1997.

[295] X. Li, W.Y. Shih, I.A. Aksay and W.-H. Shih, "Electromechanical behavior of PZT-brass unimorphs," *Journal of the American Ceramic Society*, 82(7), pp. 1733–1740, 1999.

[296] Y.L. Li, S.Y. Hu, Z.K. Liu and L.Q. Chen, "Phase-field model of domain structures in ferroelectric thin films," *Applied Physics Letters*, 78(24), pp. 3878–3880, 2001.

[297] C. Liang and C.A. Rogers, "One-dimensional thermomechanical constitutive relations for shape memory materials, *Journal of Intelligent Material Systems and Structures*, 1, pp. 207–234, 1990.

[298] M.E. Lines and A.M Glass, *Principles and Applications of Ferroelectrics and Related Materials*, Oxford University Press, Oxford, UK, 1977, Oxford Classics Series, 2001.

[299] F. Liorzou, B. Phelps and D.L. Atherton, "Macroscopic models of magnetization," *IEEE Transactions on Magnetics*, 36(2), pp. 418–428, 2000.

[300] E.A. Little, "Dynamic behavior of domain walls in barium titanate," *Physical Review*, 98(4), pp. 978–984, 1955.

[301] A.E.H. Love, *A Treatise on the Mathematical Theory of Elasticity*, Cambridge University Press, Fourth Edition, 1927.

[302] W. Lu, A. Fadeev, B. Qi, E. Smela, B. Mattes, J. Ding, G. Spinks, J. Mazurkiewicz, D. Zhou, G. Wallace, D. MacFarlane, S. Forsyth and M. Forsyth, "Use of ionic liquids for Pi-conjugated polymer electrochemical devices," *Science*, 297, pp. 983–987, 2002.

[303] W. Lu, D.-N. Fang and K-C. Hwang, "Nonlinear electric-mechanical behavior and micromechanics modelling of ferroelectric domain evolution," *Acta Materialia*, 47(10), pp. 2913–2926, 1999.

[304] W. Lu and G.J. Weng, "A self-consistent model for the stress-strain behavior of shape-memory alloy polycrystals," *Acta Materialia*, 46, pp. 5423–5433, 1998.

[305] J. Luan and F.C. Lee, "Design of a high frequency switching amplifier for smart material actuators with improved current mode control," PESC '98 Record, 29th Annual Power Electronics Specialists Conference, Volume 1, pp. 59–64, 1998.

[306] A. Ludwig and E. Quandt, "Giant magnetostrictive thin films for applications in microelectromechanical systems," *Journal of Applied Physics*, 87(9), pp. 4691–4695, 2000.

[307] D.G. Luenberger, *Optimization by Vector Space Methods*, John Wiley and Sons, New York, 1968.

[308] H.V. Ly, R. Reitich, M.R. Jolly, H.T. Banks and K. Ito, "Simulations of particle dynamics in magnetorheological fluids," *Journal of Computational Physics*, 155, pp. 160–177, 1999.

[309] C.S. Lynch, "The effect of uniaxial stress on the electro-mechanical response of 8/65/35 PLZT, *Acta Materialia*, 44(10), pp. 4137–4148, 1996.

[310] C.C. Ma, R. Wang, et. al., "Frequency response of TiNi shape memory alloy thin film micro-actuators," Proceedings of the IEEE, Thirteenth Annual International Conference on Micro Electro Mechanical Systems, pp. 370–374, 2000.

[311] J.H. Mabe, R.T. Ruggeri, E. Rosenzweig, C.-J. Yu, "Nitinol performance characterization and rotary actuator design," Smart Structures and Materials 2004, Proceedings of the SPIE, Volume 5388, pp. 95–109, 2004.

[312] J.W. Macki, P. Nistri and P. Zecca, "Mathematical models for hysteresis," *SIAM Review*, 35, pp. 94–123, 1993.

[313] J.A. Main and E. Garcia, "Design impact of piezoelectric actuator nonlinearities," *Journal of Guidance, Control, and Dynamics*, 20(2), pp. 327–332, 1997.

[314] J.A. Main and E. Garcia, "Piezoelectric stack actuators and control system design: strategies and pitfalls," *Journal of Guidance, Control, and Dynamics*. 20(3), pp. 479–485, 1997.

[315] J.A. Main, E. Garcia and D.V. Newton, "Precision position control of piezoelectric actuators using charge feedback," *Journal of Guidance, Control, and Dynamics*. 18(5), pp. 1068–73, 1995.

[316] J.A. Main, D. Newton, L. Massengil and E. Garcia, "Efficient power amplifiers for piezoelectric applications," *Smart Materials and Structures*, 5(6), pp. 766–775, 1996.

[317] N. Makris, S.A. Burton, D. Hill and M. Jordan, "Analysis and design of ER damper for seismic protection of structures," *ASCE Journal of Engineering Mechanics*, 122(10), pp. 1003–1011, 1996.

[318] S. Markuš, *The Mechanics of Vibrations of Cylindrical Shells,* Elsevier, New York, 1988.

[319] J.E. Marsden and T.S. Ratiu, *Introduction to Mechanics and Symmetry*, Second Edition, Springer–Verlag, New York, 1999.

[320] J.E. Martin and R.A. Anderson, "Electrostriction in field-structured composites: Basis for a fast artificial muscle?" *Journal of Chemical Physics*, 9(1), pp. 4278–4280, 1999.

[321] J.E. Massad, *Macroscopic Models for Shape Memory Alloy Characterization and Design*, PhD Dissertation, North Carolina State University, Raleigh, NC, 2003.

[322] J.E. Massad and R.C. Smith, "A domain wall model for hysteresis in ferroelastic materials," *Journal of Intelligent Material Systems and Structures*, 14(7), pp. 455–471, 2003.

[323] J.E. Massad and R.C. Smith, "A homogenized free energy model for hysteresis in thin-film shape memory alloys," CRSC Technical Report CRSC-TR04-26; *International Journal on the Science and Technology of Condensed Matter Films*, to appear.

[324] J.E. Massad, R.C. Smith and G.P. Carman, "A free energy model for thin-film shape memory alloys," Smart Structures and Materials 2003, Proceedings of the SPIE, Volume 5049, pp. 13–23, 2003.

[325] W.P. Mason, *Piezoelectric Crystals and Their Application to Ultrasonics*, D. Van Nostrand Company, Inc., New York, 1950.

[326] W.P. Mason, *Physical Acoustics and the Properties of Solids,* D. Van Nostrand Company, Inc., New York, 1958.

[327] W.P. Mason, "Piezoelectricity, its history and applications," *The Journal of the Acoustical Society of America,* 70, pp. 1561–1566, 1981.

[328] Y. Matsuzaki, K. Funami and H. Naito, "Inner loops of pseudoelastic hysteresis of shape memory alloys: Preisach approach," Smart Structures and Materials 2002, Proceedings of the SPIE, Volume 4699, pp. 355–364, 2002.

[329] L.D. Mauck and C.S. Lynch, "Thermo-electro-mechanical behavior of ferroelectric materials. Part I: A computational micromechanical model versus experimental results," *Journal of Intelligent Material Systems and Structures,* 14, pp. 587–602, 2003.

[330] L.D. Mauck and C.S. Lynch, "Thermo-electro-mechanical behavior of ferroelectric materials. Part II: Introduction of rate and self-heating effects," *Journal of Intelligent Material Systems and Structures,* 14, pp. 605–621, 2003.

[331] G.A. Maugin, *The Thermodynamics of Nonlinear Irreversible Behaviors,* World Scientific Publishing Company, Singapore, 1999.

[332] I.D. Mayergoyz, "Mathematical models of hysteresis," *Physical Review Letters*, 56(15), pp. 1518–1521, 1986.

[333] I.D. Mayergoyz, "Mathematical models of hysteresis," *IEEE Transactions on Magnetics*, 22(5), pp. 603–608, 1986.

[334] I.D. Mayergoyz, *Mathematical Models of Hysteresis,* Springer–Verlag, New York, 1991.

[335] I.D. Mayergoyz, *Nonlinear Diffusion of Electromagnetic Fields with Applications to Eddy Currents and Superconductivity*, Academic Press, New York, 1998.

[336] A.-M.R. McGowan, Ed., *Smart Structures and Materials 2002: Industrial and Commercial Applications of Smart Structures Technologies,* SPIE, Volume 4698, 2002.

[337] B.Z. Mei, J.I. Schienbeim and B.A. Newman, "The ferroelectric behavior of odd-numbered nylons," *Ferroelectrics*, 144, pp. 51–60, 1993.

[338] W.J. Merz, "Double hysteresis loop of $BaTiO_3$", *Physical Review*, 91(3), pp. 513–517, 1953.

[339] W.J. Merz, "Domain formation and domain wall motions in ferroelectric $BaTiO_3$ single crystals," *Physical Review*, 95(3), pp. 690–698, 1954.

[340] J.E. Miesner and J.P. Teter, "Piezoelectric/magnetostrictive resonant inchworm motor," Proceedings of SPIE, Smart Structures and Materials, Volume 2190, pp. 520–527, 1994.

[341] D.W. Miller, S.A. Collins and S.P. Peltzman, "Development of spatially convolving sensors for structural control applications," *AIAA Paper 90-1127-CP*, pp. 2283–2297, 1990.

[342] R.C. Miller, "Some experiments on the motion of 180° domain walls in $BaTiO_3$," *Physical Review*, 111(3), pp. 736–739, 1958.

[343] R.C. Miller and A. Savage, "Motion of 180° domain walls in metal electroded barium titanate crystals as a function of electric field and sample thickness," *Journal of Applied Physics*, 31(4), pp. 662–669, 1960.

[344] R.C. Miller and G. Weinreich, "Mechanism for the sideways motion of 180° domain walls in barium titanate," *Physical Review*, 117(6), pp. 1460–1466, 1960.

[345] T. Mitsui, I. Tatsuzaki and E. Nakamura, *An Introduction to the Physics of Ferroelectrics*, Gordon and Breach Science Publishers, New York, 1976.

[346] M. Mohebi, N. Jamasbi and J. Liu, "Simulation of the formation of nonequilibrium structures in magnetorheological fluids subject to an external magnetic field," *Physical Review*, 54, pp. 5407–5413, 1996.

[347] W.W. Morey, G. Meltz and J.M. Weiss, "Recent advances in fiber grating sensors for utility industry applications," Self Calibrated Intelligent Optical Sensors and Systems, Proceedings of the SPIE, Volume 2594, pp. 90–98, 1995.

[348] K.M. Mossi, R.P. Bishop, R.C. Smith and H.T. Banks, "Evaluation criteria for THUNDER actuators," Smart Structures and Materials 1999, Proceedings of the SPIE, Volume 3667, pp. 738–743, 1999.

[349] K. Mossi, Z. Ounaies, R. Smith and B. Ball, "Prestressed curved actuators: Characterization and modeling of their piezoelectric behavior," Smart Structures and Materials 2003, Proceedings of the SPIE, Volume 5053, pp. 423–435, 2003.

[350] K.M. Mossi, G.V. Selby and R.G. Bryant, "Thin-layer composite unimorph ferroelectric driver and sensor properties," *Materials Letters*, 35, pp. 39–49, 1998.

[351] A.J. Moulson and J.M. Herbert, *Electroceramics: Materials, Properties, Applications*, Chapman and Hall, New York, 1990.

[352] V. Mueller, A. Fuith, J. Fousek, H. Warhanek and H. Beige, "Spontaneous strain in ferroelastic incommensurate $[N(CH_3)_4]_2CuCl_4$ crystals," *Solid State Communications*, 104(8), pp. 455–458, 1997.

[353] I. Müller and K. Wilmanski, "A model for a pseudoelastic body," *Il Nuovo Cimento della Societa Italiana di Fisica: B*, 57, pp. 283–318, 1980.

[354] K.A. Murphy, M.F. Gunther, A.M. Vengsarkar and R.O. Claus, "Fabry-Perot fiber-optic sensors in full-scale fatigue testing on an F-15 aircraft," *Applied Optics*, 31(4), pp. 431–433, 1992.

[355] S. Nambu and D.A. Sagala, "Domain formation and elastic long-range interactions in ferroelectric perovskites," *Physical Review B*, 50(9), pp. 5838–5849, 1994.

[356] A.W. Naylor and G.R. Sell, *Linear Operator Theory in Engineering and Science*, Springer–Verlag, New York, 2000.

[357] J.M. Nealis and R.C. Smith, "Nonlinear adaptive parameter estimation algorithms for hysteresis models of magnetostrictive actuators," Smart Structures and Materials 2002, Proceedings of the SPIE, Volume 4693, pp. 25–36, 2002.

[358] J.M. Nealis and R.C. Smith, "$\mathcal{H}_\infty$ Control Design for a Magnetostrictive Transducer," Proceedings of the 42nd IEEE Conference on Decision and Control, pp. 1801–1806, 2003.

[359] J.M. Nealis and R.C. Smith, "Model-Based Robust Control Design for Magnetostrictive Transducers Operating in Hysteretic and Nonlinear Regimes," CRSC Technical Report CRSC-TR03-25; *IEEE Transactions on Control Systems Technology*, submitted.

[360] S. Nemat-Nasser, "Micromechanics of actuation of ionic polymer-metal composites," *Journal of Applied Physics*, 92(5), pp. 2899–2915, 2002.

[361] K.M. Newbury and D.J. Leo, "Electromechanical modeling and characterization of ionic polymer benders," *Journal of Intelligent Material Systems and Structures*, 13, pp. 51–60, 2002.

[362] R.E. Newnham, "Electroceramics," *Reports on Progress in Physics*, 52, pp. 123–156, 1989.

[363] R.E. Newnham, "Molecular mechanisms in smart materials," *Materials Research Society Bulletin*, Volume XXII, No. 5, pp. 20–34, 1997.

[364] V.V. Novozhilov, *Thin Shell Theory*, Second Augmented and Revised Edition, Translated from the Second Russian Edition by P.G. Lowe, Edited by J.R.M. Radok, P. Noordhoff Ltd., Groningen, The Netherlands, 1964.

[365] J.F. Nye, *Physical Properties of Crystals: Their Representation by Tensors and Matrices*, Oxford Press, London, 1957.

[366] K. Oguro, Y. Kawami and H. Takenaka, "Bending of an ion-conducting polymer film-electrode composite by an electric stimulus at low voltage," *Journal of the Micromachine Society*, 5, pp. 27–30, 1992.

[367] R.C. O'Handley, S.J. Murray, M. Marioni, H. Nembach, and S.M. Allen, "Phenomenology of giant magnetic-field-induced strain in ferromagnetic shape-memory materials," *Journal of Applied Physics*, 87(9), pp. 4712–4717, 2000.

[368] A. Ölander, *Zeitschrift für Kristallographie, Kristallgeometrie, Kristallphysik, Kristallchemie*, 83, pp. 145–148, 1932.

[369] J. Ortín, "Preisach modeling of hysteresis for a pseudoelastic Cu-Zn-Al single crystal," *Journal of Applied Physics*, 71(3), pp. 1454–1461, 1992.

[370] K. Otsuka and T. Kakeshita, "Science and technology of shape-memory alloys: New developments," *MRS Bulletin*, 27(2), pp. 91–100, 2002.

[371] Z. Ounaies, J.A. Young and J.S. Harrison, "An overview of the piezoelectric phenomenon in amorphous polymers," in *Field Responsive Polymers: Electroresponsive, Photoresponsive, and Responsive Polymers in Chemistry and Biology,* I.M. Khan and J.S. Harrison, Eds., pp. 88–103, ACS Symposium Series 726, American Ceramical Society, Washington, DC 1999.

[372] J.A. Palmer, B. Dessent, J.F. Mulling, T. Usher, E. Grant, J.W. Eischen, A.I. Kingon, "The design and characterization of a novel piezoelectric transducer-based linear motor," *IEEE/ASME Transactions on Mechatronics*, 9(2), pp. 392–398, 2004.

[373] Q. Pan and R.D. James, "Micromagnetic study of Ni_2MnGa under applied field," *Journal of Applied Physics*, 87(9), pp. 4702–4706, 2000.

[374] N. Papenfuß and S. Seelecke, "Simulation and control of SMA actuators," Smart Structures and Materials 1999, Proceedings of the SPIE, Volume 3667, pp. 586–595, 1999.

[375] M. Parthasarathy and D.J. Klingenberg, "Electrorheology: mechanisms and models," *Material Science and Engineering*, R17, pp. 57–103, 1996.

[376] R.K. Pathria, *Statistical Mechanics*, Pergamon Press, Oxford, 1972.

[377] E. Patoor and M. Berveiller, "Micromechanical modeling of the thermodynamic behavior of shape memory alloys," *Mechanics of Solids with Phase Changes*, M. Berveiller and F.D. Fischer, Eds., Springer, New York, pp. 121–188, 1997.

[378] B.F. Phelps, F. Liorzou and D.L. Atherton, "Inclusive model of ferromagnetic hysteresis," *IEEE Transactions on Magnetics*, 38(2), pp. 1326–1332, 2002.

[379] S.M. Pilgrim, M. Massuda, J.D. Prodey, and A.P. Ritter, "Electrostrictive sonar drivers for flextensional transducers," *Transducers for Sonics and Ultrasonics*, M. McCollum, B.F. Hamonic, and O.B. Wilson, Eds., Lancaster PA, Technomic, 1993.

[380] J. Pitkäranta, "The problem of membrane locking in finite element analysis of cylindrical shells," *Numerische Mathematik*, 61, pp. 523–542, 1992.

[381] K.C. Pitman, "The influence of stress on ferromagnetic hysteresis," *IEEE Transactions on Magnetics*, 26(5), pp. 1978–1980, 1990.

[382] F. Preisach, "Uber die magnetische nachwirkung," *Zeitschrift fur Physik*, 94, pp. 277–302, 1935.

[383] P.M. Prenter, *Splines and Variational Methods*, John Wiley and Sons, New York, 1975.

[384] G. Puglisi and L. Truskinovsky, "Mechanics of a discrete chain with bi-stable elements," *Journal of the Mechanics and Physics of Solids*, 48, pp. 1–27, 2000.

[385] G. Puglisi and L. Truskinovsky, "Rate independent hysteresis in a bi-stable chain," *Journal of the Mechanics and Physics of Solids*, 50, pp. 165–187, 2002.

[386] G. Puglisi and L. Truskinovsky, "A mechanism of transformational plasticity," *Continuum Mechanics and Thermodynamics*, 14(5), pp. 437–457, 2002.

[387] J. Rabinow, "The magnetic fluid clutch," *AIEE Transactions*, 67, pp. 1308–1315, 1948.

[388] J.K. Raye and R.C. Smith, "A temperature-dependent model for relaxor ferroelectric compounds," Smart Structures and Materials 2004, Proceedings of the SPIE, Volume 5383, pp. 1–10, 2004.

[389] B.D. Reddy, *Introductory Functional Analysis*, Springer–Verlag, New York, 1998.

[390] J.N. Reddy, *An Introduction to the Finite Element Method*, McGraw–Hill Book Company, New York, 1984.

[391] J.B. Restorff, "Magnetostrictive materials and devices," *Encyclopedia of Applied Physics*, Volume 9, pp. 229–244, 1994.

[392] J.B. Restorff, H.T. Savage, A.E. Clark and M. Wun-Fogle, "Preisach modeling of hysteresis in Terfenol-D," *Journal of Applied Physics*, 67(9), pp. 5016–5018, 1996.

[393] W.P. Robbins and D.E. Glumac, "A planar unimorph-based actuator with large vertical displacement capability. Part II: theory," *IEEE Transactions on Ultrasonics, Ferroelectrics and Frequency Control,* 45(5), pp. 1151–1159, 1998.

[394] G. Robert, D. Damjanovic and N. Setter, "Preisach modeling of piezoelectric nonlinearity in ferroelectric ceramics," *Journal of Applied Physics*, 89(9), pp. 5067–5074, 2001.

[395] J. Rödel and W.S. Kreher, "Modeling of domain wall contributions to the effective properties of polycrystalline ferroelectric ceramics," Smart Structures and Materials 2000, Proceedings of the SPIE, Volume 3992, pp. 353–362, 2000.

[396] N.N. Rogacheva, *The Theory of Piezoelectric Shells and Plates*, CRC Press, Boca Raton, FL, 1994.

[397] A.L. Roytburd, "Elastic domains and polydomain phases in solids," *Phase Transitions*, B45(1), pp. 1–34, 1993.

[398] D. Rugar, O. Züger, S.T. Hoen, C.S. Yannoni, H.-M. Vieth and R.D. Kendrick, "Force detection of nuclear magnetic resonance," *Science*, 264, pp. 1560–1563, 1994.

[399] K. Sadeghipour, R. Salomon and S. Neogi, "Development of a novel electrochemically active membrane and smart material based vibration sensor/damper," *Smart Materials and Structures*, 1(2), pp. 172–179, 1992.

[400] M.V. Salapaka, H.S. Bergh, J. Lai, A. Majumdar and E. McFarland, "Multimode noise analysis of cantilevers for scanning probe microscopy," *Journal of Applied Physics*, 81(6), pp. 2480–2487, 1997.

[401] S. Salapaka, A. Sebastian, J.P. Cleveland and M.V. Salapaka, "High bandwidth nano-positioner: A robust control approach," *Review of Scientific Instruments*, 73(9), pp. 3232–3241, 2002.

[402] E.K.H. Salje, *Phase Transitions in Ferroelastic and Co-elastic Crystals*, Cambridge University Press, Cambridge, UK, 1990; Student Edition, 1993.

[403] L. Sandlund, M. Fahlander, T. Cedell, A.E. Clark, J.B. Restorff and M. Wun-Fogle, "Magnetostriction, elastic moduli, and coupling factors of composite Terfenol-D," *Journal of Applied Physics*, 75(10), pp. 5656–5658, 1994.

[404] R.L. Sarno and M.E. Franke, "Suppression of flow induced pressure oscillations in cavities," *Journal of Aircraft*, 31(1), pp. 90–96, 1994.

[405] G. Schitter, P. Menold, H.F. Knapp, F. Allgower and A. Stemmer, "High performance feedback for fast scanning atomic force microscopes," *Review of Scientific Instruments*, 72(8), pp. 3320–3327, 2001.

[406] L. Schumaker, *Spline Functions: Basic Theory*, John Wiley and Sons, New York, 1981.

[407] J. Schweiger, "Aircraft control applications of smart structures," *Encyclopedia of Smart Materials*, M. Schwartz, Ed., John Wiley and Sons, New York, pp. 42–59, 2002.

[408] A. Sebastion and S. Salapaka, "H_∞ loop shaping design for nano-positioning," Proceedings of the American Control Conference, pp. 3708–3713, 2003.

[409] S. Seelecke, "Modeling the dynamic behavior of shape memory alloys," *International Journal of Non-Linear Mechanics*, 37, pp. 1363–1374, 2002.

[410] S. Seelecke, "A fully coupled thermomechanical model for shape memory alloys, Part I: theory," *Journal of the Mechanics and Physics of Solids,* submitted.

[411] S. Seelecke and O. Heintze, "A model for the strain-rate dependent inner loop behavior of NiTi," Preprint.

[412] S. Seelecke and O. Kastner, "A fully coupled thermomechanical model for shape memory alloys, Part II: numerical simulation," *Journal of the Mechanics and Physics of Solids,* submitted.

[413] S. Seelecke and I. Müller, "Shape memory alloy actuators in smart structures — Modeling and simulation," *ASME Applied Mechanics Reviews*, 57(1), pp. 23–46, 2004.

[414] D. Segalman, W. Witkowski, D. Adolf and M. Shahinpoor, "Theory of electrically controlled polymeric muscles as active materials in adaptive structures," *Smart Materials and Structures*, 1(1), pp. 95–100, 1992.

[415] T.I. Seidman and C.R. Vogel, "Well posedness and convergence of some regularisation methods for non-linear ill posed problems," *Inverse Problems*, 5, pp. 227-238, 1989.

[416] N. Setter and L.E. Cross, *Journal of Applied Physics*, "The role of B-site cation disorder in diffuse phase transition behavior of perovskite ferroelectrics," 51(8), pp. 4356–4360, 1980.

[417] A. Shadowitz, *The Electromagnetic Field*, Dover Publications, New York, 1975.

[418] M. Shahinpoor, "Conceptual design, kinematics and dynamics of swimming robotic structures using ionic polymeric gel muscles," *Smart Materials and Structures*, 1(1), pp. 91–94, 1992.

[419] M. Shahinpoor, "Continuum electromechanics of ionic polymeric gels as artificial muscles for robotic applications," *Smart Materials and Structures*, 3, pp. 367–372, 1994.

[420] M. Shahinpoor, "Micro-electro-mechanics of ionic polymeric gels as electrically controllable artificial muscles," *Journal of Intelligent Material Systems and Structures*, 6(3), pp. 307–314, 1995.

[421] M. Shahinpoor, Y. Bar-Cohen, J. Simpson and J. Smith, "Ionic polymer-metal composites (IPMCs) as biomimetic sensors, actuators and artificial muscles — a review," *Smart Materials and Structures*, 7, pp. R15–R30, 1998.

[422] I.H. Shames and C.L. Dym, *Energy and Finite Element Methods in Structural Mechanics*, Taylor and Francis, New York, 1991.

[423] N. Shankar and C.L. Hom, "An acoustic/thermal model for self-heating in PMN sonar projectors," *Journal of the Acoustical Society of America*, 108, pp. 2151–2158, 2000.

[424] J.A. Shaw, B.-C. Chang, M.A. Iadicola and Y.M. Leroy," "Thermodynamics of a 1-D shape memory alloy: modeling, experiments, and application," Smart Structures and Materials 2003, Proceedings of the SPIE, Volume 5049, pp. 76–87, 2003.

[425] J.A. Shaw and S. Kyriakides, "Thermomechanical aspects of NiTi," *Journal of the Mechanics and Physics of Solids*, 43(1), pp. 1243–1281, 1995.

[426] D.D. Shin and G.P. Carman, "Operating frequency of thin film NiTi in fluid media," Proceedings of the 2001 ASME IMECE, MEMS Symposium, pp. 221–227, 2001.

[427] M.E. Shirley and R. Venkataraman, "On the identification of Preisach measures," Smart Structures and Materials 2003, Proceedings of the SPIE, Volume 5049, pp. 326–336, 2003.

[428] T.M. Simon, F. Reitich, M.R. Jolly, K. Ito and H.T. Banks, "Estimation of the effective permeability in magnetorheological fluids," *Journal of Intelligent Material Systems and Structures*, 10, pp. 872–879, 1999.

[429] T.M. Simon, F. Reitich, M.R. Jolly, K. Ito and H.T. Banks, "On the effective magnetic properties of magnetorheological fluids," *Mathematical and Computer Modeling*, 33, pp. 273–284, 2001.

[430] R.C. Smith, "A Galerkin method for linear PDE systems in circular geometries with structural acoustic applications," *SIAM Journal on Scientific Computing*, 18(2), pp. 371–402, 1997.

[431] R.C. Smith, "Inverse compensation for hysteresis in magnetostrictive transducers," *Mathematical and Computer Modelling*, 33, pp. 285–298, 2001.

[432] R.C. Smith, C. Bouton and R. Zrostlik, "Partial and full inverse compensation for hysteresis in smart material systems," Proceedings of the 2000 American Control Conference, pp 2750–2754, 2000.

[433] R.C. Smith and M.J. Dapino, "A homogenized energy theory for ferromagnetic hysteresis," *IEEE Transactions on Magnetics*, submitted.

[434] R.C. Smith, M.J. Dapino and S. Seelecke, "A free energy model for hysteresis in magnetostrictive transducers," *Journal of Applied Physics*, 93(1), pp. 458–466, 2003.

[435] R.C. Smith and A. Hatch, "Parameter estimation techniques for nonlinear hysteresis models," Smart Structures and Materials 2004, Proceedings of the SPIE, Volume 5383, pp. 155–163, 2004.

[436] R.C. Smith, A. Hatch and T. De, "Model development for piezoceramic nanopositioners," Proceedings of the 42nd IEEE Conference on Decision and Control, pp. 2638–2643, 2003.

[437] R.C. Smith, A. Hatch, B. Mukherjee and S. Liu, "A homogenized energy model for hysteresis in ferroelectric materials: General density formulation," CRSC Technical Report CRSC-TR04-23; *Journal of Intelligent Material Systems and Structures*, to appear.

[438] R.C. Smith and C.L. Hom, "Domain wall theory for ferroelectric hysteresis," *Journal of Intelligent Material Systems and Structures*, 10(3), pp. 195–213, 1999.

[439] R.C. Smith and C.L. Hom, "A temperature-dependent hysteresis model for relaxor ferroelectrics," Smart Structures and Materials 2000, Proceedings of the SPIE, Volume 3992, pp. 267–278, 2000.

[440] R.C. Smith and H.L. Hom, "A temperature-dependent constitutive model for relaxor ferroelectrics," CRSC Technical Report CRSC-TR00-26; *Journal of Intelligent Material Systems and Structures*, to appear.

[441] R.C. Smith and J.E. Massad, "A unified methodology for modeling hysteresis in ferroic materials," Proceedings of the 2001 ASME Design Engineering Technical Conferences and Computers and Information in Engineering Conference, Vol 6, Pt B, pp. 1389–1398, 2001.

[442] R.C. Smith and Z. Ounaies, "A domain wall model for hysteresis in piezoelectric materials," *Journal of Intelligent Material Systems and Structures*, 11(1), pp. 62–79, 2000.

[443] R.C. Smith and Z. Ounaies, "Model development for high performance piezoelectric polymers," Material Research Society Symposium Proceedings Volume 698, pp. 217–222, 2002.

[444] R.C. Smith, Z. Ounaies and R. Wieman, "A model for rate-dependent hysteresis in piezoceramic materials operating at low frequencies," Smart Structures and Materials 2000, Proceedings of the SPIE, Volume 3992, pp. 128–136, 2000.

[445] R.C. Smith and M. Salapaka, "Model development for the positioning mechanisms in an atomic force microscope," in *Control and Estimation of Distributed Parameter Systems*, W. Desch, F. Kappel and K. Kunisch, Eds., *International Series of Numerical Mathematics*, Volume 143, pp. 249–259, 2002.

[446] R.C. Smith, M.V. Salapaka, A. Hatch, J. Smith and T. De, "Model development and inverse compensator design for high speed nanopositioning," Proceedings of the 41st IEEE Conference on Decision and Control, pp. 3652–3657, 2002.

[447] R.C. Smith and S. Seelecke, "An energy formulation for Preisach models," Smart Structures and Materials 2002, Proceedings of the SPIE, Volume 4693, pp. 173–182, 2002.

[448] R.C. Smith, S. Seelecke, M.J. Dapino and Z. Ounaies, "A unified model for hysteresis in ferroic materials," Smart Structures and Materials 2003, Proceedings of the SPIE, Volume 5049, pp. 88–99, 2003.

[449] R.C. Smith, S. Seelecke, M.J. Dapino and Z. Ounaies, "A unified framework for modeling hysteresis in ferroic materials," CRSC Technical Report CRSC-TR04-35; *Journal of the Mechanics and Physics of Solids*, submitted.

[450] R.C. Smith, S. Seelecke, Z. Ounaies and J. Smith, "A free energy model for hysteresis in ferroelectric materials," *Journal of Intelligent Material Systems and Structures*, 14(11), pp. 719–739, 2003.

[451] R.C. Smith and R. Zrostlik, "Inverse compensation for ferromagnetic hysteresis," Proceedings of the 38th IEEE Conference on Decision and Control, pp. 2875–2880, 1999.

[452] G.S. Smolensky, "Physical phenomena in ferroelectrics with diffused phase transition," *Journal of the Physical Society of Japan (Suppl.)*, 28, pp. 26–37, 1970.

[453] W. Soedel, *Vibrations of Shells and Plates,* Second Edition, Marcel Dekker, Inc., New York, 1993.

[454] A.N. Soukhojak and Y.-M. Chiang, "Generalized rheology of active materials," *Journal of Applied Physics,* 88(11), pp. 6902–6909, 2000.

[455] A. Sozinov, A.A. Likhachev, N. Lanska and K. Ullakko, "Giant magnetic-field induced strain in NiMnGa seven-layered martensite phase," *Applied Physics Letters*, 80, pp. 1746–1749, 2002.

[456] B.F. Spencer, Jr., S.J. Dyke, M.K. Sain and J.W. Carlson, "Phenomenological model for magnetorheological dampers," *ASCE Journal of Engineering Mechanics*, 123(3), pp. 230–238, 1997.

[457] P.N. Sreeran, G. Salvady and N.G. Naganathan, "Hysteresis prediction for a piezoceramic material system," Proceedings of the 1993 ASME Winter Annual Meeting, New Orleans, LA, Volume AD-VOL 35, pp. 35–42, 1993.

[458] A.V. Srinivasan and D.M. McFarland, *Smart Structures: Analysis and Design*, Cambridge University Press, Cambridge, UK, 2001.

[459] R. Stalmans, J. Van Humbeeck and L. Delaey, "The two way memory effect in copper-based shape memory alloys — thermodynamics and mechanisms," *Acta Metallurgica et Materialia,* 40(11), pp. 2921–2931, 1992.

[460] A. Stancu and L. Spinu, "Temperature- and time-dependent Preisach model for a Stoner–Wohlfarth particle system," *IEEE Transactions on Magnetics*, 34(6), pp. 3876–3875, 1998.

[461] H.E. Stanley, *Introduction to Phase Transitions and Critical Phenomena*, Oxford University Press, Oxford, 1971.

[462] F.L. Stasa, *Applied Finite Element Analysis for Engineers,* Holt, Rinehart and Winston, New York, 1985.

[463] C.G.F. Stenger, F.L. Scholten and A.J. Burggraaf, "Ordering and diffuse phase transitions in $Pb(Sc_{0.5}Ta_{0.5})O_3$ ceramics," *Solid State Communications*, 32, pp. 989–992, 1979.

[464] E.C. Stoner and E.P. Wohlfarth, "A mechanism of magnetic hysteresis in heterogeneous alloys," *Philosophical Transactions of the Royal Society of London*, 240A, pp. 599–642, 1948.

[465] G. Strang, *Linear Algebra and Its Applications*, 3rd Edition, Harcourt, Brace and Company, Orlando, FL, 1988.

[466] S. Strässler and C. Kittel, "Degeneracy and the order of the phase transformation in the molecular-field approximation," *Physical Review*, 139(3A), pp. A758–A760, 1965.

[467] J.K. Strelec, D.C. Lagoudas, M.A. Khan and J. Yen, "Design and implementation of a shape memory alloy actuated reconfigurable airfoil, *Journal of Intelligent Material Systems and Structures*, 14(4/5), pp. 257–273, 2003.

[468] J. Su, J.S. Harrison and T. St. Clair, "Novel polymeric elastomers for actuation," ISAF 2000: Proceedings of the 2000 12th IEEE International Symposium on Applications of Ferroelectrics, Volume 2, p. 811–814, 2000.

[469] S. Tadokoro, T. Takamori and K. Oguro, *Electroactive Polymer Actuators as Artificial Muscles,* Chapter 13, SPIE Press, Bellingham, WA, pp. 331–336, 2001.

[470] B.K. Taleghani and J.F. Campbell, "Nonlinear finite element modeling of THUNDER piezoelectric actuators," NASA report NASA/TM-1999-209322, 1999.

[471] X. Tan and J.S. Baras, "Modeling and control of hysteresis in magnetostrictive actuators," *Automatica*, 40(9), pp. 1469–1480, 2004.

[472] X. Tan, R. Venkataraman and P.S. Krishnaprasad, "Control of hysteresis: Theory and experimental results," Smart Structures and Materials 2001, Proceedings of the SPIE, Volume 4326, pp. 101–112, 2001.

[473] K. Tanaka, S. Kobayashi and Y. Sato, "Thermodynamics of transformation, pseudoelasticity and shape memory effect in alloys," *International Journal of Plasticity*, 2, pp. 59–72, 1986.

[474] K. Tanaka and S. Nagaki, "A thermomechanical description of materials with internal variables in the process of phase transitions," *Ingenieur Archiv* (later *Archive of Applied Mechanics*), 51, pp. 287–299, 1982.

[475] K. Tanaka, F. Nishimura and H. Tobushi, "Phenomenological analysis on subloops in shape memory alloys due to incomplete transformations," *Journal of Intelligent Material Systems and Structures*, 5(4), pp. 487–493, 1994.

[476] J. Tani, T. Takagi and J. Qiu, "Intelligent material systems: Application of functional materials," *Applied Mechanics Reviews*, 51(8), pp. 505–521, 1998.

[477] G. Tao and P.V. Kokotović, "Adaptive control of plants with unknown hysteresis," *IEEE Transactions on Automatic Control*, 40(2), pp. 200–212, 1995.

[478] G. Tao and P.V. Kokotović, *Adaptive Control of Systems with Actuator and Sensor Nonlinearities*, John Wiley and Sons, New York, 1996.

[479] V. Thomée, *Galerkin Finite Element Methods for Parabolic Problems*, Springer–Verlag, Berlin, 1997.

[480] S. Timoshenko and S. Woinowsky-Krieger, *Theory of Plates and Shells,* Second Edition, McGraw-Hill Book Company, Inc., New York, 1987.

[481] M. Tokuda, S. Sogino, T. Inaba and P. Sittner, "Two-way shape memory effect obtained by training of combined cyclic loading," in *IUTAM Symposium on Mechanics of Martensitic Phase Transformation in Solids*, Q.P. Sun, Ed., Kluwer Academic Publishers, Dordrecht, pp. 55–62, 2002.

[482] J.-C. Tolédano and P. Tolédano, *The Landau Theory of Phase Transitions: Application to Structural, Incommensurate, Magnetic and Liquid Crystal Systems*, World Scientific Publishing, Singapore, 1987.

[483] R.D. Turner, T. Valis, W.D. Hogg, and R.M. Measures, "Fiber-optic strain sensors for smart structures," *Journal of Intelligent Material Systems and Structures*, 1, 1990, pp. 26–49.

[484] J.A. Tuszyński, B. Mróz, H. Kiefte and M.J. Clouter, "Comments on the hysteresis loop in ferroelastic $LiCsSO_4$," *Ferroelectrics*, 77, pp. 111–120, 1988.

[485] H.S. Tzou, *Piezoelectric Shells: Distributed Sensing and Control of Continua*, Kluwer Academic Publishers, Boston, 1993.

[486] H.S. Tzou and M. Gadre, "Theoretical analysis of a multi-layered thin shell coupled with piezoelectric actuators for distributed vibration controls," *Journal of Sound and Vibration*, 132(3), pp. 433–450, 1989.

[487] H.S. Tzou and R.V. Howard, "A piezothermoelastic thin shell theory applied to active structures," *ASME Journal of Vibration and Acoustics*, 116(3), pp. 295–302, 1994.

[488] H.S. Tzou and C.I. Tseng, "Distributed piezoelectric sensor/actuator design for dynamic measurement/control of distributed parameter systems: A piezoelectric finite element approach," *Journal of Sound and Vibration*, 138(1), pp. 17–34, 1990.

[489] K. Uchino, *Piezoelectric Actuators and Ultrasonic Motors*, Kluwer Academic Publishers, Boston, 1997.

[490] E. Udd, "Fiber optic smart structures," Chapter 14 in *Fiber Optic Sensors: An Introduction for Engineers and Scientists,* E. Udd, Ed., pp. 439–467, John Wiley and Sons, New York, 1991.

[491] T.D. Usher, A. Sim, G. Ashford, G. Camargo, A. Cabanyog and K. Ulibarri, Jr., "Modeling and applications of new piezoelectric actuator technologies," Smart Structures and Materials 2004, Proceedings of the SPIE, Volume 5383, pp. 31–38, 2004.

[492] R. Valenzuela, *Magnetic Ceramics*, Cambridge University Press, Cambridge, UK, 1994.

[493] J. Van Humbeeck and R. Stalmans, "Characteristics of shape memory alloys," in *Shape Memory Materials*, K. Otsuka and C.M. Wayman, Eds., Cambridge University Press, Cambridge, UK, pp. 149–183, 1998.

[494] J. Van Humbeeck and R. Stalmans, "Shape memory alloys, types and functionalities," in *Encyclopedia of Smart Materials,* M. Schwartz, Ed., John Wiley and Sons, New York, pp. 951–964, 2002.

[495] R. Venkataraman, "A hybrid actuator," MS Thesis, University of Maryland, College Park, MD, 1995.

[496] R. Venkataraman and P.S. Krishnaprasad, "Qualitative analysis of a bulk ferromagnetic hysteresis model," Proceedings of the 37th IEEE Conference on Decision and Control, pp. 2443–2448, 1998.

[497] D. Viehland, S.J. Jang and L.E. Cross, "Freezing of the polarization fluctuations in lead magnesium niobate relaxors," *Journal of Applied Physics*, 68(6), pp. 2916–2921, 1990.

[498] D. Viehland, S.J. Jang, L.E. Cross, M. Wuttig, "Deviation from Curie–Weiss behavior in relaxor ferroelectrics," *Physical Review*, 46(13), pp. 8003–8006, 1992.

[499] C.R. Vogel, *Computational Methods for Inverse Problems,* SIAM, Philadelphia, PA, 2002.

[500] W. Voigt, *Lehrbuch der Kristallphysik*, B.G. Teubner, Leipzig, Berlin, 1928.

[501] S.V. Vonsovskii, *Magnetism: Volume Two*, Translated from Russian by Ron Hardin, John Wiley and Sons, New York, 1974.

[502] V.K. Wadhawan, *Introduction to Ferroic Materials*, Gordon and Breach Science Publishers, Amsterdam, 2000.

[503] T.T. Wang, J.M. Herbert and A.M. Glass, Eds., *The Applications of Ferroelectric Polymers*, Blackie and Son, Ltd., Glasgow, 1988.

[504] G. Webb, A. Kurdila and D.C. Lagoudas, "Adaptive hysteresis model for model reference control with actuator hysteresis," *Journal of Guidance, Control and Dynamics*, 23(3), pp. 459–465, 2000.

[505] R. Weinstock, *Calculus of Variations with Applications to Physics and Engineering*, Dover, New York, 1974.

[506] P. Weiss, "La variation du ferromagnétisme avec la temperature," *Comptes Rendus Hebdomadaires des Séances de L'Académie des Sciences,* (later *Comptes Rendus de L'Académie des Sciences*), 143, pp. 1136–1139, 1906.

[507] P. Weiss, "L'hypothèse du champ moléculaire et de la propriété ferromagnétique," *Journal de Physique Théorique et Appliquée*, Series 4, Volume 6, pp. 661–690, 1907.

[508] F.B. Weissler, "Semilinear evolution equations in Banach spaces," *Journal of Functional Analysis*, 32, 1979, pp. 277–296.

[509] R. Wieman, R.C. Smith, T. Kackley, Z. Ounaies and J. Bernd, "Displacement models for THUNDER actuators having general loads and boundary conditions," Smart Structures and Materials 2001, Proceedings of the SPIE, Volume 4326, pp. 252–263, 2001.

[510] K. Wilde, P. Gardoni and Y. Fujino, "Base isolation system with shape memory alloy device for elevated highway bridges," *Engineering Structures*, 22, pp. 222–229, 2000.

[511] E.J. Williams, "The effect of thermal agitation on atomic arrangement in alloys III," *Proceedings of the Royal Society of London*, A152, pp. 231–252, 1935.

[512] W. Williams and D.J. Dunlop, "Three-dimensional micromagnetic modelling of ferromagnetic domain structure," *Nature*, 337, pp. 634–637, 1989.

[513] J.M. Wiltse and A. Glezer, "Manipulation of free shear flows using piezoelectric actuators," *Journal of Fluid Mechanics*, 249, pp. 261–285, 1993.

[514] W.M. Winslow, "Method and means for translating electrical impulses into mechanical force," U.S. Patent 2,417,850, 1947.

[515] W.M. Winslow, "Induced fibration of suspensions," *Journal of Applied Physics*, 20, pp. 1137–1140, 1949.

[516] S.A. Wolf, D.D. Awschalom, R.A. Buhrman, J.M. Daughton, S. von Molnár, M.L. Chtchelkanova and D.M. Teger, "Spintronics: A spin-based electronics vision for the future," *Science*, 294, pp. 1488–1495, 2001.

[517] J.L. Woolman, K.P. Mohanchandra and G.P. Carman, "Composition and annealing effects on the mechanical properties of superelastic thin film nickel titanium," Smart Structures and Materials 2003, Proceedings of the SPIE, Volume 5053, pp. 230–238, 2003.

[518] M. Wun-Fogle, J.B. Restorff, A.E. Clark and J.F. Lindberg, "Magnetization and magnetostriction of dendritic [112] $Tb_xDy_yHo_zFe_{1.95}$ $(x+y+z=1)$ rods under compressive stress." *Journal of Applied Physics*, 83(11), pp. 7279–7281, 1998.

[519] M. Wuttig, L. Liu, K. Tsuchiya and R.D. James, "Occurrence of ferromagnetic shape memory alloys," *Journal of Applied Physics*, 87(9), pp. 4712–4717, 2000.

[520] Y. Xiao and K. Bhattacharya, "Modeling electromechanical properties of ionic polymers," Smart Structures and Materials 2001, Proceedings of the SPIE, Volume 4329, pp, 292–300, 2001.

[521] D. Xu, L. Wang, G. Ding, Y. Zhou, A. Yu and B. Cai, "Characteristics and fabrication of NiTi/Si diaphragm micropump," *Sensors and Actuators A: Physical*, 93(1), pp. 87–92, 2001.

[522] M. Yamakita, N. Kamanichi, Y. Kaneda, K. Asaka and Z-W. Luo, "Development of artificial muscle actuator using ionic polymer with its application to biped walking robots," Smart Structures and Materials 2003, Proceedings of the SPIE, Volume 5051, pp, 301–308, 2003.

[523] I.V. Yannas and A.J. Grodzinsky, "Electromechanical energy conversion with collagen fibers in an aqueous medium," *Journal of Mechanochemistry and Cell Motility*, 2, pp. 113–125, 1973.

[524] J.M. Yeomans, *Statistical Mechanics of Phase Transitions*, Clarendon Press, Oxford, UK, 1992.

[525] J-H. Yoo, J. Sirohi, N.M. Wereley and I. Chopra, "A magnetorheological hydraulic actuator driven by a piezopump," Smart Structures and Materials 2003, Proceedings of the SPIE, Volume 5056, pp, 444–456, 2003.

[526] J.A. Young, B.L. Farmer and J.A. Hinkley, "Molecular modeling of the poling of piezoelectric polyimides," *Polymer*, 40, pp. 2787–2795, 1999.

[527] O.C. Zienkiewicz and R.L. Taylor, *The Finite Element Method*, Butterworth–Heinemann, Boston, 2000.

[528] D. Zwillinger, Editor-in-Chief, *CRC Standard Mathematical Tables and Formulae*, 30th Edition, CRC Press, Boca Raton, 1996.

Index